Radar Signals

An Introduction to Theory and Application

Radar Signals

An Introduction to Theory and Application

Charles E. Cook

Marvin Bernfeld

Artech House
Boston • London

Library of Congress Cataloging Data
Cook, Charles E. (Charles Emerson), 1926–
Radar Signals: An Introduction to Theory and Application
[by] Charles E. Cook and Marvin Bernfeld
Includes bibliographical references and index.
xvi, 531 p., illustrated, 24 cm
ISBN 0-89006-733-3
TK6575.C65   66-30803
621.3848   MARC

International Standard Book Number: 0-89006-733-3
Library of Congress Card Catalog Number: 66-30803

Originally published by Academic Press, 1967, 1987

10 9 8 7 6 5 4 3 2 1

Preface

This book is devoted to the development of the basic theory and application of radar signals that are designated as large time-bandwidth or pulse-compression waveforms. This class of signals provides one of the cornerstones for modern radar technology, and yet there has been no single treatment of this subject at an introductory level from which the graduate student and practicing engineer can obtain the understanding and background necessary for participation in this branch of radar system engineering. In writing this book, the authors have attempted to present in a unified manner the theoretical and practical aspects of radar signal processing applications that are contained throughout various sources in the published literature. The authors have felt for some time the need for having this information more readily available. To meet this need for their co-workers at the Sperry Gyroscope Company, they organized and prepared the notes for an after-hours graduate level course covering the theory, application, and design of large time-bandwidth radar signals. This present work is an outgrowth of that activity. We note that, although no examples are specifically designated in the various chapters, problem assignments and examinations could be developed directly from the text material and fitted into the major interests of the students in the course.

The initial impetus to research and development of complex radar waveforms was obtained from the needs of military radar applications during and after World War II. During this period, rapid strides were made in developing and understanding the requirements for optimum (linear) signal processing. At the same time, engineering developments were paving the way for practical implementations based on this broader understanding, with numerous proposals being advanced for achieving what are now known as pulse-compression, or coded waveform, systems. The work of Ville, Woodward and Davies, and Woodward provided a unified basis from which sprang a second phase of concentrated effort in defining the constraints imposed on radar signal ambiguity distributions and the principles for realizing specific radar objectives as a result of proper choice of the radar signal. For some time there persisted the hope that a single large time-bandwidth signal might be discovered that would be optimum for a number of different applications. Over a decade of effort and a much clearer understanding of the problem have shown that this hope was largely illusory. In this respect, the statement of Woodward "that the basic question of what (radar signal) to transmit remains substantially unanswered" was somewhat prophetic. Despite the great progress since that time (1953), this statement still contains an element of truth unless a very clear definition of the nature of the radar reflecting environment can be made. Under these conditions, it may be possible that a practically implemented radar signal can be optimally matched to its environment. However, prognostications of the distributions and motions of reflecting surfaces are prone to more or less uncertainty. In this event, the radar designer is usually forced to a compromise type of radar signal (or signals) that he *thinks* (based on incomplete analysis, simplified environmental models, insight, etc.) will perform adequately over a given number of possible situations. In this context, the intent of the authors is to place before the reader the theory and principles that will aid his own efforts in attacking a particular set of problems, and to indicate where the weight of logic should direct his attention. This is supported by examples of different types of radar signals that have been widely investigated for different applications. Among these, the linear FM signal has been found to be suitable (and easily implemented) for many tasks. Thus, this waveform is used as an illustrative example in many parts of the text that treat the practical problems associated with large time-bandwidth radar systems. It is interesting to note that large time-bandwidth signals are now found in such diversified uses as meteorology, seismology, ionospheric sounding, harbor navigation, airport traffic control, flaw detection in metals, and the experimental analysis of boundary layers between different media. It is anticipated that this list will increase with passing time, and it is our hope that this volume will contribute to the wider use of this class of signals.

C. E. COOK
M. BERNFELD

February, 1967

Acknowledgments

Many different laboratories and individuals have made contributions to the practical realization of modern radar waveform and signal processing techniques. Some of these are mentioned in the opening chapter. Initial investigation and development of pulse-compression and matched-filter systems at the Sperry Gyroscope Company began in early 1952 under the direction of W. W. Mieher and C. E. Brockner. Early contributors to this program were J. E. Chin, G. R. Latham, and L. R. Sadler. An experimental X-band pulse-compression system was constructed in 1955. The efforts of M. Buchbinder and the late V. F. Ragni were instrumental in the success of this part of the program. This was followed by a demonstration in a high power radar under the direction of A. E. Hylas. Since that time, a number of individuals at Sperry have made significant contributions in the study and implementation of matched-filter radar systems. In particular, we are indebted to our colleagues S. E. Bogotch and C. A. Palmieri for, respectively, supplying data on implementation techniques and developing the Sperry Analog Waveform Correlator. This latter device, an active correlator, was used to obtain many of the waveform examples in Chapters 4, 8, and 9. Mr. J. F. Cerar has been the source of numerous ideas in the experimental phase of these efforts. At various times we have benefitted from discussions with our co-workers in the Advanced Radar Studies Department, as well as with others at Sperry. Mr. J. Paolillo made significant contributions to the development of the material in Section 6.6 and Chapter 7, and in addition had made helpful suggestions in other areas. We wish to thank L. Susman and S. G. McCarthy for commenting on various portions of the manuscript, along with those students who survived our first attempt at teaching the material and who pointed out a number of errors in the original notes. When we first proposed to undertake this project, we received the enthusiastic support of D. M. Skidmore, I. A. Paul, and F. W. Ziegler. A special note of thanks is due Mrs. M. Pace, F. J. Ash, J. N. Bannister, S. Tagliaferro, and the ladies of the Sperry Gyroscope Company Radiation Division typing pool for their help given in various phases of manuscript and photographic artwork preparation.

The authors are equally indebted to their many associates at other laboratories for their advice, critical comments, and help in assembling material for several of the chapters. In particular, Dr. A. H. Meitzler and his co-workers at the Bell Telephone Laboratories very graciously offered to review the material in Chapter 13, and Mr. H. C. Wing of Sanders Associates provided material for inclusion in that chapter. Mr. G. F. Miller of the Canadian General Electric Company, Ltd., and Professor H. Miyakawa of the University of Tokyo made possible the use of several interesting photographs of short code waveforms and ambiguity functions that appear in Chapter 8, for which we wish to express our gratitude. Part of the material in Chapters 5, 8–10 was originally developed on Contract No. AF 30 (602)-3352. Our thanks go to the United States Air Force for permission to publish this data, and to Mr. T. Maggio and Mr. C. F. Silfer of the Rome Air Development Center who were responsible for this program. Thanks are also due the Editors of the *Proceedings* and *Transactions* of the Institute of Electrical and Electronics Engineers and of the *Microwave Journal* for permission to reproduce figures from these journals.

Mr. E. N. Fowle of the MITRE Corporation and Dr. W. L. Rubin and Mr. J. V. DiFranco of the Sperry Gyroscope Company have been friendly critics and sources of information for many years, and we would be remiss in not expressing a warm note of thanks for their interest in and encouragement of our work. The help and advice of Academic Press in the final stages of manuscript preparation have been invaluable. Finally, our appreciation for the patience and understanding of our wives and families during the preparation of this volume cannot be expressed in words. Their quiet and heartfelt support was a significant factor in our undertaking and completing this work.

Part of the material in this volume was developed through the aid of Sperry Gyroscope Company sponsored research programs. The authors wish to express their appreciation to the Sperry Rand Corporation for permission to publish this manuscript.

Contents

Chapter 1.   The Basic Elements of Matched Filtering and Pulse Compression

Chapter 2.   Optimum Predetection Processing—Matched-Filter Theory

Chapter 3.   Matched-Filter Requirements for Arbitrary FM Pulse-Compression Signals

Chapter 4.   The Radar Ambiguity Function

Chapter 5.   Parameter Estimation

Chapter 6.   The Linear FM Waveform and Matched Filter

Chapter 7.   Matched-Filter Waveform Considerations—Range Sidelobe Reduction

Chapter 8.   Discrete Coded Waveforms

Chapter 9.   The Measurement Accuracies of Matched-Filter Radar Signals— Waveform Design Criteria

Chapter 10. Waveform Design Criteria for Multiple and Dense Target Environments

Chapter 11. Effects of Distortion on Matched-Filter Signals

Chapter 12. The Design of Dispersive Delay Functions—I

Chapter 13. The Design of Dispersive Delay Functions—II Ultrasonic Delay Lines

Chapter 14.   **Microwave and Optical Matched-Filter Techniques**

List of Symbols

α	rms half-duration (radian measure of signal)	Φ_0	Filter phase constant
β	rms half-bandwidth (radian measure) of signal	$\chi(\tau, \phi)$	Signal response function
		$\|\chi(\tau, \phi)\|^2$	Signal ambiguity function
$\beta_f(\omega)$	Filter phase response, $\beta_f(\omega) = -\Phi(\omega)$ for matched-filter case	$\chi_p(\tau', \phi)$	Response function of a component pulse of a discrete coded signal
$\beta_f'(\omega)$	Filter group delay $d\beta_f(\omega)/d\omega$	$\chi(k\delta, \phi)$	Response function of discrete coded signal at integer values of δ
γ	Propagation constant (used in Chapters 13 and 14)		
δ	Component subpulse width of discrete coded signals	$\psi(t)$	Complex signal
		$\Psi(f)$	Spectrum of complex signal
δ_r	rms range measurement error	ω	Radian frequency, $2\pi f$
δ_v	rms velocity measurement error	ω_d	radian Doppler shift, $2\pi\phi$
$\theta(t)$	Radian phase modulation	$a(t)$	Signal envelope function
$\theta_e(t)$	Radian phase modulation error	$\{a_n\}$	Amplitude sequence for discrete coded signals
$\theta'(t)$	Frequency modulation, $d\theta(t)/dt$	$\{b_n\}$	Huffman code combined amplitude and phase sequence
θ_0	Phase constant of phase modulation		
θ_1, θ_2	Generalized signal parameters (used only in Chapter 5)	$\{c_n\}, \{d_n\}$	Alternate representations of discrete code phase sequence
$\Theta(\tau, \phi)$	Symmetrical ambiguity function	$\{\theta_n\}$	Phase sequence for discrete coded signals
λ	Wavelength		
μ	Linear FM radian modulation rate	$\{\omega_n\}$	Radian frequency sequence for discrete coded signals
ρ	Range-Doppler correlation or coupling factor	$F_r(0)$	Frequency resolution constant
		$F_r(\tau)$	Generalized frequency resolution constant
σ_ϕ	rms frequency measurement error		
σ_ϕ^2	Frequency measurement error variance	$H(\omega)$	Complex filter transfer characteristic
σ_τ	rms time measurement error	E	Energy of real signal $s(t)$
σ_τ^2	Time measurement error variance	M	Number of subpulses $P_n(t)$ in a discrete coded signal
ϕ	Doppler shift	N	Number of possible subpulse positions in a discrete coded signal
$\Phi(\omega)$	Spectrum phase function		

N_0	Power density, white noise
$N(\omega)$	Power density, nonwhite noise
$P_n(t)$	Unity height pulse of fixed duration δ
$P(D)$	Polynomial in the unit delay operator D describing shift register operation
$P(s)$	Huffman polynomial
$s(t)$	Real signal
$S(f)$	Spectrum of $s(t)$
T	Signal time extent
$T\Delta f$	Time-bandwidth product ($=$ compression ratio for linear FM case)
Δf	Spectrum bandwidth extent ($\equiv 1/T$ for uncoded pulse)
$T_r(0)$	Time resolution constant
$T_r(\phi)$	Generalized time resolution constant
$u(t)$	Complex envelope function, $a(t)\exp[j\theta(t)]$
$U(\omega)$	Spectrum of $u(t)$
$w(t)$	Time weighting function for sidelobe reduction
$W(\omega)$	Frequency weighting function for sidelobe reduction

CHAPTER 1

The Basic Elements of Matched Filtering and Pulse Compression

1.1 Introduction

When radar was in its earliest stages of development, it was generally accepted that radar systems fell into two basic categories. Thus, an operational radar was either a continuous wave (cw) system that had inherently very good velocity (or Doppler shift) measuring capability, or a pulsed system that had good range measuring and resolution capability. In a pulsed radar system the dimensions of the transmitted pulse usually represented a compromise between the desired range resolution (i.e., as small a pulse width as possible) and the desired maximum detection range (i.e., maximizing the energy per pulse by using as long a pulse width as allowable). The effort to balance off the sometimes conflicting requirements of range resolution and maximum detection range often forced compromises in other areas of the radar design. One example is that of a decreased rate of volume search by the radar antenna in order to obtain more reflected pulses from an object so that the maximum detection range could be increased by use of pulse-to-pulse integration techniques. The design of a radar system[1] normally begins by examining the constraints imposed by the radar range equation, given by

$$P_r = \frac{P_t G^2 \lambda^2 \sigma}{(4\pi)^3 R^4} \tag{1-1a}$$

or

$$R_{max} = \left[\frac{P_t G^2 \lambda^2 \sigma}{(4\pi)^3 S_{min}} \right]^{1/4} \tag{1-1b}$$

---

[1] See Bibliography at the end of this chapter for a listing of recent texts on general radar system design considerations.

where P_t is the transmitted power, P_r the received power, G the gain of transmitting and receiving antenna, λ the transmitted wavelength, σ the effective
reflecting area of the object, R the distance from radar to reflecting object,
$R_{\max}$ the maximum detection range, and $S_{\min}$ the minimum detectable signal.

Making use of (1-1a) or (1-1b), the radar designer could weigh the merits
of the various possible compromises that could achieve a desired combination of results.

As radar development progressed, and emphasis changed from merely
getting things to work to getting things to work in an optimum or near
optimum manner, newer concepts came into being that laid the foundations
of waveform design as an integral part of the radar system development.
In one particular area, pulsed Doppler systems represented an effort to
obtain good velocity measurement or discrimination and range resolution
together. This was the forerunner of later techniques of waveform design
based on the use of complex pulse train functions. However, it remained
for a stimulating monograph by Woodward [1] to bring a unification of
thought to the various aspects of radar waveform design that had begun to
develop in the time period following World War II. It came to be seen that
waveform *shape* could be an added dimension in the radar design, and that
such characteristics of the radar system as range resolution need not enter
into the other important considerations, such as the average transmitted
power and the transmitted pulse length. In considering the radar equation
(1-1a) or (1-1b), the basic ideas developed by Woodward indicated that the
transmitted pulse could be designed to be as wide as necessary to meet the
energy requirements of the system or to take fullest advantage of the
available transmitter power tubes, and that *after* the detectability requirements had been satisfied the range resolution conditions could be met by
coding the transmitted signal with wideband modulation information. One
of the most important contributions by Woodward to the development
of modern radar technology was to point out that range resolution and
accuracy were functions of the signal bandwidth, and not of the transmitted
pulse width.

The extraction of the wideband information contained in the type of
signal implied above requires the use of a more complex receiving system
than that needed for a simple pulse radar. These complex receiver systems
are designated as matched-filter signal processing systems. The use of
the term signal processing generally implies operations performed upon
the received signal in the rf or i.f. portions of the radar receiver, and is
distinguished from the use of the term data processing, which normally
implies operations upon the detected version of the radar signal.

The freedom to design various characteristics into the radar signal has
been an important factor in the development of modern radar systems,

which have become closely tied to the use of advanced signal processing techniques. The waveform design concepts associated with modern radar systems have been largely designated as pulse-compression, matched-filter, or coded-waveform techniques [2–4]. Some of the basic practical reasons for the development of these techniques are listed here, although these are by no means all inclusive:

1. More efficient use of the average power available at the radar transmitter and, in some cases, avoidance of peak power problems in the high power sections of the radar transmitter.
2. Increased system resolving capability, both in range and velocity. In the case of range resolution, the generation of extremely fast rise time and high peak power signals is bypassed when pulse-compression techniques are used.
3. Reduction of vulnerability to certain types of interfering signals that do not have the same properties as the coded waveform.
4. Extraction of information from the signals present at the receiver input to yield estimates of important parameters associated with the individual signals, such as range, velocity, and possibly acceleration. This aspect of radar signal processing is referred to as parameter estimation.

Figure 1.1 illustrates in simple form a radar system that makes use of waveform coding techniques to obtain the operating results just outlined. The purpose of this volume will be to develop the concepts and techniques that are necessary to understand and apply radar signal processing that takes the form of pulse compression matched filtering. Thus, the theory and implementation of the parts of Fig. 1.1 that are designated as "coding source" and "decoding device" will be examined in detail.

The topics covered fall into three general areas. The first five chapters introduce the broad theoretical aspects related to matched-filter techniques. The matched-filter concept and the historical development of pulse compression are reviewed, and the various criteria of system performance that lead to the matched-filter requirement are presented. Among the subjects treated in detail are the important principle of stationary phase, used to derive the approximate relationship between a frequency modulated coded waveform and its matched filter; the radar ambiguity function and its application as a radar waveform design criterion; the theory of parameter estimation and the relationship of the coded waveform characteristics to radar measurements. Chapters 6 through 10 consider topics related to specific radar waveforms. Particular emphasis is given to the linear FM signal, since this is in many respects *the* canonic pulse-compression matched-

1. BASIC ELEMENTS OF MATCHED FILTERING

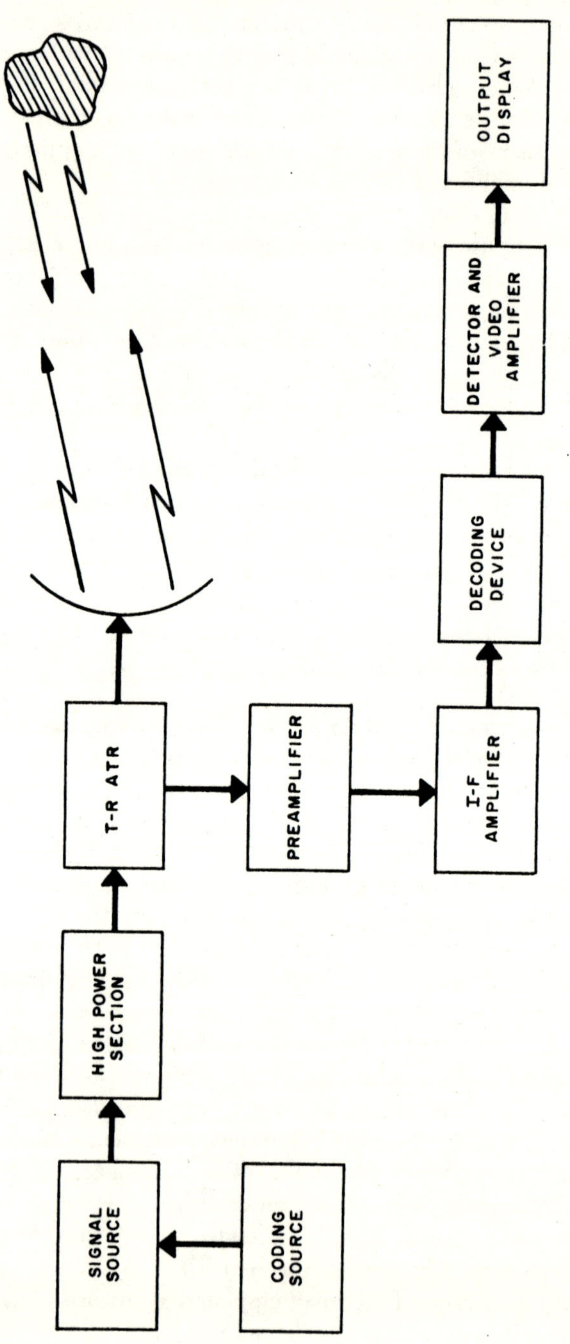

Fig. 1.1   A coded waveform radar system.

filter waveform. The linear FM signal has found the most widespread application of all of the large time-bandwidth signals that have been suggested for radar applications, and has received the most attention in the area of component development. The various discrete coded waveforms, as opposed to the continuous FM codes, are treated as a unified class of signals in Chapter 8. Some of these signals represent discrete approximations to well-known FM signals, while others have no relationship at all to the FM class. Parameter estimation theory as a waveform design criterion is explored in Chapter 9, where the theoretical relationships derived in Chapter 5 are applied to a number of the matched-filter signals discussed in other chapters. Chapter 10 considers waveform design criteria for several cases of multiple and dense reflecting environments. The final four chapters consider various practical problems associated with matched-filter implementations. These include the effect of distortions, the design of linear delay filters with lumped constant and ultrasonic delay line techniques, and the application of optical and microwave methods to general matched-filter designs.

The topics covered are intended to give the systems engineer and student a basic understanding of radar waveform design principles, as well as to provide the interested individual with the necessary background to proceed to more advanced topics in signal processing techniques and applications.

1.2 The Matched-Filter Concept

The basic concept of matched filtering evolved from the effort to obtain a better theoretical understanding of the factors leading to optimum performance of radar systems. The goal was to translate this sounder theoretical understanding into developments that could be instrumented by practical engineers. The technique of matched filtering constitutes the optimum linear processing of radar signals. This form of signal processing transforms the raw radar data, available at the receiver input and assumed to be corrupted by white Gaussian noise, into a form that is suitable for performing optimum detection decisions (i.e., target or no target), or for estimating target parameters (i.e., range, velocity, etc.) with minimum rms errors, or for obtaining maximum resolution among a group of targets.

The characteristics of matched filters can be designated by either a frequency response function or a time response function, each being related to the other by a Fourier transform operation. In the frequency domain the matched-filter transfer function, $H(\omega)$, is the complex conjugate function of the spectrum of the signal that is to be processed in an optimum fashion. Thus, in general terms

$$H(\omega) = kS^*(\omega)\exp[-j\omega T_d] \qquad (1\text{-}2)$$

where $S(\omega)$ is the spectrum of the input signal $s(t)$, and T_d is a delay constant required to make the filter physically realizable. The normalizing factor, k, and the delay constant are generally ignored in formulating the underlying significant relationship, usually expressed as

$$H(\omega) = S^*(\omega) \qquad (1\text{-}3)$$

The corresponding time domain relationship between the signal to be operated upon and the matched filter is obtained from the inverse Fourier transform of $H(\omega)$. This leads to the result that the impulse response of the matched filter is a replica of the time inverse of the known signal function. Thus, if $h(t)$ represents the matched-filter impulse response, the general relationship equivalent to Eq. (1-2) is

$$h(t) = ks(T_d - t) \qquad (1\text{-}4)$$

As above, the arbitrary delay term, T_d, can be ignored[1] to point out the basic important relationship

$$h(t) = ks(-t) \qquad (1\text{-}5)$$

North [5] is generally credited as being the first to derive the properties of the optimum receiver, for the case of white Gaussian noise, based on the signal spectrum parameters (Eq. 1-2). Because of this matched filters are also referred to as North filters; however, Van Vleck and Middleton [6] were apparently the first to use the designation "matched filter" in the context of optimizing the signal-to-noise ratio of pulsed signals. The derivation of the matched-filter requirements is included in Chapter 2 for the sake of completeness, and to give the interested reader a deeper insight into matched-filter systems. Figure 1.2 illustrates the relationships given by Eqs. (1-3) and (1-5).

The basis for the derivation of the condition for optimum detection performance is shown in Fig. 1.3, where a simplified receiving system is presented. The output waveform at (b) is composed of a combination of signal and noise. The goal for the system designer is to attempt to optimize the probability of detecting the signal during some observation interval. This interval might represent range gating, or merely the fact that the attention of the observer has been directed to a particular point either by chance or by a prior knowledge. The observation threshold might be either an explicit threshold, as in an automatic alarm device, or an implied threshold resulting from the bias of a human observer to discount the larger

---

[1] In this case the matched-filter constant will be retained in the notation, since it represents the factor necessary to determine a unity gain filter response.

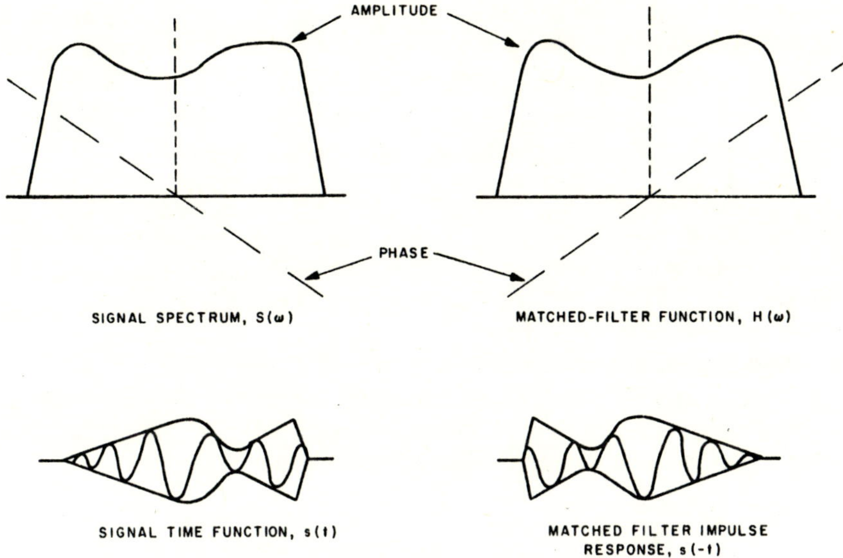

FIG. 1.2   Signal and matched-filter relations. (From Bernfeld *et al.* [4].)

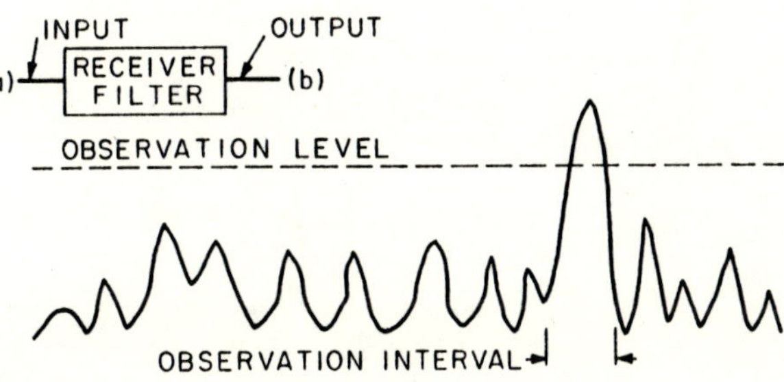

FIG. 1.3   Detection performance criterion.

noise spikes as not being true signals. The actual statistics involved in the detection process will depend on many factors, such as threshold level and existence of prior information as to the location of a signal. Regardless of these influencing factors, it can be seen from Fig. 1.3 that the logical procedure is to try to maximize the peak of the signal relative to the noise. Since the signal is, presumably, seldom present (a continuous signal would, by definition, convey no useful information), the continual observation of the random fluctuations of the noise signal will focus attention on these momentary deviations from the long term average or rms value of the noise.

From this point of view comes the logical conclusion that obtaining a maximum peak signal relative to the rms noise will lead to the optimization being sought, or

$$\left(\frac{S}{N}\right)_{max} = \frac{\text{maximum instantaneous output signal power}}{\text{output noise power}} \qquad (1\text{-}6)$$

The reader interested in the derivation of the conditions for maximizing the ratio given by Eq. (1-6) will find this developed in Chapter 2. Chapter 2 also considers the statistical basis for optimizing system detection capability. Both of these approaches lead to the matched-filter conditions described by Eqs. (1-3) and (1-5). When a matched filter, as given by these expressions, is employed it is found that the maximum signal-to-noise ratio at the filter output for the case of white Gaussian noise is given by

$$\left(\frac{S}{N}\right)_{max} = \frac{2 \times \text{received signal energy}}{\text{noise spectral density (watts/cycle/sec)}} \qquad (1\text{-}7)$$

The relationship given by (1-7) can be derived heuristically for the simple pulsed radar signal on the basis of the parameters shown in Fig. 1.4. The

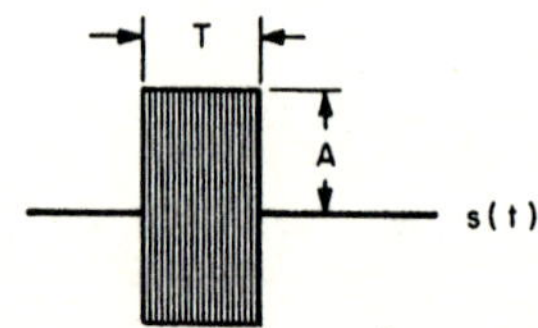

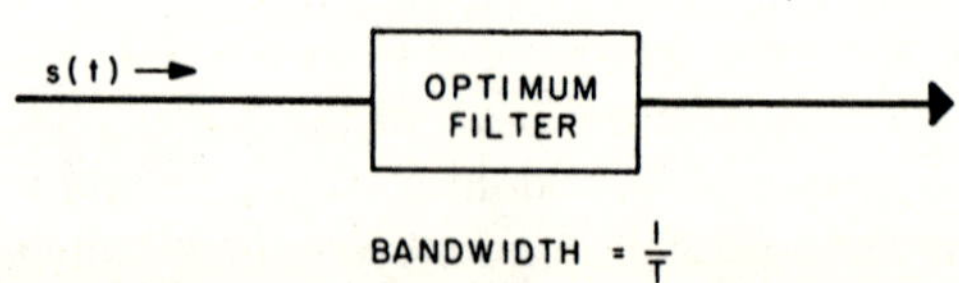

FIG. 1.4 Pulsed signal parameters.

energy of the received signal is

$$E = \tfrac{1}{2}A^2 T \tag{1-8}$$

and the noise power is given by

$$P_N = N_0\,\Delta f \tag{1-9}$$

where N_0 is the noise power density and Δf the effective filter bandwidth.

The peak signal-to-noise power ratio in terms of the signal energy is

$$\left(\frac{S}{N}\right)_{\text{max}} = \frac{A^2}{N_0\,\Delta f}$$

$$= \frac{2E}{T N_0\,\Delta f} \tag{1-10}$$

For optimum results it is known that the filter bandwidth is related to the pulse duration, such that

$$\Delta f = k/T \tag{1-11}$$

where k is always near unity. Assuming that $k = 1$, then the peak signal-to-noise ratio equation (1-10) reduces to

$$\left(\frac{S}{N}\right)_{\text{max}} = \frac{2E}{N_0} \tag{1-12}$$

This result implies to the radar designer that, so long as a matched filter is used in the predetection stages of a receiving system, the ability to detect a radar signal is only a function of the energy it contains and depends in no way on the shape or form of the signal as it enters the receiver. In order to obtain the optimum output signal-to-noise ratio a matched filter must be used. However, the theory provides a considerable region of flexibility with respect to this point. Hence, if it is impractical to build the exact matched filter, it is usually possible to employ a reasonable approximation with very little effect on the radar system detection capability.

1.3   The Pulse-Compression Concept—Historical Background

The development of the matched-filter concept represented an effort to define, independent of practical limitations, theoretical performance criteria for pulsed-radar systems. At much the same interval of time engineers faced with the very real shortcomings of World War II radars were considering methods of improving the practical results that could be achieved (this group included some of the world's best known physicists and network mathematicians, who had wartime assignments in the development of usable radar systems). As the war progressed, and radar receiver techniques became more advanced, it became obvious that the major impediment to radically improved radar performance lay in the power limitations of the transmitters being used. The problem here was twofold. The peak power available in transmitter tubes was limited, and in addition, even if greater peak power had been available, many of the transmitter components could not have operated under higher power conditions than prevailed at the time. A straightforward solution would have been to take advantage of the greater average power capabilities of transmitter tubes by going to wider pulse widths. However, there was a conflicting requirement to achieve even greater resolution for such purposes as ground mapping and separation of aircraft targets in massed formations. Thus, the obvious solution to overcome system power restrictions was not acceptable where the need was greatest.

A proposed method to avoid the dilemma described above was thought of by several individuals [7–11], serving on both sides during the war. This essentially provided that a wide pulse be transmitted, during which the carrier frequency was linearly swept.[1] Intuitively, this was seen to yield a correlation between time and frequency that could be exploited in the radar receiver. The means of exploitation was proposed as a filter having a linear time delay vs frequency characteristic of such a sense that it would apparently delay one end of the received wide pulse by a greater amount than the other, thus causing the signal to compress in time and increase in peak amplitude (see Fig. 6.19, Chapter 6). Any judgment as to priority of concept of the basic pulse-compression idea is beyond the scope, or intent, of this present work. The fact that such widely separated scientists demonstrated the insight to arrive at a more or less identical solution to a common problem is remarkable enough in itself.

The conception of pulse compression came too late to be a factor in World War II. In addition, when it seemed that special type tubes, such as high power klystrons, would be required to implement the technique interest lagged, since such tubes did not then exist. Seemingly pulse compression, as a radar technique, was buried as a curiosity in the patent files left as a legacy of the war.

Eventually, the necessary transmitter components were developed, and with this progress came a renewed interest in advanced concepts such as pulse compression and matched filters. By the early 1950's several major laboratories, representing government and private industry, had initiated research programs based on bringing the pulse-compression concept to fruition [12–15].

---

[1] Dicke [10] discusses the requirements for processing a nonlinear FM signal, but mainly in the sense of being faced with a distorted linear FM function.

1.4 The Pulse-Compression Concept—A Heuristic Development of the Significant Parameters

The basic reasoning of the patents cited in the previous section can be seen by referring to Fig. 1.5. Here are shown a transmitted pulse of duration T (Fig. 1.5(a)) in which the carrier frequency is linearly swept (Fig. 1.5(b)). A pulse-compression filter with the time delay vs frequency characteristic given in Fig. 1.5(c) is used to delay one end of the received pulse relative to the other, thus producing at the filter output a narrower pulse of greater peak amplitude (Figs. 1.5(d) and 1.5.(e)). The linear time delay characteristic of the filter would act to delay the high frequency components at the start of the pulse more than the low frequency components at the end of the pulse, with frequency components inbetween experiencing a proportional delay. The net result would be a time compression of the pulse. Since a passive linear filter is postulated, the principle of conservation of energy applies and the buildup in peak power of the compressed pulse would be

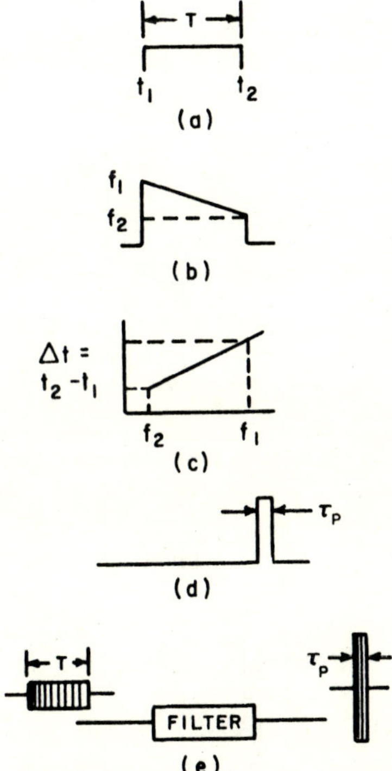

FIG. 1.5 Idealized pulse-compression characteristics. (a) Wide-pulse envelope. (b) Carrier-frequency modulation. (c) Filter time delay characteristics. (d) Compressed-pulse envelope. (e) Input-output waveforms of compression filter. (From Cook, Ref. 3, Chapter 6.)

proportional to the ratio of the pulse widths of the filter input and output signals. Thus,

$$\hat{P}_0/\hat{P}_i = T/\tau_P \tag{1-13}$$

where $\hat{P}_i$ is the peak power of the input pulse and $\hat{P}_0$ the peak power of the compressed pulse.

If the pulse width τ_P represents the desired resolution, it can be seen that the feasibility of this technique permits the use of a radar pulse of width T, representing an increase in the average power that can be transmitted. The associated frequency modulation contains the information necessary to construct the desired compressed pulse of greater effective peak power. However, the actual peak power limitations of a pulsed radar system are bypassed, making it possible for the system designer to achieve both the required pulse energy necessary for a stipulated detection range and the required resolving capability upon which the tactical mission of the system is based.

The fundamental relationships outlined above can be further developed on a heuristic level to yield a good approximation to the important characteristics associated with a pulse-compression system. Thus, the transmitted pulse is to have a rectangular envelope (this need not necessarily be so, as Chapter 3 points out, but is an assumption based on a high power transmitter) and a carrier frequency that is of the form

$$\omega = \omega_0 + \mu t, \qquad |t| \leqslant T/2 \tag{1-14}$$

The phase angle of the transmitted frequency becomes, when envelope contributions are ignored

$$\theta(t) = \int \omega \, dt = \omega_0 t + \frac{\mu t^2}{2} + C_1 \tag{1-15}$$

Thus, the phase angle θ is seen to contain a square-law term

$$\tfrac{1}{2}\mu t^2 \tag{1-16}$$

Further, if the product of the transmitted pulse width T and the frequency deviation $\Delta f = \Delta\omega/2\pi = f_2 - f_1$ is large, the linear progression of the carrier frequency between f_2 and f_1 should result in an essentially rectangular spectrum amplitude distribution. Figure 1.6 plots the essential features of the pulse derived by this method of reasoning.

The compression filter is to have a linear time delay vs frequency characteristic of the opposite sense to the linear FM sweep. Functionally this may be expressed as

$$t_d = 2K(\omega - \omega_1) + b \tag{1-17}$$

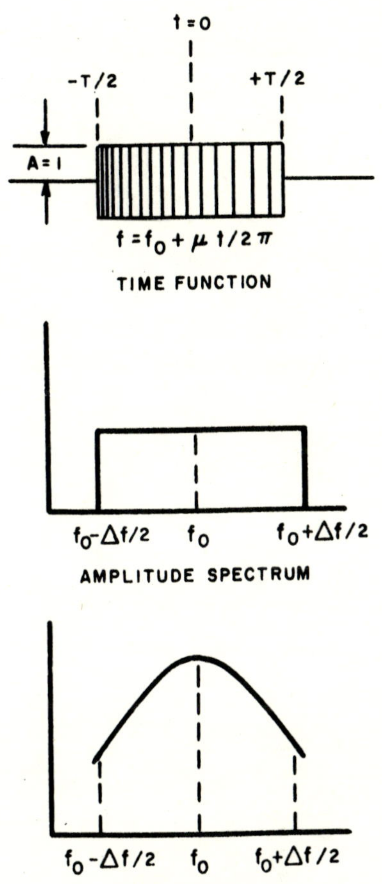

FIG. 1.6 Wide-pulse waveform parameters and assumed amplitude and phase spectra. (From Cook, Ref. 3, Chapter 6.)

Since the filter is being used in a band-pass application, the associated filter phase shift is

$$\beta_f(\omega) = \int t_d \, d\omega = K(\omega - \omega_1)^2 + b\omega + C_2 \qquad (1\text{-}18)$$

It must be realized that in a practical filter design only the portion of the phase function that corresponds to a positive time delay can be synthesized. The relationships of (1-17) and (1-18) are plotted in Fig. 1.7.

If the constants μ and K are properly matched, the spectrum at the compression-filter output is assumed to consist of a rectangular amplitude

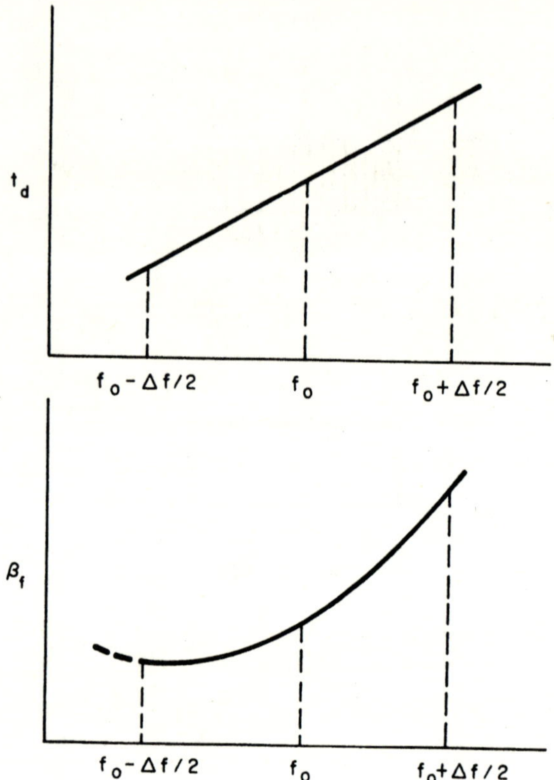

FIG. 1.7 Compression filter time delay and phase shift.

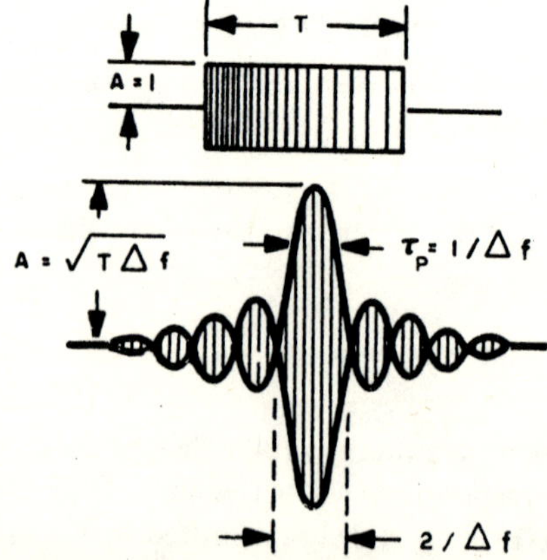

FIG. 1.8 Wide-pulse and compressed-pulse waveforms derived by heuristic analysis. (From Cook, Ref. 3, Chapter 6.)

distribution and a flat or linear phase component. The time function of the compressed pulse is readily recognized from the spectrum parameters as having a $(\sin x)/x$ envelope, the pulse width being $\tau_P = 1/\Delta f$ when measured 4 db down from the peak amplitude. The spacing between the first zeros of this envelope is $2/\Delta f$. The carrier frequency of the compressed pulse under the above assumption is a constant, f_0, and the peak amplitude is $\sqrt{T\Delta f}$ as shown in Fig. 1.8.

No loss of generality will result if the compression filter is assumed to have a rectangular bandwidth of Δf centered at f_0. Thus, this allows the pulse-compression operation to be thought of as matched filtering in the North sense, as described in Section 1.2.

The type of reasoning developed above ignores a number of second order effects that can be important to the radar designer. These will be more fully examined in subsequent chapters. The important results obtained by this approach are the relationship between the peak powers of the input and output waveforms and the pulse-compression parameter $T\Delta f$, or time-bandwidth product as it is designated in matched-filter terminology, yielding

$$\hat{P}_0/\hat{P}_i = T/\tau_P = T\Delta f \qquad (1\text{-}19)$$

A further result of significant interest is that the shape of the compressed pulse is $(\sin x)/x$ and not a truncated function as implied by the referenced patents. This particular result is substantiated by the rigorous analysis given in Chapter 6, and is of considerable concern for those radar applications in which many targets, that have a large dynamic range of reflecting cross sections, must be examined. This has given rise to various methods of mismatching the matched-filter characteristics to reduce the range sidelobes, as they are called, of the $(\sin x)/x$ time function. In addition, various types of nonlinear FM coded pulse waveforms have been considered as a means for implementing matched-filter pulse-compression signals with more desirable pulse shapes. The next section indicates that the heuristic approach can be placed on a more formal basis to examine the properties of the general class of frequency modulated pulse-compression functions.

1.5 The Matched-Filter Characteristics for a General FM Pulse-Compression Signal

In the previous section the approximate matched-filter requirements for the linear FM pulse-compression signal were derived by an analytical approach that ignored second order effects and concentrated on the more obvious waveform and matched-filter characteristics. An important result was the implied relationship between the linear frequency modulation function and the spectrum amplitude distribution for the case of the rectangular pulse envelope. Thus, the linear progression of the carrier frequency was assumed to trace out a rectangular spectrum amplitude function. Since the total bandwidth depends on the rate of the frequency sweep (radians/sec^2) and the total time duration of the signal, T, the relationship between the spectrum power density distribution and the signal phase θ for a fixed duration T can be expressed as

$$|A(\omega_k)|^2 \sim 1/\theta''(t) \tag{1-20}$$

Since

$$\theta(t) = \omega_0 t + \tfrac{1}{2}\mu t^2 + C_1$$

then

$$\theta''(t) = \mu \tag{1-21}$$

and

$$|A(\omega_k)|^2 \sim 1/\mu$$

In this outline the designation ω_k indicates the frequency parameter for which $A(\omega)$ takes on a value associated with a particular value of θ. The reasoning indicated by the preceding discussion can be placed on a much more formal basis that permits the derivation of the approximate spectrum and matched-filter parameters for a general frequency-modulated pulse-compression waveform that is expressed by

$$s(t) = a(t)\cos[\omega_0 t + \theta(t)] \tag{1-22}$$

where $d\theta(t)/dt$ is the general frequency modulation function. The method of analysis that yields these results is based on Lord Kelvin's principle of stationary phase [16], and has been applied to the investigation of pulse-compression matched-filter signals by Key *et al.* [17] and Watters [18]. This topic is discussed in Chapter 3.

REFERENCES

1. P. M. Woodward, "Probability and Information Theory, with Applications to Radar." Pergamon Press, Oxford, 1953.

2. G. L. Turin, An introduction to matched filters, *IRE Trans. Inform. Theory* **IT-6**, 311–329 (1960).

3. E. N. Fowle, The design of radar signals, The MITRE Corporation, Bedford, Massachusetts, Spec. Rept. SR-98 (1963).

4. M. Bernfeld, C. E. Cook, J. Paolillo, and C. A. Palmieri, Matched filtering, pulse compression, and waveform design, Parts I–IV, *Microwave J.* **7**, No. 10, 57–64, No. 11, 81–90, No. 12, 70–76 (1964); **8**, No. 1, 73–81 (1965).

5. D. O. North, Analysis of factors which determine signal-to-noise discrimination in radar, RCA Laboratories, Princeton, New Jersey, Rept. PTR-6c (June, 1943).

6. J. H. Van Vleck and D. Middleton, A theoretical comparison of visual, aural, and meter reception of pulsed signals in the presence of noise, *J. Appl. Phys.* **17**, 940–971 (1946).

7. E. Huttman, German Patent No. 768,068, March 22, 1940.

8. W. A. Cauer, German Patent No. 892,772, December 19, 1950.

9. D. O. Sproule and H. J. Hughes, British Patent No. 604,429, July 5, 1948.

10. R. H. Dicke, U.S. Patent No. 2,624,876, January 6, 1953.

11. S. Darlington, U.S. Patent No. 2,678,997, May 18, 1954.

12. C. E. Cook, Analysis and experimental verification of the performance of a network for time compressing a frequency modulated pulse. Unpublished research notes (1954).

13. C. E. Cook, Modification of pulse-compression waveforms, *Proc. Natl. Electron. Conf.* **14**, 1058–1067 (1958).

14. J. E. Chin and C. E. Cook, The mathematics of pulse compression—a problem in systems analysis, *Sperry Eng. Rev.* **12**, 11–16 (1959).

15. J. R. Klauder, A. C. Price, S. Darlington, and W. J. Albersheim, The theory and design of chirp radars, *Bell Syst. Tech. J.* **39**, 745–808 (1960).

16. S. Goldman, "Frequency Analysis, Modulation and Noise." McGraw-Hill, New York, 1948.

17. E. L. Key, E. N. Fowle, and R. D. Haggarty, A method of pulse compression employing nonlinear frequency modulation, M.I.T. Lincoln Laboratory, Lexington, Massachusetts, Tech. Rept. 207 (1959).

18. E. C. Watters, A note on the design of coded pulses, *Proc. Pulse Compress. Symp.*, Rome Air Development Center, New York, Tech. Rept. TR-59-161 (1959).

Bibliography of Texts on Radar System Analysis

1. M. I. Skolnik, "Introduction to Radar Systems." McGraw-Hill, New York, 1962.

2. D. K. Barton, "Radar System Analysis." Prentice-Hall, Englewood Cliffs, New Jersey, 1964.

3. R. S. Berkowitz (ed.), "Modern Radar: Analysis, Evaluation, and System Design." Wiley, New York, 1965.

4. S. Ye. Falkovich, "Reception of Radar Signals against a Background of Fluctuating Interference." Soviet Radio Publishing House, Moscow, 1961.

CHAPTER 2

Optimum Predetection Processing—Matched-Filter Theory

2.1 Introduction

In Chapter 1 it was stated that optimum predetection processing in a radar system for the case of white Gaussian noise interference consists of filtering the received signal with a network for which the impulse response is

$$h(t) = s(-t) \tag{2-1}$$

and the corresponding transfer function $H(\omega)$ is

$$H(\omega) = S^*(\omega) \tag{2-2}$$

where $s(t)$ is the signal at the input to the radar receiver and $S(\omega)$ is its corresponding Fourier transform. It is this processing technique that is referred to as matched filtering.

The properties of the matched filter just cited can be derived on the basis of several different criteria. These are:

(1) Signal-to-noise criterion
(2) Likelihood ratio criterion
(3) Inverse probability criterion

All three criteria yield the same result that a matched filter, as characterized above, is required in the receiver to extract the desired information from the signal most efficiently.

The signal-to-noise criterion assumes that the optimum predetection system maximizes the signal-to-noise ratio at one instant of time. This approach was investigated by several individuals, notably North [1] and Van Vleck and Middleton [2]. The likelihood ratio criterion defines the optimum predetection filter as one that processes the received signal in such a way as to provide an evaluation of the ratio of two conditional signal probabilities (i.e., signal plus noise or noise only). This criterion was formulated from considerations of statistical decision theory developed by Wald [3], Neyman and Pearson [4], and others. Some of the earliest applications of statistical decision theory to radar reception were by Marcum and Swerling [5], Lawson and Uhlenbeck [6], Middleton [7], Middleton and Van Meter [8, 9], and Peterson and Birdsall [10]. The inverse probability criterion describes the ideal receiver as one that processes the received signal so as to produce an *a posteriori* probability distribution (or an equivalent representation) as an output. This approach is based on Shannon's information theory, and was first developed for radar system analysis by Woodward and Davies [11], and later expanded by Woodward [12].

The next section presents the derivation of the matched-filter properties from the point of view of maximizing the signal-to-noise ratio. The remaining sections discuss in outline form the other aspects of detection theory that are based on statistical concepts.

2.2 Signal-to-Noise Criterion [1, 2]

The signal-to-noise ratio to which attention is most often directed in a radar system is defined as follows:

$$\frac{S}{N} = \frac{\text{peak instantaneous output signal power}}{\text{output noise power}} \tag{2-3}$$

The objective of this section will be to determine the properties of the linear system (filter) that maximizes the ratio defined in (2-3).

By definition, the peak instantaneous output signal power is, assuming a normalized output impedance of 1 ohm, the square of the maximum signal output voltage in the absence of noise. This signal voltage is thus

$$g(t) = \frac{1}{2\pi} \int_{-\infty}^{\infty} S(\omega)H(\omega)\exp[\,j\omega t\,]\,d\omega \tag{2-4}$$

where $S(\omega)$ is the signal spectrum and $H(\omega)$ is the system transfer function.

Considering the time T_d when $g(t)$ is desired to be a maximum, where T_d is equal to or greater than the signal duration, then

$$g(T_d) = \frac{1}{2\pi} \int_{-\infty}^{\infty} S(\omega)H(\omega)\exp[\,j\omega T_d\,]\,d\omega \tag{2-5}$$

In addition, the normalized noise power at the filter output is expressed as

$$\sigma^2 = \frac{N_0}{2} \int_{-\infty}^{\infty} |H(\omega)|^2 \frac{d\omega}{2\pi} \tag{2-6}$$

where N_0 is the one-sided noise power density, in watts/cycle/second, at the filter input. Thus, the ratio that is to be maximized is, from Eq. (2-3), expressed as follows:

$$S/N = g^2(T_d)/\sigma^2 \tag{2-7}$$

Since the optimization has been stated in terms of finding the best $H(\omega)$ when all other parameters are fixed (i.e., input signal and N_0), this can be approached through the calculus of variations, by means of which the assumption can be made that $H(\omega)$ can vary about its optimum function (if it exists).

As a matter of convenience it is helpful to work with the logarithm of the signal-to-noise ratio, as North did, so that the expression to be used is

$$\log S/N = 2 \log g(T_d) - \log \sigma^2 \tag{2-8}$$

The effect on signal-to-noise due to the variational $\delta H(\omega)$ can be expressed in terms of $g(T_d)$ and σ^2 in which $H(\omega)$ is implicit, as follows:

$$\delta(\log S/N) = \frac{\partial \log S/N}{\partial g(T_d)} \delta[g(T_d)] + \frac{\partial \log S/N}{\partial \sigma^2} \delta[\sigma^2] \tag{2-9}$$

From Eq. (2-8) one can obtain the indicated partial derivatives,[1] so that

$$\frac{\partial \log S/N}{\partial g(T_d)} = \frac{2}{g(T_d)} \tag{2-10a}$$

$$\frac{\partial \log S/N}{\partial \sigma^2} = -\frac{1}{\sigma^2} \tag{2-10b}$$

Substituting these into Eq. (2-9) yields the result

$$\delta(\log S/N) = \frac{2\delta[g(T_d)]}{g(T_d)} - \frac{\delta[\sigma^2]}{\sigma^2} \tag{2-11}$$

The additional substitution of Eqs. (2-5) and (2-6) for $g(T_d)$ and σ^2, respectively, changes the expression (2-11) to

$$\delta(\log S/N) = \frac{2\int_{-\infty}^{\infty} S(\omega)\, \delta[H(\omega)]\, \exp[\,j\omega T_d]\, d\omega/2\pi}{g(T_d)}$$

$$- \frac{(N_0/2)\int_{-\infty}^{\infty} \delta[H(\omega)H^*(\omega)]\, d\omega/2\pi}{\sigma^2} \tag{2-12}$$

---

[1] Variations, $\delta F(x, y)$, are governed by the same rules as are total differentials in ordinary calculus, except that no limits on size are imposed [15].

Since the variational of a product is

$$\delta[H(\omega)H^*(\omega)] = H^*(\omega)\,\delta[H(\omega)] + H(\omega)\,\delta[H^*(\omega)]$$

$$= 2\,\mathrm{Re}\,H^*(\omega)\,\delta[H(\omega)] \tag{2-13}$$

then, if the constraint is imposed that

$$\int_{-\infty}^{\infty} \delta[H(\omega)]\exp[j\omega t]\,d\omega = \text{real time function} = \lambda(t) \tag{2-14}$$

it can be shown that

$$\int_{-\infty}^{\infty} H^*(\omega)\,\delta[H(\omega)]\frac{d\omega}{2\pi} = \int_{-\infty}^{\infty} h(t)\lambda(t)\,dt = \int_{-\infty}^{\infty} H(\omega)\,\delta[H^*(\omega)]\frac{d\omega}{2\pi} \tag{2-15}$$

and therefore

$$2\,\mathrm{Re}\int_{-\infty}^{\infty} H^*(\omega)\,\delta[H(\omega)]\frac{d\omega}{2\pi} = 2\int_{-\infty}^{\infty} H^*(\omega)\,\delta[H(\omega)]\frac{d\omega}{2\pi} \tag{2-16}$$

Making use of the development of Eqs. (2-13)–(2-16) in the expression for $\delta(\log\ S/N)$, it is possible to factor out the variational term, $\delta[H(\omega)]$, in Eq. (2-12) to obtain

$$\delta(\log S/N) = \int_{-\infty}^{\infty}\left[\frac{2S(\omega)\exp[j\omega T_d]}{g(T_d)} - \frac{N_0 H^*(\omega)}{2\sigma^2}\right]\delta[H(\omega)]\frac{d\omega}{2\pi} \tag{2-17}$$

By setting this last relation equal to zero, one can obtain the conditions for the stationary point with respect to signal-to-noise ratio. For this zero value to exist and be independent of the variational change in $H(\omega)$, the following condition must be satisfied:

$$\frac{2S(\omega)\exp[j\omega T_d]}{g(T_d)} = \frac{N_0 H^*(\omega)}{2\sigma^2} \tag{2-18}$$

Since $g(T_d)$, N_0, and σ^2 are constants, this reduces to

$$H(\omega) = kS^*(\omega)\exp[-j\omega T_d] \tag{2-19}$$

which was the condition stated in Chapter 1.

An alternate derivation, showing that the condition of (2-19) indeed produces a maximum, can be obtained by making use of the following equivalent expressions for $g(T_d)$ and σ^2, where $h(t)$ is the impulse response of the filter $H(\omega)$:

$$g(T_d) = \int_0^{T_d} s(T_d - \tau)h(\tau)\,d\tau \tag{2-20a}$$

and

$$\sigma^2 = \frac{N_0}{2} \int_0^\infty h^2(\tau)\, d\tau \tag{2-20b}$$

where this last expression can be derived from Eq. (2-6) by the application of Parseval's theorem relating the total energy content in the spectrum $H(\omega)$ and the time function $h(t)$. The application of these to Eq. (2-3) leads to the following restatement of the signal-to-noise ratio:

$$\frac{S}{N} = \frac{\left[\int_0^{T_d} s(T_d - \tau) h(\tau)\, d\tau\right]^2}{(N_0/2)\int_0^\infty h^2(\tau)\, d\tau} \tag{2-21}$$

Using Schwarz's inequality[1] to obtain

$$\frac{S}{N} \leq \frac{\int_0^\infty s^2(T_d - \tau)\, d\tau \int_0^\infty h^2(\tau)\, d\tau}{(N_0/2)\int_0^\infty h^2(\tau)\, d\tau} \tag{2-22}$$

it follows that the signal-to-noise ratio will be a maximum when

$$h(t) = ks(T_d - t) \tag{2-23}$$

which is simply the Fourier transform of $H(\omega)$, yielding the alternate definition of the matched-filter requirement.

From the above it is evident that the maximum signal-to-noise ratio can be expressed as

$$\left(\frac{S}{N}\right)_{max} = \frac{\int_0^{T_d} s^2(T_d - \tau)\, d\tau}{N_0/2} \tag{2-24}$$

It is noted that the numerator in expression (2-24) represents the total signal energy, E, of the signal. This leads to the interesting conclusion that the detection capability of a particular signal depends only on its energy content, and not on the time structure of the signal. However, to obtain this condition in practice it is necessary to process the signal through a matched filter.

---

[1] One form of Schwarz's inequality states

$$\left(\int_{-\infty}^\infty f(x)g(x)\, dx\right)^2 \leq \int_{-\infty}^\infty f^2(x)\, dx \int_{-\infty}^\infty g^2(x)\, dx$$

By substituting the derived conditions for the optimum signal-to-noise ratio at the filter output into the alternate expressions given for $g(t)$, the following are obtained:

$$g(t) = \int_0^t s(\tau - t)s(\tau - T_d)\,d\tau \qquad (2\text{-}25)$$

and

$$g(t) = \int_{-\infty}^{\infty} |S(\omega)|^2 \exp[\,j\omega t\,]\,d\omega/2\pi \qquad (2\text{-}26)$$

The first of these expressions can be recognized as the autocorrelation function of $s(t)$, hence the equivalency of matched filtering and correlation processing. The relations derived for the matched-filter condition are graphically illustrated in Fig. 1.2 of Chapter 1.

In order to justify the signal-to-noise criterion one can examine the effect of this ratio on system performance. In this case, one of the best measures of system performance is the probability that an error will be made during the detection process. For example, consider the system which is designed to communicate a random binary message by means of on–off signaling. At the point of reception it is necessary to make some decision about the two possible states of the received message. The presence of noise, however, will cause some of the decisions to be incorrect. The probability that an error will be made is given by

$$P_E = \frac{1}{2}\left[1 - \frac{1}{2}\left\{\text{erf}\left(\frac{A - b}{\sqrt{2}\sigma}\right) + \text{erf}\left(\frac{b}{\sqrt{2}\sigma}\right)\right\}\right] \qquad (2\text{-}27)$$

where

$$\text{erf }x = \frac{2}{\sqrt{\pi}}\int_0^x \exp[\,-x^2\,]\,dx$$

and where b is a threshold voltage (decision criterion), $A^2/2\sigma^2$ the signal-to-noise ratio, and A the receiver signal amplitude, representing one of the message states. This relationship is based on the assumption that the noise in the system is Gaussian, that each message state is equally probable (i.e., $P_0 = P_1 = \frac{1}{2}$), that the decision is based on a single sample of the received signal, and that the decision criterion is a preestablished threshold level. It is apparent from this equation that the larger the signal-to-noise ratio, $A^2/2\sigma^2$, becomes the smaller the probability of error is. The conclusion is that the processing of the received signal should maximize the signal-to-noise ratio.

If the interfering Gaussian noise is not white, but can be described by a power density spectrum $N(\omega)$, then Eq. (2-6) is written

$$\sigma^2 = \frac{1}{2}\int_{-\infty}^{\infty} N(\omega)|H(\omega)|^2 \frac{d\omega}{2\pi} \tag{2-6a}$$

A similar development to that given by Eqs. (2-9)–(2-19) yields the general optimization of $H(\omega)$ described by

$$H(\omega) = \frac{kS^*(\omega)\exp|{-j\omega T_d}|}{N(\omega)} \tag{2-19a}$$

2.3   The Likelihood Criterion—Statistical Decision Theory [3–5]

It is evident from Eq. (2-27) that the probability of error also depends on the decision criterion as well as the signal-to-noise ratio. For

$$A^2/2\sigma^2 = \text{const} \tag{2-28}$$

it is possible to minimize the probability of making an error (P_E). For the conditions stated in the previous section, this minimum of P_E occurs when $b = A/2$, yielding

$$P_E = \frac{1}{2}\left[1 - \text{erf}\left(\frac{A}{2\sqrt{2}\sigma}\right)\right] \tag{2-29}$$

It can be shown that a decision based on examining a single sample of the received signal and comparing it to the criterion $b = A/2$ is equivalent to making a decision based on the criterion:

$$\ell = \frac{P_1(v)}{P_0(v)} \geq 1; \qquad \text{decide } A \text{ plus noise}$$

$$\tag{2-30}$$

$$\ell = \frac{P_1(v)}{P_0(v)} < 1; \qquad \text{decide noise only}$$

Here, $P_1(v)$ is the conditional probability that the received signal voltage (v) will lie in the range of values dv, given that the received signal consists of signal plus noise. Similarly, $P_0(v)$ is the conditional probability that the received signal voltage will lie in the same range of values if it consists of noise only. This ratio, denoted by ℓ, is the so-called likelihood ratio; it is one of the foundations of statistical decision theory and parameter estimation theory.

In general, the probability of error will always be a minimum if the decision is in favor of A plus noise when

$$\ell \geqslant \frac{P_0}{P_1} \qquad (2\text{-}31)$$

and is in favor of noise only when

$$\ell < \frac{P_0}{P_1} \qquad (2\text{-}32)$$

where P_0 is the probability that the message state contains noise only and P_1 is the probability that the message state contains signal plus noise. This criterion is known as the Bayes decision rule, and is based on the viewpoint of Siegert's "ideal" observer [6], who desires a minimum overall probability of error. However, since the values of P_0 and P_1 are usually not known at the receiver, as for the case of the radar problem, the ratio P_0/P_1 is replaced by a decision criterion that is obtained through the Neyman–Pearson theory of testing statistical hypotheses [4]. Briefly, this theory is based on maintaining a fixed probability of error for the case of the signal being noise only. This fixed probability is known as the probability of false alarm. The decision criterion is then determined by minimizing the probability of error for the case when the signal is signal plus noise, this being equivalent to maximizing the probability of detection when the false alarm probability is fixed. The statistical tests that are specified as a result of this procedure, however, also make use of the likelihood ratio and a Bayes type decision rule. Extensive discussion of this may be found in Middleton [7].

Up to this point, attention has been directed to applying the hypothesis tests to a single sample of the received signal in order to facilitate the development of the important statistical concepts. Optimum predetection processing in this case consists of supplying the received signal in its raw form to an observer (human or otherwise). In general, however, more than one statistically independent sample is used. If the message state is constant between the interval at which the samples are taken it can be shown that the optimum predetection system is an integrator. However, if the message varies as a function of time during this interval then it can be shown that the optimum predetection receiver is a correlator. This will now be examined.

For the case of multiple samples, when the exact form of the message signal is known, the likelihood ratio is expressed as follows:

$$\ell(z = v_1, v_2, \ldots, v_{2BT_0}) = \frac{P_1(z = v_1, v_2, \ldots, v_{2BT_0})}{P_0(z = v_1, v_2, \ldots, v_{2BT_0})} \qquad (2\text{-}33)$$

The symbol $(z = v_1, v_2, \ldots, v_{2BT_0})$ can be visualized as a point lying in multidimensional space consisting of $2BT_0$ coordinates, where B is the entire bandwidth of the noise and T_0 is the observation interval. The dimensions of this space consist of $2BT_0$ observations of the received signal taken at intervals $1/B$ to ensure statistical independence. For the case of rf signals, two orthogonally related samples are taken at each observation point, yielding $2BT_0$ samples. Figure 2.1 illustrates the method of sampling for the case of video signals. For this case the signal is sampled at intervals of $1/2B$ in order to obtain statistical independence. A comprehensive discussion of sampling theory can be found in Woodward [12].

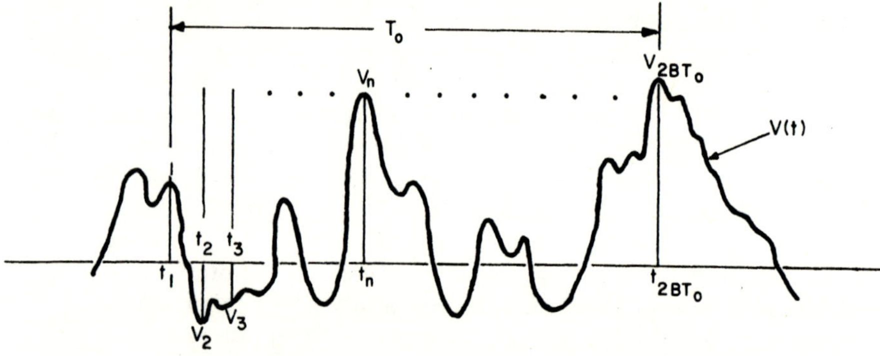

FIG 2.1   Illustration of a typical observation interval of the envelope of a radar echo which is sampled at $t_1, t_2, \ldots, t_n$. The question to be answered is: does this represent signal plus noise or is this noise only? (From Bernfeld *et al.*, Ref. 4, Chapter 1.)

Corresponding to each sample point z there is a single value of the multidimensional probability density functions, p_1 and p_0, and a corresponding value of the likelihood ratio. For example, the two dimensional probability density function corresponding to two observations of Gaussian noise would appear as shown in Fig. 2.2.

Through the application of the sampling theorem it has been shown that the received signal is sufficiently processed, for an output that represents the likelihood ratio, when it is correlated with a replica of the message signal $s(t)$, or:

$$c_T = \int_0^{T_0} v(t)s(t)\, dt \tag{2-34}$$

When the received signal $v(t)$ contains $s(t)$, then the correlated message signal is described by

$$\int_0^{T_0} v(t)s(t)\, dt = \int_0^{T_0} s^2(t)\, dt + \int_0^{T_0} (\text{noise})\, s(t)\, dt \tag{2-35}$$

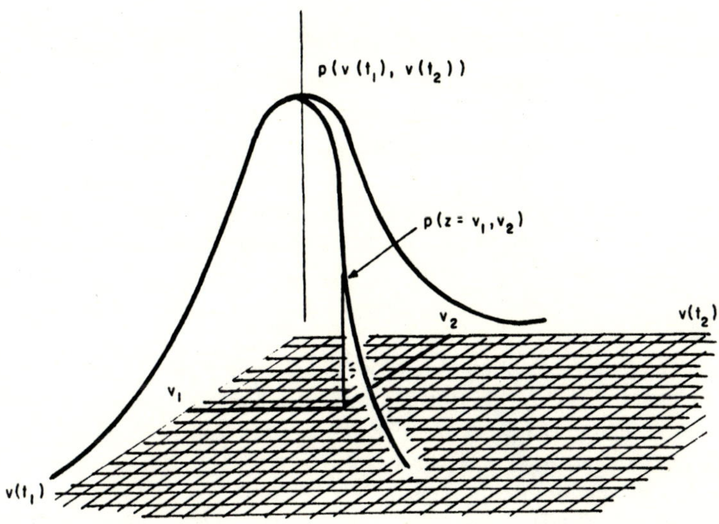

FIG. 2.2 Illustration of the right quarter plane of a joint probability density function $p(v(t_1), v(t_2))$ for the case of only two samples of (Gaussian) noise. The point $z = v_1, v_2$ represents the magnitude of the noise samples at the times t_1 and t_2, respectively. (From Bernfeld et al., Ref. 4, Chapter 1.)

This expression represents the energy in the message signal plus a small contribution from the correlation between noise and the message signal. However, the output of a matched filter is given by

$$y(t) = \int_0^t v(t - \tau)s(T_d - \tau)\, d\tau \qquad (2\text{-}36)$$

Therefore, it is seen that the correlated received signal equals the output of a matched filter when $t = T_d$.[1] In addition, at this instant the output signal $y(t)$ has reached a maximum.

2.4 The Likelihood Criterion—Parameter Estimation Theory [11–14]

It is evident that matched filtering is more than necessary when the message signal is known exactly. However, the message signal often carries information in the form of some parameter of the signal. For example, in the radar application this parameter may be the delay of the received signal

---

[1] In general, the implementation of a matched filter is less complex than a correlation receiver. A single matched filter yields an output for all possible values of range delay of a received signal. A correlation receiver must provide an undistorted delayed replica signal for each range delay of interest, and thus may require many such signals if a large range interval is to be searched.

referenced to the time of the transmitted signal. When these parameters are important, provisions must be taken to prevent this information from being destroyed while processing the received signal for optimum detection decisions. Woodward, and others, have shown that signal delay information is preserved by correlating the signal $v(t)$ with all possible received signals $s(t - t_x)$, where t_x is a delay parameter. On this basis, the delay information can be extracted while at the same time preserving the requirements for optimum detection decisions. To achieve this result it is necessary to process the received signals through a matched filter. In this case the matched filter only represents sufficient processing, and thus is the ideal predetection system.

When the tactical application of the radar system requires the estimation of other parameters in addition to that of delay, a single matched filter is no longer sufficient. In this case, the theory of matched filtering shows that a bank of matched filters, as shown in Fig. 2.3, suitably modified to cover the extent of the unknown parameters is necessary. For example, when Doppler shift is a desired parameter the bank of matched filters must have center frequencies that include the range of all the expected shifts in frequency of the received signal. The basic theory of parameter estimation and the fundamental radar signal relationships derived from it are developed in greater detail in Chapter 5, with applications to specific radar signals being discussed in Chapter 9.

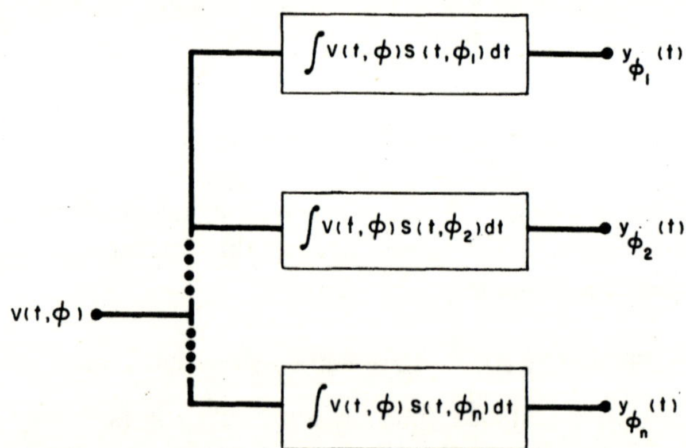

FIG. 2.3 Illustration of a typical bank of optimum (matched) predetection filters which yield output signals from which parameters such as Doppler shift can be estimated. (From Bernfeld *et al.*, Ref. 4, Chapter 1.)

2.5 Inverse Probability [11, 12]

During the early 1950's Woodward and Davies applied Shannon's information theory to the solution of the ideal receiver problem. They defined the ideal receiver as one that would supply at the output all the available information existing in the received signal about the radar parameters. This information takes the form of a probability distribution, $p_v(x)$, for each desired parameter, given that a signal has been received. It was argued that it was not necessary to consider the observer as part of the reception system, as is done in statistical decision theory, since the observer can do no better than determine $p_v(x)$. This probability distribution is referred to as the inverse, or *a posteriori*, probability. Again, however, this distribution can be described in terms of the likelihood function. Therefore, except for some differences in constants of proportionality, the structure of the ideal receiver is essentially the same as before.

The relationship between the inverse probability and the likelihood ratio can be derived by means of the Bayes rule for conditional probability. To develop this concept it is most convenient to consider the binary case, letting the symbol "1" indicate signal plus noise and the symbol "0" noise only. The respective probabilities of signal plus noise, or noise only are then expressed by

$$P_v(1) = \frac{P_1 P_1(v)}{p(v)\, dv}, \qquad P_v(0) = \frac{P_0 P_0(v)}{p(v)\, dv} \tag{2-37}$$

where P_1 and P_0 are the *a priori* probabilities defined in conjunction with the likelihood inequalities of Eqs. (2-31) and (2-32):

$$p(v)\, dv = P_1 P_1(v) + P_0 P_0(v) \tag{2-38}$$

so that by making the appropriate substitutions one obtains

$$P_v(1) = \frac{P_1(v)/P_0(v)}{P_1(v)/P_0(v) + P_0/P_1} = \frac{\ell}{\ell + (P_0/P_1)} \tag{2-39}$$

and

$$P_v(0) = \frac{P_0(v)/P_1(v)}{P_0(v)/P_1(v) + P_1/P_0} = \frac{1}{1 + \ell(P_1/P_0)} \tag{2-40}$$

Once the receiver has evaluated $P_v(1)$ and $P_v(0)$ it has completed all the processing that is required. The values of these two probabilities represent the maximum information available about the two possible message states that can be extracted from the received signal. Knowing these probabilities, the decision process becomes automatic, since the most probable signal state is always selected. It is of interest to note that this criterion is equivalent to the Bayes decision rule. This can be seen by taking the ratio of $P_v(1)/P_v(0)$, and assuming that the decision criterion is unity. Thus,

$$\frac{P_v(1)}{P_v(0)} = \ell\, \frac{P_1}{P_0} \tag{2-41}$$

and when

$$\ell \geq P_0/P_1, \qquad \text{choose state "1"}$$

$$\ell < P_0/P_1, \qquad \text{choose state "0"}$$

The application of the method of inverse probability to the general radar problem can be illustrated by the example of the radar system that must perform the measurement of the delay, or range, of the received signal. From the viewpoint of Woodward [12], the ideal receiver can do no more than compute the conditional probability $p_v(\tau)$ that a signal has arrived at time τ when it is given that the signal $v(t)$ is present. Making use of the Bayes rule, one can express the joint probability of τ and $v(t)$ as

$$p(\tau, v) = p_\tau(v)p(\tau) = p_v(\tau)p(v) \tag{2-42}$$

where $p(\tau)$ is the prior probability distribution of the signal being delayed by τ seconds, $p(v)$ the prior probability distribution of the occurrence of the signal $v(t)$, $p_\tau(v)$ the conditional probability distribution of $v(t)$ given τ, and $p_v(\tau)$ the conditional probability distribution of τ given $v(t)$.

The signal $v(t)$ is

$$v(t) = s(t - t_d) + n(t) \tag{2-43}$$

where t_d is the delay of the signal and $s(t)$ is the transmitted signal. The conditional probability $p_\tau(v)$ can be expressed in terms of the likelihood function, which, for $v(t)$ a continuous ensemble of independent random Gaussian variables, becomes

$$p_\tau(v) = k \exp\left[-\frac{1}{N_0}\int_0^T [v(t) - s(t - \tau)]^2 \, dt\right] \tag{2-44}$$

The constant k is a normalizing factor determined by

$$\int p_\tau(v) \, dv = 1 \tag{2-45}$$

Since the radar receiver is required to compute $p_v(\tau)$, one obtains from (2-42)

$$p_v(\tau) = \frac{p_\tau(v)p(\tau)}{p(v)} = \frac{kp(\tau)}{p(v)}\exp\left[-\frac{1}{N_0}\int_0^T [v(t) - s(t - \tau)]^2 \, dt\right] \tag{2-46}$$

Once a signal is present $p(v)$ is represented by some constant. If it is also assumed that $p(\tau)$ is a uniform distribution, then this is also a constant. Both of these constants can be absorbed into k, and (2-46) becomes

$$p_v(\tau) = k \exp\left[-\frac{1}{N_0}\int_0^T [v(t) - s(t - \tau)]^2 \, dt\right] \tag{2-47}$$

or

$$p_v(\tau) = k \exp\left[-\frac{1}{N_0}\int_0^T v^2(t)\,dt\right]\exp\left[-\frac{1}{N_0}\int_0^T s^2(t-\tau)\,dt\right]$$

$$\times \exp\left[\frac{2}{N_0}\int_0^T v(t)s(t-\tau)\,dt\right] \tag{2-48}$$

The first two exponentials are functions, respectively, of the energy of the received signal $v(t)$ and the transmitted signal $s(t)$, and thus are constants that can also be absorbed into k. It can be seen, then, that the ideal receiver is one that performs the operation expressed by the integral

$$\int_0^T v(t)s(t-\tau)\,dt = \int_0^T s(t-t_d)s(t-\tau)\,dt + \int_0^T n(t)s(t-\tau)\,dt \tag{2-49}$$

This integral will be a maximum when $\tau = t_d$, thus yielding the maximum value of $p_v(\tau)$, or the most probable value of τ. Woodward points out that it is not actually necessary to compute $p_v(\tau)$, since it is sufficient, from the point of view of the information process, to just perform the operation indicated by (2-49). The sufficient receiver that computes (2-49) performs the same irreversible operation on $v(t)$ that would be performed by the ideal receiver in computing $p_v(\tau)$. Irreversibility of the receiver operation implies that the desired information in the signal is retained, and that unwanted information is destroyed. Further, in the case of the ideal receiver, once $p_v(\tau)$ has been determined one cannot deduce the structure of $v(t)$ from this function. From the earlier sections of this chapter, it can be noted that (2-49) represents a correlation receiver, or equivalently a receiver that contains a matched filter as described in Section 2.2. The sufficient receiver maximizes the signal-to-noise ratio for Gaussian white noise, which is equivalent to obtaining the full posterior distribution of τ. Woodward's main point is that this is all that can be expected of the ideal receiver from the information theory viewpoint. This approach does not take into account explicitly such things as the decision rules or error probabilities of the system. However, since the structure of the receiving system that maximizes the information at the receiver output is identical to that for the other criteria discussed, it is clear that there is no incompatibility among the various criteria for optimizing the operation of the radar system.

Figure 2.4 summarizes the three general theoretical points of view concerning optimum predetection filtering.

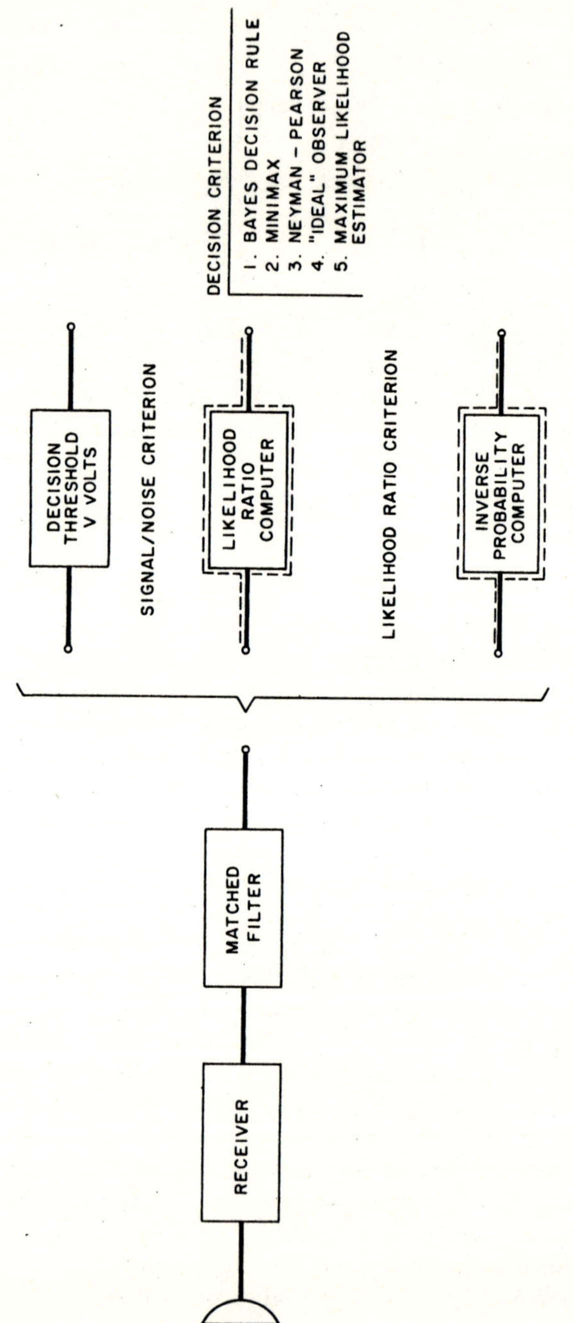

FIG. 2.4 Summary of the theoretical aspects of optimum predetection filtering. (From Bernfeld *et al.*, Ref. 4, Chapter 1.)

REFERENCES

1. D. O. North, An analysis of the factors which determine signal-to-noise discrimination in radar, RCA Lab. Princeton, New Jersey, Rept. PTR-6c (1943); and *Proc. IEEE* **51**, 1016–1027 (1963).

2. J. H. Van Vleck and D. Middleton, A theoretical comparison of visual, aural, and meter reception of pulsed signals in the presence of noise, *J. Appl. Phys.* **17**, 940–971 (1946).

3. A. Wald, "Statistical Decision Functions." Wiley, New York, 1950.

4. J. Neyman and E. S. Pearson, The problem of the most efficient tests of statistical hypothesis, *Phil. Trans. Roy. Soc.* Ser. A **231**, 289–333 (1933).

5. J. I. Marcum and P. Swerling, Studies of target detection by pulsed radar, *IRE Trans.* **IT-6**, 59–308 (1960). See also: Rand Research Memos: RM 754 (December, 1947), RM 753 (July, 1948), RM-1217 (March, 1954).)

6. J. L. Lawson and G. E. Uhlenbeck, "Threshold Signals." McGraw-Hill, New York, 1950.

7. D. Middleton, Statistical criteria for the detection of pulsed carriers in noise, *J. Appl. Phys.* **24**, 371–378, 379–391 (1953).

8. D. Middleton and D. Van Meter, Modern statistical approaches to reception in communication theory, *IRE Trans.* **IT-4**, 119–145 (1954).

9. D. Middleton and D. Van Meter, Detection and extraction of signals in noise from the point of view of statistical decision theory, *J. Soc. Ind. Appl. Math.* **3**, 192–253 (1955); **4**, 88–119 (1956).

10. W. W. Peterson and T. G. Birdsall, The theory of signal detectability, Electronic Defense Group, Dept. EE, Univ. of Michigan, Ann Arbor, Michigan, Tech. Rept. 13 (June, 1953).

11. P. M. Woodward and I. L. Davies, A theory of radar information, *Phil. Mag.* **41**, 1001 (1950).

12. P. M. Woodward, "Probability and Information Theory with Applications to Radar." Pergamon Press, Oxford, 1953.

13. R. Manasse, Range and velocity accuracy from radar measurements, Lincoln Lab., M.I.T., Lexington, Massachusetts, Group Rept. 312–26 (February, 1955).

14. E. J. Kelly, I. S. Reed, and W. L. Root, The detection of radar echoes in noise, *J. Soc. Ind. Appl. Math.* **8**, 309–341, 481–507 (1960).

15. L. A. Pipes, "Applied Mathematics for Engineers and Physicists," pp. 292–298. McGraw-Hill, New York, 1946.

CHAPTER 3

Matched-Filter Requirements for Arbitrary FM Pulse-Compression Signals

3.1 Introduction

A general line of reasoning was indicated in Chapter 1 in relating the frequency modulation function to the spectrum amplitude distribution for the linear FM signal. The principle of stationary phase can be used to place the heuristic concepts developed in Chapter 1 on a more logical basis. This permits the intuitive ideas discussed there to be extended to a large class of general FM pulse compression matched-filter signals described by

$$s(t) = a(t)\cos[\omega_0 t + \theta(t)] \tag{3-1}$$

where $a(t)$ is the transmitted envelope function, and $\omega_0 + \theta'(t)$ the radian frequency modulation function. In most radar applications $a(t)$ is assumed to be a rectangular pulse envelope function, although the general analysis is not restricted to this particular case.

Historically, the nonlinear FM pulse-compression signals were suggested as a means for improving the compressed-pulse waveform that had a $(\sin x)/x$ form for the linear FM matched-filter function. However, this approach leads to some important restrictions that must be placed on the use of nonlinear FM waveforms. Other techniques of more general applicability for improving the shape of the matched-filter output signal are discussed in Chapter 7. However, the analysis of the general class of signals described by Eq. (3-1) provides important insights concerning the design of FM pulse-compression signals.

3.2 The Principle of Stationary Phase

In developing some of the aspects of the analysis of time and frequency functions it is often expedient to make use of the exponential complex representation of a signal. If the real signal is given by

$$s(t) = a(t)\cos[\omega_0 t + \theta(t)] \qquad (3\text{-}2)$$

an equivalent complex representation of the modulation is

$$u(t) = a(t)\exp[\,j\theta(t)] \qquad (3\text{-}3)$$

The Fourier transform of this complex time function is expressed as follows:

$$U(\omega) = \int_{-\infty}^{\infty} a(t)\exp[\,j\{-\omega t + \theta(t)\}]\,dt \qquad (3\text{-}4)$$

where $\omega = 2\pi f$. $U(\omega)$ and $u(t)$ are related by Parseval's theorem, such that

$$\int_{-\infty}^{\infty} |u(t)|^2\,dt = \frac{1}{2\pi}\int_{-\infty}^{\infty} |U(\omega)|^2\,d\omega = \int_{-\infty}^{\infty} |U(f)|^2\,df = 2E \qquad (3\text{-}5)$$

where E is the energy in the real signal of Eq. (3-2).[1] The presence of the imaginary exponential term in (3-4) makes the integrand in this equation an oscillating function, varying at the rate

$$\frac{d}{dt}[\omega t - \theta(t)] \qquad (3\text{-}6)$$

It can be seen that the major contribution to the Fourier spectrum given by Eq. (3-4) will occur when the rate of change of oscillation is minimal. This condition is embodied in what is referred to as the stationary phase point, which is given by

$$\frac{d}{dt}[\omega t - \theta(t)] = 0 \qquad \text{or} \qquad \omega - \theta'(t) = 0 \qquad (3\text{-}7)$$

This expression is a parametric one, since it relates two independent variables, ω and t. Thus, a particular value of t that meets the condition (3-7) can only be established after a value of ω has been assumed. Unless $\theta(t)$ is a constant, the time t when the phase is stationary will be different for each frequency ω_k. The two components of the exponent in Eq. (3-4) can be represented, as shown in Fig. 3.1, as two independently rotating vectors. It can be seen that the maximum effect of these two vectors in this integral equation for $U(\omega)$ will occur when the vectors are rotating at the same rate. For a particular frequency ω_k indicated in Fig. 3.1, this will prevail at a time t that produces the constant phase difference $\omega_k t - \theta(t)$. Since ωt is a vector rotating at a constant angular velocity, and $\theta(t)$ is in general rotating at a varying rate, there will be some time t_k when the rate of change

---

[1] The use of the radian frequency $\omega = 2\pi f$ is a convenience that makes notation in the complex formulation more compact, although this requires the $(2\pi)^{-1}$ factor in integral expressions such as given by Eq. (3-5).

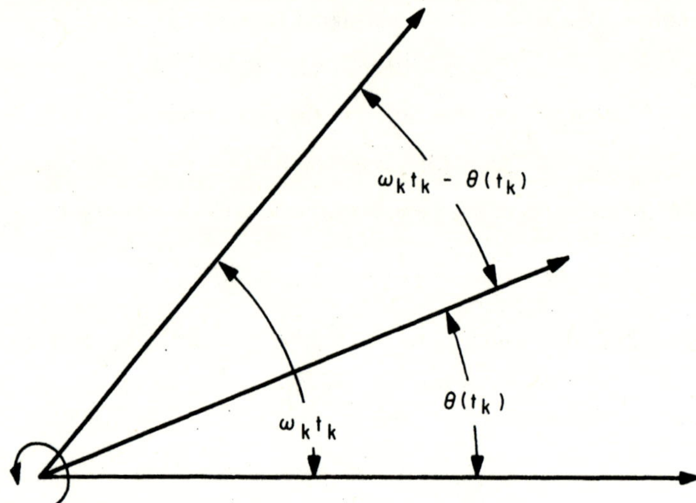

FIG. 3.1   Phasor representation of imaginary components of the spectrum integral of Eq. (3-4). (From Bernfeld *et al.*, Ref. 4, Chapter 1.)

of the phase difference is zero. This is identically the condition described by Eq. (3-7). Figure 3.2 illustrates how the phase associated with each of these two vectors builds up as a function of time. The stationary point occurs at a time t_k when the tangent to the function $\theta(t)$ parallels the slope of $\omega_k t$. Since the slope of this straight line curve is ω_k, the frequency parameter, it is again seen that the time t_k at which phase stationarity is obtained is related to ω_k.

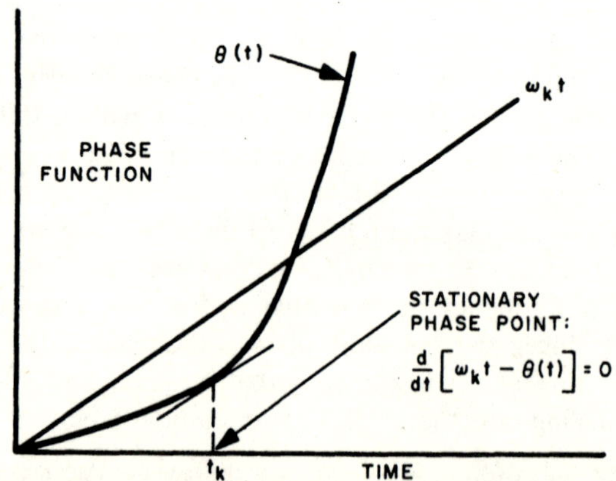

FIG. 3.2   Stationary phase criterion for frequency ω_k. (From Bernfeld *et al.*, Ref. 4, Chapter 1.)

Referring again to Fig. 3.2, the difference between $\omega_k t$ and $\theta(t)$ can be expressed in a Taylor series about the time t_k yielding

$$\omega_k t - \theta(t) = \omega_k t_k - \theta(t_k) + [\omega_k - \theta'(t_k)](t - t_k)$$

$$- \frac{\theta''(t_k)}{2!}(t - t_k)^2 + \cdots \tag{3-8}$$

If in (3-8) the time difference $(t - t_k) = \pm\delta$ is kept small this equation can be accurately represented by the first three terms. Thus, in the small interval around t_k

$$\omega_k t - \theta(t) = \omega_k t_k - \theta(t_k) + [\omega_k - \theta'(t_k)](t - t_k)$$

$$- \frac{\theta''(t_k)}{2!}(t - t_k)^2 \tag{3-9}$$

Since, by the condition of Eq. (3-7), $\omega_k - \theta'(t_k) = 0$, (3-9) reduces to, in the interval around t_k

$$\omega_k t - \theta(t) = \omega_k t_k - \theta(t_k) - \frac{\theta''(t_k)}{2}(t - t_k)^2 \tag{3-10}$$

By substituting this last relationship into the Fourier transform, $U(\omega)$, and considering $a(t)$ to be relatively constant when compared to the rate of change of phase at t_k, the spectrum for the frequency can be written as

$$U(\omega_k) = a(t_k)\int_{t_k - \delta}^{t_k + \delta} \exp\left[-j\left(\omega_k t_k - \theta(t_k) - \frac{\theta''(t_k)}{2}(t - t_k)^2\right)\right] dt \tag{3-11}$$

It is this latter result that will be made use of in deriving the approximate relationship between a known coded waveform function and the appropriate matched filter.

3.3 Application of Principle of Stationary Phase to General Pulse-Compression Signals

The principle of stationary phase can be applied to oscillating functions that have continuous first derivatives. Among such functions are a large class of pulse-compression matched-filter signals that are based on frequency modulation techniques. The linear FM signal represents a special case of this general class of waveforms.

The spectrum of the signal given by Eq. (3-1) is

$$S(\omega) = \int_{-\infty}^{\infty} a(t)\cos[\omega_0 t + \theta(t)]\exp[-j\omega t]\, dt \tag{3-12}$$

This can be further rearranged to

$$S(\omega) = \tfrac{1}{2}\int_{-\infty}^{\infty} a(t)\exp[\,j\{(\omega_0 - \omega)t + \theta(t)\}]\,dt$$

$$+ \tfrac{1}{2}\int_{-\infty}^{\infty} a(t)\exp[-j\{(\omega_0 + \omega)t + \theta(t)\}]\,dt$$

$$= S_+(\omega) + S_-(\omega) \tag{3-13}$$

$S_+(\omega)$ and $S_-(\omega)$ are essentially the positive and negative frequency portions of the spectrum. It will be assumed that the respective center frequencies of $S_+(\omega)$ and $S_-(\omega)$ are sufficiently far removed from each other that the effect of $S_-(\omega)$ in the positive frequency region is vanishingly small. This condition is shown in Fig. 3.3. When this is the case the signal is said to be narrow band, and the exponential complex representation of the signal may be used. From (3-4) one obtains $U(\omega) = 2S_+(\omega + \omega_0)$. Thus, the development of the previous section can be applied to the analysis of narrow band frequency modulated signals. The complex spectrum of the signal given by Eq. (3-1) is

$$U(\omega) = \int_{-\infty}^{\infty} a(t)\exp[\,j\{-\omega t + \theta(t)\}]\,dt \tag{3-14}$$

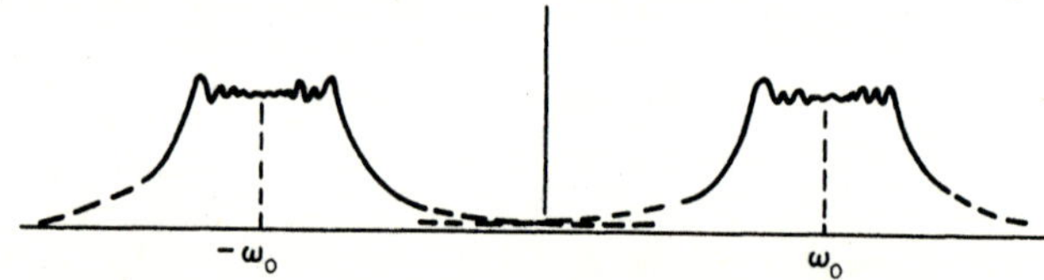

FIG. 3.3   Positive and negative spectrum contributions.

The stationary phase condition for Eq. (3-14) is given by Eq. (3-7), and the development leading to Eq. (3-11) can be employed. In Eq. (3-11) t_k represents a constant time point. Therefore, terms depending on t_k alone may be taken outside of the integral. This results in

$$U(\omega) = a(t_k)\exp[\,j\{-\omega t_k + \theta(t_k)\}]\int_{t_k-\delta}^{t_k+\delta} \exp\!\left[j\frac{\theta''(t_k)}{2}(t - t_k)^2\right]dt \tag{3-15}$$

By making the following changes of variables

$$t - t_k = \mu \qquad \text{and} \qquad \frac{\theta''(t_k)}{2}\mu^2 = \frac{\pi y^2}{2}$$

so that

$$d\mu = \sqrt{\pi}\,[\theta''(t_k)]^{-1/2}\,dy$$

the spectrum contribution about the point t_k becomes

$$U(\omega) = 2\sqrt{\pi}\,\frac{a(t_k)}{\sqrt{|\theta''(t_k)|}}\,\exp[j\{-\omega t_k + \theta(t_k)\}] \int_0^{(|\theta''(t_k)|/\pi)^{1/2}\delta} \exp\left[j\frac{\pi y^2}{2}\right]dy \quad (3\text{-}16)$$

The integral in Eq. (3-16) is one form of the Fresnel integral, and if the upper limit is large can be approximated by

$$\frac{1}{\sqrt{2}}\exp\left[j\frac{\pi}{4}\right] \tag{3-17}$$

Substituting (3-17) into Eq. (3-16) results in the contribution to the spectrum at t_k being expressed as

$$U(\omega) = \sqrt{2\pi}\,\frac{a(t_k)}{\sqrt{|\theta''(t_k)|}}\,\exp\left[j\left(-\omega t_k + \theta(t_k) + \frac{\pi}{4}\right)\right] \tag{3-18}$$

Under the assumption that Eq. (3-17) holds at all values of t_k, the following relationship is obtained

$$|U(\omega_t)|^2 \doteq 2\pi\,\frac{a^2(t)}{|\theta''(t)|} \tag{3-19}$$

where ω_t is used to indicate the dependence of the frequency variable on t. The development that leads to (3-19) is similar to that of Key *et al.* [1, 2], in which the point of view is adopted that the transmitted signal represents the output of one of a pair of matched filters (one in the transmitter and one in the receiver, as shown by Fig. 3.4a), and in complex form is expressed by

$$u(t) = \frac{1}{2\pi}\int_{-\infty}^{\infty} |U(\omega)|\,\exp[j(\Phi(\omega) + \omega t)]\,d\omega \tag{3-20}$$

where $\Phi(\omega)$ is the spectrum phase function. The stationary phase point for Eq. (3-20) is given by

$$\Phi'(\omega) = -t \tag{3-21}$$

The corollary relationship, equivalent to Eq. (3-19), that can be obtained by the stationary phase analysis just outlined is

$$a^2(t_\omega) \doteq \frac{1}{2\pi}\,\frac{|U(\omega)|^2}{|\Phi''(\omega)|} \tag{3-22}$$

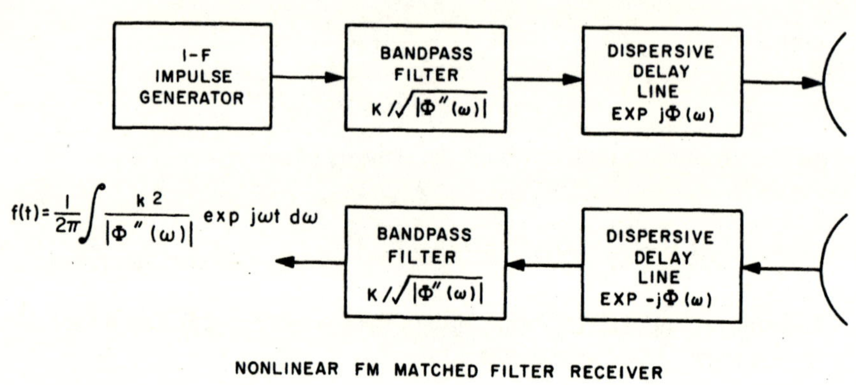

$$f(t) = \frac{1}{2\pi} \int \frac{k^2}{|\Phi''(\omega)|} \exp j\omega t \, d\omega$$

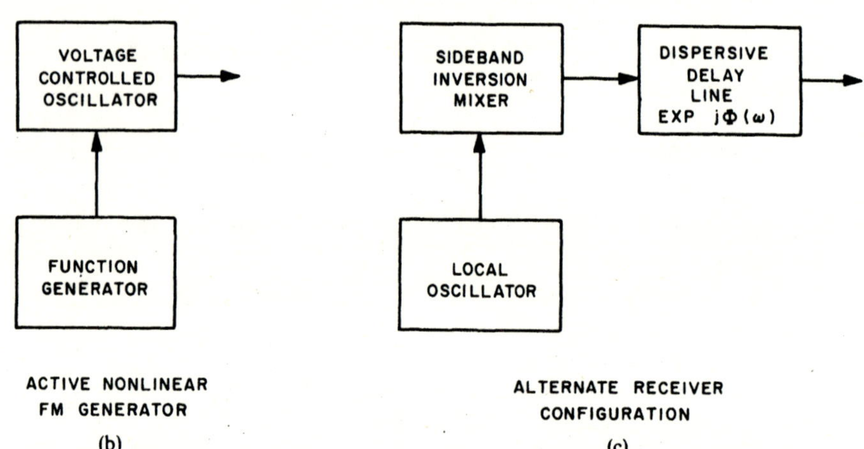

Fig. 3.4 Matched-filter transmitter and receiver techniques. (a) Top: Nonlinear FM matched-filter transmitter. Bottom: Nonlinear FM matched-filter receiver. (b) Active nonlinear FM generator. (c) Alternate receiver configuration. (From Bernfeld *et al.*, Ref. 4, Chapter 1.)

By replacing t_k by t in Eq. (3-18), it can be noted that the spectrum phase $\Phi(\omega)$ is approximately given by

$$\Phi(\omega) \doteq -\omega t + \theta(t) + \pi/4 \tag{3-23}$$

Differentiating with respect to ω, one obtains

$$\Phi'(\omega) \doteq -t - \left(\omega - \frac{d\theta(t)}{dt}\right)\frac{dt}{d\omega} \tag{3-24}$$

Applying the stationary phase condition given by Eq. (3-7), (3-24) becomes

$$\Phi'(\omega) \doteq -t \qquad\qquad (3\text{-}25)$$

This approximate relationship from the stationary phase analysis of the Fourier transform of the time function agrees with the stationary phase point defined by Eq. (3-21) for the Fourier transform of the spectrum function. This indicates the equivalence of the relationships given by Eqs. (3-19) and (3-22). These represent a set of parametric equations that can be used to obtain approximate expressions for $|U(\omega)|$ (or $|S(\omega)|$) and $\Phi(\omega)$ when the signal function is given, or expressions for $a(t)$ and $\theta(t)$ when the spectrum $U(\omega)$, including the phase function $\Phi(\omega)$, is given. In a majority of cases these approximate expressions can be quite accurate.

In the usual radar case $a(t)$ is a constant over the duration of the signal, and Eq. (3-22) can be used directly to solve for the matched-filter time delay function required for a specified spectrum modulus $|U(\omega)|$. The matched-filter transmitter and receiver system for this case is shown in Fig. 3.4a. The impulse response of the transmitter filter is the desired time expanded and frequency modulated signal.[1] The transmitter and receiver filters could be interchanged, hence the need for the absolute magnitude signs in Eqs. (3-19) and (3-22). The two possible functions for $\Phi(\omega)$ that meet the requirement of Eq. (3-22) are conjugate functions. An active FM generator can be used at the transmitter, as shown in Fig. 3.4b, to replace one of the pair of matched filters. The required frequency modulation function for this case is obtained from Eq. (3-19). For the special case of $\theta'(t)$ possessing odd symmetry, the same matched-filter function can be used at both the transmitter and receiver if a sideband inversion technique is used, as indicated by the alternate receiver configuration of Fig. 3.4c. This and other transmitter/receiver matched-filter implementations for FM pulse-compression signals are discussed further in Chapter 6.

From the linear FM example outlined in Chapter 1 it appeared intuitively obvious that the frequency modulation function $\theta'(t)$ and the matched-filter group time delay function $\Phi'(\omega)$ had an inverse relationship, as should be the case if both Eqs. (3-19) and (3-22) yield good approximate relationships for the same signal. Following the development given by Fowle [4] this can be shown to hold for the general case by considering the set of stationary phase conditions given by Eqs. (3-7) and (3-21). Thus,

$$\omega = \theta'(t) \qquad \text{and} \qquad -t = \Phi'(\omega) \qquad (3\text{-}26a)$$

---

[1] From the practical viewpoint the requirement for the transmitter filter driving function is that it be a pulse with a spectrum wider than the band-pass response of the matched filter, and essentially flat over that region.

Let

$$\theta'(t) = \omega(t) \qquad \text{and} \qquad \Phi'(\omega) = -T(\omega)$$

Then the conditions of (3-26a) become

$$\omega = \omega(t) \qquad t = T(\omega) \tag{3-26b}$$

From the above one obtains $\omega = T^{-1}(t)$, and thus

$$\omega(t) = T^{-1}(t) \tag{3-27}$$

Equation (3-27) shows that the instantaneous frequency function $\omega(t)$ and the group time delay function $T(\omega)$ are inverse functions. This is illustrated in Fig. 3.5. This relationship can also be expressed as

$$f(t) = T^{-1}(t) \tag{3-28}$$

where $\omega(t) = 2\pi f(t)$ and $T(f) = T(\omega/2\pi)$.

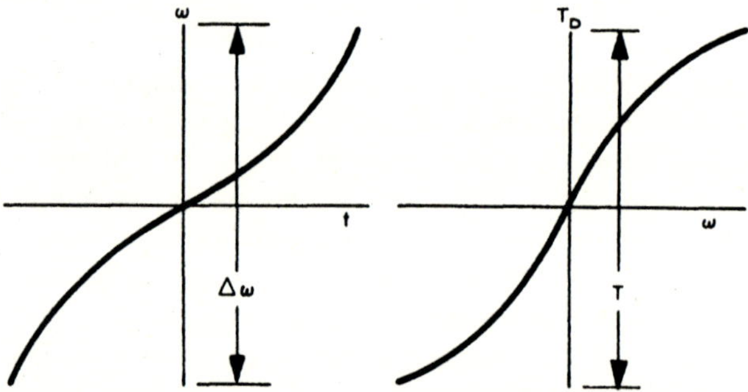

FIG. 3.5   Matched-filter time delay and frequency modulation relationship for nonlinear FM signal. (From Bernfeld *et al.*, Ref. 4, Chapter 1.)

It should be pointed out that the equality implied by (3-27) or (3-28) is actually an approximation, as shown by Eq. (3-24), that depends on the functions

$$u(t) = a(t)\exp[j\theta(t)]$$

and

$$U(\omega) = |U(\omega)|\exp[j\Phi(\omega)]$$

being a Fourier pair. Since the stationary phase analysis leads to approximate relationships among $a(t)$, $|U(\omega)|$, $\Phi(\omega)$, and $\theta(t)$, the Fourier pair relationship is also an approximate one, given by

$$a(t) \exp[j\theta(t)] \doteq \frac{1}{2\pi} \int_{-\infty}^{\infty} |U(\omega)| \exp[j\{\Phi(\omega) + \omega t\}]\, d\omega \qquad (3\text{-}29a)$$

and

$$|U(\omega)| \exp[j\Phi(\omega)] \doteq \int_{-\infty}^{\infty} a(t) \exp[-j\{\omega t - \theta(t)\}]\, dt \qquad (3\text{-}29b)$$

For most FM signals these approximations improve as the product of the time extent of the signal and the frequency extent of the spectrum increases. The equality implied by Eq. (3-28) is met with increasing accuracy for large time-bandwidth product signals. Just how large the time-bandwidth product must be depends on the type of signal. This is discussed further in Section 3.4.

In certain radar applications it may be possible to amplitude modulate the transmitted signal. In a high power system this would imply a sacrifice in the energy transmitted per pulse. However, in situations where detectability is not the most important problem, this approach has been advocated as a useful technique in the design of pulse-compression signals [5]. In addition, there are nonradar applications of waveform design in which the freedom to both amplitude modulate and frequency modulate the transmitted signal is useful. The design of pulse-compression matched-filter signals in which the pulse envelope $a(t)$ and the spectrum modulus $|U(\omega)|$ are independently specified has been developed in detail by Fowle [4]. The general procedure for this case is now outlined.

From the stationary phase condition of Eq. (3-7) one obtains, by differentiation,

$$|\Phi''(\omega)| = -\frac{dt}{d\omega} \qquad (3\text{-}30)$$

and thence by substitution into Eq. (3-22)

$$a^2(t)\, dt = \frac{1}{2\pi} |U(\omega)|^2\, d\omega \qquad (3\text{-}31)$$

Making use of the relationship between $a(t)$ and $U(\omega)$ given by Eq. (3-5) (Parseval's theorem), the following integral expressions are obtained[1]:

$$\int_{-\infty}^{t} a^2(\tau)\, d\tau = \frac{1}{2\pi} \int_{-\infty}^{\omega} |U(\Omega)|^2\, d\Omega \qquad (3\text{-}32)$$

or

$$\int_{-\infty}^{t} a^2(\tau)\, d\tau = \frac{1}{2\pi} \int_{\omega}^{\infty} |U(\Omega)|^2\, d\Omega \qquad (3\text{-}33)$$

---

[1] The limits in (3-32) and (3-33) are not independent. By implication from the previous discussions $t = t(\omega)$, or $\omega = \omega(t)$.

Equations (3-32) and (3-33) each define a possible solution for the phase functions $\theta(t)$ and $\Phi(\omega)$. As will be shown, the two solutions for each phase function have a conjugate relationship. Fowle makes the following definitions:

$$\int_{-\infty}^{t} a^2(\tau)\, d\tau = P(t) \tag{3-34}$$

$$\frac{1}{2\pi} \int_{-\infty}^{\omega} |U(\Omega)|^2 \, d\Omega = Q(\omega) \tag{3-35}$$

$$\frac{1}{2\pi} \int_{\omega}^{\infty} |U(\Omega)|^2 \, d\Omega = 2E - Q(\omega) \tag{3-36}$$

where $P(-\infty) = Q(-\infty) = 0$, and $Q(\infty) = 2E$, the energy associated with the complex signal (see Eq. (3-5)). From (3-32) one obtains

$$t = P^{-1}[Q(\omega)] \tag{3-37}$$

Invoking Eq. (3-21) yields

$$\Phi'(\omega) = -P^{-1}[Q(\omega)] \tag{3-38}$$

and

$$\Phi_1(\omega) = -\int P^{-1}[Q(\omega)]\, d\omega + C_1 \tag{3-39}$$

The use of (3-33) gives

$$\Phi'(\omega) = -P^{-1}[2E - Q(\omega)] \tag{3-40}$$

and

$$\Phi_2(\omega) = -\int P^{-1}[2E - Q(\omega)]\, d\omega + C_2 \tag{3-41}$$

where C_1 and C_2 are constants of integration.

Similar solutions for $\theta(t)$ can be obtained by the same procedure, resulting in

$$\theta_1(t) = \int Q^{-1}[P(t)]\, dt + C_3 \tag{3-42}$$

$$\theta_2(t) = \int Q^{-1}[2E - P(t)]\, dt + C_4 \tag{3-43}$$

The solutions for $\theta(t)$ and $\Phi(\omega)$ permit the construction of the two approximate Fourier pairs

$$|U(\omega)| \exp[\,j\Phi_1(\omega)] \doteq \mathscr{F}\{a(t) \exp[\,j\theta_1(t)]\} \tag{3-44a}$$

$$|U(\omega)| \exp[\,j\Phi_2(\omega)] \doteq \mathscr{F}\{a(t) \exp[\,j\theta_2(t)]\} \tag{3-44b}$$

and the matched-filter condition results in the envelope of the matched-filter output autocorrelation function being given by

$$g(t) = \frac{1}{2\pi} \int_{-\infty}^{\infty} |U(\omega)|^2 \exp[-j\omega t]\, d\omega \tag{3-45}$$

The exactness of the approximate Fourier transform relationships given by Eqs. (3-44a) and (3-44b) depends on the type of signal and the time-bandwidth product of the signal. The following section offers examples of the application of the stationary phase results to the analysis and design of pulse-compression matched-filter signals.

3.4 Waveform Design Applications to FM Pulse-Compression Signals

The waveform design examples in this section are taken from Fowle [4], Key *et al.* [1, 2], and Cook [6], and illustrate the application of the various stationary phase relationships to the analysis and design of some fundamental types of pulse-compression signals.

Example 1[1]

The signal envelope and the complex spectrum modulus are both specified as Gaussian functions given by

$$a(t) = \frac{1}{\sqrt{T}} \exp\left[-\left(\frac{2t}{T}\right)^2\right] \tag{3-46}$$

$$|U(f)| = \frac{1}{\sqrt{\Delta f}} \exp\left[-\left(\frac{2f}{\Delta f}\right)^2\right] \tag{3-47}$$

The time duration and bandwidth of these functions are defined as the interval between their e^{-1} points, or T and Δf, respectively. The time-bandwidth product of this signal is $T\Delta f$.[2] The pulse envelope and spectrum modulus are illustrated in Fig. 3.6, as is the matched-filter output envelope. This waveform is the Fourier transform of Eq. (3-47), given by

$$g(t) = \exp\left[-\frac{\pi^2 \Delta f^2 t^2}{8}\right] \tag{3-48}$$

The use of Eq. (3-32) yields

$$\frac{1}{T}\int_{-\infty}^{t} \exp\left[-8\left(\frac{\tau}{T}\right)^2\right] d\tau = \frac{1}{\Delta f}\int_{-\infty}^{f} \exp\left[-8\left(\frac{\Omega}{2\pi\,\Delta f}\right)^2\right]\frac{d\Omega}{2\pi} \tag{3-49}$$

---

[1] From Fowle [4].

[2] The definitions of time duration and bandwidth differ from that used by Fowle. The notation used in the examples of this chapter conform to the standard usage employed in the other chapters.

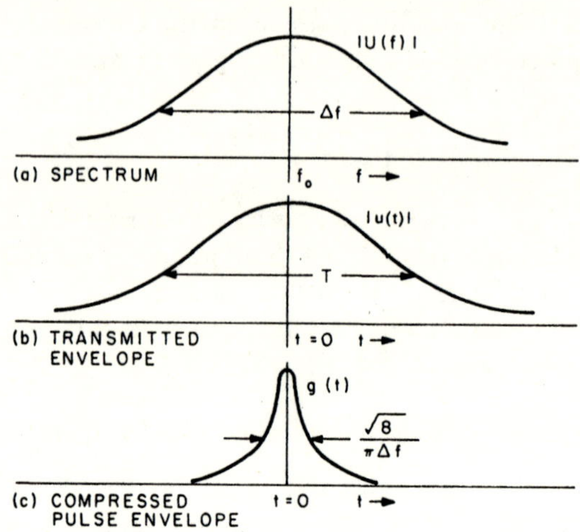

FIG. 3.6   Characteristics of the Gaussian envelope, Gaussian spectrum pulse-compression matched-filter signal.

Let $2\sqrt{2}(\tau/T) = y$ and $2\sqrt{2}(\Omega/2\pi\,\Delta f) = z$; Eq. (3-49) becomes

$$\int_{-\infty}^{2\sqrt{2}(t/T)} \exp[-y^2]\,dy = \int_{-\infty}^{2\sqrt{2}(f/\Delta f)} \exp[-z^2]\,dz \tag{3-50}$$

From the limits of (3-50) one obtains

$$\frac{t}{T} = \frac{f}{\Delta f} = \frac{\omega}{\Delta\omega} \tag{3-51}$$

where $\omega = 2\pi f$. The use of Eq. (3-26a) yields

$$-\Phi'(\omega) = \beta_f'(\omega) = \frac{T\omega}{\Delta\omega} \tag{3-52}$$

and

$$\Phi(\omega) = -\frac{1}{2}\frac{T}{\Delta\omega}\omega^2 + C_1 \tag{3-53}$$

Similarly, the frequency modulation function becomes

$$\theta'(t) = \frac{\Delta\omega}{T}t \tag{3-54}$$

and

$$\theta(t) = \frac{1}{2}\frac{\Delta\omega}{T}t^2 + C_3 \tag{3-55}$$

Equations (3-52) and (3-55) show that the matched filter has a linear delay vs frequency characteristic and that the waveform modulation is linear FM. The amplitude response of the matched filter is, of course, a Gaussian function matched to Eq. (3-47). Fowle has compared these results with those obtained by taking the exact Fourier transform of

$$u(t) = \frac{1}{\sqrt{\Delta f}} \exp\left[-\left(\frac{2t}{T}\right)^2\right]\exp\left[-j\left(\frac{\Delta\omega}{2T}t^2 + C_3\right)\right] \qquad (3\text{-}56)$$

The comparison shows that for $T\Delta f = 40/\pi$ (as defined here) the spectrum modulus and phase given by Eqs. (3-47) and (3-53) are within a fraction of 1% of those of the exact Fourier transform of (3-56). Thus for $T\Delta f$ greater than approximately 10, (3-46), (3-47), (3-53), and (3-55) can be used to define a Fourier pair quite accurately. As a general rule for large time-bandwidth-signals, the specification of functionally equivalent pulse envelope and spectrum modulus functions yields a $\theta'(t)$ that is a linear FM, and a group delay $\beta'(\omega) = -\Phi'(\omega)$ that is a linear delay as a function of frequency.

Example 2[1]

The previous example required both $a(t)$ and $|U(\omega)|$ to be smooth, continuous functions. This second example requires $a(t)$ to be a rectangular (i.e., discontinuous) function and $|U(\omega)|$ to be a smooth, continuous function, given by

$$a(t) = \frac{1}{\sqrt{T}}, \qquad 0 < t < T \qquad (3\text{-}57)$$

$$= 0, \qquad \text{elsewhere}$$

$$|U(\omega)| = \frac{\sqrt{2}}{\sqrt{\Delta\omega/2}} \frac{1}{\sqrt{1 + (2\omega/\Delta\omega)^2}} \qquad (3\text{-}58)$$

These functions are illustrated in Fig. 3.7. The matched-filter autocorrelation waveform, defined by the Fourier transform of Eq. (3-58), is

$$g(t) = \exp[-\pi\,\Delta f\,|t|] \qquad (3\text{-}59)$$

Application of Eq. (3-33) gives

$$\frac{1}{T}\int_0^t d\tau = \frac{1}{\pi}\int_\omega^\infty \frac{1}{1 + (2\omega/\Delta\omega)^2} \frac{d\omega}{(\Delta\omega/2)} \qquad (3\text{-}60)$$

---

[1] From Key *et al.* [1, 2] and Fowle [4].

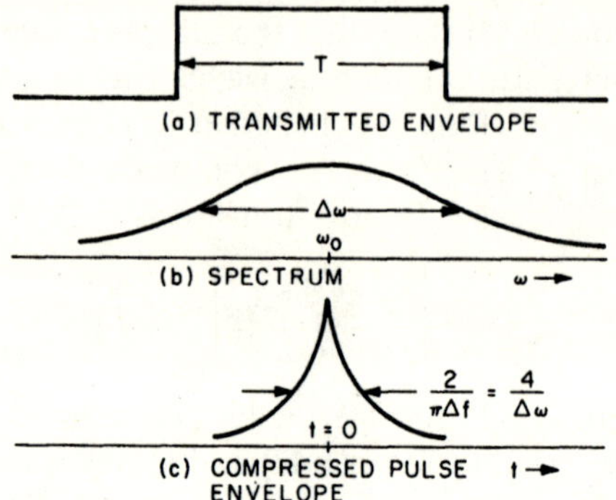

FIG. 3.7   Characteristics of signal associated with the inverse tangent group delay matched filter. (After Key *et al.* [2].)

Integration of Eq. (3-60) yields

$$\frac{t}{T} = \frac{1}{\pi}\left(\frac{\pi}{2} - \tan^{-1}\frac{2\omega}{\Delta\omega}\right) \tag{3-61}$$

from which can be obtained

$$-\Phi_1'(\omega) = \beta_1'(\omega) = \frac{T}{2}\left(1 - \frac{2}{\pi}\tan^{-1}\frac{2\omega}{\Delta\omega}\right) \tag{3-62}$$

Making use of Eq. (3-32) leads to the alternate formulation

$$\frac{1}{T}\int_0^t d\tau = \frac{1}{\pi}\int_{-\infty}^{\omega} \frac{1}{1 + (2\omega/\Delta\omega)^2}\frac{d\omega}{\Delta\omega/2} \tag{3-63}$$

and

$$-\Phi_2'(\omega) = \beta_2'(\omega) = \frac{T}{2}\left(\frac{2}{\pi}\tan^{-1}\frac{2\omega}{\Delta\omega} + 1\right) \tag{3-64}$$

The group time delay functions $-\Phi_1'(\omega)$ and $-\Phi_2'(\omega)$ are plotted in Fig. 3.8.

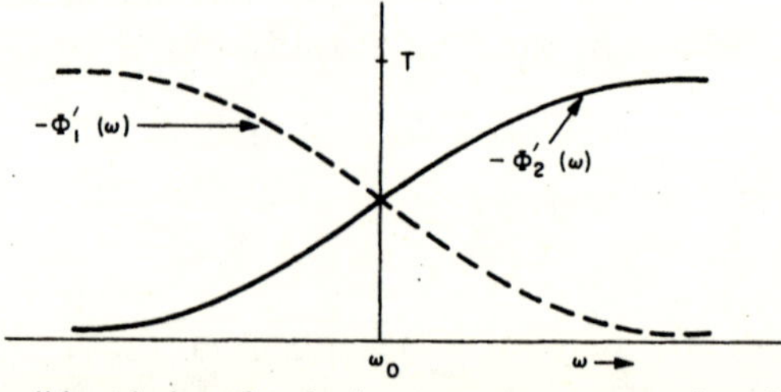

FIG. 3.8   Two possible solutions for the inverse tangent group time delay matched-filter function.

It can be seen by inspection that they result in a $\Phi_1(\omega)$ and a $\Phi_2(\omega)$ that are conjugate functions, where

$$\Phi_1(\omega) = -\frac{T}{\pi} \int \left(\frac{\pi}{2} - \tan^{-1} \frac{2\omega}{\Delta\omega} \right) d\omega + C_1 \qquad (3\text{-}65)$$

Using the relationship indicated in (3-61), the FM function is given by

$$\Theta_1'(t) = \frac{\Delta\omega}{2} \tan \left(\frac{\pi t}{T} - \frac{\pi}{2} \right) \qquad (3\text{-}66)$$

$$\Theta_1(t) = \frac{\Delta\omega}{2} \int \tan \left(\frac{\pi t}{T} - \frac{\pi}{2} \right) dt + C_3 \qquad (3\text{-}67)$$

Key *et al.* [2] have numerically computed the $\theta'(t)$ and $a(t)$ for the spectrum modulus and phase given by (3-58) and (3-65) for values of $T\Delta f$ of $10/\pi$ and $100/\pi$. These are shown in Fig. 3.9. Based on this data, time-bandwidth products $T\Delta f$ greater than 20 or 30 are required to construct a Fourier pair with good accuracy for signals that have the general characteristics of the one used in this example. This covers the class of pulse-compression signals for which either $a(t)$ or $|U(\omega)|$ are to be discontinuous, and the other smooth and continuous.

If both $a(t)$ and $|U(\omega)|$ are required to be rectangular functions, as indicated by the discussion of the rectangular envelope linear FM signal described in Chapter 1, then $|U(\omega)|$ does not become a good approximation to the rectangular distribution unless $T\Delta f > 100$, as can be seen from the spectra shown in Fig. 6.6. This follows from the restriction that both the envelope and spectrum cannot be simultaneously truncated. However, for $T\Delta f$ sufficiently large, simultaneous truncation of both the pulse envelope and spectrum can be closely approximated. It should be noted that if the rectangular spectrum is desired in order to obtain a $\sin x/x$ matched-filter output waveform, then a good approximation to this is obtained with a linear FM signal for $T\Delta f > 20$. Thus, the rectangular spectrum does not need to be closely approximated to achieve this waveform.

Example 3[1]

The matched-filter output waveforms of both the preceding examples exhibit smooth continuous curves, being Gaussian and exponential, respectively. In the case of the Gaussian compressed-pulse function, the skirt of the waveform is down only $-26\,\mathrm{db}$ at $t = \pm 1.8/\Delta f$. The exponential compressed pulse derived from the nonlinear FM function of Example 2 requires a matched-filter group time delay function that must be controlled

---

[1] From Cook [6].

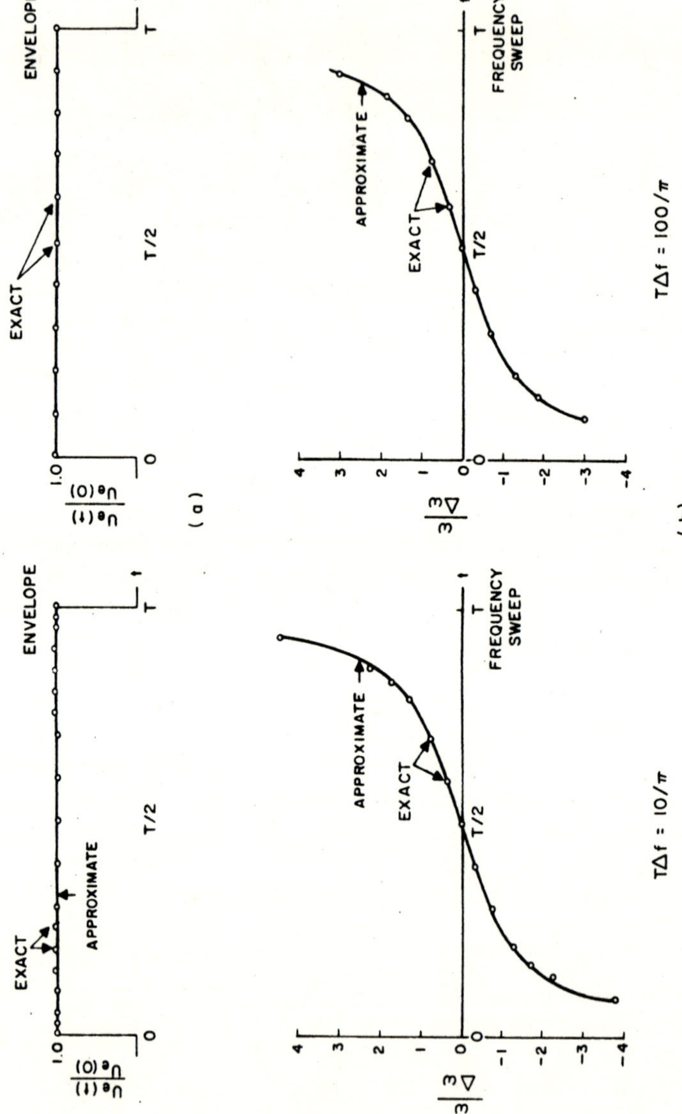

Fig. 3.9   Calculated expanded-pulse envelope and frequency modulation function for the inverse tangent group delay matched filter. (a) Calculated pulse envelope functions. (b) Calculated frequency vs time functions. (From Fowle [4]).

out to several times the basic bandwidth Δf. The tail of the compressed
pulse of this signal also falls off fairly slowly relative to the inverse of the
bandwidth required for the matched-filter design. In many applications it
is desired that the compressed pulse fall off rapidly as a function of time
to a certain level, and then remain below that level thereafter. Most waveforms that fit into this category exhibit a time sidelobe structure, similar
to that of antenna sidelobes, rather than a smoothly falling waveform skirt
characteristic. Generally, there is a trade-off between the rate at which the
compressed-pulse function falls off and the desired maximum level of the
time sidelobes. For a given available bandwidth in which the matched-filter
group time delay must be contained, more efficient use of this bandwidth
in designing an FM pulse-compression signal can usually be achieved with
signals of the type described above. One such class of signals has a rectangular envelope function and spectrum moduli given by

$$|U_n(\omega)|^2 = A_n \cos^n \frac{\pi\omega}{\Delta\omega_n}, \qquad |\omega| \le \frac{\Delta\omega_n}{2} \tag{3-68}$$

Making use of Eq. (3-22), coupled with an envelope duration T, the reader
may verify by direct integration that the following group time delay func-
tions $T_n(\omega) = \beta'_n(\omega)$ are obtained (A_n is obtained by evaluating $T_n(\omega)$ at one
of its end points):

$$n = 1 \qquad T_1(\omega) = \frac{T}{2}\sin\frac{\pi\omega}{\Delta\omega_1} \tag{3-69}$$

$$n = 2 \qquad T_2(\omega) = T\left[\frac{\omega}{\Delta\omega_2} + \frac{1}{2\pi}\sin\frac{2\pi\omega}{\Delta\omega_2}\right] \tag{3-70}$$

$$n = 3 \qquad T_3(\omega) = \frac{T}{4}\left[\sin\frac{\pi\omega}{\Delta\omega_3}\left\{\cos^2\frac{\pi\omega}{\Delta\omega_3} + 2\right\}\right] \tag{3-71}$$

$$n = 4 \qquad T_4(\omega) = T\left[\frac{\omega}{\Delta\omega_4} + \frac{1}{2\pi}\sin\frac{2\pi\omega}{\Delta\omega_4} + \frac{2}{3\pi}\cos^3\frac{\pi\omega}{\Delta\omega_4}\sin\frac{\pi\omega}{\Delta\omega_4}\right] \tag{3-72}$$

This family of group time delay functions is plotted in Fig. 3.10. Relative
to the $(\sin x)/x$ waveform of the rectangular spectrum that fills the interval
$\Delta\omega$, the compressed-pulse widths (-3 db level) for each spectrum defined by
Eq. (3-68) are as follows: $n = 1$, 1.34 times the $(\sin x)/x$ pulse width; $n = 2$,
1.62 times the $(\sin x)/x$ pulse width; $n = 3$, 1.87 times; $n = 4$, 2.09 times.
A denormalizing factor can be applied to the $\Delta\omega_n$ so that the matched-filter
output waveforms have equal -3 db pulse widths. This is shown in Fig. 3.11,
where the denormalization is taken with respect to $|U(\omega)|^2$ for $n = 1$. The
peak sidelobe levels for the compressed pulse associated with each spectrum

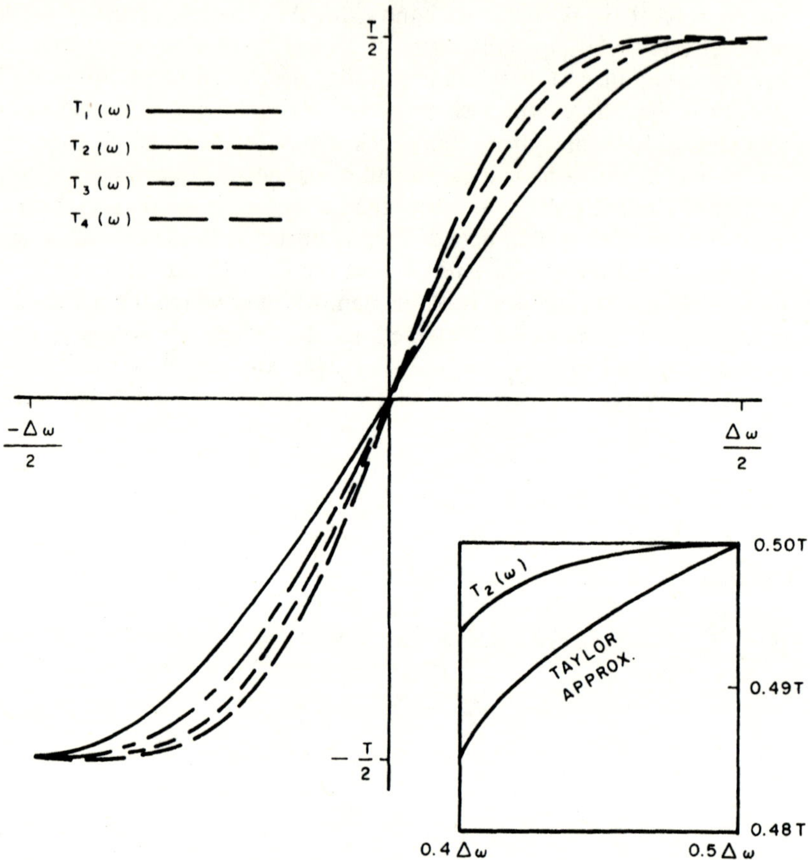

FIG. 3.10   Group time delays for cos^n power spectra. (From Cook [6].)

distribution are also shown in this figure. It can be seen that to achieve low time sidelobe levels for approximately equal spectrum bandwidths requires control over the spectrum tail response for a considerable interval beyond the region of major spectral energy. This, in turn, implies control over $T_n(\omega)$ over the same frequency interval. In order to reduce this interval, and still obtain a matched-filter compressed-pulse waveform with low time sidelobes, one can use (3-22) to obtain a $T(\omega)$ for a cosine-squared on pedestal spectrum given by

$$|U_k(\omega)|^2 = A_k\left(k + (1 - k)\cos^2\frac{\pi\omega}{\Delta\omega_k}\right)$$
(3-73)

This function approximates a Taylor weighted spectrum (see Chapter 7)

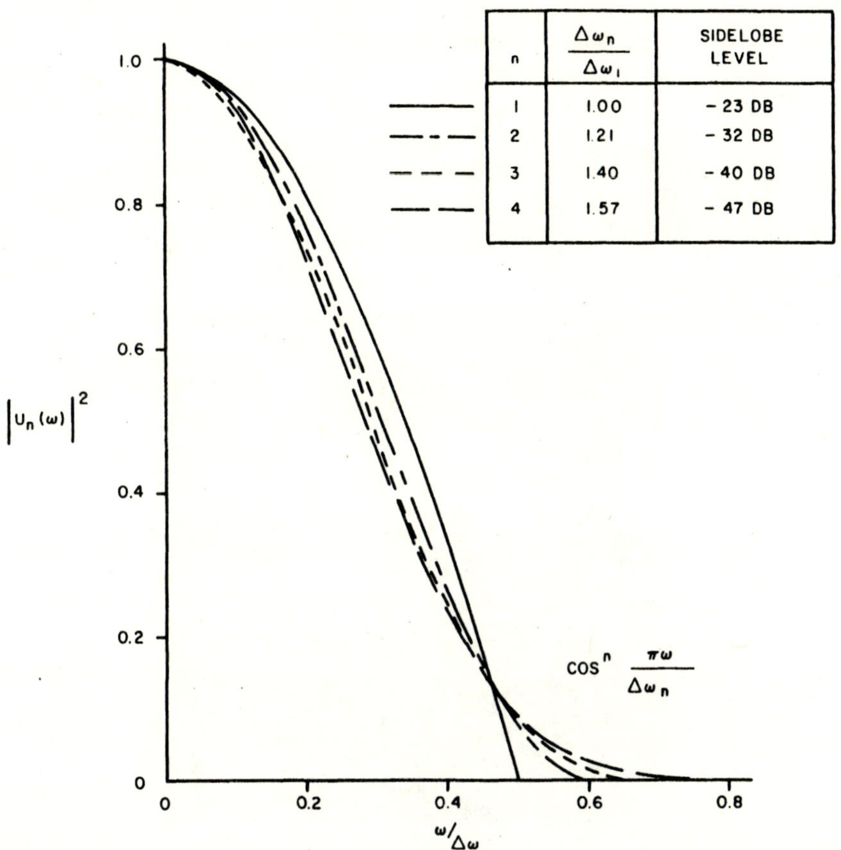

n	$\dfrac{\Delta\omega_n}{\Delta\omega_1}$	SIDELOBE LEVEL
1	1.00	− 23 DB
2	1.21	− 32 DB
3	1.40	− 40 DB
4	1.57	− 47 DB

FIG. 3.11   Cos^n power spectra for $T_n(\omega)$ denormalized to yield equal pulse widths. (From Cook [6].)

for $k = 0.088$. The group time delay derived from (3-73) is

$$T_k(\omega) = T\left(\frac{\omega}{\Delta\omega_k} + \frac{1-k}{2\pi}\sin 2\pi\frac{\omega}{\Delta\omega_k}\right) \qquad (3\text{-}74)$$

The detail of (3-74) near $\Delta\omega/2$ for the Taylor approximation is shown in the inset of Fig. 3.10. For $k = 0.088$ and $\Delta\omega_k = 1.1\,\Delta\omega_1$ one obtains the same compressed-pulse width as the cosine-squared spectrum and -40 db peak sidelobes. The analytic functions for $\theta'(t)$ are not simply expressed for the class of group time delay functions of this example. $\theta'(t)$ is evaluated most directly in cases such as this by a graphical technique. This involves rotating the axes of Fig. 3.10, letting the time variable equal the delay variable and interchanging the plus and minus values of frequency. Figure 3.12 illustrates the impulse response (i.e., expanded pulse) for a nonlinear delay matched filter for $|U(\omega)|^2$ an approximate cosine-squared spectrum. The expanded pulse from a linear group time delay matched filter for the same input impulse is considerably less rectangular in shape, and the compressed pulse has higher level sidelobes. The overshoot at each end of the nonlinear FM expanded pulse results from departures from the ideal matched-filter time delay and amplitude response near the band edges of the filter.

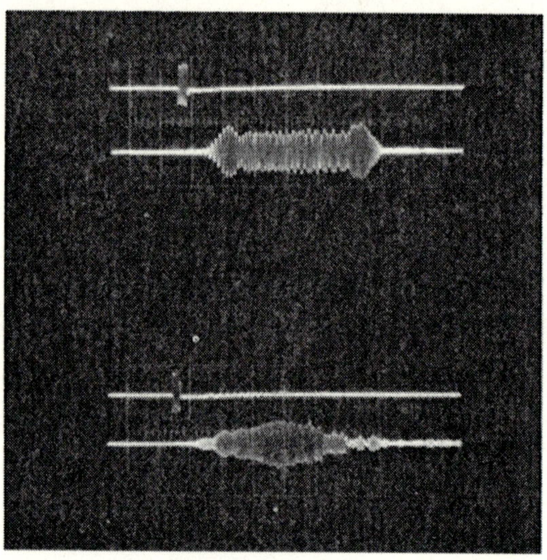

FIG. 3.12 Pulse expansion (matched-filter) response of nonlinear and linear delay filters. Top: Driving impulse and response of nonlinear delay filter. Bottom: Driving impulse and response of linear delay filter.

Example 4

Examples 1, 2, and 3 cover cases for $a(t)$ a rectangular envelope and $\theta'(t)$ a nonlinear FM function, or $a(t)$ amplitude modulated and $\theta'(t)$ a linear FM function. An example of both an amplitude modulated envelope and a nonlinear FM characteristic is given by

$$a(t) = \left(\frac{\pi}{2T}\cos\frac{\pi t}{T}\right)^{1/2}, \qquad -\frac{T}{2} < t < \frac{T}{2} \qquad (3\text{-}75)$$

$$|U(f)| = \sqrt{\frac{2}{\Delta f}\cos\frac{\pi f}{\Delta f}}, \qquad -\frac{\Delta f}{2} < f < \frac{\Delta f}{2} \qquad (3\text{-}76)$$

The use of Eq. (3-32) gives

$$\frac{\pi}{2T}\int_{-T/2}^{t} \cos\frac{\pi\tau}{T}\,d\tau = \frac{2}{\Delta f}\int_{-\Delta f/2}^{f} \cos^2\frac{\pi f}{\Delta f}\,df \tag{3-77}$$

Let $y = \pi\tau/T$ and $z = \pi f/\Delta f$; Eq. (3-77) becomes

$$\frac{1}{2}\int_{-\pi/2}^{\pi t/T} \cos y\,dy = \frac{2}{\pi}\int_{-\pi/2}^{\pi f/\Delta f} \cos^2 z\,dz \tag{3-78}$$

Performing the designated integrations, one obtains

$$\frac{1}{2}\left[\sin\frac{\pi t}{T} + 1\right] = \frac{f}{\Delta f} + \frac{1}{2\pi}\sin\frac{2\pi f}{\Delta f} + \frac{1}{2} \tag{3-79}$$

and

$$t = \frac{T}{\pi}\sin^{-1}\left[\frac{f}{\Delta f} + \frac{1}{2\pi}\sin\frac{2\pi f}{\Delta f}\right] \tag{3-80}$$

The relationship $-\Phi'(\omega) = t$ results in the group time delay function shown in Fig. 3.13. As in the previous example, $\theta'(t)$ is determined most easily by the graphical technique. The curvature of this frequency modulation function lies between that of the linear FM and the nonlinear FM required to obtain a cosine-squared power spectrum with a rectangular pulse envelope.

The motivating reason behind the investigation of nonlinear FM signals was based on the desire to shape the signal spectrum so as to improve the matched-filter compressed-pulse waveform (autocorrelation function) without sacrificing any of the system power advantages of transmitting a rectangular pulse envelope. As it turns out, the cost of reducing the time (or range) sidelobes of the matched-filter autocorrelation function by this approach is the increase of the time sidelobe level of the matched-filter output signal when the received input signal has been frequency shifted (moving reflector case). Because of this the application of nonlinear FM pulse-compression signals usually would be confined to situations in which the expected velocity spread of the received signals to be examined is relatively small. An example of this is discussed in Chapter 7.

The important conclusions of this section are that given the general FM matched-filter input signal $u(t)$, the following parameters can be obtained, to a good approximation, with the stationary phase analysis:

(1) $|U(\omega)|$, the signal spectrum magnitude and matched-filter amplitude response.

(2) $\Phi(\omega)$, the spectrum phase and conjugate of the matched-filter phase response.

(3) $-\Phi'(\omega)$, the matched-filter group time delay function, as a result of (2).

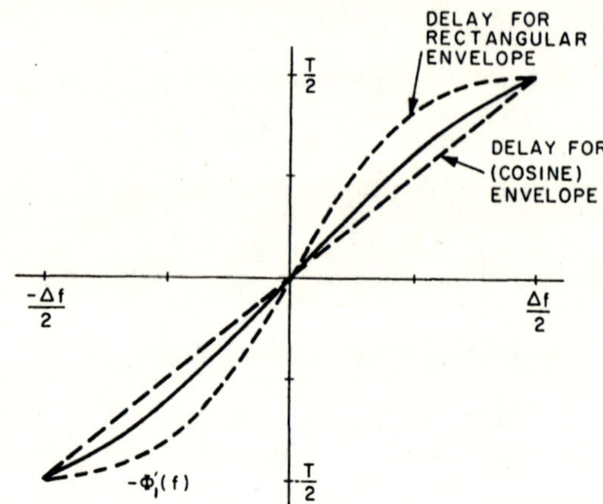

FIG. 3.13   Group time delay function for amplitude modulated and nonlinear FM matched-filter signal of Example 4.

In addition, other related functions of the matched-filter signal can be obtained. These are briefly described in the next section.

3.5   The Matched-Filter Ambiguity Function

The magnitude of the spectrum of the matched-filter signal is one of the results that can be readily obtained from the stationary phase analysis. Knowing the complex spectrum magnitude $|U(\omega)|$, the matched-filter auto-correlation function can be calculated by

$$g(t) = \frac{1}{2\pi} \int_{-\infty}^{\infty} |U(\omega)|^2 \exp[\,j\omega t\,]\,d\omega \tag{3-81}$$

From the impulse response relationship shown in Chapter 1, an alternate form for the matched-filter waveform is

$$g(t) = \int_{-\infty}^{\infty} u(\tau)u(\tau - t)\,d\tau \tag{3-82}$$

These two equations can be put in more general terms so that

$$g(t, \omega_d) = \frac{1}{2\pi} \int_{-\infty}^{\infty} U(\omega + \omega_d)U^*(\omega) \exp[\,j\omega t\,]\,d\omega \tag{3-83}$$

$$= \int_{-\infty}^{\infty} u(\tau)u^*(\tau - t) \exp[\,j\omega_d\tau\,]\,d\tau = \chi(-t, \omega_d) \tag{3-84}$$

The function $g(t, \omega_d)$ describes the response of the matched filter to a frequency shifted input signal. This more general matched-filter output function is useful in determining the filter response (for the narrow band case) when the input signals represent returns from moving objects. $|\chi|$ is denoted as the matched-filter response function or radar uncertainty function [7, 8]. The ambiguity function of the radar signal is derived from (3-83) or (3-84), being denoted by $|\chi(t, \omega_d)|^2$ or $|\chi(t, \phi)|^2$, where $\phi = \omega_d/2\pi$.

As applied to the design and analysis of matched-filter waveforms, the uncertainty, or ambiguity, function defined above is a criterion that can be used to determine the usefulness of a particular matched-filter signal for determining either the *simultaneous, one pulse* measurement of velocity and range of a moving object, or for resolving velocity and range among several moving objects. From this point of view it represents an important concept in certain tactical applications of radar systems.

The ambiguity function is often represented in a three dimensional form. The $|\chi|$ function or $|\chi|^2$ function is plotted as a surface above a plane in which the coordinate axes are time (or range) and frequency shift (or velocity). Figure 3.14 illustrates this concept, showing typical cuts through this surface. The peak of the ambiguity function always occurs at ($t = 0$, $\omega_d = 0$). Slices of the ambiguity function parallel to the time or range axis represent the squared magnitude of the matched-filter outputs for different frequency shift values, ω_d. The slices parallel to the frequency axis represent

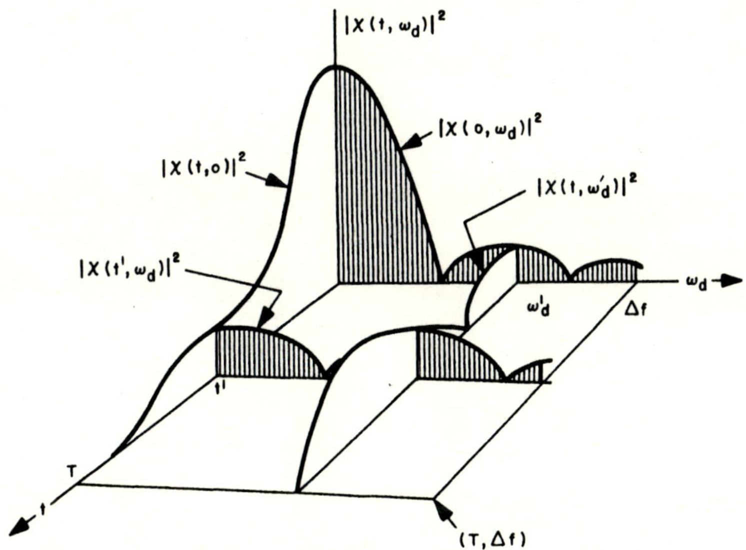

FIG. 3.14   Cross sections of the ambiguity function. (From Bernfeld *et al.*, Ref. 4, Chapter 1.)

the variation of this output magnitude at a fixed moment of time, and are related to the basic velocity measuring capability of the waveform, $u(t)$. If the $|\chi|^2$ function is sharply peaked about the origin, $u(t)$ is said to have good simultaneous range and velocity resolution capability. If the ambiguity function is broadly distributed above the time-frequency plane $u(t)$ does not have this capability, and range and velocity cannot be determined independently of each other by examining a single target return at the matched-filter output. This aspect of matched-filter signals will be examined in detail in the chapters dealing with general waveform design topics. For the present the uncertainty or ambiguity function will be taken as being descriptive of the matched-filter waveform degradation under conditions of frequency shifted inputs. A basic constraint associated with the ambiguity function $|\chi|^2$ is that the volume under its surface is constant for all input waveforms normalized to have equal energy. This is an important consideration in the general matched-filter waveform design problem. The ambiguity function is of central importance to the radar designer in understanding the advantages and disadvantages of specific radar signal design approaches. The next chapter is devoted to a fuller discussion of this important function and its properties.

REFERENCES

1. E. L. Key, E. N. Fowle, and R. D. Haggarty, A method of pulse compression employing nonlinear frequency modulation, Lincoln Lab., M.I.T., Lexington, Massachusetts, Tech. Rept. 207 (September, 1959).
2. E. L. Key, E. N. Fowle, and R. D. Haggarty, A method of designing signals of large time-bandwidth product, *IRE Intern. Conv. Record Pt. 4*, 146–155 (1961).
3. E. N. Fowle, A general method for controlling the time and frequency envelopes of FM signals, Lincoln Lab., M.I.T., Lexington, Massachusetts, Group Rept. 41G-0008 (June, 1961).
4. E. N. Fowle, The design of FM pulse-compression signals, *IEEE Trans.* **IT-10**, 61–67 (1964).
5. E. N. Fowle, Private communication.
6. C. E. Cook, A class of nonlinear FM pulse-compression signals, *Proc. IEEE (Correspond.)* **52**, 1369–1371 (1964).
7. P. M. Woodward, "Probability and Information Theory with Applications to Radar." Pergamon Press, Oxford, 1953.
8. W. M. Siebert, A radar detection philosophy, *IRE Trans.* **IT-2**, 204–219 (1956).

CHAPTER 4

The Radar Ambiguity Function

4.1 Introduction

Chapter 2 reviewed the problem of the optimum processing of radar signals required to determine the presence or absence of a reflected waveform. It was shown that the optimum receiver for white Gaussian noise interference was the class of linear filters that are characterized by the pair of equations:

$$H(\omega) = kS^*(\omega)$$
$$\updownarrow \qquad \updownarrow$$
$$h(t) \;\; = ks(-t)$$

$(4\text{-}1)$

where $H(\omega)$ and $h(t)$ are the transfer function and the impulse response of the optimum filter, and $S^*(\omega)$ and $s(-t)$ are the spectrum conjugate and the mirror image (or time inverse) corresponding to the waveform $s(t)$. These equations constitute the conditions for matched filtering. It is evident that a particular waveform for transmission must first be specified before an optimum detection system can be implemented. Because of this requirement the radar designer has the important preliminary task of finding a suitable waveform for each different radar system application. To the designer concerned with resolution and accuracy, this waveform will leave the least amount of ambiguity to contend with after the matched-filtering process.

This chapter discusses in greater detail a two-dimensional waveform amplitude distribution function in time delay τ and frequency displacement ϕ that was introduced briefly in Chapter 3. This function is the basis in modern radar technology for the systematic search for the best waveform for a particular application and is commonly referred to as the *ambiguity function*. The ambiguity function is defined as $\chi(\tau, \phi)\chi^*(\tau, \phi)$, where [1]

$$\chi(\tau, \phi) = \int_{-\gamma}^{\infty} u(t)u^*(t + \tau) \exp[-j2\pi\phi t]\, dt \qquad (4\text{-}2a)$$

The waveform $s(t)$ is represented by $u(t)$ in (4-2a), a complex representation that is discussed more fully in Section 4.2. Although this chapter considers only the parameters of time delay and frequency shift, the concept of ambiguity can be extended to include more than just two parameters, or different pairs of parameters [2]. It can be shown through Parseval's theorem[1] that $\chi(\tau, \phi)$ can be expressed, in addition to the form given by (4-2a), in terms of $U(f)$, the Fourier transform of $u(t)$; thus,

$$\chi(\tau, \phi) = \int_{-\infty}^{\infty} U^*(f)U(f + \phi)\exp[-j2\pi f\tau]\, df \qquad (4\text{-}2\text{b})$$

In this volume, $\chi(\tau, \phi)$ will be referred to as the complex time/frequency response function or simply the response function. The reason for this designation will become clear as the treatment in this chapter develops. This response function was introduced by Ville [3], while its importance with respect to radar was pointed out by Woodward [1]. Siebert [4] and Wilcox [5] have made important contributions to the understanding of this function in deriving most of its fundamental properties and associated theorems. Additional contributions to the understanding of the properties of $\chi(\tau, \phi)$ and $|\chi(\tau, \phi)|^2$ have been made by other workers [6–10]. Some of these contributions involve the general properties of $\chi(\tau, \phi)$, and some involve particular waveform considerations.[2]

This chapter is intended to provide a fundamental understanding of the radar ambiguity function. The complex signal representation and the Doppler approximation are discussed in Sections 4.2 and 4.3. Following this, the general ambiguity function formulation is presented in Section 4.4; the two points of view that lead to the derivation of the ambiguity function are discussed there. Sections 4.5 through 4.11 present the important general properties of $\chi(\tau, \phi)$ and $|\chi(\tau, \phi)|^2$ and associated theorems. Finally, Sections 4.12 and 4.13 discuss the ambiguity function properties of some specific waveforms and the application of the principle of stationary phase to the analysis of ambiguity functions.

4.2 Complex Waveform Representation

The class of radar waveforms consists of all real waveforms that can be described by

$$s(t) = a(t)\cos[2\pi f_0 t + \theta(t)] \qquad (4\text{-}3)$$

where the envelope $a(t)$ and the phase modulation $\theta(t)$ are relatively slowly

---

[1] Parseval's theorem is given by $\int v(t)u^*(t)\, dt = \int V(f)U^*(f)\, df$; this important theorem will be invoked at other places in this chapter.

[2] Klauder [10] uncovered an analogy with the characteristic function for the Wigner distribution, a quantity of interest in the field of quantum mechanics.

varying time functions in contrast with $\cos[2\pi f_0 t]$. These waveforms have the common property that their Fourier transform $S(f)$ is concentrated mostly around f_0, the carrier frequency. Moreover, since they are real functions of time it can be shown that

$$S(f) = S^*(-f) \tag{4-4}$$

For theoretical analysis involving only linear operations on $s(t)$, or some time weighted version of $s(t)$ such as $h(\tau - t)s(t)$, it is convenient to describe $s(t)$ as follows

$$s(t) = \text{Re}\{\psi(t)\} \tag{4-5}$$

where $\psi(t)$ is some complex waveform which necessarily has a real part, implied by the operator $\text{Re}\{\ \}$, equal to $s(t)$. These analyses can generally be accomplished using less steps when $s(t)$ is replaced by $\psi(t)$ and the real part operator is invoked following the completion of all the linear operations on $\psi(t)$. In principle this is a much simpler procedure, but unfortunately in practice it is difficult, if not impossible, to derive the correct $\psi(t)$ corresponding to a particular waveform.

The general form of $\psi(t)$ is often expressed as

$$\psi(t) = a(t)\exp[\,j\{2\pi f_0 t + \theta(t)\}] \tag{4-6}$$

However, unless $s(t)$ is a "narrow band waveform"[1] the form given by equation (4-6) is equally as cumbersome to employ as $s(t)$. The correct form of $\psi(t)$ for more convenient linear systems analysis is given by

$$\psi(t) = s(t) + j\frac{1}{\pi}\int_{-\infty}^{\infty} \frac{-s(\tau)}{t-\tau}\,d\tau \tag{4-7}$$

This form is credited to Gabor [11]. It is seen that the real part consists of $s(t)$ as required, while the imaginary part consists of

$$\hat{s}(t) = \frac{1}{\pi}\int_{-\infty}^{\infty} \frac{s(\tau)}{t-\tau}\,d\tau \tag{4-8}$$

---

[1] A narrow band waveform by definition is one which has the property that

$$S_+(f) = 0, \qquad 0 > f$$
$$S_-(f) = 0, \qquad 0 < f$$

where

$$S_+(f) = \int_{-\infty}^{\infty} \frac{\psi(t)}{2}\exp[-j2\pi ft]\,dt$$

$$S_-(f) = \int_{-\infty}^{\infty} \frac{\psi^*(t)}{2}\exp[-j2\pi ft]\,dt$$

where $\hat{s}(t)$ is the Hilbert transform of $s(t)$. The imaginary part of $\psi(t)$ given by (4-7) should be compared with the imaginary part of (4-6) which is simply

$$\text{Im}\{\psi(t)\} = a(t)\sin[2\pi f_0 t + \theta(t)] \qquad (4-9)$$

where $\text{Im}\{\ \}$ indicates the imaginary part of $\psi(t)$. In general, Eqs. (4-6) and (4-7) will not be equal.

The convenience that results from using the Gabor complex representation can be illustrated by considering the calculation of the energy contained in $s(t)$. The energy of $s(t)$ is given by

$$E = \int_{-\infty}^{\infty} s^2(t)\, dt \qquad (4-10)$$

The reader can demonstrate for himself how difficult it is to compute E from this equation by inserting (4-3) for $s(t)$ in (4-10). In general, the analysis will involve the essential difference frequency term as well as the sum frequency term. It can easily be shown that the energy given in terms of the general complex representation is

$$E = \tfrac{1}{2}\text{Re}\left\{\int_{-\infty}^{\infty} [\psi(t)\psi(t) + \psi^*(t)\psi(t)]\, dt\right\} \qquad (4-11)$$

This expression can be just as difficult to evaluate as (4-10). However, when $\psi(t) = s(t) + j\hat{s}(t)$ is inserted, expression (4-11) reduces to

$$E = \tfrac{1}{2}\int_{-\infty}^{\infty} \psi(t)\psi^*(t)\, dt \qquad (4-12)$$

The operator $\text{Re}\{\ \}$ is omitted from (4-12) since the quantity E is always real. This expression is clearly much more convenient than (4-11). It reduces to this simple form since

$$\int_{-\infty}^{\infty} \psi(t)\psi(t)\, dt = 0 \qquad (4-13)$$

when $\psi(t) = s(t) + j\hat{s}(t)$. This result also implies that

$$\int_{-\infty}^{\infty} s^2(t)\, dt = \int_{-\infty}^{\infty} \hat{s}^2(t)\, dt \qquad (4-14)$$

and

$$\iint_{-\infty}^{\infty} \frac{s(t)s(\tau)}{t - \tau}\, dt\, d\tau = 0 \qquad (4-15)$$

To prove (4-13), Parseval's theorem is applied to obtain

$$\int_{-\infty}^{\infty} \psi(t)\psi(t)\, dt = \int_{-\infty}^{\infty} \Psi(f)\Psi(-f)\, df \qquad (4-16)$$

where $\Psi(f)$ is the Fourier transform corresponding to $\psi(t)$. The right side of this equation clearly vanishes when

$$\Psi(f) = 0, \qquad 0 > f \tag{4-17}$$

The proof is completed by noting the following Fourier transform pair

$$s(t) + j\hat{s}(t) \Rightarrow \Psi(f) = \begin{cases} 2S(f), & 0 < f \\ 0, & 0 > f \end{cases} \tag{4-18}$$

In general, for other linear operations on $s(t)$, such as the convolution integral, it can be shown that

$$\int_{-\infty}^{\infty} s(\tau)h(t-\tau)\,dt = \tfrac{1}{2}\,\mathrm{Re}\left\{\int_{-\infty}^{\infty} [\psi(\tau)\psi_h(\tau,t) + \psi(\tau)\psi_h^*(\tau,t)]\,d\tau\right\} \tag{4-19}$$

where $\psi(\tau)$ and $\psi_h(\tau)$ are complex representations for $s(\tau)$ and $h(-\tau)$, respectively. When $\psi(\tau)$ and $\psi_h(\tau)$ are Gabor complex representations, Eq. (4-19) reduces in a manner similar to (4-12) to yield

$$\int_{-\infty}^{\infty} s(\tau)h(t-\tau)\,d\tau = \tfrac{1}{2}\,\mathrm{Re}\left\{\int_{-\infty}^{\infty} \psi(\tau)\psi_h^*(\tau,t)\,d\tau\right\} \tag{4-20}$$

It is seen that when $h(t-\tau) = s(\tau+t)$ (4-20) describes the autocorrelation function of $s(t)$.

A previous statement in this section alluded to the difficulty of deriving the Gabor waveform; this is because it is usually not possible to express the Hilbert transform in closed form. In practice, therefore, for the analysis of specific waveforms, the Gabor complex representation is frequently approximated by (4-6). However, this approximation becomes an exact representation when $s(t)$ is a "narrow band waveform." In general, the approximation error $\delta(t)$ is given as a function of time by

$$\delta(t) = 2\,\mathrm{Im}\left\{\int_{-\infty}^{0} S_+(f)\exp[j2\pi ft]\,dt\right\} \tag{4-21}$$

where

$$S_+(f) = \int_{-\infty}^{\infty} a(t)\exp[-j\{2\pi(f-f_0)t - \theta(t)\}]\,dt \tag{4-22}$$

Equation (4-21) clearly depends only on the spillover of $S_+(f)$ into the negative frequency region; practical pulsed radar waveforms will always have some spillover since they are always confined to a finite duration. However, as long as $\Delta f/f_0$ is small, where Δf denotes the signal bandwidth, the error will be small. For most radar waveforms this condition will be satisfied so that $\delta(t)$ can be neglected; this is not the case for sonar waveforms and more care must be taken in treating the complex representation for this class of signals, as well as for some very wideband radar signals.

For the remainder of this chapter, the waveforms will be treated in terms of their complex representation. So as to prevent any confusion, it will be assumed for the particular waveforms that are considered in this chapter that their complex representation is approximated, where necessary, by (4-6). However, for the general treatment of waveforms that appears in this chapter as well as the following chapter, the exact Gabor representation will be employed. The following convention will be used to describe the complex representation

$$\psi(t) = u(t)\exp[\,j2\pi f_0 t]$$ (4-23)

where

$$2Ef_0 = \int_{-\infty}^{\infty} f|\Psi(f)|^2\, df$$ (4-24)

$$u(t) = a(t)\exp[\,j\theta(t)]$$ (4-25)

4.3   The Doppler Approximation

Although it is common practice to approximate the Doppler effect as a frequency translation, the true Doppler effect appears as a scale factor on the time variable. Thus, if the waveform is $\psi(t)$ before transmission, then the waveform after being reflected from a moving target is $\sqrt{\alpha}\psi(\alpha t)$. The scale factor $\alpha = 1 \pm 2v/c$, where the plus $(+)$ and minus $(-)$ are for closing and opening velocities v, and c is the velocity of propagation. The multiplier $\sqrt{\alpha}$ is necessary in order to account for the fact that the energy does not change under the transformation caused by the Doppler effect.

To derive the preceding result, a simple but frequently used model is postulated. This model consists of a point target traveling at constant velocity v along a radial path away from (or toward) the radar. If $\psi(t)$ is transmitted, then the reflected waveform is $\psi(t - t_d)$. The delay time t_d is itself a function of time given by

$$t_d = \frac{2R(t)}{c \pm v}$$ (4-26)

For the model chosen, the range function $R(t)$ is

$$R(t) = R_0 \pm vt$$ (4-27)

where R_0 is the initial range of the target at time $t = 0$. By substituting for t_d and collecting terms, the result obtained is

$$\psi(t - t_d) = \psi\left(\alpha t - \frac{2R_0}{c \pm v}\right)$$ (4-28)

where $2R_0/(c \pm v)$ is the nominal delay to and from the target of $\psi(t)$.

An important Fourier transform pair is

$$\sqrt{\alpha}\psi(\alpha t) \Rightarrow \frac{1}{\sqrt{|\alpha|}}\Psi(f/\alpha) \qquad (4\text{-}29)$$

The implication of this transform pair is that the correct matched filter for the received waveform is $k\Psi^*(f/\alpha)$ and not the usually implemented $k\Psi^*(f \pm \phi)$. However, it can be shown that as long as $(2v/c)T\Delta f < 1$ the mismatch that develops is inconsequential. Moreover, under the same condition the analysis of the matched-filter response for Doppler shifted signals can also be treated by using $k\Psi^*(f \pm \phi)$. Throughout this book the Doppler effect is approximated by the simple frequency translation, except where specifically noted.

4.4 The General Ambiguity Function Formulation

From our every day visual experiences we know that the differences in color, size, profile, etc., are what make one object distinguishable from another. So too for radar, the ability to distinguish between two targets depends on the differences that are superimposed on the reflected radar waveform. This section discusses the waveform differences derived from time displacement (denoted by τ) and frequency shift (denoted by ϕ). The measure of difference that will be used for this purpose is the "mean square departure" of a waveform and its shifted self [1]; this is given by

$$\int_{-\infty}^{\infty} |\psi(t) - \psi(t + \tau, f + \phi)|^2 \, dt \qquad (4\text{-}30)$$

where $\psi(t + \tau, f + \phi)$ and $\psi(t)$ are the shifted and unshifted waveforms. This measure of difference has an alternate form, derived by invoking Parseval's theorem, given by

$$\int_{-\infty}^{\infty} |\Psi(f) - \Psi(f + \phi, t + \tau)|^2 \, df \qquad (4\text{-}31)$$

where $\Psi(f)$ is the Fourier transform of $\psi(t)$.

Each of the equations (4-30) and (4-31) can be reduced, respectively, to the general ambiguity function formulations given by expressions (4-2a) and (4-2b). However, to demonstrate this it is necessary to consider only one of these expressions since, without any loss in generality, Parseval's theorem can be invoked on the result of this reduction to obtain the alternate ambiguity function expression.

Considering (4-30), the following expansion is obtained:

$$\int_{-\infty}^{\infty} |\psi(t) - \psi(t + \tau, f + \phi)|^2 \, dt = k - 2 \operatorname{Re}\left\{\int_{-\infty}^{\infty} \psi(t)\psi^*(t + \tau, f + \phi) \, dt\right\}$$

$$(4-32)$$

where k is a constant related to the signal energy given by

$$k = \int_{-\infty}^{\infty} |\psi(t)|^2 \, dt + \int_{-\infty}^{\infty} |\psi(t + \tau, f + \phi)|^2 \, dt \qquad (4-33)$$

Since the integral that appears in the expansion given by (4-32) is the only term that varies with τ and ϕ, it is therefore the significant term. This integral exists as long as the two waveforms $\psi(t)$ and $\psi(t + \tau, f + \phi)$ are sufficiently alike; it is the general correlation function formulation for simultaneous shifts in time and frequency. In general, this term tends to become smaller for values of τ and ϕ removed from the origin located at $\tau = \phi = 0$. The origin for this integral can be placed, as shown in Fig. 4.1,

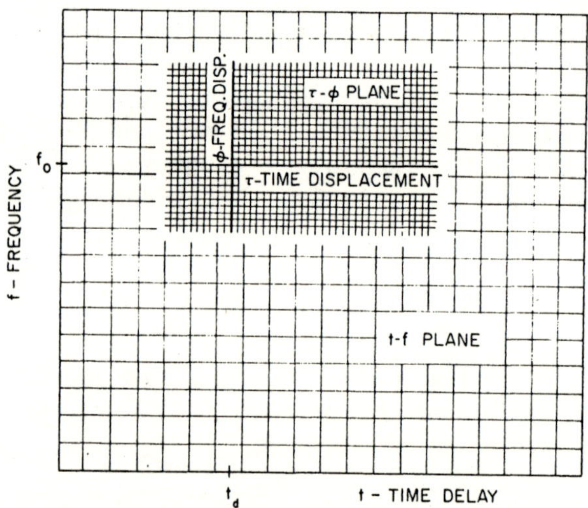

FIG. 4.1   Ambiguity function coordinates: τ-ϕ and t-f planes.

anywhere on a quarter plane having the coordinates of absolute range delay t_d and frequency f_0. The location of the origin is determined by the reference time delay and frequency that are prescribed for $\psi(t)$. The τ-ϕ plane can be treated as the time-frequency space in which the radar signals of interest exist. The principle objective in the context of radar waveform design is to control the behavior of this two-dimensional correlation function. The next section discusses the constraints of this problem.

It will also be shown in the next section that this function achieves its greatest value at $\tau = \phi = 0$. It is possible, however, for this term to become peaked at locations away from the origin. The implication of these peaks is that they cannot be distinguished from the displaced $\psi(t)$'s of other signals which correspond to these points. The ambiguity reflected by such pop-ups is clear; not only do they make estimations of time delay and frequency highly ambiguous, but, in addition, for situations that involve an unknown number of targets (suppose these targets can only exist at the discrete points where the pop-ups occur) there is no way of being certain how many targets are actually being observed by the radar. Whether or not these potential ambiguities are important in the search for a suitable waveform depends on how likely it is that such waveform displacements in time and frequency will occur. This knowledge depends upon the ability of the radar designer to postulate an accurate model of the environment in which the radar will operate. This subject is discussed further in Chapter 10.

By applying (4-21), it can be shown that

$$\int_{-\infty}^{\infty} \psi(t)\psi^*(t + \tau, f + \phi)\,dt = \exp[-j2\pi(f_0 + \phi)\tau]$$

$$\times \int_{-\infty}^{\infty} u(t)u^*(t + \tau)\exp[-j2\pi\phi t]\,dt \qquad (4\text{-}34)$$

The integral on the right side of (4-34) will be recognized as $\chi(\tau, \phi)$; Parseval's theorem will produce the alternate form of $\chi(\tau, \phi)$. The ambiguity function is defined as $\chi(\tau, \phi)\chi^*(\tau, \phi)$ as noted earlier.

As just derived, the ambiguity function is a mathematical concept which hopefully provides a good measure of ambiguity inherent in a radar signal. The "mean square departure" criterion is only one of many possible measures for the difference between two waveforms and is generally considered to be the best for this purpose. In order to attach a meaningful physical significance to $\chi(\tau, \phi)$ for the systems engineer an alternate derivation of this function will now be developed. This approach involves the viewpoint that $\chi(\tau, \phi)$ is the time-frequency response for a matched filter.

Suppose $\Psi^*(f)$ is the transfer function of a matched filter for $\psi(t)$. It is readily shown by referring to the definition previously given for $U(f)$ in Section 4.1 that

$$\Psi^*(f) = U^*(f - f_0) \qquad (4\text{-}35)$$

where f_0 is the reference frequency shown in Fig. 4.1. In addition, suppose the input waveform is a Doppler shifted replica of $\psi(t)$. The Fourier transform for this input is represented by

$$\Psi(f + \phi) = U(f - f_0 + \phi) \qquad (4\text{-}36)$$

It can then be shown that the output from this filter is given by

$$\Gamma(\tau, \phi) = \exp[j2\pi f_0 \tau] \int_{-\infty}^{\infty} U^*(f)U(f + \phi) \exp[j2\pi f\tau] \, df \qquad (4\text{-}37)$$

It is noted that

$$\Gamma(\tau, \phi) = \exp[j2\pi f_0 \tau]\chi(-\tau, \phi) \qquad (4\text{-}38)$$

This equation appears to imply that the output from a matched filter is the time inverse of the ambiguity function. However, by labeling τ as τ_{af} and τ_{mf}, respectively, for the ambiguity function and the response function it is seen that

$$\tau_{af} = -\tau_{mf} \qquad (4\text{-}39)$$

This difference can be seen more clearly by referring to Fig. 4.2.

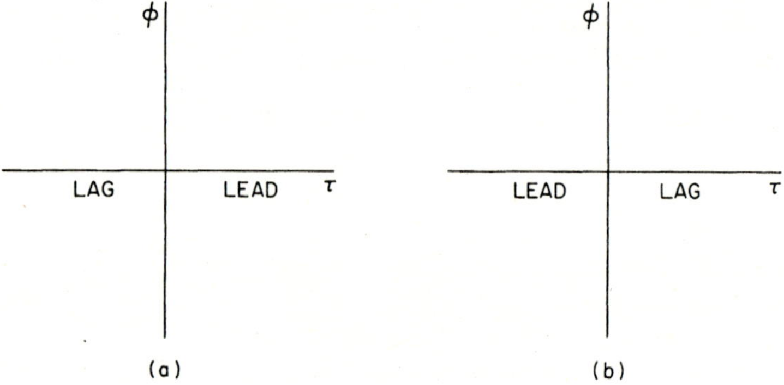

FIG. 4.2 τ-ϕ planes. (a) Ambiguity function. (b) Time/frequency response function.

4.5   Properties of the Ambiguity Function

The general formulation for $\chi(\tau, \phi)$ has been an attractive study area ever since it was first introduced as a basis for radar waveform design. A prime objective of these studies has been the hope of finding the necessary and sufficient conditions for a two-variable function to be an ambiguity function. Although the results achieved have fallen short of this goal, much has been learned about the general properties of $\chi(\tau, \phi)$. These properties are reviewed in this section.[1] However, for this purpose it will often be convenient to present these properties in terms of $\Theta(\tau, \phi)$ which is defined by

$$\chi(\tau, \phi) = \exp[j\pi\phi\tau]\Theta(\tau, \phi) \qquad (4\text{-}40)$$

This equation is obtained by introducing the new variable $s = t + \tau/2$ to

---

[1] The principle references for these properties are Siebert [4] and Wilcox [5]. In addition, other sources have been drawn on and are cited at the appropriate points in this chapter.

expression (4-2a) to obtain the symmetrical form

$$\Theta(\tau, \phi) = \int_{-\infty}^{\infty} u\left(t - \frac{\tau}{2}\right) u^*\left(t + \frac{\tau}{2}\right) \exp[-j2\pi\phi t]\, dt \qquad (4\text{-}41)$$

or by introducing $p = f + \phi/2$ into expression (4-2b) to obtain

$$\Theta(\tau, \phi) = \int_{-\infty}^{\infty} U^*\left(f - \frac{\phi}{2}\right) U\left(f + \frac{\phi}{2}\right) \exp[-j2\pi f\tau]\, df \qquad (4\text{-}42)$$

These transformations are inconsequential in the context of waveform design since it is the magnitude of $\chi(\tau, \phi)$ that usually interests the designer and not the phase structure, which is the only thing affected as a result of employing the Θ-function.

Property I: Symmetry about the Origin

It is readily shown from Eqs. (4-41) and (4-42) that

$$\Theta(\tau, \phi) = \Theta^*(-\tau, -\phi) \qquad (4\text{-}43)$$

Therefore, it follows that

$$|\Theta(\tau, \phi)|^2 = |\Theta(-\tau, -\phi)|^2 \qquad (4\text{-}44)$$

Moreover, Stutt [6] has also pointed out that

$$|\Theta(\tau, \phi)|\, \exp[j\gamma(\tau, \phi)] = |\Theta(-\tau, -\phi)|\, \exp[-j\gamma(-\tau, -\phi)] \qquad (4\text{-}45)$$

Hence, the phase function $\gamma(\tau, \phi)$ is *odd* symmetric.

Property II: The Greatest Value of $\Theta(\tau, \phi)$

The greatest value of the ambiguity function always appears above the origin $\tau = \phi = 0$; that is,

$$|\Theta(\tau, \phi)|^2 \leqslant |\Theta(0, 0)|^2 \qquad (4\text{-}46)$$

The proof follows from the application of the Schwarz inequality.[1] Thus, by applying this inequality it is seen that

$$|\Theta(\tau, \phi)|^2 = \left| \int_{-\infty}^{\infty} u\left(t - \frac{\tau}{2}\right) u^*\left(t + \frac{\tau}{2}\right) \exp[-j2\pi\phi t]\, dt \right|^2$$

$$\leqslant \int_{-\infty}^{\infty} \left| u\left(t - \frac{\tau}{2}\right) \right|^2 dt \int_{-\infty}^{\infty} \left| u\left(t + \frac{\tau}{2}\right) \right|^2 dt \qquad (4\text{-}47)$$

---

[1] The Schwarz inequality is

$$\left[\int f^*(z)g(z)\, dz \right]^2 \leqslant \int f(z)f^*(z)\, dz \int g(z)g^*(z)\, dz$$

But since

$$\int_{-\infty}^{\infty} \left| u\left(t - \frac{\tau}{2}\right) \right|^2 dt = \int_{-\infty}^{\infty} \left| u\left(t + \frac{\tau}{2}\right) \right|^2 dt = |\chi(0,0)| \qquad (4\text{-}48)$$

the inequality given in (4-46) is obtained.

Property III: Time Scaling

If $\Theta(\tau, \phi)$ corresponds to $u(t)$, then $(1/|a|)\Theta(a\tau, \phi/a)$ corresponds to $u(at)$. The proof involves a straightforward substitution in (4-41).

Property IV: Quadratic Phase

b. Frequency

If $\Theta(\tau, \phi)$ corresponds to $U(f)$ then $\Theta(\tau + h\phi/\pi, \phi)$ corresponds to $U(f)\exp[jhf^2]$. The proof involves a straightforward substitution in (4-42).

$$= 0, \qquad |t| > T/2$$

The new waveform $A \exp[jkt^2]$ is the linear FM pulse.

b. Frequency

If $\Theta(\tau, \phi)$ corresponds to $u(t)$ then $\Theta(\tau + h\phi/\pi, \phi)$ corresponds to $U(f) \exp[jhf^2]$. The proof involves a straightforward substitution in (4-42).

Property V: Volume Invariance

If $|\chi(\tau, \phi)|^2$ is an ambiguity function then it can be shown that

$$\iint_{-\infty}^{\infty} |\chi(\tau, \phi)|^2 \, d\tau \, d\phi = |\chi(0,0)|^2 \qquad (4\text{-}50)$$

This property is usually referred to as the "radar uncertainty principle" [4], or sometimes as the "law of the conservation of ambiguity" [5]. It leads to the conclusion that the total potential ambiguity is the same for all signals that possess the same energy. In addition, it also represents a necessary condition for a two-variable function to be a $|\chi(\tau, \phi)|^2$ function. It is perhaps the most important ambiguity function constraint since it implies that all signals are equally good (or bad) as long as they are not compared against a specific radar environment. Fortunately, each different waveform yields a new distribution of ambiguity. Figure 4.3 illustrates the ambiguity distribution of some important radar waveforms. It is this variation in the ambiguity distribution, coupled with at least a basic understanding of the radar

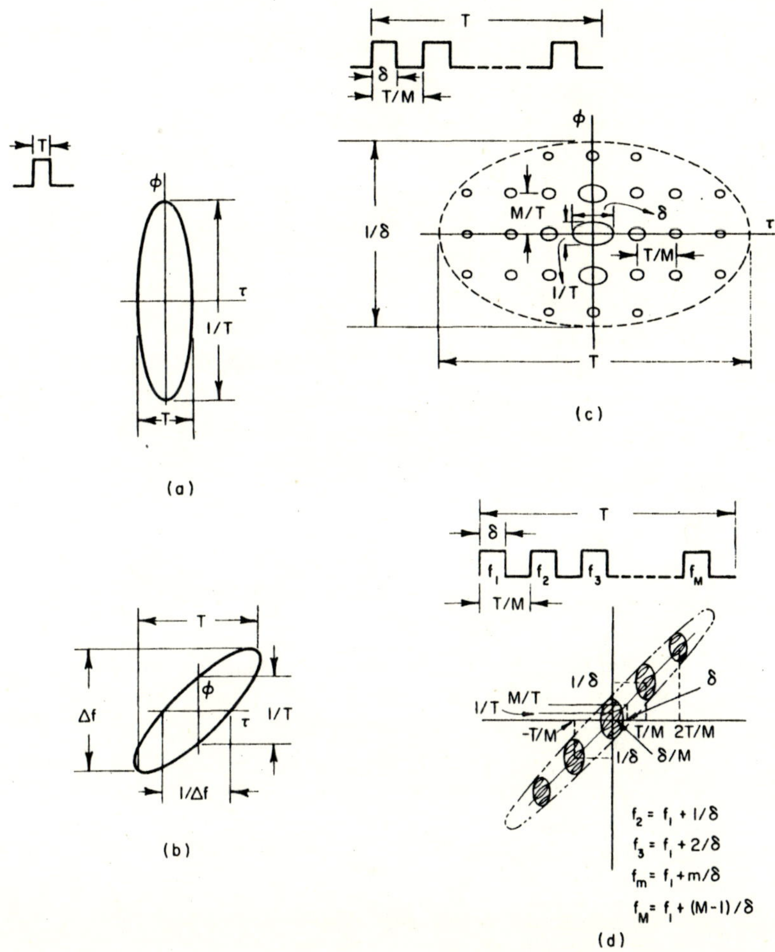

FIG. 4.3 Examples of ambiguity function distributions—Contours at a constant amplitude level relative to $|\chi(0, 0)|$. (a) Short monotone pulse. (b) Linear FM pulse. (c) Uniform pulse train (see Chapter 8). (d) Pulse-to-pulse stepped FM pulse train (see Chapter 8). (From Bernfeld *et al.*, Ref. 4, Chapter 1.)

system application, that enables the radar designer to systematically search for a suitable waveform for transmission. It follows, as shown by Price and Hofstetter[12], that a condition for Gaussian envelope signals is

$$\iint_{-\infty}^{\infty} |\chi(\tau, \phi)|^{2p}\, d\tau\, d\phi = \frac{1}{p}|\chi(0, 0)|^2 \tag{4-51}$$

where p is any integer > 1.

The proof of Eq. (4-50) will now be furnished; the proof for (4-51) can be found in Price and Hofstetter [12]. It is seen that

$$\chi^*(\tau, \phi) = \int_{-\infty}^{\infty} u^*(t)u(t + \tau)\exp[\,j2\pi\phi t]\,dt$$

$$= \int_{-\infty}^{\infty} U(f)U^*(f + \phi)\exp[\,j2\pi f\tau]\,df \tag{4-52}$$

Consequently,

$$\chi(\tau, \phi)\chi^*(\tau, \phi) = |\chi(\tau, \phi)|^2$$

$$= \iint_{-\infty}^{\infty} u(t)u^*(t + \tau)U(f)U^*(f + \phi)\exp[\,j2\pi(f\tau - \phi t)]\,dt\,df \tag{4-53}$$

and

$$\iint_{-\infty}^{\infty} |\chi(\tau, \phi)|^2\,d\tau\,d\phi$$

$$= \iiiint_{-\infty}^{\infty} u(t)u^*(t + \tau)U(f)U^*(f + \phi)\exp[\,j2\pi(f\tau - \phi t)]\,dt\,df\,d\tau\,d\phi \tag{4-54}$$

The latter equation can now be reduced by applying the following Fourier transform pairs:

$$\int_{-\infty}^{\infty} u^*(t + \tau)\exp[\,j2\pi f\tau]\,d\tau = \exp[\,-j2\pi ft]U^*(f) \tag{4-55}$$

and

$$\int_{-\infty}^{\infty} U^*(f + \phi)\exp[\,-2\pi\phi t]\,d\phi = \exp[\,j2\pi ft]u^*(t) \tag{4-56}$$

By making these substitutions into the right side of (4-54) it is seen that

$$\iint_{-\infty}^{\infty} |\chi(\tau, \phi)|^2\,d\tau\,d\phi = \iint_{-\infty}^{\infty} u(t)u^*(t)U(f)U^*(f)\,df\,dt$$

$$= \int_{-\infty}^{\infty} |u(t)|^2\,dt \int_{-\infty}^{\infty} |U(f)|^2\,df \tag{4-57}$$

Since

$$\int_{-\infty}^{\infty} |u(t)|^2\,dt = \int_{-\infty}^{\infty} |U(f)|^2\,df = |\chi(0, 0)| \tag{4-58}$$

the proof has been completed.

An alternate approach to the proof of Property V is now considered. This will lead to some important integral pairs. For this approach the following Fourier transform pair is needed:

$$\int_{-\infty}^{\infty} u(t)\exp[-j2\pi(f+\phi)t]\,dt = U(f+\phi) \tag{4-59}$$

When this transform pair, together with (4-55), is applied to (4-54) the result obtained is

$$\iint_{-\infty}^{\infty} |\chi(\tau,\phi)|^2\,d\tau\,d\phi = \iint_{-\infty}^{\infty} |U(f)|^2|U(f+\phi)|^2\,df\,d\phi \tag{4-60}$$

It is seen that the inner integrals are equal. This relationship is discussed further in Section 4.11. By applying Parseval's theorem, it is also seen that

$$\int_{-\infty}^{\infty} |U(f)|^2|U(f+\phi)|^2\,df = \int_{-\infty}^{\infty} |\chi(\tau,0)|^2\exp[-j2\pi\phi\tau]\,d\tau \tag{4-61}$$

Another approach to the proof of Property V will now be considered. For this the following Fourier transform is needed:

$$\int_{-\infty}^{\infty} U(f)\exp[j2\pi f(t+\tau)]\,df = u(t+\tau) \tag{4-62}$$

When this transform pair, together with (4-56), is applied to (4-54) the result obtained is

$$\iint_{-\infty}^{\infty} |\chi(\tau,\phi)|^2\,d\phi\,d\tau = \iint_{-\infty}^{\infty} |u(t)|^2|u(t+\tau)|^2\,dt\,d\tau \tag{4-63}$$

Once more, it is observed that the inner integrals are equal and by applying Parseval's theorem the following is obtained

$$\int_{-\infty}^{\infty} |u(t)|^2|u(t+\tau)|^2\,dt = \int_{-\infty}^{\infty} |\chi(0,\phi)|^2\exp[j2\pi\phi\tau]\,d\phi \tag{4-64}$$

This result is also discussed further in Section 4.11.

The volume invariance property also applies to mismatched ambiguity functions; that is, when the receiver signal processor is not matched to the transmitted signal. The linear delay mismatch and range sidelobe reduction techniques discussed, respectively, in Chapters 6 and 7 are examples of this. How the receiver mismatch affects the ambiguity distribution in local regions of the ambiguity plane will depend on the exact nature of the mismatch involved.

Property VI: The Self Transform Property

If $|\chi(\tau,\phi)|^2$ is an ambiguity function then

$$\iint_{-\infty}^{\infty} |\chi(\tau,\phi)|^2\exp[j2\pi(\phi v - \tau\sigma)]\,d\tau\,d\phi = |\chi(v,\sigma)|^2 \tag{4-65}$$

The proof for this property is essentially the same as the proof for the preceding property. Thus, the left side of Eq. (4-65) is expanded by substituting the alternate integral forms for $\chi(\tau, \phi)$ in its place, yielding

$$\iiiint_{-\infty}^{\infty} u(t)u^*(t + \tau)U(f)U^*(f + \phi)\exp[\,j2\pi\{(v - t)\phi - (\sigma - f)\tau\}]\,dt\,df\,d\tau\,d\phi$$

$$(4\text{-}66)$$

This integral can now be reduced by applying the following Fourier transform pairs:

$$\int_{-\infty}^{\infty} U^*(f + \phi)\exp[\,j2\pi\phi(v - t)]\,d\phi = u^*(t - v)\exp[\,j2\pi f(t - v)] \qquad (4\text{-}67a)$$

and

$$\int_{-\infty}^{\infty} u^*(t + \tau)\exp[-j2\pi\tau(\sigma - f)]\,d\tau = U^*(f - \sigma)\exp[-j2\pi(f - \sigma)t] \qquad (4\text{-}67b)$$

On introducing the variables $p = t - v$ and $q = f - \sigma$, the integral in (4-66) reduces to the following:

$$\iint_{-\infty}^{\infty} u^*(p)u(p + v)U^*(q)U(q + \sigma)\exp[\,j2\pi(\sigma p - qv)]\,dp\,dq \qquad (4\text{-}68)$$

However, it is readily seen that this integral is the product

$$\chi(\sigma, v)\chi^*(\sigma, v) = |\chi(\sigma, v)|^2 \qquad (4\text{-}69)$$

Property VII: Rotational Invariance

The property of being an ambiguity function is invariant under the transformation that rotates the $\tau - \phi$ plane through the angle θ such that

$$\begin{bmatrix} \tau' \\ \phi' \end{bmatrix} = \begin{bmatrix} \cos\theta & -\sin\theta \\ \sin\theta & \cos\theta \end{bmatrix} \begin{bmatrix} \tau \\ \phi \end{bmatrix} \qquad (4\text{-}70)$$

Thus, if $\chi(\tau, \phi)$ is the ambiguity function that corresponds to $u(t)$ then $\chi(\tau', \phi')$ is an ambiguity function also and it corresponds to

$$v(t) = |\cos\theta|^{1/2} \int_{-\infty}^{\infty} df \exp[-j2\pi f t]\cos\theta$$

$$\times \int_{-\infty}^{\infty} u(x)\exp\left[-j2\pi\left\{fx - \frac{\tan\theta}{2}x^2 + \frac{\sin 2\theta}{4}\right\}\right]dx \qquad (4\text{-}71)$$

For a class of waveforms, called Hermite waveforms, it is found that

$$v(t) = u(t) \tag{4-72}$$

Additional information regarding this class of waveforms is furnished in Section 4.12 of this chapter, where some important ambiguity function examples are presented.

The property that an ambiguity function transforms into another ambiguity function when the coordinate axis are rotated has been generalized by Reis [8]. Thus, as long as

$$\begin{vmatrix} a_{11} & a_{12} \\ a_{21} & a_{22} \end{vmatrix} = \pm 1 \tag{4-73}$$

then $\chi(a_{11}\tau + a_{12}\phi, a_{21}\tau + a_{22}\phi)$ corresponds to

$$v(t) = |a_{11}|^{1/2} \int_{-\infty}^{\infty} df \exp\left[j2\pi a_{11}f\left(t - \frac{a_{12}}{2}f\right)\right]$$

$$\times \int_{-\infty}^{\infty} u(x) \exp\left[-j2\pi x\left(f - \frac{a_{21}}{2a_{11}}x\right)\right] dx \tag{4-74}$$

The new ambiguity function $\chi(\tau', \phi')$ can be derived by applying in sequence the properties II, IVa, and IVb. Thus, if $u(t)$ corresponds to $\chi(\tau, \phi)$ then $z(t) = u(at)\exp[jkt^2]$ corresponds to $(1/|a|)\chi\{a\tau, (1/a)[\phi + k\tau/\pi]\}$. Now if $Z(f)$ is the Fourier transform of $z(t)$ then the ambiguity function that corresponds to $V(f) = Z(f)\exp[jhf^2]$ is $(1/|a|)\chi\{a\tau + ah\phi/\pi, k\tau/a\pi + (\phi/a)[1 + kh/\pi^2]\}$ where the characteristic determinant is

$$\begin{vmatrix} a & \dfrac{ah}{\pi} \\ \dfrac{k}{a\pi} & \dfrac{1}{a}\left(1 + \dfrac{kh}{\pi^2}\right) \end{vmatrix} = 1 \tag{4-75}$$

Property VIII: Product Rule

If $\chi_u(\tau, \phi)$ is the ambiguity function corresponding to $u(t)$ and if $\chi_v(\tau, \phi)$ is the ambiguity function that corresponds to $v(t)$, then for $z(t) = u(t)v(t)$ the corresponding ambiguity function $\chi_{uv}(\tau, \phi)$ is given by the convolution of $\chi_u(\tau, \phi)$ and $\chi_v(\tau, \phi)$ along the Doppler axis

$$\chi_{uv}(\tau, \phi) = \int_{-\infty}^{\infty} \chi_u(\tau, \lambda)\chi_v(\tau, \phi - \lambda)\, d\lambda \tag{4-76}$$

Similarly, if $Z(f) = U(f)V(f)$ then

$$\chi_Z(\tau, \phi) = \int_{-\infty}^{\infty} \chi_U(t, \phi)\chi_V(\tau - t, \phi)\, dt \tag{4-77}$$

4.6 The Uniqueness Theorem

A necessary and sufficient condition for some function of two variables to be a $\Theta(\tau, \phi)$ function is that it transform as follows [4, 5]:

$$u\left(x - \frac{\tau}{2}\right)u^*\left(x + \frac{\tau}{2}\right) = \int_{-\infty}^{\infty} \Theta(\tau, \phi)\exp[j2\pi\phi x]\,d\phi \qquad (4\text{-}78)$$

By introducing the new variables $\rho = x - \tau/2$ and $\sigma = x + \tau/2$, this condition is conveniently expressed as

$$u(\rho)u^*(\sigma) = \int_{-\infty}^{\infty} \Theta(\sigma - \rho, \phi)\exp[j\pi\phi(\rho + \sigma)]\,d\phi \qquad (4\text{-}79)$$

It can also be shown that if (4-78) holds for a function of two variables then

$$U(\Omega)U^*(\beta) = \int_{-\infty}^{\infty} \Theta(\tau, \Omega - \beta)\exp[j\pi(\Omega + \beta)\tau]\,d\tau \qquad (4\text{-}80)$$

As a result, if $\Theta_u(\tau, \phi)$ corresponds to $u(t)$ and $\Theta_v(\tau, \phi)$ corresponds to $v(t)$ and if $\Theta_u(\tau, \phi) = \Theta_v(\tau, \phi)$ then it follows from (4-78) that $u(t)$ and $v(t)$ can only differ by a constant multiplier c where $|c| = 1$ (i.e., $v(t) = cu(t)$).

This uniqueness property derives from the fact that as long as $\Theta_u(\tau, \phi) = \Theta_v(\tau, \phi)$ then (4-78) provides

$$u(\rho)u^*(\sigma) = v(\rho)v^*(\sigma) \qquad (4\text{-}81)$$

For this equality to hold, it is necessary that

$$u(\rho)u^*(\sigma) = cu(\rho)c^*u^*(\sigma) \qquad (4\text{-}82)$$

Therefore,

$$[|c|^2 - 1]u(\rho)u^*(\sigma) = 0 \qquad (4\text{-}83)$$

and since $u(t)$ is not identically zero it follows that $|c|^2 = 1$.

A similar condition for a function of two variables to be a $|\Theta(\tau, \phi)|^2$ has so far not been discovered.

4.7 Volume Free Area and Average Sidelobe Level

A conjecture made by P. E. Green stimulated Price and Hofstetter [12] to prove that the maximum possible volume free area that can surround a spike located at $\tau = \phi = 0$ is 4. This result implies that the maximum range-Doppler area that can be unambiguously explored is unity. If $V(A)$ denotes the volume under $|\chi(\tau, \phi)|^2$ within the boundaries of a centrally located convex, symmetric area A, and if $c(A)$ denotes the volume free area within these boundaries and V_0 denotes the volume of the spike at $\tau = \phi = 0$, then Price and Hofstetter's result is given as follows:

$$V(A) \geqslant \tfrac{1}{4}c(A)V_0 \tag{4-84}$$

The equality for (4-84) holds as long as $c(A) < 4$.

This result is illustrated in Fig. 4.4 for a train of impulses given by

$$u(t) = \sum_{n=-\infty}^{\infty} \delta(t - nT) \tag{4-85}$$

where $\delta(t)$ denotes an impulse waveform. It is seen in Fig. 4.4 that the

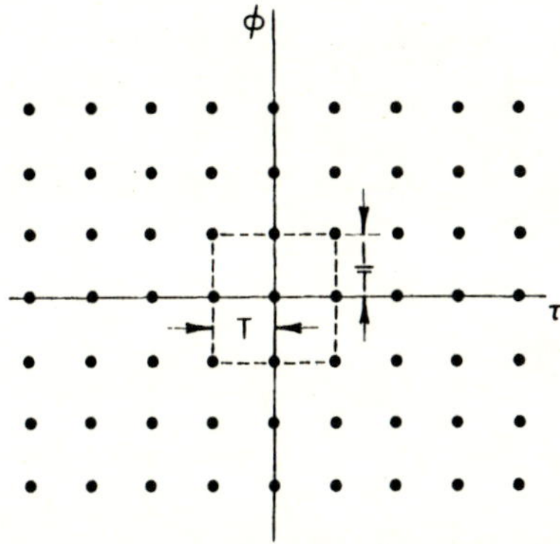

FIG. 4.4   "Bed of nails" ambiguity function of an infinite train of impulses.

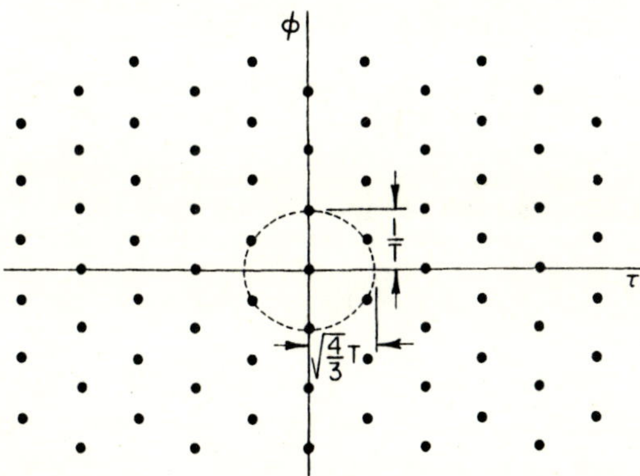

FIG. 4.5   Ambiguity function of a rotated "bed of nails."

ambiguity function for (4-85) consists of a "bed of nails" having a clear rectangular area around the central spike which is equal to 4. Figure 4.5 illustrates a volume free elliptical area obtained by multiplying (4-85) by

$\exp[jkt^2]$ to give a waveform described by

$$u(t) = \sum_{n=-\infty}^{\infty} \delta(t - nT)\exp[jk(nT)^2] \tag{4-86}$$

Since the ambiguity function for (4-85) is given by

$$\chi(\tau, \phi) = \sum_{m=-\infty}^{\infty}\sum_{n=-\infty}^{\infty} \delta\left(\tau - nT, \phi - \frac{m}{T}\right) \tag{4-87}$$

then, recalling Property IV of Section 4.5, it is seen that the ambiguity function for (4-87) is given by

$$\chi\left(\tau, \phi + \frac{1}{2nT}\right) = \sum_{m=-\infty}^{\infty}\sum_{n=-\infty}^{\infty} \delta\left(\tau - nT, \phi - \frac{m}{T} + \frac{kmT}{\pi}\right) \tag{4-88}$$

Let $k = (\pi/2)(1/T)^2$. This causes the impulses of (4-87), along constant τ lines defined by $n = 2p - 1$, to shift an amount $\phi = 1/2T$ to yield

$$\chi\left(\tau, \phi + \frac{1}{2nT}\right) = \sum_{n=-\infty}^{\infty} \delta\left(\tau - nT, \phi - \frac{1}{2nT}\right) \tag{4-89}$$

It is seen in Fig. 4.5 that the volume free area for (4-89) is bounded by an ellipse whose half axes are $\sqrt{4/3}T$ along the time axis and $1/T$ along the frequency axis and whose area is 3.6.

An extension of the result concerning volume free areas is obtained by considering similar areas for which $|\chi(\tau, \phi)|^2 \leqslant \eta$. Price and Hofstetter have shown that for this case the inequality becomes

$$V_0 + \eta c(A) \geqslant V(A) \geqslant \frac{c(A)}{4}V_0 \tag{4-90}$$

This inequality can hold only as long as $\eta < V_0/4$ and

$$c(A) \leqslant V_0\left(\frac{V_0}{4} - 1\right)^{-1} \tag{4-91}$$

In general, it is seen that the greater η becomes the larger $c(A)$ becomes.

Price and Hofstetter have shown for an ambiguity function containing an impulse at $\tau = \phi = 0$ that for an average taken over an area many times 4 the average sidelobe level must exceed $\sqrt{V_0/2}$, where V_0 is the volume of the centrally located impulse. This average sidelobe level is given by

$$S^2 = \frac{V(A) - V_0}{c(A)} \tag{4-92}$$

For the ideal "thumbtack" ambiguity function, which consists of an impulse of volume V_0 located at the origin and surrounded by a level plateau of height V_0, it is found that the average sidelobe level is $\sqrt{V_0}$.

4.8 An Expansion Theorem for Ambiguity Functions [5, 6]

Consider the sequence of functions

$$f_0(t), \quad f_1(t), \quad f_2(t), \quad \ldots \quad , f_n(t) \tag{4-93}$$

If the members of this sequence are integrable square between $-\infty < t < \infty$ and if they are orthonormal in accordance with the definition

$$\int_{-\infty}^{\infty} f_n(t) f_m^*(t)\, dt = \delta_{mn} \tag{4-94}$$

where δ_{mn} is the Kronecker delta, then it follows if the sequence $f_n(t)$ is complete that $B_{mn}(\tau, \phi)$, which is defined by

$$B_{mn}(\tau, \phi) = \int_{-\infty}^{\infty} f_n\left(t - \frac{\tau}{2}\right) f_m^*\left(t + \frac{\tau}{2}\right) \exp[-j2\pi\phi t]\, dt \tag{4-95}$$

is also a member of a complete orthonormal sequence, where

$$\iint_{-\infty}^{\infty} B_{mn}(\tau, \phi) B_{pq}^*(\tau, \phi)\, d\tau\, d\phi = \delta_{np}\, \delta_{mq} \tag{4-96}$$

If $u(t)$ is integrable square then

$$u(t) = \sum_{n=0}^{\infty} a_n f_n(t) \tag{4-97}$$

where

$$a_n = \int_{-\infty}^{\infty} u(t) f_n^*(t)\, dt \tag{4-98}$$

and

$$\sum_{n=0}^{\infty} |a_n|^2 = \int_{-\infty}^{\infty} |u(t)|^2\, dt \tag{4-99}$$

The ambiguity function for $u(t)$ is given by

$$\Theta(\tau, \phi) = \sum_{m=0}^{\infty} \sum_{n=0}^{\infty} a_n a_m B_{nm}(\tau, \phi) \tag{4-100}$$

where

$$c_{nm} = a_n a_m = \iint_{-\infty}^{\infty} \Theta(\tau, \phi) B_{nm}^*(\tau, \phi)\, d\tau\, d\phi \tag{4-101}$$

and

$$\sum_{n=0}^{\infty} \sum_{m=0}^{\infty} |c_{nm}|^2 = \left(\sum_{k=0}^{\infty} |a_k|\right)^2 \tag{4-102}$$

4.9 The Ambiguity Function Close to the Origin

For values of τ and ϕ close to the origin, the ambiguity function can be described by the first two terms of the Taylor series expansion of $|\chi(\tau, \phi)|^2$. Thus

$$\frac{|\chi(\tau, \phi)|^2}{|\chi(0, 0)|^2} = 1 + \frac{1}{|\chi(0, 0)|^2}\left[\tau^2 \frac{\partial^2 \chi}{\partial \tau^2} + 2\tau\phi \operatorname{Re}\left\{\frac{\partial^2 \chi}{\partial \tau\, \partial \phi}\right\} + \phi^2 \frac{\partial^2 \chi}{\partial \phi^2}\right]_{\tau=\phi=0} \tag{4-103}$$

It is possible to express (4-103) in terms of defined measures of waveform bandwidth, time duration, and a quantity, which is discussed more fully in Chapter 5, called the "error coupling coefficient." Proceeding to derive this result, the coefficient of τ^2 in (4-103) is considered first. It is readily shown that

$$\left.\frac{\partial^2 \chi}{\partial \tau^2}\right|_{\tau=\phi=0} = \int_{-\infty}^{\infty} u(t)[u^*(t)]'' \, dt \tag{4-104}$$

It follows from the application of Parseval's theorem that

$$\int_{-\infty}^{\infty} u(t)[u^*(t)]'' \, dt = (2\pi j)^2 \int_{-\infty}^{\infty} f^2 |U(f)|^2 \, df \tag{4-105}$$

It is seen that the expression on the right side of Eq. (4-105) contains the second moment of $|U(f)|^2$. This expression is used to define the bandwidth of $u(t)$. Specifically, the bandwidth, denoted here by β, is defined as follows:

$$2E\beta^2 = (2\pi)^2 \int_{-\infty}^{\infty} f^2 |U(f)|^2 \, df \tag{4-106}$$

where E denotes the energy in $\operatorname{Re}\{u(t) \exp[j\omega_0 t]\}$ and is given by

$$E = \tfrac{1}{2} \int_{-\infty}^{\infty} u(t)u^*(t) \, dt \tag{4-107}$$

The coefficient of τ^2 can now be expressed as follows:

$$\left.\frac{\partial^2 \chi}{\partial \tau^2}\right|_{\tau=\phi=0} = -2E\beta^2 \tag{4-108}$$

Considering next the coefficient of ϕ^2 in (4-103), the following is obtained:

$$\left.\frac{\partial^2 \chi}{\partial \phi^2}\right|_{\tau=\phi=0} = (2\pi j)^2 \int_{-\infty}^{\infty} t^2 |u(t)|^2 \, dt \tag{4-109}$$

The integral in the expression on the right side of (4-109) is clearly the second moment of $|u(t)|^2$; this has been used to define the time duration,[1] denoted here by α, of the waveform $u(t)$ as follows:

$$2E\alpha^2 = (2\pi)^2 \int_{-\infty}^{\infty} t^2 |u(t)|^2 \, dt \tag{4-110}$$

Therefore, the coefficient of ϕ^2 can be expressed as follows:

$$\left. \frac{\partial^2 \chi}{\partial \phi^2} \right|_{\tau=\phi=0} = -2E\alpha^2 \tag{4-111}$$

Finally, determining the coefficient of $2\tau\phi$ in (4-103) yields

$$\operatorname{Re}\left\{ \frac{\partial^2 \chi}{\partial \tau \, d\phi} \right\}_{\tau=\phi=0} = 2\pi \operatorname{Im}\left\{ \int_{-\infty}^{\infty} tu(t)[u^*(t)]' \, dt \right\} \tag{4-112}$$

where $\operatorname{Im}\{\ \}$ denotes "the imaginary part of." It will be shown in Chapter 5 that a waveform parameter called the "error coupling coefficient," denoted by ρ, can be expressed in terms of the right side of Eq. (4-112). Thus,

$$2E\rho = -\frac{2\pi}{\alpha\beta} \operatorname{Im}\left\{ \int_{-\infty}^{\infty} tu(t)[u^*(t)]' \, dt \right\} \tag{4-113}$$

Therefore, the coefficient of $2\tau\phi$ is

$$\operatorname{Re}\left\{ \frac{\partial^2 \chi}{\partial \tau \, \partial \phi} \right\}_{\tau=\phi=0} = -2E\alpha\beta\rho \tag{4-114}$$

By noting that $|\chi(0,0)| = 2E$, the Taylor series expansion expressed in terms of the waveform parameters α, β, and ρ is

$$\frac{|\chi(\tau, \phi)|^2}{|\chi(0, 0)|^2} = 1 - [\beta^2\tau^2 + 2(\alpha\beta)\rho\tau\phi + \alpha^2\phi^2] \tag{4-115}$$

It is seen that the curve formed by the intersection of $|\chi(\tau, \phi)|^2$ and a level plane close to the maximum of $|\chi(0, 0)|^2$ is

$$\beta^2\tau^2 + 2(\alpha\beta)\rho\tau\phi + \alpha^2\phi^2 = \gamma^2 \tag{4-116}$$

This curve is always an ellipse [5]. By letting $\gamma^2 = N_0/8E$, Eq. (4-116) describes the so called "uncertainty ellipse" [13] which is illustrated in

---

[1] The definitions of bandwidth and time duration given by (4-106) and (4-110) are rms values, and do not necessarily conform to the usual concept of 3 db bandwidth or envelope duration.

Fig. 4.6. The width of this ellipse along the τ-axis is

$$\Delta\tau = \frac{1}{\beta\sqrt{2E/N_0}} \qquad (4\text{-}117)$$

The width of the ellipse along the ϕ-axis is

$$\Delta\phi = \frac{1}{\alpha\sqrt{2E/N_0}} \qquad (4\text{-}118)$$

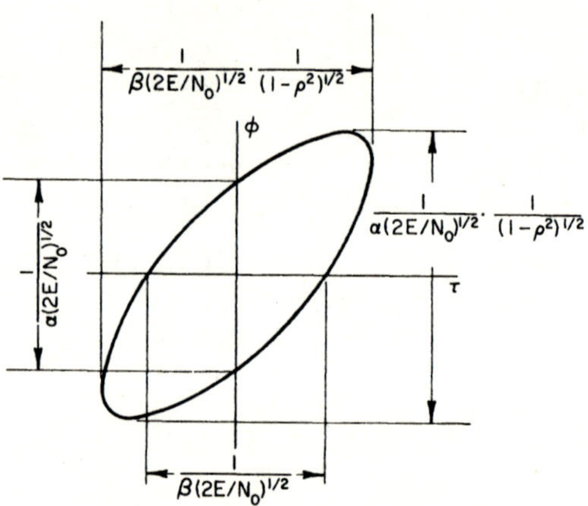

FIG. 4.6   The uncertainty ellipse.

The farthest points in both directions on the uncertainty ellipse are

$$\Delta\tau_{\text{max}} = \pm\frac{1}{2\beta\sqrt{2E/N_0}} \cdot \frac{1}{\sqrt{1-\rho^2}} \qquad (4\text{-}119)$$

and

$$\Delta\phi_{\text{max}} = \pm\frac{1}{2\alpha\sqrt{2E/N_0}} \cdot \frac{1}{\sqrt{1-\rho^2}} \qquad (4\text{-}120)$$

The quantities given by (4-117) through (4-120) will take on more significance in Chapter 5, which treats the problem of estimating the time delay and the Doppler shift of a received signal. It is shown in Chapter 5 that these equations express the smallest possible estimation error in an rms sense.

4.10   Generalized Waveform Uncertainty Principle

The reciprocal relationship between the time duration of a waveform and the bandwidth of its Fourier transform leads one to suspect that the product of time duration α and bandwidth β cannot be made indefinitely small. Gabor [11] was the first to express the lower bound on this product by the inequality

$$\alpha\beta \geqslant \pi \tag{4-121}$$

This inequality constitutes the waveform uncertainty principle. A more general expression of waveform uncertainty is the following inequality:

$$\alpha^2\beta^2[1 - \rho^2] \geqslant \pi^2 \tag{4-122}$$

This inequality develops from consideration of the integral on the right side of (4-112). The imaginary part of this integral is given by (4-113). It is readily shown that the real part of this integral is

$$\text{Re}\left\{\int_{-\infty}^{\infty} tu(t)[u^*(t)]' \, dt\right\} = -\tfrac{1}{2}\chi(0, 0) \tag{4-123}$$

Therefore, the combination of the real and imaginary parts yields

$$\pi^2 + (\alpha\beta\rho)^2 = \frac{2\pi}{|\chi(0, 0)|^2}\left|\int_{-\infty}^{\infty} tu(t)[u^*(t)]' \, dt\right|^2 \tag{4-124}$$

By applying the Schwarz inequality and then Parseval's theorem, Eq. (4-124) reduces to (4-122). It is seen from (4-122) that the lower bound on the product $\alpha\beta$ depends on the coupling coefficient ρ.

4.11   The Time Resolution Constant

The first quantitative basis for judging the potential resolution quality of a radar waveform was given by Woodward [1]. He defined a quantity, called the "time resolution constant," for use in situations involving no Doppler shift, given by

$$T_R(0) = \frac{\displaystyle\int_{-\infty}^{\infty} |\chi(\tau, 0)|^2 \, d\tau}{|\chi(0, 0)|^2} \tag{4-125}$$

The waveform which furnished the smallest value for $T_R(0)$ was assumed to have the greatest potential for resolving between two signals in time.[1]

It is seen that $T_R(0)$ is related to the reciprocal of the "occupied" bandwidth of a waveform. To understand what is meant by the term occupied bandwidth, consider the following example. Suppose the spectrum

---

[1] Section 11.8 presents some examples of waveforms where this assumption is questionable.

corresponding to a waveform $u(t)$ is given by (see Fig. 4.7)

$$U(f) = U_0, \qquad |f - f_n| \leq \frac{\Delta f_n}{2}$$

$$= 0, \qquad f_n + \frac{\Delta f_n}{2} \leq f \leq f_{n+1} - \frac{\Delta f_{n+1}}{2} \qquad (4\text{-}126)$$

$$= 0, \qquad f_1 - \frac{\Delta f_1}{2} > f; \quad f_N + \frac{\Delta f_N}{2} < f$$

where $n = 1, 2, \ldots, N$. The occupied bandwidth for this waveform is then

$$\sum_{n=1}^{N} \Delta f_n \qquad (4\text{-}127)$$

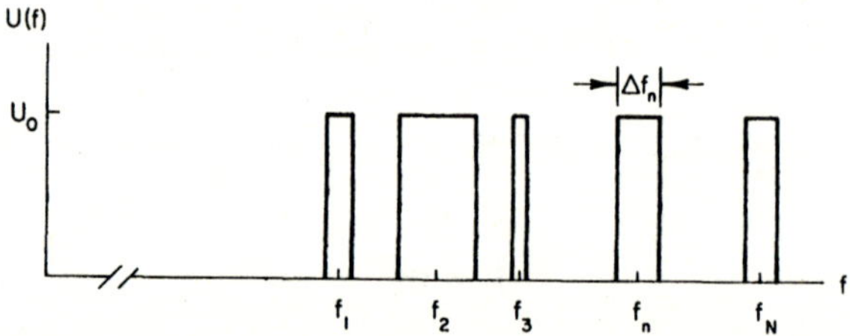

FIG. 4.7   Occupied and unoccupied spectral bands.

In order to demonstrate the relationship between $T_R(0)$ and the sum of the occupied frequency bands, Parseval's theorem is applied to (4-125) to obtain

$$T_R(0) = \frac{\int_{-\infty}^{\infty} |U(f)|^4 \, df}{\left[\int_{-\infty}^{\infty} |U(f)|^2 \, df\right]^2} \qquad (4\text{-}128)$$

Substituting $U(f)$, given by (4-126) into this equation yields

$$T_R(0) = \frac{U_0^4 \sum_{n=1}^{N} \Delta f_n}{[U_0^2 \sum_{n=1}^{N} \Delta f_n]^2} \qquad (4\text{-}129)$$

Hence, it is seen that

$$T_R(0) = \frac{1}{\sum_{n=0}^{N} \Delta f_n} \qquad (4\text{-}130)$$

By regrouping the factors in (4-129) an alternate expression for $T_R(0)$ is obtained

$$T_R(0) = \frac{U_0^2}{2E} \qquad (4\text{-}131)$$

where E stands for energy as defined by (4-107). For equal energy waveforms that have a fixed overall bandwidth equal to the sum of the interlaced spans of occupied and unoccupied frequencies, the quantity U_0^2 will clearly attain its smallest value when the entire spectral span is fully occupied. Accordingly, it is seen from (4-131) that this will also result in the smallest ambiguity. A well-known and important conclusion derives from this result: when it is known, *a priori*, that the waveforms will be unaffected by a Doppler displacement, then single pulse transmissions are less ambiguous than a train of pulses. By allowing only the shape of the spectral envelope to vary within the limits of the fixed band of frequencies, it is found, by means of the calculus of variations, that $T_R(0)$ will be smallest when the spectrum is filled and also has a uniform height. The linear FM pulse that has a large time-bandwidth product (say >20) tends to exhibit such a spectrum. Therefore, from the viewpoint of $T_R(0)$ the linear FM pulse approximates the optimum bandlimited waveform for resolving among several signals as long as it is known that the received waveforms will never be displaced from any arbitrary reference frequency.[1] This conclusion was first given by Manasse [14].

The concept of the time resolution constant, although apparently not referred to as such, was later extended by Westerfield *et al.* [15]. These authors showed the importance of the time resolution constant expressed as a function of Doppler shift ϕ as follows:

$$T_R(\phi) = \frac{\int_{-\infty}^{\infty} |\chi(\tau, \phi)|^2 \, d\tau}{|\chi(0, 0)|^2} \tag{4-132}$$

They also introduced the integral relationships

$$\int_{-\infty}^{\infty} |\chi(\tau, \phi)|^2 \, d\tau = \int_{-\infty}^{\infty} |U(f)|^2 |U(f + \phi)|^2 \, df \tag{4-133}$$

$$\int_{-\infty}^{\infty} |\chi(\tau, \phi)|^2 \, d\tau = \int_{-\infty}^{\infty} |\chi(\tau, 0)|^2 \exp[-j2\pi\phi\tau] \, d\tau \tag{4-134}$$

These relationships were derived in Section 4.5 of this chapter.

The following observations are made concerning the different forms for $T_R(\phi)$. The first form given by Eq. (4-132) can be interpreted as a summary of the ambiguities, observed at the matched-filter output, as a function of Doppler shift. This is the interpretation which is an extension of Woodward's time resolution constant. The latter form, given by Eq. (4-134), clearly indicates that $T_R(\phi)$ is related to the zero-Doppler ambiguity profile through the Fourier transform. Finally, the remaining form given by (4-133)

---

[1] This restriction can be removed if the system is implemented to take advantage of the range prediction property of the linear FM signal, as discussed in Section 9.5.

lends itself to a more physical interpretation. By viewing $|U(f)|^2$ as the energy response of a matched filter, this last form may be regarded as the energy response of off-Doppler waveforms at the matched-filter output. For the special case where there is a single waveform corresponding to every Doppler shift in the interval $-\infty < \phi < \infty$, it is seen by integrating $T_R(\phi)$ between these limits that the combined energy is equal to $|\chi(0, 0)|^2$. The quantity $T_R(\phi)$ will receive further consideration in Chapter 10 as a waveform design criterion in dense and multiple target environments.

Woodward also introduced a frequency resolution constant given by

$$F_R(0) = \frac{\int_{-\infty}^{\infty} |\chi(0, \phi)|^2 \, d\phi}{|\chi(0, 0)|^2} \tag{4-135a}$$

The definition for the frequency resolution constant has since been extended to include variations with time delay τ [15]. The generalized definition for the frequency resolution constant is then given by

$$F_R(\tau) = \frac{\int_{-\infty}^{\infty} |\chi(\tau, \phi)|^2 \, d\phi}{|\chi(0, 0)|^2} \tag{4-135b}$$

Equations (4-63) and (4-64) provide alternate forms for $F_R(\tau)$.

A measure of the combined resolving capability of a radar waveform is furnished by the product $T_R(0)F_R(0)$. For a single tone Gaussian pulse $T_R(0)F_R(0) = 1$ and for a single tone rectangular pulse $T_R(0)F_R(0) = 2/3$. Therefore, because it yields the smaller of the two products, the rectangular pulse seems to be inherently more suited for radar situations requiring a combined (time delay-Doppler shift) resolving capability.

4.12 Ambiguity Function Examples

The preceding sections of this chapter discussed the significance of the ambiguity function and described its general behavior. The principles developed in these sections establish the framework for deriving suitable waveforms for transmission by the radar system. By disregarding the contents of these sections, in particular the section on properties, it is very easy to go seriously astray. For example, one might be tempted to search for a waveform—a panacea perhaps—that would yield an impulse at the origin of the ambiguity plane. However, it is immediately evident from the integral relationships stemming from the property on volume invariance that it is theoretically impossible to find such a waveform.

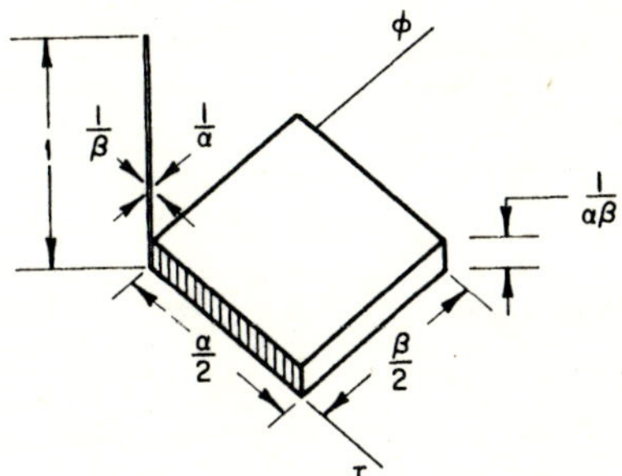

FIG. 4.8 A thumbtack ambiguity function.

The so-called thumbtack ambiguity function (see Fig. 4.8) is about the closest realization of a single impulse ambiguity surface. It consists of a narrow spike surrounded by a uniformly low pedestal. The dimensions of the spike are given as $1/\beta$ along the time axis and $1/\alpha$ along the frequency axis. The major portion of the volume, however, lies under the pedestal. And, according to the same integral relationships just cited, this volume is confined in essence to α-units along the time axis and β-units along the frequency axis. The height of this volume is obtained by invoking the volume invariance property and is $|\chi(0, 0)|^2/\alpha\beta$. Although many waveforms come close, the specific waveform has not yet been discovered that would furnish such an ambiguity surface. In any case, this ambiguity surface is far from being useful for all applications. Like all other ambiguity surfaces, its utility must be carefully analyzed. This is discussed further in Chapter 10.

Some additional examples of ambiguity surfaces are presented in Table 4-I [16]. For each of these examples, it is assumed that the corresponding waveforms are normalized according to

$$\int_{-\infty}^{\infty} u(t)u^*(t)\,dt = 1 \tag{4-136}$$

Therefore, all the ambiguity surfaces presented in Table 4-I have unity height at the origin. Table 4-I includes the description of $u(t)$, $U(f)$, and $|\chi(\tau, 0)|$. The principle contours of $|\chi(\tau, \phi)|$ along the time axis and frequency axis are obtained respectively by setting $\phi = 0$ and $\tau = 0$. The level contours (i.e., for $|\chi(\tau, \phi)| \cong 0.7$) of several of these examples are illustrated in Fig. 4.3. The reader should recall from Section 4.9 that close to the peak of $|\chi(\tau, \phi)|^2$ these contours are usually ellipses. It is only when lower levels are considered that different resolution contours are observed.

Some general observations are made below concerning the ambiguity surfaces presented in Table 4-I.

TABLE 4-I

	$u(t)$	$U(f)$

1. Gaussian Envelope

Monotone $\quad \left(\dfrac{2k^2}{\pi}\right)^{1/4} \exp[-k^2 t^2]$ $\qquad \left(\dfrac{2\pi}{k^2}\right)^{1/4} \exp\left[\dfrac{-\pi^2 f^2}{k^2}\right]$

Linear FM $\quad \left(\dfrac{2k^2}{\pi}\right)^{1/4} \exp[-(k^2 - jb)t^2]$ $\qquad \left(\dfrac{2k^2\pi}{k^4 + b^2}\right)^{1/4} \exp\left[\dfrac{-\pi^2 f^2}{k^2 - jb} + j\dfrac{1}{2}\tan^{-1}\dfrac{k^2}{b}\right]$

Parabolic FM[b] $\quad \left(\dfrac{2k^2}{\pi}\right)^{1/4} \exp[-(k^2 - jct)t^2]$ $\qquad \left(\dfrac{k^2}{3cf}\right)^{1/4} \exp\left[-\dfrac{2\pi k^2 f}{3c}\right] \cos\left[\dfrac{2}{3}\dfrac{(2\pi f)^{3/2}}{\sqrt{3c}} - \dfrac{\pi}{2}\right]$

where $f > 0$

2. Rectangular Envelope

Monotone $\quad \begin{cases} \sqrt{\dfrac{1}{T}}, & |t| < T/2 \\ 0, & |t| > T/2 \end{cases}$ $\qquad \sqrt{T}\,\dfrac{\sin \pi Tf}{\pi Tf}$

Linear FM $\quad \begin{cases} \sqrt{\dfrac{1}{T}}\exp[jbt^2], & |t| < T/2 \\ 0, & |t| > T/2 \end{cases}$ $\qquad \sqrt{\dfrac{1}{2\Delta f}}[(C_1 + C_2)^2 + (S_1 + S_2)^2]^{1/2}$

$$\times \exp\left[-j\left(\dfrac{\pi^2 f^2}{b^2} - \tan^{-1}\dfrac{S_1 + S_2}{C_1 + C_2}\right)\right]$$

where

$$C_n = \int_0^{x_n} \cos \pi y^2 \, dy; \quad S_n = \int_0^{x_n} \sin \pi y^2 \, dy$$

$(n = 1, 2)$

and

$$\left.\begin{matrix} x_1 \\ x_2 \end{matrix}\right\} = (2T\Delta f)^{1/2}\left(\dfrac{1}{2} \mp \dfrac{f}{\Delta f}\right)$$

Ambiguity Function Properties[a]

$|\chi(\tau, 0)|$	$|\chi(0, \phi)|$	$|\chi(\tau, \phi)|$

$\exp\left[-\tfrac{1}{2}k^2\tau^2\right]$

$\exp\left[-\dfrac{1}{2}\left(k^2 + \dfrac{b^2}{k^2}\right)\tau^2\right]$

$\left(\dfrac{1}{1 + (3c/2k^2)^2\tau^2}\right)^{1/4}\exp\left[-\tfrac{1}{2}k^2\tau^2\right]$

$\exp\left[-\dfrac{1}{2}\dfrac{\pi^2\phi^2}{k^2}\right]$

$\exp\left[-\dfrac{1}{2}\dfrac{\pi^2\phi^2}{k^2}\right]$

$\exp\left[-\dfrac{1}{2}\dfrac{\pi^2\phi^2}{k^2}\right]$

$\exp\left[-\dfrac{1}{2}\left(k^2\tau^2 + \dfrac{\pi^2\phi^2}{k^2}\right)\right]$

$\exp\left[-\dfrac{1}{2}\left\{k^2\tau^2 + \dfrac{\pi^2}{k^2}\left(\phi + \dfrac{b\tau}{\pi}\right)^2\right\}\right]$

$\left(\dfrac{1}{1 + (3c/2k^2)^2\tau^2}\right)^{1/4}$
$\times \exp\left[-\left(\dfrac{k^2\tau^2}{2} + \dfrac{2k^2\pi^2\phi^2}{4k^4 - 9c^2\tau^2}\right)\right]$

$\begin{cases} 1 - \dfrac{|\tau|}{T}, & |\tau| < T \\ \\ 0, & |\tau| > T \end{cases}$

$\dfrac{\sin \pi T\phi}{\pi T\phi}$

$\begin{cases} \left(1 - \dfrac{|\tau|}{T}\right)\dfrac{\sin \pi\phi T(1 - |\tau|/T)}{\pi\phi T(1 - |\tau|/T)}, \\ \qquad\qquad\qquad |\tau| < T \\ 0, \quad |\tau| > T \end{cases}$

$\begin{cases} \left(1 - \dfrac{|\tau|}{T}\right)\dfrac{\sin b\tau[T - |\tau|]}{b\tau[T - |\tau|]}, & |\tau| < T \\ \\ 0, & |\tau| > T \end{cases}$

$\dfrac{\sin \pi T\phi}{\pi T\phi}$

$\begin{cases} \left(1 - \dfrac{|\tau|}{T}\right)\dfrac{\sin[b\tau - \pi\phi][T - |\tau|]}{[b\tau - \pi\phi][T - |\tau|]}, \\ \qquad\qquad\qquad |\tau| < T \\ 0, \quad |\tau| > T \end{cases}$

TABLE 4-I

$u(t)$	$U(f)$

Rectangular Envelope (Continued)

Parabolic FM[b]

$$u(t) = \begin{cases} \sqrt{\dfrac{1}{T}}\,\exp[\,jct^3\,], & |t| < T/2 \\[2ex] 0, & |t| > T/2 \end{cases}$$

$$\left(\dfrac{1}{6T^2cf}\right)^{\frac{1}{4}} \cos\left[\dfrac{2}{3}\dfrac{(2\pi f)^{3/2}}{\sqrt{3c}} - \dfrac{\pi}{4}\right]$$

$$\text{where } 0 < f < \dfrac{3cT^2}{8\pi}$$

Cosine FM

$$u(t) = \begin{cases} \sqrt{\dfrac{1}{T}}\,\exp\left[\,j\Delta\theta \sin\dfrac{2\pi t}{T}\right], & |t| < T/2 \\[2ex] 0, & |t| > T/2 \end{cases}$$

$$[T]^{1/2} \sum_{n=-\infty}^{\infty} J_n(x)\dfrac{\sin \pi(f - n/T)T}{\pi(f - n/T)T}$$

$$\text{where } x = 2\Delta\theta \sin\dfrac{\pi\tau}{T}$$

3. Hermite Waveforms

$$\dfrac{[2]^{1/4}}{\sqrt{n!}}\,\exp[\,-\pi t^2\,]H_n(2\sqrt{\pi}t)$$

$$(n = 0, 1, 2, \ldots) \qquad (-j)^n u(f)$$

where

$$H_n(x) = (-1)^n \exp\left[\dfrac{x^2}{2}\right]$$

$$\times \dfrac{d^n}{dx^n}\left(\exp\left[-\dfrac{x^2}{2}\right]\right)$$

[a] From Bernfeld [16].
[b] Spectrum calculated using principle of stationary phase.

(continued)

$|\chi(\tau, 0)|$	$|\chi(0, \phi)|$	$|\chi(\tau, \phi)|$

$$\frac{1}{T}\sqrt{\frac{\pi}{6c\tau}}[(C_1 + C_2)^2 + (S_1 + S_2)^2]^{1/2}$$

$$\frac{\sin \pi T\phi}{\pi T\phi}$$

$$\begin{cases} \dfrac{1}{T}\sqrt{\dfrac{\pi}{6c\tau}}[(C_1 + C_2) \\ \qquad + (S_1 + S_2)^2]^{1/2}, & |\tau| < T \\ 0, & |\tau| > T \end{cases}$$

where
$$C_n = \int_0^{x_n} \cos \pi y^2 \, dy$$

$$S_n = \int_0^{x_n} \sin \pi y^2 \, dy, \quad n = 1, 2$$

$$\left.\begin{matrix} x_1 \\ x_2 \end{matrix}\right\} = \pm\frac{1}{2}\sqrt{\frac{6c\tau}{\pi}}(T - |\tau|)$$

$$\left.\begin{matrix} x_1 \\ x_2 \end{matrix}\right\} = -\left(\frac{\pi}{6c\tau}\right)^{1/2}\phi$$

$$\pm\frac{1}{2}\left(\frac{6c\tau}{\pi}\right)^{1/2}(T - |\tau|)$$

$$\begin{cases} \dfrac{1}{T}\left|\displaystyle\sum_{n=-\infty}^{\infty}(j)^n J_n(x)(T - |\tau|) \right. \\ \left. \times \dfrac{\sin(\pi n/T)(T - |\tau|)}{(\pi n/T)(T - |\tau|)}\right|, & |\tau| < T \\ 0, & |\tau| > T \end{cases}$$

$$\frac{\sin \pi T\phi}{\pi T\phi}$$

$$\left|\frac{1}{T}\sum_{n=-\infty}^{\infty}(j)^n J_n(x)|T - |\tau||\right.$$
$$\left. \times \frac{\sin \pi(\phi - n(T)(T - |\tau|)}{\pi(\phi - n/T)(T - |\tau|)}\right|, \quad |\tau| < T$$

where $x = 2\,\Delta\theta \sin\dfrac{\pi\tau}{T}$

$$\exp\left[-\frac{\pi}{2}\tau^2\right]L_n(\pi\tau)^2$$

$$\exp\left[-\frac{\pi}{2}\phi^2\right]$$
$$\times L_n(\pi\phi^2)$$

$$\exp\left[-\frac{\pi}{2}(\tau^2 + \phi^2)\right]L_n(\pi|\tau^2 + \phi^2|)$$

$$(n = 0, 1, 2, \ldots)$$

where

$$L_n(x) = \frac{1}{n!}\exp|x|\frac{d^n}{dx^n}(x^n \exp|-x|)$$

4.12.a GAUSSIAN ENVELOPE SINGLE TONE AND LINEAR FM WAVEFORMS

It is seen that the vertical profiles along the time and frequency axes, as well as all profiles taken parallel to these axes, are Gaussian in shape. In fact, along any line through the origin the shape of the profile will also be Gaussian. For the single tone Gaussian pulse, the level contour (see Fig. 4.9) will be a circle when the coefficient $k^2 = \pi$, otherwise it will be an ellipse with major axis aligned with the τ-axis when $k^2 > \pi$ and major axis aligned

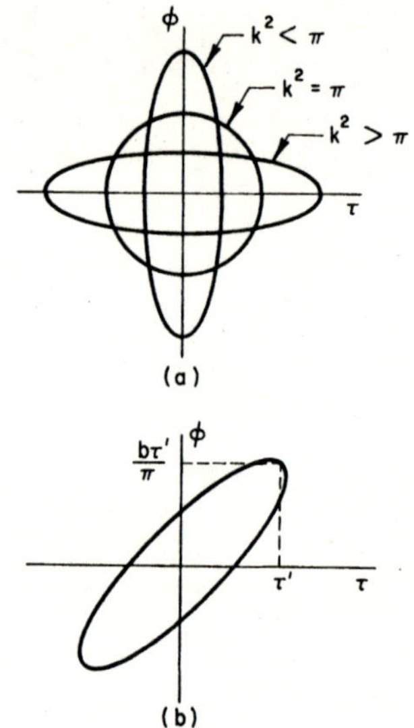

FIG. 4.9 Level ambiguity contours for Gaussian envelope signals. (a) Monotone pulse. (b) Linear FM pulse.

with the ϕ-axis when $k^2 < \pi$. The level contour for the linear FM waveform will be an ellipse whose major axis is aligned with a line given by

$$\phi = \pm \frac{b\tau}{\pi} \tag{4-137}$$

where $2b$ is the FM rate (see Fig. 4.9).

4.12.b RECTANGULAR ENVELOPE WAVEFORMS

For all FM waveforms that are enclosed in rectangular envelopes the form of $\chi(0,\phi)$, is given by

$$\chi(0, \phi) = \frac{\sin \pi T\phi}{\pi T\phi} \tag{4-138}$$

where T is the time duration of the envelope. This is clearly seen from the Fourier transform relationship

$$\chi(0, \phi) = \int_{-T/2}^{T/2} |u(t)|^2 \exp[-j2\pi\phi t]\, dt \qquad (4\text{-}139)$$

where

$$|u(t)| = \frac{1}{\sqrt{T}}, \qquad -\frac{T}{2} < t < \frac{T}{2}$$

4.12.c RECTANGULAR ENVELOPE SINGLE TONE AND LINEAR FM WAVEFORMS

For these waveforms, $\chi(\tau, \phi)$ has a $(\sin x)/x$ profile along every line parallel to the ϕ-axis. The profiles attain their peak values above the axis for the single tone pulse (see Fig. 4.10) and above a line through the origin

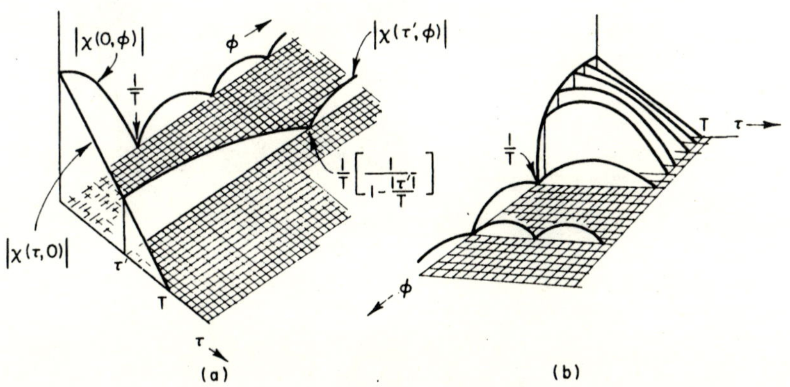

FIG. 4.10 Time/frequency response function for rectangular envelope monotone pulse. (a) Constant τ profiles. (b) Constant ϕ profiles.

for the linear FM case given by Eq. (4-137) (see Fig. 4.11). It is seen that these maxima are weighted by

$$\mathscr{R}(\tau) = 1 - \frac{|\tau|}{T} \qquad (4\text{-}140)$$

The general ambiguity structures for these waveforms are illustrated in Figs. 4.12 and 4.13. The linear FM waveform (rectangular envelope case) is discussed further in Chapter 6.

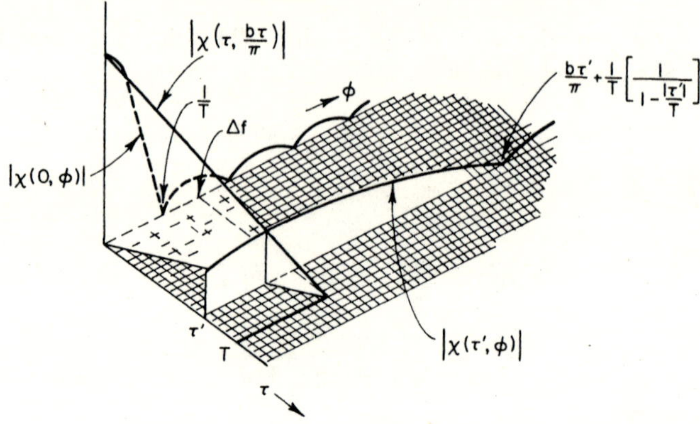

FIG. 4.11   Response function for rectangular envelope linear FM pulse.

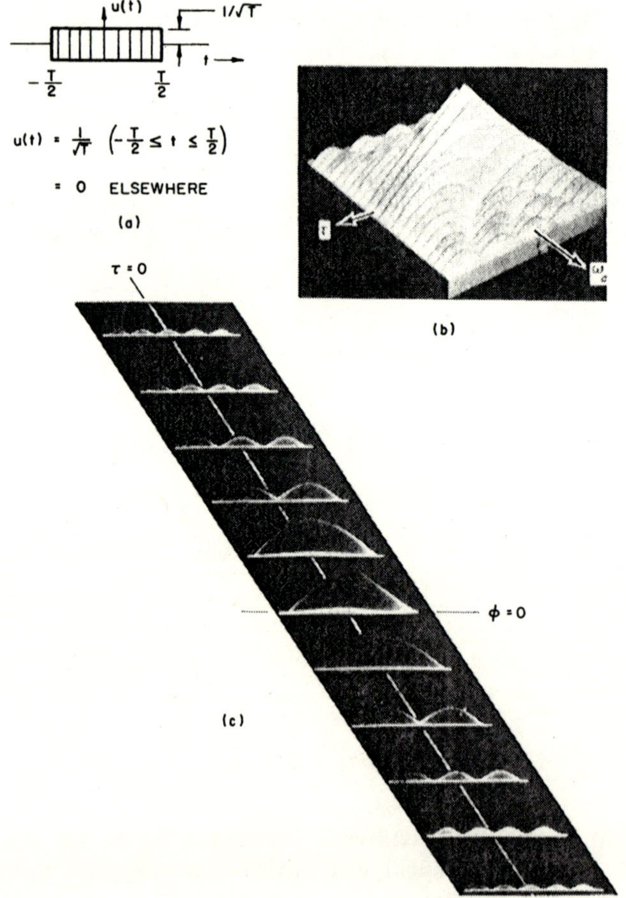

FIG. 4.12   Constant frequency pulsed signal and response surface. (a) Waveform. (b) Calculated composite surface. (c) Experimental composite surface.

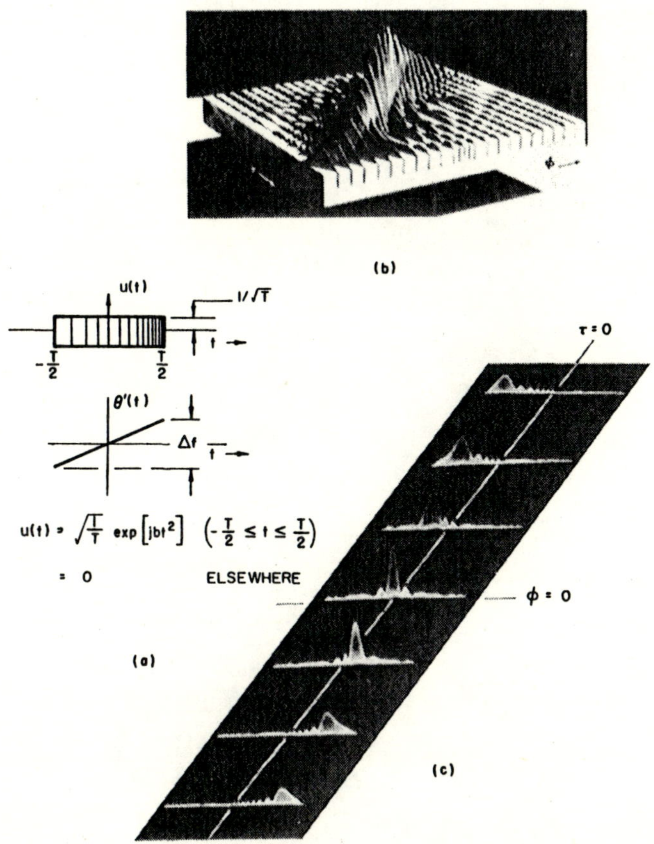

FIG. 4.13   Linear FM pulse waveform and response function properties $T\Delta f = 10$. (a) Waveform. (b) Calculated surface. (c) Experimental composite surface.

4.12.d   RECTANGULAR ENVELOPE HIGHER ORDER FM WAVEFORMS

The $(\sin x)/x$ generating function found to develop along all lines parallel to the ϕ-axis, for the preceding examples, deforms for the higher order FM waveforms. For the case of the parabolic FM waveform the ambiguity function (see Table 4-I and Fig. 4.14) is characterized by Fresnel integrals. In general, it is not possible to describe the structures of the ambiguity function in terms of simple generating functions. It is often helpful for these cases to apply the principle of stationary phase to obtain gross indications of the ambiguity function characteristics. This application is discussed in the next section.

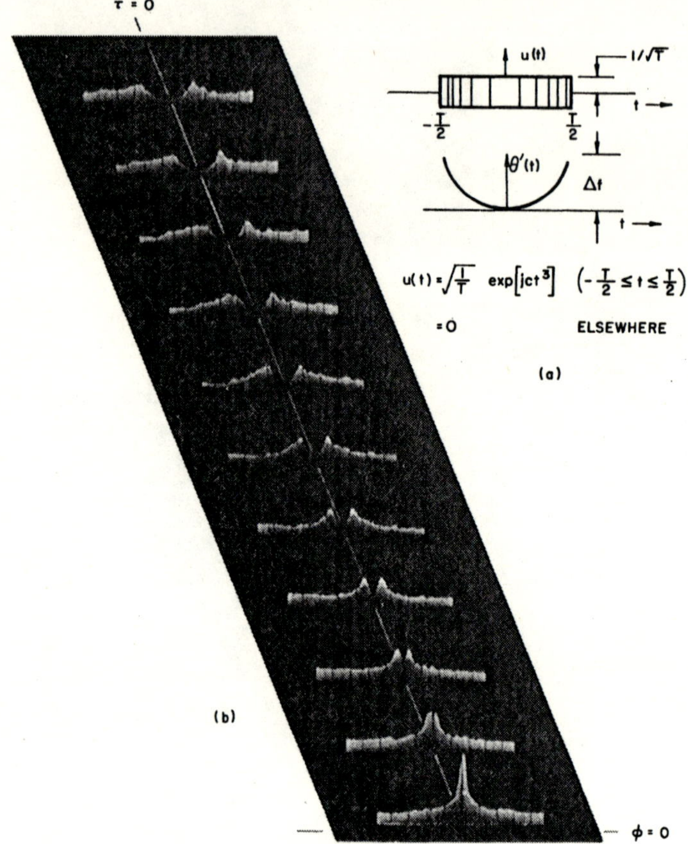

FIG. 4.14   Parabolic FM pulse waveform and response function properties, $T\Delta f = 200$. (a) Waveform. (b) Experimental response function surface.

4.12.e   THE LINEAR V-FM PULSE

The linear V-FM waveform and its corresponding ambiguity function is illustrated in Fig. 4.15. For the downward V illustrated there, the waveform is given by

$$u(t) = u_1(t) + u_2(t), \qquad -T < t < T$$
$$= 0, \text{ elsewhere} \tag{4-141}$$

where

$$u_1(t) = \frac{1}{\sqrt{2T}} \exp[-jbt^2], \qquad -T < t < 0$$

$$u_2(t) = \frac{1}{\sqrt{2T}} \exp[jbt^2], \qquad 0 < t < T \tag{4-142}$$

The ambiguity function for this waveform consists of four parts; two

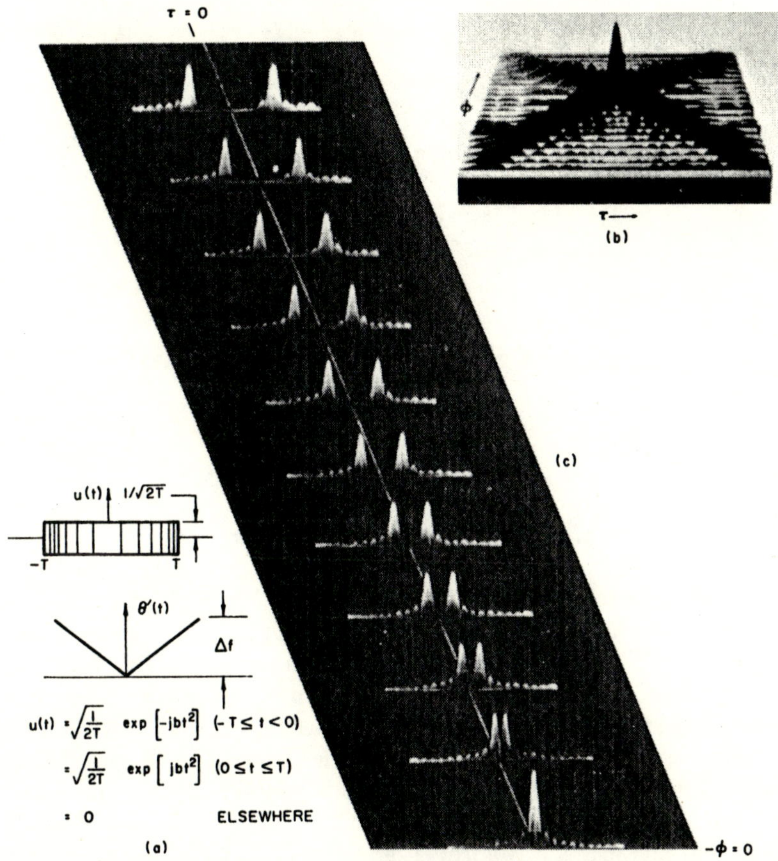

FIG. 4.15   V-FM pulse waveform and response function properties, $2T\Delta f = 200$. (a) Waveform. (b) Calculated composite response surface (note pedestal not included). (c) Experimental composite response surface.

autocorrelation terms corresponding to $u_1(t)$ and $u_2(t)$, and two cross-correlation terms corresponding to the two cross products of $u_1(t)$ and $u_2(t)$. Expressed in terms of these components, the entire ambiguity function is given by

$$\chi(\tau, \phi) = \chi_{11}(\tau, \phi) + \chi_{22}(\tau, \phi) + \chi_{12}(\tau, \phi) + \chi_{21}(\tau, \phi) \tag{4-143}$$

The principle contributions to the function are furnished by $\chi_{11}(\tau, \phi)$ and $\chi_{22}(\tau, \phi)$, the autocorrelation terms. These are related as follows[1]:

$$\chi_{22}(\tau, \phi) = \chi_{11}^*(-\tau, \phi) \tag{4-144}$$

[1] Equations (4-144) and (4-145) apply to all waveforms that have even envelope functions and even frequency modulation functions.

$$\chi_{22}(\tau, \phi) = \chi_{11}(\tau, -\phi) \exp[j2\pi\phi\tau] \qquad (4\text{-}145)$$

where

$$\chi_{11}(\tau, \phi) = \frac{1}{2}\left[1 - \frac{|\tau|}{T}\right] \frac{\sin\{\pi[\phi - b\tau/\pi][T - |\tau|]\}}{\pi[\phi - b\tau/\pi][T - |\tau|]} \exp[j(\pi\phi[T + \tau] - b\tau T)],$$

$$-T < \tau < T$$

$$(4\text{-}146)$$

Clearly, since $\chi_{12}(0, \phi) = \chi_{21}(0, \phi) = 0$, then

$$\chi(0, \phi) = 2 \operatorname{Re}\{\chi_{11}(0, \phi)\}$$

$$= \frac{\sin 2\pi T\phi}{2\pi T\phi} \qquad (4\text{-}147)$$

In general, the sum consisting of the principle contributions rides on a low level pedestal represented by the crosscorrelation terms $\chi_{12}(\tau, \phi)$ and $\chi_{21}(\tau, \phi)$. These terms can be shown to be characterized by the Fresnel integral. Each is confined to one half of the $\tau - \phi$ plane between the time delay limits $0 < |\tau| < 2T$. The heights of $\chi_{12}(\tau, \phi)$ and $\chi_{21}(\tau, \phi)$ vary about the level $\frac{1}{2}[\pi/bT^2]^{1/2}$. This can be expressed in terms of the time-bandwidth product $T\,\Delta f$ of each linear FM component if, by definition, $b = \pi\,\Delta f/T$, where Δf is the swept frequency difference shown in Fig. 4.15. It is also noted that only the pedestal functions exist in the intervals $T < |\tau| < 2T$. The waveforms shown in Fig. 4.16 illustrate the V-FM autocorrelation function $\chi(\tau, 0)$ for different time-bandwidth products. The basic matched-filter implementation for this waveform is shown in Fig. 4.17. The matched filter for any bidirectional FM signal would be similar, with the pulse-compression filter in each channel being matched to the corresponding segment of the frequency modulation function.

The importance of the linear V-FM signal is that it is the most basic FM pulse-compression function that yields uncorrelated measurements of range and velocity (i.e., $\rho = 0$). This can be related to the splitting up of the matched-filter output into two distinct signals when a Doppler shift causes the received signal to be mismatched to the matched filter. This Doppler effect is shown by the waveforms of Fig. 4.15. It can be seen that the on-Doppler signal is readily recognized. However, if more than one signal is present, there is an increased amount of uncertainty as to the number of targets and their actual ranges and velocities. A method for reducing the large value of off-Doppler ambiguity associated with the V-FM class of signals is to add curvature to each branch of the FM function. This causes a reduction in the peak off-Doppler outputs. The parabolic FM waveform

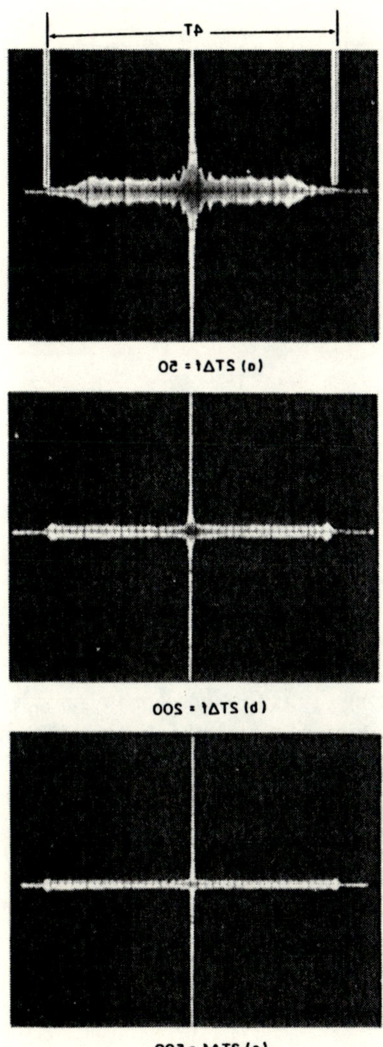

FIG. 4.16 V-FM autocorrelation functions (note pedestal level as a function of time-bandwidth product).

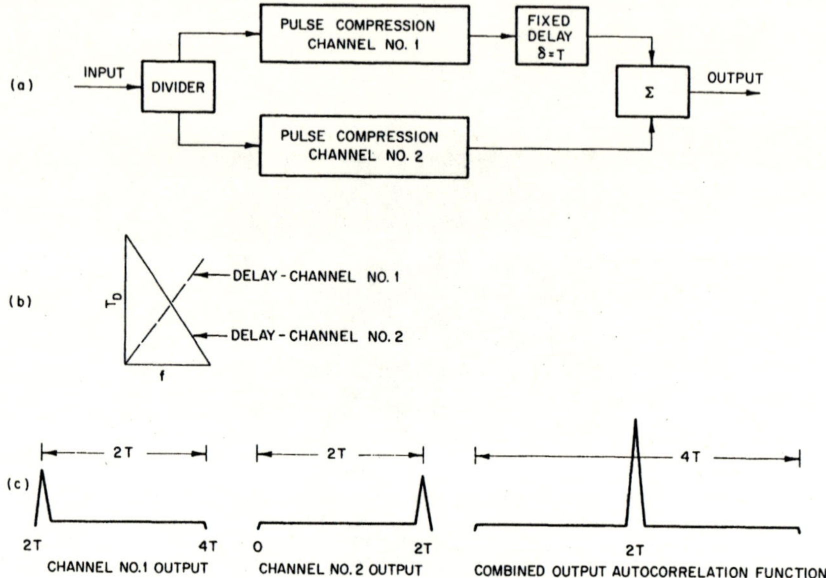

FIG. 4.17   Block diagram of matched filter for V-FM signal of Fig. 4.15 (from Bernfeld *et al.*, Ref. 4, Chapter 1).

is one example of this. The reduced off-Doppler peaks for this signal can be observed in Fig. 4.14. However, a concurrent cost of this lowering of peak off-Doppler ambiguity is the increase in the sidelobe responses of the on-Doppler and off-Doppler matched-filter outputs. This offers a practical example of the volume invariance property of ambiguity functions.

4.12.f   HERMITE WAVEFORMS

A class of waveforms that are rotationally invariant on the ambiguity plane (see Section 4.5, Property VI) are given by [5, 10]

$$u_n(t) = \frac{2^{1/4}}{\sqrt{n!}} \exp[-\pi t^2] H_n(2\sqrt{\pi}\, t) \tag{4-148}$$

where $H_n(x)$ is an nth order Hermite polynomial. The Hermite polynomials are defined by

$$H_n(x) = (-1)^n \exp\left(\frac{x^2}{2}\right) \frac{d^n}{dx^n} \exp\left(-\frac{x^2}{2}\right) \tag{4-149}$$

where

$$H_0(x) = 1$$

$$H_1(x) = x$$

$$H_2(x) = x^2 - 1$$

$$H_3(x) = x^3 - 3x$$

$$H_4(x) = x^4 - 6x^2 - 3$$

$$\cdot \qquad \cdot \qquad \cdot \qquad \cdot$$

These waveforms have the property that they are self-transformable; i.e.,

$$u_n(t) \Rightarrow U_n(f) = (-j)^n u_n(f) \tag{4-150}$$

The ambiguity function for these waveforms is readily shown to be [10]

$$\chi_n(\tau, \phi) = \exp\left[-\frac{\pi}{2}(\tau^2 + \phi^2)\right] L_n[\pi(\tau^2 + \phi^2)] \tag{4-151}$$

where $L_n(x)$ is the nth order Laguerre polynomial. The Laguerre polynomials are defined by

$$L_n(x) = \frac{1}{n!} e^x \frac{d^n}{dx^n} (x^n e^{-x}) \tag{4-152}$$

$$= \sum_{k=0}^{n} \frac{n!(-x)^k}{(n-k)!k!} \tag{4-153}$$

Examples of these waveforms are illustrated by Jahnke and Emde [17] as well as by Klauder. Moreover, Klauder also illustrates the ambiguity function for several orders of n as shown in Figs. 4.18 and 4.19. The side-lobes of these ambiguity functions fall off approximately as $t^{-1/2}$ in contrast with the linear FM sidelobes which fall off as t^{-1}.

4.13 The Principle of Stationary Phase Applied to Ambiguity Function Analysis

The ambiguity surface properties described in the preceding section can be explained by regarding the general ambiguity functions expression as a Fourier transform operating on the product $u(t)u^*(t + \tau)$ with τ taken as a parameter. It is seen from the principle of stationary phase that

$$|\chi(\tau, \phi)| \doteq \sqrt{2\pi} \left. \frac{u(t)u^*(t + \tau)}{\sqrt{|\Delta\theta''(t)|}} \right|_{t = f(\phi)} \tag{4-154}$$

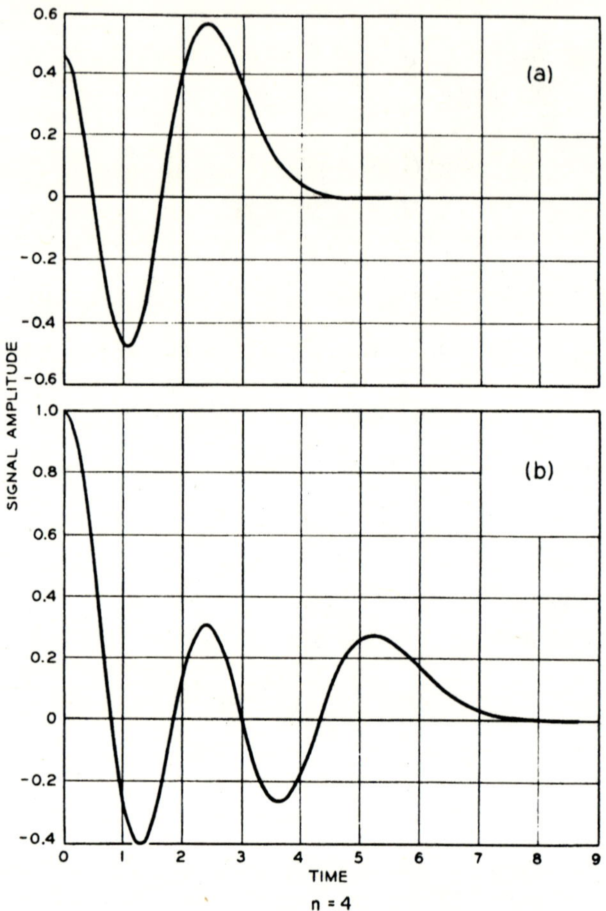

Fig. 4.18a

$$u(t) = \left(\frac{1}{\pi}\right)^{1/4}\frac{1}{\sqrt{n!\,2^n}}H_n(t)e^{-t^2/2} \qquad\qquad \chi(\tau, 0) = \frac{1}{n!}L_n\left(\frac{\tau^2}{2}\right)e^{-\tau^2/4}$$

where   where

$$H_n(t) = (-1)^n e^{t^2}\frac{d^n}{dt^n}e^{-t^2} \qquad\qquad L_n(x) = e^x\frac{d^n}{dx^n}x^n e^{-x}$$

Figs. 4.18a–c   Hermite signal functions, denoted by (a), and the corresponding correlation function $\chi(\tau, 0)$, denoted by (b). (Courtesy of J. R. Klauder [10] and Bell Syst. Tech. Journal.)

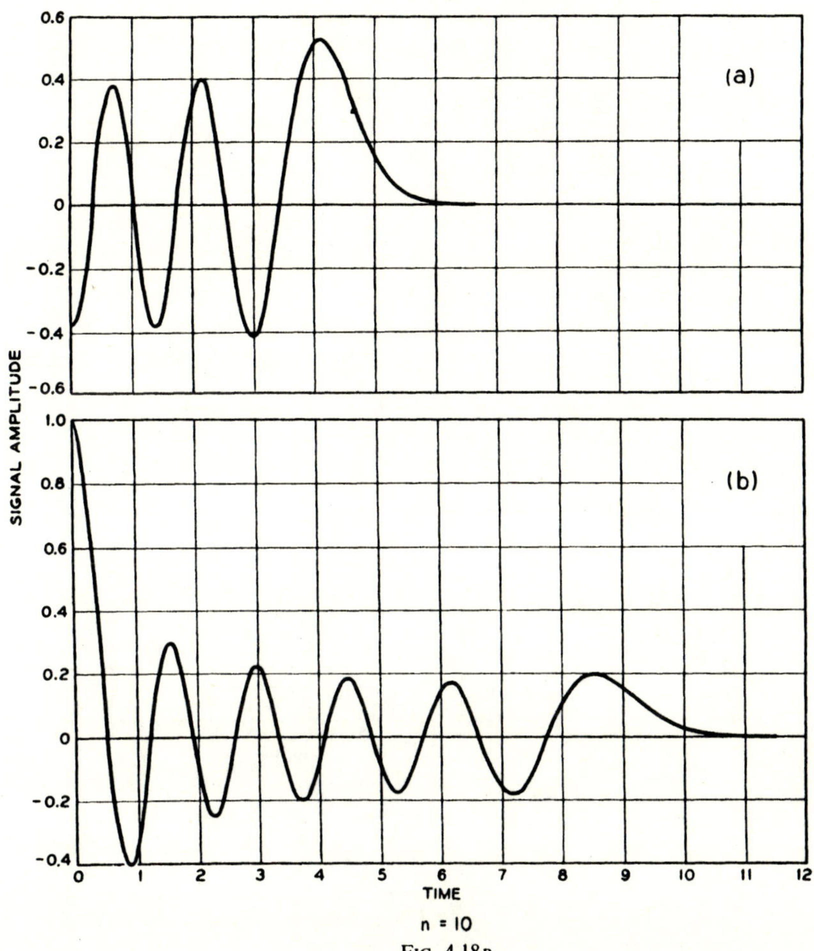

FIG. 4.18B

where $\Delta\theta''(t) = \theta''(t) - \theta''(t + \tau)$ is the difference between the FM rates of the time shifted and unshifted waveforms $u(t + \tau)$ and $u(t)$. The solution for $t = f(\phi)$ as well as the characterization for the principle of stationary phase is obtained from

$$2\pi\phi - \Delta\theta'(t) = 0 \qquad (4\text{-}155)$$

It is easily shown, for a single tone pulse, that $\Delta\theta'(t) = 0$. Since this is clearly independent of time, the principle of stationary phase indicates that $\chi(\tau, \phi)$ is concentrated for the most part above the τ-axis, and the shape of $\chi(\tau, \phi)$ for constant τ is determined by the Fourier transform of $|u(t)u^*(t + \tau)|$.

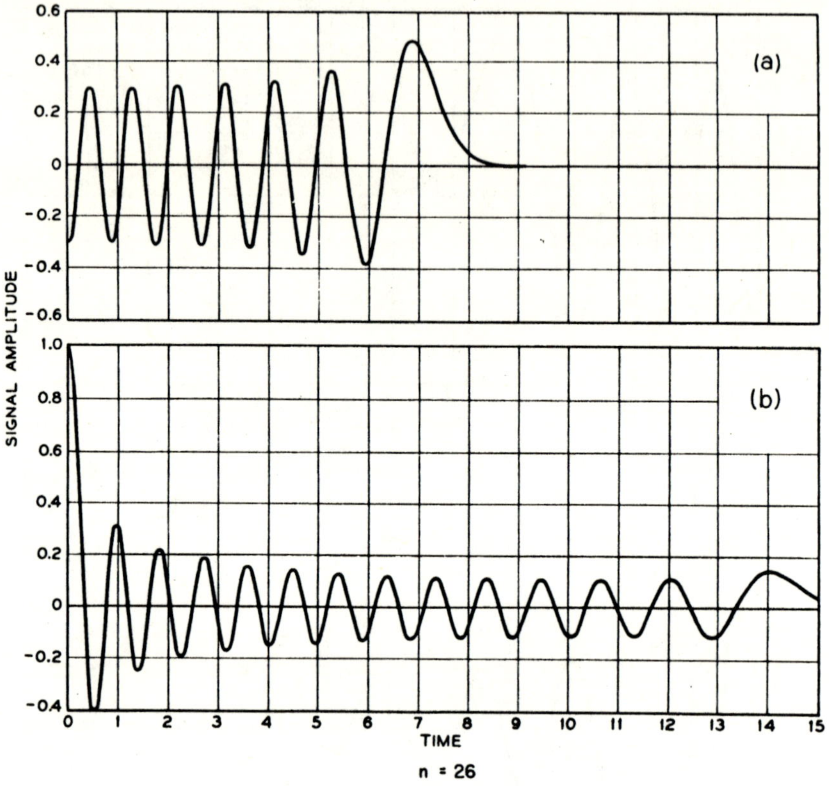

FIG. 4.18c

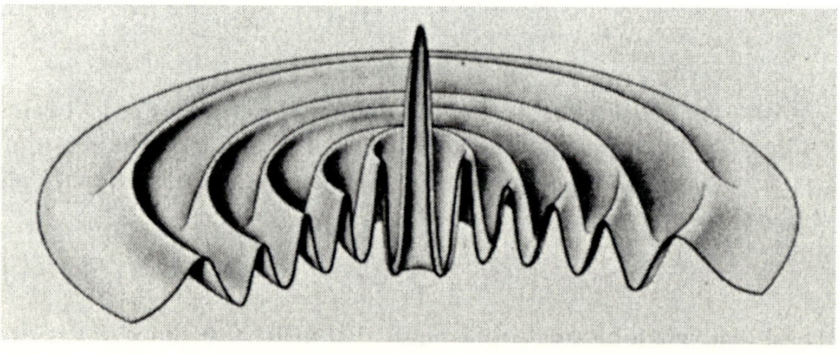

FIG. 4.19 Composite response surface $\chi(\tau, \phi)$ for Hermite waveform, $n = 10$ (courtesy of J. R. Klauder [10] and Bell Syst. Techn. Journal).

Similarly, for the linear FM pulse $\Delta\theta'(t) = \pm 2b\tau$ and $\chi(\tau, \phi)$ is concentrated, for constant τ, along $\phi = \pm b\tau/\pi$. It is evident that heuristic predictions of $\chi(\tau, \phi)$ taken along constant τ lines can be made once the form of the frequency modulation is specified. A further example of the application of the principle of stationary phase to the analysis of ambiguity functions is given by the ambiguity function for the parabolic FM waveform, for which $\theta(t) = ct^3$. For this waveform $\Delta\theta'(t) = \pm 3c[\tau^2 + 2t\tau]$. By applying Eq. (4-153) to the rectangular envelope parabolic FM waveform illustrated in Fig. 4.14 it is seen that

$$|\chi(\tau, \phi)| = \sqrt{2\pi}\,\frac{1}{T\sqrt{6c\tau}} \tag{4-156}$$

The close resemblance between this result and the exact form given in Table 4-I should be noted.

A word of caution is offered regarding the application of the principle of stationary phase. It can be safely applied as long as $|u(t)u^*(t + \tau)|$ does not vary rapidly in contrast with $\exp|\,j\{\theta(t) - \theta(t + \tau)\}\,|$. For rectangular envelope waveforms, this means that it will not be accurate for values of τ close to the origin as well as near $\tau = \pm T$. However, since the ambiguity function close to the origin is near unity while it is very small near $\tau = \pm T$, this restriction does not seriously affect the usefulness of the principle of stationary phase for ambiguity function analysis. It will be found that for a great majority of waveforms that are analyzed, exact closed form descriptions of $\chi(\tau, \phi)$ will not be possible. Therefore, the principle of stationary phase is a very important tool for waveform analysis and design.

4.14 Summary

In modern radar technology the ambiguity function provides the basis for a systematic search for the best waveform for a particular system application. This chapter has presented a complete review, including important derivations, of the many facets of the ambiguity function and its characteristics that are of interest to the radar systems engineer. Although comments have been made in different sections about the importance of evaluating waveforms in the context of specific radar requirements, this chapter has not attempted to explore these possibilities in any depth. Discussions of different examples of radar environments and waveform applications are treated in later chapters.

It has been shown that the ambiguity function may be taken, point for point, as a measure of the difference, in a mean square sense, between a waveform $s(t)$ and a time and frequency shifted version of the same waveform. This was the viewpoint initially taken by Woodward. As an important alternate viewpoint, for those desiring a more physically satisfying picture of the analytical methods employed, it has also been shown that the ambiguity function corresponds to the time/frequency response function observed at the matched-filter output. From both of these aspects, the ambiguity function is clearly a measure of the ability of the radar system to distinguish similar coded waveforms at the receiver that differ in time of arrival and/or frequency shift. Consequently, the ambiguity function can be taken also as a measure of (1) how accurately and unambiguously the range and velocity of a single target can be estimated, and (2) how reliably two targets can be resolved in terms of these parameters.

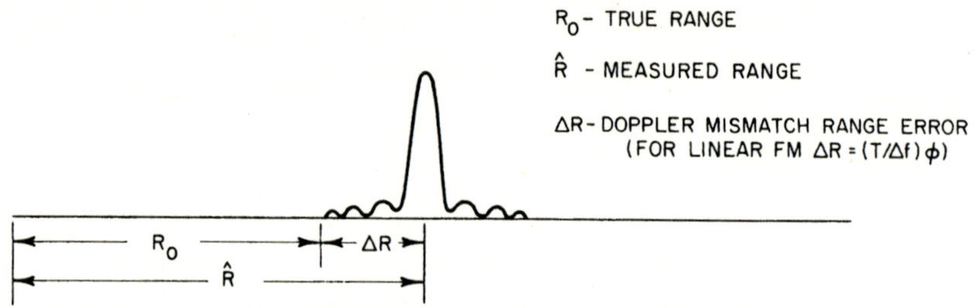

FIG. 4.20   Range measurement ambiguity caused by Doppler mismatch.

Ambiguity can be interpreted in different ways. Figure 4.20 indicates the problem associated with unidirectional FM signals that have unknown Doppler shifts. The effect of the Doppler shift is to introduce a time displacement (or range error) of the compressed pulses at the matched-filter output. The amount of time displacement from the true signal location is related to the Doppler shift/bandwidth ratio and the type of FM function employed. The resulting effect is the so-called range-Doppler ambiguity, or range-Doppler coupling. The relationship between the radar waveform modulation characteristics and the theoretical measurement errors associated with this type of ambiguity is examined in the next chapter, with examples for specific waveforms being given in Chapter 9.

A second type of ambiguity results from the relative magnitude and time distribution of spurious responses that appear at the matched-filter output. These spurious responses can be caused by system distortions, as discussed in Chapter 11, or by the response of the matched-filter to Doppler shifted inputs. If many received signals, having a large dynamic range, are present the spurious responses associated with the larger signals can be falsely identified as "true" signals of interest. These types of false target ambiguities can be more detrimental than normal receiver noise false alarms, because they will often appear as deterministic signals.

A third form of ambiguity is related to the amount of energy contained in the range sidelobe structure of the matched-filter output. If the sidelobe energy is relatively large and extends over an appreciable time interval, then smaller signals not at the same range location as larger signals can be obscured, or masked, by the interfering sidelobes. This is of particular importance in the multiple and dense scatterer environments, and is discussed in Chapter 10.

The most significant property of the ambiguity function is that the total potential ambiguity, as measured by the volume under the surface generated by this function, is independent of the type of radar signal employed as long as all signals are normalized to have the same energy content. This result is well known as the "radar uncertainty principle" [4], or the "law of conservation of ambiguity" [5]. This implies, to the radar designer with absolutely no prior knowledge concerning the particular radar application or environment, that all signals are equally good (or bad). However, each different waveform that can be considered will yield a different *distribution* of ambiguity, as shown by the ambiguity distributions of some basic signals in Fig. 4.3, as well as the other examples considered in Section 4.12. It is this variation in the ambiguity distribution, coupled with at least a fundamental appreciation of the radar system application, that holds forth the hope that the radar designer may, as a result of a systematic search, find a suitable waveform.

A preferable procedure would be to align the ambiguity distribution so that regions of maximum potential ambiguity occur where target parameters (range delay or frequency shift) are not expected, and then to derive the transmitted waveform from this distribution by applying the uniqueness theorem.[1] Unfortunately, this would require a knowledge of the phase structure of $\chi(\tau, \phi)$. Since this information usually will not be available, it would have to be assumed. In that case, there is no guarantee that the resulting derived waveform will be physically realizable. Therefore, the procedure that is normally adopted is to examine the ambiguity functions of a variety of waveforms, choosing that one which yields the best fit (after appropriate scaling of its duration and bandwidth) for a given desired ambiguity function distribution. The degree to which this can be done is an indication of how well the radar system will perform in the projected environment. As with any type of nonideal situation, compromises will undoubtedly be called for. These will usually be arrived at by performance and cost studies of the entire system.

---

[1] The regions of potential ambiguity also locate the regions of range and Doppler shift where the interfering energy from other signals will have the most serious effect on detection performance.

REFERENCES

1. P. M. Woodward, "Probability and Information Theory, with Applications to Radar." Pergamon Press, Oxford, 1953.
2. H. Urkowitz, C. A. Hauer, and J. F. Koval, Generalized resolution in radar systems, *Proc. IRE* **50**, No. 10, 2093–2105 (1962).
3. J. Ville, Theory and application of the notion of complex signal, *Cables Transmission* **2**, 61–74, 1948. [English Transl.: I. Selin, Rand Rept. T-92 (August, 1958).]
4. W. M. Siebert, Studies of Woodward's uncertainty function, *Mass. Inst. Technol. Research Lab. Electronics, Quart. Progr. Rept.* (April, 1958).
5. C. H. Wilcox, The synthesis problem for radar ambiguity functions, Mathematical Research Center, U.S. Army, University of Wisconsin, Madison, Wisconsin, Rept. 157 (April, 1960). (See also: A mathematical analysis of waveform design problem, General Electric Co., Schenectady, New York, Rept. R57EMH54 (October, 1957).
6. C. A. Stutt, The application of time/frequency correlation functions to the continuous waveform encoding of message symbols, *WESCON*, Sect. 9/1 (1961).
7. N. DeClaris and E. Titlebaum, Transformation of radar signals and ambiguity functions. Cornell University, Ithaca, New York, Research Rept. EERL 26 (October, 1964).
8. F. B. Reis, A linear transformation of the ambiguity function plane, *IRE Trans.* (*Correspond.*) **IT-8**, 59 (1962).
9. C. S. Sorkin, The theoretical design of waveforms for pulsed radar, General Atronics Corporation, Bala Cynwyd, Pennsylvania, Rept. 770-207-11 (July, 1960).
10. J. R. Klauder, The design of radar signals having both high range resolution and high velocity resolution, *Bell System Tech. J.* **39**, 809–820 (1960).
11. D. Gabor, Theory of communications, *J. IEE* (*London*) **93**, 429–457 (1946).
12. R. Price and E. M. Hofstetter, Bounds on the volume and height distributions of the ambiguity function, *IEEE Trans.* **IT-11**, 207–214 (1965).
13. C. W. Helstrom, "Statistical Theory of Signal Detection." Pergamon Press, Oxford, 1960.
14. R. Manasse, The use of pulse coding to discriminate against clutter, *Mass. Inst. Technol., Lincoln Lab., Lexington, Massachusetts,* Group Rept. 312-12 (August, 1957).
15. E. C. Westerfield, R. H. Prager, and J. L. Stewart, Processing gains against reverberation (clutter) using matched filters, *IRE Trans.* **IT-6**, 342–348 (1960).
16. M. Bernfeld, Selection of appropriate signals by their ambiguity functions. Unpublished notes.
17. E. Jahnke and F. Emde, "Tables of Functions." Dover, New York, 1945.

CHAPTER 5

Parameter Estimation

5.1 Introduction

It was stated in Chapter 1, and shown in Chapter 2, that matched filtering constitutes the best way to process the raw (input) radar data, that is combined with white Gaussian noise, in order to obtain a waveform suitable for optimum detection decisions (absence or presence). It was further indicated that matched filtering is also the best way to process the same input signals in order to obtain maximum accuracy estimates (measurements) of range, velocity, etc., as well as effective multiple target resolution. Measures of resolution were discussed in Chapter 4. The problem of parameter estimation is treated in this chapter, making use of concepts derived from the field of statistical estimation. The Cramér–Rao inequality is used to develop the bounds on minimum measurement errors in terms of certain basic properties of the radar signal, such as duration, bandwidth, and the frequency or phase modulation function. Interestingly, it is not necessary to invoke the requirement of a specific waveform processing technique to establish the minimum error bounds. However, the method of maximum likelihood, which was shown in Chapter 2 to result in a matched-filter implementation, leads to measurement errors that approach the Cramér–Rao bound for large signal-to-noise ratios at the matched-filter output. This latter condition is usually assumed in parameter estimation problems.

As will be discussed in later sections, the Cramér–Rao derivation depends on conditions of regularity being met by the waveform. This is not the case for a majority of matched-filter waveforms that have either discontinuous envelopes (i.e., rectangular) or phase modulation functions. The treatment of this problem in regard to practical waveforms is discussed in Chapter 9, where the results derived in this chapter are applied to a number of frequency and phase modulated pulse-compression matched-filter signals. When used with care, the concepts of classical parameter estimation can be a valuable tool to obtain comparative measurement accuracies among various types of waveforms.

5.2 The Radar Parameters

An equally important function of a radar system, in addition to the detection of target echo signals, is the extraction of information from the radar echo signal that can be used to determine the line of sight distance to a target, called the range, the range-rate of the target, and sometimes the target acceleration. The ability to determine these target parameters by referring to the echo signal is derived from the effect they have on the transmitted waveform after it is reflected by the target.

The target range is determined by measuring the time between the transmission of the radar waveform and the reception of the radar echo signal. If $s(t)$ denotes the radar waveform which is transmitted, starting at time $t = 0$, then $s(t - t_0)$ denotes the radar echo signal, where $t = t_0$ is the time of propagation between the transmission and reception. The relationship between the range R and $t = t_0$ is given by

$$R = ct_0/2 \qquad (5\text{-}1)$$

where c is the velocity of electromagnetic propagation. The factor 2 appears in this relationship because of the two way trip of the electromagnetic energy.

The ability to determine the velocity is derived from the Doppler effect on the waveform. For the class of narrow band waveforms, to which radar waveforms generally belong, this effect is usually approximated by frequency translation with respect to the transmitted carrier frequency; there are some second order effects on the waveform, discussed in Chapter 4. These will be assumed negligible for the majority of general radar applications. The narrow band waveform can be described by

$$s(t) = a(t)\cos[\omega_0 t + \theta(t)] \qquad (5\text{-}2)$$

where the envelope $a(t)$ and the phase function $\theta(t)$ are slowly varying in contrast to $\cos \omega_0 t$. The carrier frequency is denoted by f_0 where $f_0 = \omega_0/2\pi$. The effect of a moving target on this waveform is therefore described by

$$s_d(t) = a(t)\cos[2\pi(f_0 + \phi)t + \theta(t)] \qquad (5\text{-}3)$$

where ϕ is the Doppler frequency translation. Alternately, if the spectrum of the transmitted signal, $s(t)$, is denoted by $S(f)$, then $S(f + \phi)$ denotes the spectrum of the radar echo signal. The relationship between the velocity, v, and the Doppler shift, ϕ, is given by

$$v = c\phi/2f_0 \qquad (5\text{-}4)$$

It is noted that when the target is moving toward the radar the Doppler shift is positive, and when it is moving in the opposite sense, this shift is negative.

In contrast with the previous two parameters, the ability to determine the acceleration of a target cannot be derived in terms of a single waveform parameter. The effect of target acceleration is to distort the structure of the waveform. The type and degree of this distortion will depend on the structure of the waveform during transmission as well as on the amount of target acceleration to which it is exposed. For example, for the case of the simple monotone pulse, the distortion appears partially in the form of an additional linear FM term.

For most conceivable situations, the amount of distortion on a single waveform echo as a result of target acceleration will be relatively small. Therefore, it is usually necessary to process more than one received waveform from the accelerating target. In this case it is better, perhaps, to indirectly extract the acceleration data by means of computations which are based on samples of range or range-rate. In this chapter we are concerned with the theoretical evaluation of the target parameters, based on the information that can be extracted from a single waveform echo. Therefore, the further discussion of the extraction of acceleration will not be considered.

5.3 The Problem of Parameter Estimation

The presence of noise at the radar receiver will always interfere with the precise evaluation of the time and frequency position of waveform echo signals. At the very best, the values of these parameters that are extracted (from a noisy background) will consist of estimates. With everything else fixed, these estimates will vary randomly about the true parametric values, as illustrated in Fig. 5.1. It is the function of the radar system to extract the

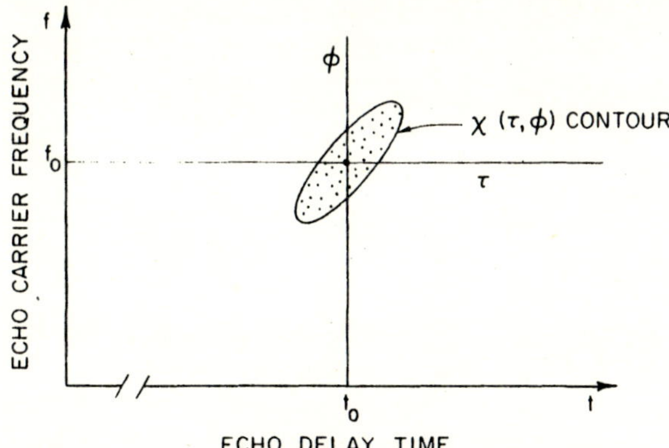

FIG. 5.1 An example of the dispersion of simultaneous range and velocity estimates of the true values t_0 and f_0.

most information about these parameters from the raw radar data so that the best estimates can be obtained. The measure of goodness usually consists of the error variance; that is, the word "best" is interpreted to mean the smallest error variance about the true parameter value. In this chapter the error variance will be assumed to be the measure of the goodness of an estimate. However, it is noted that there are other viewpoints of how to interpret the word "best"; the most general is perhaps Wald's approach with regard to average costs. For a discussion of this approach the reader is referred to Helstrom [1] or Middleton [2]. Another viewpoint is contained in the information theory interpretation by Woodward and Davies [3] and Woodward [4], which was briefly outlined in Chapter 2.

It is possible to derive an expression for the smallest possible error variances in terms of effective measures of such waveform parameters as the bandwidth and the time duration. This can be done without regard to the type of waveform processing that will furnish the most information about the time and frequency position of a radar echo. These variances, as will be shown later, also depend on the signal-to-noise ratio at the time of estimation, as well as on the statistical coupling between the errors. The increasing emphasis on accuracy, especially with regard to space probes and missiles, has made these expressions for the ultimate error variances an important guide in the design considerations of radar signals.

5.4   The Cramér–Rao Inequality [5]

The Cramér–Rao inequality states the lower bound for the error variance of parameter estimates, regardless of the method of estimation employed. In this section the derivation of this inequality for a single unknown parameter will be discussed. Later, a more general result due to Cramér will be applied for the specific case of the simultaneous estimation of the radar parameters of range and range-rate.

For the case of a single unknown parameter θ, the inequality is given by

$$E\{(\hat{\theta} - \theta)^2\} \geq \frac{\left[\dfrac{dm(\theta)}{d\theta}\right]^2}{\displaystyle\int_{-\infty}^{\infty}\!\!\!\!\cdots\!\int \left(\frac{\partial \log p(\bar{x};\theta)}{\partial \theta}\right)^2 p(\bar{x};\theta)\,d\bar{x}} \tag{5-5}$$

where $E\{\ \}$ denotes the expectation, and θ and $\hat{\theta}$ denote, respectively, the unknown parameter and the parameter estimator,[1] which is some

---

[1] The general parameter θ is not to be confused with $\theta(t)$, the notation for radian phase modulation.

function of the statistical samples $\bar{x} = x_1, x_2, \ldots, x_n$. Moreover, $m(\theta)$ is the expected value of $\hat{\theta}$ which in general will be some function of the unknown parameter; when $m(\theta) = \theta$ the estimator is said to be unbiased. The notation $p(\bar{x}; \theta)$ represents the joint probability distribution of the random variable $\bar{x}$, given the unknown parameter. It is called the likelihood function and denoted by $L(\bar{x}; \theta)$ when $\bar{x}$ is given and θ is allowed to be the variable. In fact, the inequality, given by (5-5), is correctly stated in terms of the likelihood function.

In order to present the conditions under which the inequality (5-5) holds, and to describe the various classifications of parameter estimates, the essential steps of Cramér's proof of this inequality will be given. Let the statistical description of the estimator $\hat{\theta}$ be denoted by $p(\hat{\theta}; \theta)$; then

$$m(\theta) = \int_{-\infty}^{\infty} \hat{\theta} p(\hat{\theta}; \theta) \, d\hat{\theta} \tag{5-6}$$

This can be restated as follows:

$$m(\theta) = \theta + \int_{-\infty}^{\infty} (\hat{\theta} - \theta) p(\hat{\theta}; \theta) \, d\hat{\theta} \tag{5-7}$$

The integral part of this expression represents the bias of the estimator. As stated previously, when this is zero the estimator is *unbiased*. Now, subject to certain conditions of regularity, the derivative of $m(\theta)$ is

$$\frac{dm(\theta)}{d\theta} = \int_{-\infty}^{\infty} (\hat{\theta} - \theta) \frac{\partial p(\hat{\theta}; \theta)}{\partial \theta} \, d\hat{\theta} \tag{5-8}$$

By noting that

$$\frac{d \log_e v}{dx} = \frac{1}{v} \frac{dv}{dx} \tag{5-9}$$

(5-8) becomes

$$\frac{dm(\theta)}{d\theta} = \int_{-\infty}^{\infty} (\hat{\theta} - \theta) p(\hat{\theta}; \theta) \frac{\partial \log p(\hat{\theta}; \theta)}{\partial \theta} \, d\hat{\theta} \tag{5-10}$$

where the subscript, e, indicating the natural logarithm has been dropped for convenience. Equation (5-10) can be rewritten as follows:

$$\frac{dm(\theta)}{d\theta} = \int_{-\infty}^{\infty} (\hat{\theta} - \theta) \sqrt{p} \, \frac{\partial \log p}{\partial \theta} \sqrt{p} \, d\hat{\theta} \tag{5-11}$$

where p stands for $p(\hat{\theta}; \theta)$. By squaring both sides of this equation and applying the Schwarz inequality, it can be shown that

$$\left[\frac{dm(\theta)}{d\theta}\right]^2 \leq \int_{-\infty}^{\infty} (\hat{\theta} - \theta)^2 p \, d\hat{\theta} \int_{-\infty}^{\infty} \left(\frac{\partial \log p}{\partial \theta}\right)^2 p \, d\hat{\theta} \qquad (5\text{-}12)$$

According to the Schwarz inequality, the equality holds if and only if some quantity $K(\theta)$ exists such that

$$\frac{\partial \log p}{\partial \theta} = K(\theta)(\hat{\theta} - \theta) \qquad (5\text{-}13)$$

Now, if the probability density $p(\bar{x};\theta)$ can be uniquely and continuously described in terms of a new set of random variables, such that

$$p(\bar{x};\theta)|J| = p(\hat{\theta};\theta) \, p(\bar{y}/\hat{\theta};\theta) \qquad (5\text{-}14)$$

where $|J|$ is the Jacobian of the transformation, and $\hat{\theta}, \bar{y} = y_1, y_2, \ldots, y_{n-1}$ are the new set of random variables, then, subject to certain conditions of regularity, it can be shown that

$$\int_{-\infty}^{\infty} \left(\frac{\partial \log p(\bar{x};\theta)}{\partial \theta}\right)^2 p(\bar{x};\theta) \, d\bar{x} = \int_{-\infty}^{\infty} \left(\frac{\partial \log p(\hat{\theta};\theta)}{\partial \theta}\right)^2 p(\hat{\theta};\theta) \, d\hat{\theta} \qquad (5\text{-}15)$$

when $p(\bar{y}/\hat{\theta};\theta)$ is independent of θ for all points such that $p(\hat{\theta};\theta) > 0$. Substituting this result in (5-12), and noting that the first integral on the right hand side of (5-12) is $E\{(\hat{\theta} - \theta)^2\}$ completes the proof of (5-5).

The estimator $\hat{\theta}$ is said to be a regular estimate of θ, when $p(\bar{x};\theta)$ can be described in terms of the new set of random variables $\hat{\theta}, \bar{y}$ and satisfies certain conditions of regularity. Moreover, the estimator $\hat{\theta}$ is said to be a *sufficient* estimate of θ when in addition the condition is satisfied that $p(\bar{y}/\hat{\theta};\theta)$ is independent of θ for all points such that $p(\hat{\theta};\theta) > 0$. The estimator, in this case, is said to summarize all the relevant information contained in a sample $\bar{x} = x_1, x_2, \ldots, x_n$ with respect to the unknown parameter. Finally, the estimator is said to be an *efficient* estimator when it is *unbiased* and the conditions for the equality are satisfied. These conditions are (1) the estimator is sufficient, and (2), the quantity $K(\theta)$ exists such that $\partial \log p(\hat{\theta};\theta)/\partial \theta = K(\theta)(\hat{\theta} - \theta)$. An *efficient* estimator provides the smallest error variance. However, the restrictive conditions for the existence of this estimator are rarely satisfied. Therefore, there are two further distinctions regarding the classification of estimators. Namely, an estimator that converges in probability to the unknown parameter as the number of independent samples approaches infinity is said to be a consistent estimate of θ. This implies that, for infinitely increasing samples, the probability density $p(\hat{\theta};\theta)$ is concentrated around θ such that for any $\delta > 0$ the probability that $|\hat{\theta} - \theta| < \delta$ tends to unity. An even stronger class of estimators are those for which $p(\hat{\theta}; \theta)$ is a asymptotically normal for large n. Among this class of estimators, the estimators for which the error variance is given by the lower bound of (5-5) are said to be *asymptotically* efficient estimates of θ. It has been shown that the parameter estimation approach known as the method of maximum likelihood very frequently yields an *asymptotically efficient* estimator. The method of maximum likelihood has the property that when an efficient estimator exists it can be found by this approach. The classifications discussed above, as presented by Cramér [5], were introduced by Fisher [6] who is credited with founding the theory of estimation.

5.5 Sample Space

For the following application of the Cramér–Rao inequality, and also in the succeeding discussion of the method of maximum likelihood, it will be assumed that the noise has (1) a probability distribution which is Gaussian, and (2) a one-sided power density spectrum which is uniform in the interval $|f - f_0| < W/2$ and equal to N_0. In conjunction with these assumptions, it will also be assumed that a bandlimited signal can be accurately represented in the interval $t_0 \leqslant t \leqslant t_0 + T$ by a finite number of observations on the interval spaced $1/W$ apart. Since the following discussion will deal with rf type signals, each sample time will yield two samples for a total of $2WT$ samples in the T interval. These samples will be statistically uncorrelated and, because the noise is Gaussian, they will also be statistically independent.

This particular sample space allows the convenient application of Shannon's sampling theorem [7] for the following discussions. In particular, Shannon's sampling technique will be invoked to transform from solutions which appear in the form of discrete summations involving the sampled data, to integral formulations. In the strict sense of rigorous analysis, these transformations are not correct, unless T is allowed to approach infinity, because they assume the existence of signals which are simultaneously limited in time duration as well as in bandwidth—a theoretical impossibility. For the sake of convenience, however, the inconsistency is ignored. For more rigorous analyses [8–10], it is theoretically possible to employ the Karhunen–Loéve expansion which involves the eigenfunctions of the noise covariance function.

5.6 Joint Radar Parameter Estimation Errors as Seen by an Unbiased Estimator

As seen from the viewpoint of an unbiased, joint efficient estimator, the mean square error in the joint estimation of two statistical parameters $\theta_j, j = 1, 2$, where θ_j represents time delay, τ, and Doppler shift ϕ, is given by [5]

$$E\{(\hat{\theta}_j - \theta_j)^2\} = \cfrac{1}{E\left\{\left(\cfrac{\partial}{\partial \theta_j} \log L(\bar{x}; \theta_1, \theta_2)\right)^2\right\}} \; \cfrac{1}{1 - \rho^2(\hat{\theta}_1, \hat{\theta}_2)} \qquad (5\text{-}16)$$

where $\rho(\hat{\theta}_1, \hat{\theta}_2)$ is a correlation coefficient, and represents the statistical coupling of the estimators $\hat{\theta}_1$ and $\hat{\theta}_2$. The correlation coefficient is given by

$$\rho(\hat{\theta}_1, \hat{\theta}_2) = -\cfrac{E\left\{\displaystyle\prod_{j=1}^{2} \left(\cfrac{\partial \log L(\bar{x}; \theta_1, \theta_2)}{\partial \theta_j}\right)\right\}}{\left[\displaystyle\prod_{j=1}^{2} E\left\{\left(\cfrac{\partial \log L(\bar{x}; \theta_1, \theta_2)}{\partial \theta_j}\right)^2\right\}\right]^{1/2}} \qquad (5\text{-}17)$$

In the next section of this chapter it will be shown that the method of maximum likelihood approaches an unbiased, asymptotically efficient estimator for the radar parameters τ and ϕ as the signal-to-noise ratio becomes increasingly large ($S/N \gg 1$).

The Likelihood Function

For the particular example which will be considered below, a point target will be assumed to be traveling along a linear path in line with the radar. For this case, the signal at the radar receiver input will consist of noise and the waveform that was reflected from the target. As a rule, the description of the waveform will involve "stray" parameters as well as the parameters τ and ϕ. These stray parameters which include the amplitude and the phase of the waveform, will be ignored for the present; this is equivalent to the assumption that they are known and cannot affect the estimates of the other parameters. Consequently, the general form of the receiver input can be described by

$$x(t) = n(t) + s(t; \theta_1, \theta_2) \qquad (5\text{-}18)$$

where $x(t)$ is the receiver input and constitutes the raw radar data, $n(t)$ is the Gaussian, white, bandlimited noise described in Section 5.5, and $s(t; \theta_1, \theta_2)$ is the reflected waveform. Specifically, $s(t; \theta_1, \theta_2)$ is given by

$$s(t; \theta_1, \theta_2) = a(t - \tau) \cos[2\pi(f_0 + \phi)(t - \tau) + \theta(t - \tau)] \qquad (5\text{-}19)$$

where θ_1 and θ_2 have been replaced by τ and ϕ.

The functions $x(t)$, $n(t)$, and $s(t)$ are assumed to be real valued. For convenience, however, this example is developed in terms of their respective complex representations after Gabor [11]. Thus, the following substitutions will be made

$$x(t) \rightarrow \gamma(t), \qquad n(t) \rightarrow v(t), \qquad s(t) \rightarrow \psi(t) = u(t) \exp[j2\pi f_0 t]$$

in which case, the likelihood function, expressed in terms of WT complex samples of the radar input data, is given by

$$L(\bar{x}; \theta_1, \theta_2) = \left[\frac{1}{2\pi N}\right]^{2WT/2} \exp\left[-\frac{1}{2N}\right] \sum_{i=1}^{WT} |\gamma_i - \psi_i(\theta_1, \theta_2)|^2 \quad (5\text{-}20)$$

where $N = N_0 W$. By taking $\log L$ and applying the partial derivative to this with respect to θ_j, one obtains

$$\frac{\partial \log L}{\partial \theta_j} = \mathrm{Re}\,\frac{1}{N} \sum_{i=1}^{WT} (\gamma_i - \psi_i)\frac{\partial \psi_i^*}{\partial \theta_j} \quad (5\text{-}21)$$

where Re denotes the "real part of," and the asterisk denotes the complex conjugate. It is necessary to evaluate (5-21) at the true value of the parameter θ_j. Since the factor $(\gamma_i - \psi_i)$ can be replaced by v_i, the expression (5-21) becomes

$$\left.\frac{\partial \log L}{\partial \theta_j}\right|_{\theta_j^0} = \mathrm{Re}\,\frac{1}{N} \sum_{i=1}^{WT} v_i \left.\frac{\partial \psi_i^*}{\partial \theta_j}\right|_{\theta_j^0} \quad (5\text{-}22)$$

where the vertical bar to which the subscript θ_j^0 is attached implies that the derivatives in (5-21) have been evaluated at the true value of the parameter. Since

$$E\{v_i v_j\} = 0, \qquad \text{for all } i \text{ and } j$$

$$E\{v_i v_j^*\} = \begin{cases} 0, & i \neq j \\ 2N, & i = j \end{cases}$$

consequently, by straightforward development

$$E\left\{\left(\left.\frac{\partial \log L}{\partial \theta_j}\right|_{\theta_j^0}\right)^2\right\} = \frac{1}{N} \sum_{i=1}^{WT} \left|\left.\frac{\partial \psi_i}{\partial \theta_j}\right|_{\theta_j^0}\right|^2 \quad (5\text{-}23)$$

Expression (5-23) is a fundamental result; it appears that it was first presented by Slepian [12]. An integral form for (5-23) can be derived by invoking Shannon's sampling theorem. Thus, if $\psi(t)$ is band limited such that

$$\Psi(f) = \begin{cases} 2S(f), & f_0 - \dfrac{W}{2} \leqslant f \leqslant f_0 + \dfrac{W}{2} \\[2mm] 0, & |f - f_0| > \dfrac{W}{2} \end{cases} \quad (5\text{-}24)$$

where $S(f)$ is the Fourier transform of $s(t)$, then by the sampling theorem

$$E\left\{\left(\left.\frac{\partial \log L}{\partial \theta_j}\right|_{\theta_j^0}\right)^2\right\} = \frac{1}{N_0}\int_0^T \left[\frac{\partial \psi(t; \theta_1^0, \theta_2^0)}{\partial \theta_j}\right]\left[\frac{\partial \psi(t; \theta_1^0, \theta_2^0)}{\partial \theta_j}\right]^* dt \quad (5\text{-}25)$$

where the derivatives are evaluated at the true value of the parameter.

Minimum Error Variance for Time Delay

Expression (5-25) can now be employed to derive the minimum error variance for the measurement of time delay. It can also be applied to derive the minimum error variance for the measurement of Doppler shift; this will be done in the next paragraph. The reader should realize that the application of (5-25) alone, without regard for the factor $\rho(\hat{\theta}_1, \hat{\theta}_2)$ which appears in the Cramér–Rao inequality, assumes that $\rho(\hat{\theta}_1, \hat{\theta}_2) = 0$, or that the parameter not being investigated is known. The latter of these conditions is less likely to be true than the former; it implies complete *a priori* knowledge about the particular parameter. The conditions for $\rho(\hat{\theta}_1, \hat{\theta}_2) = 0$ will be discussed in Chapter 9.

To determine the minimum error variance for time delay, it is first necessary to note that the time delayed form of $\psi(t)$ is $\psi(t - \tau)$. Consequently, by invoking the chain rule, it is found that

$$\frac{\partial \psi(t - \tau)}{\partial \tau} = -\psi'(t - \tau) \tag{5-26}$$

By using this result in (5-25), one obtains

$$E\left\{ \left(\frac{\partial \log L}{\partial \tau} \right)^2 \right\} = \frac{1}{N_0} \int_{-\tau}^{T-\tau} \psi'(t)[\psi^*(t)]' \, dt \tag{5-27}$$

After suitable application of Parseval's theorem, the integral on the right side of (5-27) becomes

$$\int_0^T \psi'(t)[\psi^*(t)]' \, dt = -[2\pi j]^2 \int_0^\infty f^2 \Psi(f)\Psi^*(f) \, df \tag{5-28}$$

where $\Psi(f)$ was defined by (5-24). Introducing a change of variable such that $f = \eta + f_0$, where f_0 is the centroid of $|\Psi|^2$, (5-28) becomes

$$\int_0^T \psi'(t)[\psi^*(t)]' \, dt = [2\pi]^2 \int_{-f_0}^\infty \eta^2 \Psi(\eta + f_0)\Psi^*(\eta + f_0) \, d\eta + [2\pi f_0]^2 2E \tag{5-29}$$

The first term of the expression on the right side of this equation is used to define the rms bandwidth β in units of radians per second, for radar waveforms [4]. Thus, (5-28) finally becomes

$$\int_0^T \psi'(t)[\psi^*(t)]' \, dt = 2E\beta^2 + [2\pi f_0]^2 2E \tag{5-30}$$

and the following is obtained:

$$E\{(\hat{\tau} - \tau)^2\} = \frac{1}{(2E/N_0)\{\beta^2 + [2\pi f_0]^2\}} \tag{5-31}$$

The estimator that would yield (5-31) would require *a priori* knowledge of the carrier phase in the radar return waveform, as assumed in the first paragraphs of this section. Since this information, as a rule, is not available, the error variance will be much greater than (5-31) indicates, unless the estimator is averaged with respect to the "stray" carrier phase. This is equivalent to envelope detection after the signal has been processed.[1] In this case, the error variance becomes

$$E\{(\hat{\tau} - \tau)^2\} = \frac{1}{(2E/N_0)\beta^2} \tag{5-32}$$

Minimum Error Variance for Doppler Frequency

To determine the minimum error variance for the Doppler frequency measurement. the form of $\psi(t)$ that is now used is

$$\psi(t) = u(t)\exp[j2\pi(f_0 + \phi)t] \tag{5-33}$$

where $u(t)$ is called the waveform "pre-envelope" [13]. Consequently,

$$\frac{\partial\psi(t)}{\partial\phi} = j2\pi t\psi(t) \tag{5-34}$$

By using this result, (5-25) becomes

$$E\left\{\left(\frac{\partial \log L}{\partial\phi}\right)^2\right\} = \frac{(2\pi)^2}{N_0}\int_{-\tau}^{T-\tau} t^2\psi(t)\psi^*(t)\,dt \tag{5-35}$$

The expression on the right side of (5-35) is used [4] to define the rms time duration, α, in units of radian-seconds, where

$$2E\alpha^2 = (2\pi)^2\int_{-\tau}^{T-\tau} t^2\psi(t)\psi^*(t)\,dt \tag{5-36}$$

Thus,

$$E\left\{\left(\frac{\partial \log L}{\partial\phi}\right)^2\right\} = \frac{1}{(2E/N_0)\alpha^2} \tag{5-37}$$

Equation (5-37) is similar to its counterpart (5-32).

The Error Coupling

The estimation errors will, in general, not be uncoupled, nor will the parameter not under investigation be known from *a priori* knowledge. Consequently, in contrast with the error variances (5-32) and (5-37), the actual error variances will generally be larger by virtue of the fact that $\rho(\hat{\theta}_1, \hat{\theta}_2) \neq 0$.

---

[1] This is discussed further at the end of the Section 5.8.

Proceeding now to express $\rho(\hat{\theta}_1, \hat{\theta}_2)$ in terms of the waveform, it is recalled that

$$\rho(\hat{\theta}_1, \hat{\theta}_2) = -\frac{E\left\{\dfrac{\partial \log L}{\partial \theta_1} \cdot \dfrac{\partial \log L}{\partial \theta_2}\right\}}{\left[E\left\{\left(\dfrac{\partial \log L}{\partial \theta_1}\right)^2\right\} \cdot E\left\{\left(\dfrac{\partial \log L}{\partial \theta_2}\right)^2\right\}\right]^{1/2}} \tag{5-38}$$

Since the denominator of (5-38) has already been obtained [see (5-32) and (5-37)], the problem that remains is to reduce the numerator of this expression so that it is expressed in terms of the waveform. By referring to (5-22), it is possible to obtain in a straightforward manner

$$E\left\{\left.\frac{\partial \log L}{\partial \theta_1}\right|_{\theta_1^0} \cdot \left.\frac{\partial \log L}{\partial \theta_2}\right|_{\theta_2^0}\right\} = \frac{1}{N}\operatorname{Re}\sum_{i=1}^{WT} \left.\frac{\partial \psi_i^*}{\partial \theta_1}\right|_{\theta_1^0} \cdot \left.\frac{\partial \psi_i}{\partial \theta_2}\right|_{\theta_2^0} \tag{5-39}$$

Again, invoking the sampling theorem yields

$$E\left\{\frac{\partial \log L}{\partial \tau} \cdot \frac{\partial \log L}{\partial \phi}\right\} = \frac{1}{N_0}\operatorname{Re}\int_0^T \frac{\partial \psi^*(t)}{\partial \tau} \cdot \frac{\partial \psi(t)}{\partial \phi}\, dt \tag{5-40}$$

The notations which indicate that the derivatives should be evaluated at the true values of τ and ϕ have been dropped from (5-40) for convenience; however, it can be assumed that this operation is implied by (5-40).

By referring to the two forms which were previously given for $\psi(t)$ expression (5-40) finally becomes

$$E\left\{\frac{\partial \log L}{\partial \tau} \cdot \frac{\partial \log L}{\partial \phi}\right\} = -\frac{\pi}{N_0}\int_{-\tau}^{T-\tau} jt[\psi(t)\{\psi^*(t)\}' - \psi^*(t)\psi'(t)]\, dt \tag{5-41}$$

Now, the general form for $\psi(t)$ is

$$\psi(t) = a(t)\exp[\,j\{2\pi f t + \theta(t)\}] \tag{5-42}$$

Using (5-42) in (5-41) finally yields

$$E\left\{\frac{\partial \log L}{\partial \tau} \cdot \frac{\partial \log L}{\partial \phi}\right\} = -\frac{2\pi}{N_0}\int_{-\tau}^{T-\tau} t|\psi|^2[2\pi f + \theta'(t)]\, dt \tag{5-43}$$

where the prime indicates differentiation. This result can be expanded into two parts. Thus

$$E\left\{\frac{\partial \log L}{\partial \tau} \cdot \frac{\partial \log L}{\partial \phi}\right\} = -\frac{2\pi}{N_0}\int_{-\tau}^{T-\tau} t\theta'|\psi|^2\, dt - \frac{(2\pi)^2 f}{N_0}\int_{-\tau}^{T-\tau} t|\psi|^2\, dt \tag{5-44}$$

The second term on the right side of (5-44) is recognized as a first moment of $|\psi(t)|^2$. By choosing the time reference such that the first moment disappears, expression (5-44) becomes

$$E\left\{\frac{\partial \log L}{\partial \tau} \cdot \frac{\partial \log L}{\partial \phi}\right\} = -\frac{2\pi}{N_0} \int_{-\tau}^{T-\tau} t\theta'|\psi|^2 \, dt \qquad (5\text{-}45)$$

By inserting this result in (5-38), along with (5-32) and (5-37), one obtains

$$\rho(\tau, \phi) = \frac{\pi}{E\alpha\beta} \int_{-\tau}^{T-\tau} t\theta'|\psi|^2 \, dt \qquad (5\text{-}46)$$

An alternate form for (5-46) can be derived by invoking Parseval's theorem. Thus, alternately,

$$\rho(\tau, \phi) = \frac{\pi}{E\alpha\beta} \int_{-f_0}^{\infty} f\Phi'|\Psi(f + f_0)|^2 \, df \qquad (5\text{-}47)$$

where $\Phi'(f + f_0)$ is the derivative of the phase function of the Fourier transform of $\psi(t)$.

Since the results that were developed in this section are important in the context of waveform design, they will be tabulated for ready reference below:

1. *Carrier frequency:*

$$2Ef_0 = \int_{-\infty}^{\infty} f\Psi(f)\Psi^*(f) \, df \qquad (5\text{-}48)$$

2. *Bandwidth:*

$$2E\left[\frac{\beta}{2\pi}\right]^2 = \int_{-\infty}^{\infty} f^2|\Psi(f + f_0)|^2 \, df \qquad (5\text{-}49)$$

3. *Time duration:*

$$2E\left[\frac{\alpha}{2\pi}\right]^2 = \int_{-\infty}^{\infty} t^2|\psi(t)|^2 \, dt \qquad (5\text{-}50)$$

4. *Minimum range measurement error variance:*

$$\sigma_R^2 = \frac{c^2}{4(2E/N_0)\beta^2(1 - \rho^2)} \qquad (5\text{-}51)$$

5. *Minimum velocity measurement error variance:*

$$\sigma_v^2 = \frac{(c/f_0)^2}{4(2E/N_0)\alpha^2(1 - \rho^2)} \qquad (5\text{-}52)$$

6. *Correlation coefficient of estimation errors*:

$$\rho(\tau, \phi) = \frac{\pi}{E\alpha\beta} \int_{-\infty}^{\infty} t\theta'(t)|\psi(t)|^2 \, dt \tag{5-53}$$

These quantities will be the subject of further discussion in Chapter 9, where they will be applied to a number of specific waveform examples.

5.7 The Theoretical rms Bandwidth for Signals That Exhibit Step Discontinuous Amplitude and/or Phase Characteristics

The definition of the theoretical rms bandwidth β of a signal was given in Section 5.6 by the expression

$$\beta^2 = \frac{(2\pi)^2}{2E} \int_{-\infty}^{\infty} f^2 U U^* \, df \tag{5-54}$$

where U represents the spectrum for the complex signal modulation $u(t) = a(t)\exp[j\theta(t)]$. Using Parseval's theorem, the following alternate formulation is obtained:

$$\beta^2 = -\frac{1}{2E} \int_{-\infty}^{\infty} u^* u'' \, dt \tag{5-55}$$

This can be expanded after noting that

$$u^* u'' = [a''a - a^2\theta^2] + j[a^2\theta'' + 2aa'\theta'] \tag{5-56}$$

and that β is a real valued quantity. Thus,

$$\beta^2 = -\frac{1}{2E} \int_{-\infty}^{\infty} [a''a - a^2(\theta')^2] \, dt \tag{5-57}$$

Finally, using the method of integration by parts, the expression for β^2 becomes

$$\beta^2 = \frac{1}{2E} \int_{-\infty}^{\infty} \{[a']^2 + [a\theta']^2\} \, dt \tag{5-58}$$

This result implies that the rms bandwidth becomes infinite for signals which exhibit discontinuous amplitude and/or phase characteristics. It is noted that the rectangular pulse always falls into this category, regardless of the form of FM employed. The problem that this causes in comparing the measurement accuracies of matched-filter signals is discussed in Chapter 9.

5.8 The Method of Maximum Likelihood [5]

The Cramér–Rao inequality has yielded expressions for the ultimate accuracy of simultaneous radar parameter estimations in terms of effective measures of such waveform design parameters as bandwidth and time duration. The derivations of these were isolated from the consideration of the method of estimation. Therefore the problem remains to justify the application of matched filtering for "best" parameter estimation. Since the word "best" has been previously interpreted in this chapter to mean the smallest error variance, this justification will be obtained from an approach based on the method of maximum likelihood, inasmuch as this method of estimation yields an *asymptotically efficient* estimator (see Section 5.4) for large signal-to-noise ratios, where the noise is Gaussian.

The method of maximum likelihood is perhaps the most important general approach for estimating the unknown parameters that remain to be determined in order to complete the statistical description of an infinite population of random quantities. Since the general problem of estimation involves several unknown parameters, the method of maximum likelihood requires the solution of as many simultaneous equations as there are parameters of the form

$$\frac{\partial L(\bar{x};\bar{\theta})}{\partial \theta_i} = 0, \qquad i = 1, 2, \ldots, m \tag{5-59}$$

where L denotes the likelihood function (see Section 5.3) which describes the variation of a given joint probability distribution $p(\bar{x};\theta)$ at a given sample point $\bar{x}' = x_1', x_2', \ldots, x_n'$ in terms of the set of parameters $\bar{\theta} = \theta_1, \theta_2, \ldots, \theta_i, \ldots, \theta_m$. Since $\log L$ and L attain their maximum values for the same value of θ, then for convenience the method of maximum likelihood requires that

$$\frac{\partial \log L}{\partial \theta} = 0 \tag{5-60}$$

A graphic illustration at this point will help the reader, perhaps, to develop some insight about the method of maximum likelihood. Suppose, for example, it is desired to estimate the mean, denoted by m, of an infinite population of random numbers that have a Gaussian probability distribution. If one number is drawn as a sample of the population, then according to the method of maximum likelihood, this number should be employed as an estimate of the mean, since the maximum point of the Gaussian probability distribution can be made to appear at the sample point. The reader will note that the mean of a Gaussian distribution appears under the maximum point. The error variance for this estimate of the mean is clearly equal to the variance σ^2 of the population of numbers.

It is possible to reduce the estimation error variance by using additional samples of the population, as a basis for the estimate of the mean. Suppose, now, two samples, designated x_1' and x_2', are drawn from the population. These samples can be treated as samples of the random variables x_1 and x_2, respectively. The random variables have a joint probability distribution denoted by $p(x_1, x_2 ; m)$. This joint probability distribution can be visualized as a surface above a plain that has the coordinates x_1 and x_2. Moreover, the two sample x_1' and x_2' are the coordinates of a point on this plane. Since the same mean, m, is common to both random variables, the maximum point on this Gaussian surface is subject to the constraint that it must appear above a straight line given by $x_1 = x_2$.

According to the method of maximum likelihood, the estimate of the mean is derived from the set of coordinates (m, m) for the surface maximum which maximize the joint probability distribution, $p(x_1, x_2 ; m)$ of the sample point (x_1', x_2'). Since the Gaussian surface has rotational symmetry, the coordinates of the surface maximum which maximize $p(x_1', x_2' ; m)$ will lie, as shown in Fig. 5.2, at the intersection of the normal drawn from x_1', x_2' to the line $x_1 = x_2$.

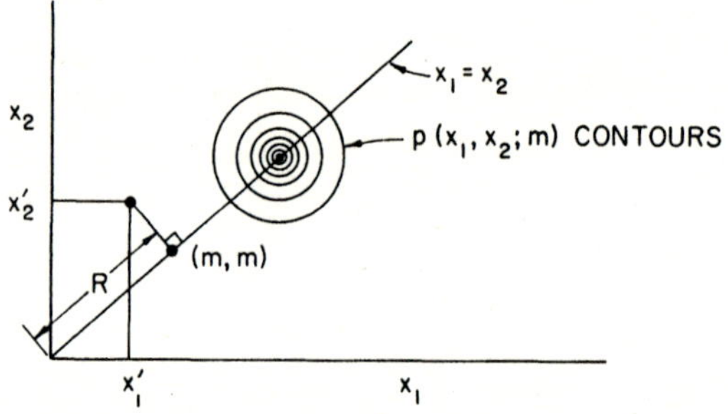

FIG. 5.2   A graphic illustration of a two-sample maximum-likelihood estimation.

Now, it is possible to express the coordinates of the surface maximum which maximize $p(x_1', x_2' ; m)$ in terms of the sample (x_1', x_2'). To do this it is simply necessary to express the length of the line $x_1 = x_2$ to the point (m, m) in terms of x_1' and x_2' and alternately in terms of m. The estimator is obtained by equating these lengths. The length of this line in terms of x_1' and x_2' is

$$R = \frac{1}{\sqrt{2}}(x_1' + x_2') \tag{5-61}$$

where the factor $1/\sqrt{2}$ equals the direction cosines of the line. Alternately, the length of the line in terms of m is

$$R = \sqrt{2}\,m \tag{5-62}$$

Therefore, the estimate of the mean is

$$\hat{m} = \frac{x'_1 + x'_2}{2} \tag{5-63}$$

This result, of course, is well known. The error variance of this estimate of the mean is easily found to be $\sigma^2/2$.

Note for this particular example, the method of "least squares" will also yield the same estimate of the mean. The estimate of the mean based on the method of least squares is derived by minimizing the distance between the sample point (x'_1, x'_2) and the coordinates of the surface maximum (m, m). The two methods of estimation, however, will not always yield the same estimators; this depends on the structure of the joint probability.

To complete this graphic illustration of the method of maximum likelihood the number of samples will be increased to n values. For this case the constraint on the location of the hypersurface maximum, for the corresponding multidimensional Gaussian probability distribution, will be $x_1 = x_2 = \cdots = x_n$. Accordingly, the length of this multidimensional line will be

$$R = \frac{1}{\sqrt{n}}[x'_1 + x'_2 + \cdots + x'_n] \tag{5-64}$$

where, as before, $1/\sqrt{n}$ is equal to the direction cosines with respect to the line. Alternately,

$$R = \sqrt{n}\, m \tag{5-65}$$

Therefore,

$$\hat{m} = \frac{x'_1 + x'_2 + \cdots + x'_n}{n} \tag{5-66}$$

and the error variance is readily determined to be σ^2/n.

Before continuing to the specific radar problem, a formal derivation for this particular estimator will be given. However, in order to align this derivation with the following discussion, the statement of the problem will be rephrased. It will now be supposed that it is desired to estimate the unknown level of a constant voltage which appears in the presence of band-limited white Gaussian noise. It is seen that the two problems are equivalent. For this case, however, the n samples will consist of independent observations of a time series.[1]

Let V denote the unknown level of the constant voltage; then for independent observations of the time series, denoted by $\bar{x} = x_1, x_2, \ldots, x_i, \ldots, x_n$ and taken in the sequence that they appear, the likelihood function

---

[1] A time series is defined as one manifestation of a stochastic process.

is

$$L(\bar{x}; V) = \prod_{i=1}^{2WT} \frac{1}{\sqrt{2\pi}\,\sigma} \exp\left[-\frac{(x_i - V)^2}{2\sigma^2}\right] \qquad (5\text{-}67)$$

where $\sigma^2 = N_0 W$ and N_0 and W are, respectively, the noise power density and bandwidth. By taking $\log L$ and its derivative with respect to V, in sequence, this yields

$$\frac{\partial \log L}{\partial \theta} = \frac{1}{\sigma^2} \sum_{i=1}^{2WT} (x_i - V) \qquad (5\text{-}68)$$

By setting this equation equal to zero and solving for V, the estimator is obtained. Thus

$$\hat{V} = \frac{1}{2WT} \sum_{i=1}^{2WT} x_i \qquad (5\text{-}69)$$

The variance of this estimator is readily found to be

$$E\{(\hat{V} - V)^2\} = \frac{\sigma^2}{n} = \frac{N_0}{2T} \qquad (5\text{-}70)$$

Now, returning to the specific radar estimation problem, it will be assumed that the voltage V is some given function of time which is completely known except for its time and frequency positions. This voltage will be denoted by $s(t; \theta_1, \theta_2)$ and said to represent the radar echo signal. As in the previous section, the complex representation for $s(t; \theta_1, \theta_2)$ will be adopted later for convenience.

As viewed graphically, the raw input data yields a sample point in $2WT$-dimensional space. And, for the case of a single unknown parameter, the locus of points which constitute the constraint on the hypersurface maximum form a multidimensional arc. Each point on this arc represents a unique parametric value. For the case of two parameters the locus of points which constitute the constraint on the hypersurface maximum generate a surface. Each point on this surface represents a unique set of parametric values. It is possible to extend this concept to any number of unknown parameters.

Unlike the previous case, the estimators for the radar parameters cannot be derived explicitly in terms of the $2WT$ samples of the raw input data. The estimates for these parameters are obtained by noting the position (along an arc, a surface or hypersurface, as the case may be) of the maximum of the multidimensional Gaussian distribution that maximizes the likelihood function. In essence, it is necessary to correlate the set of samples taken from the raw input data with the different sample sets taken from all possible echo signals in the absence of noise.

The argument given in the last paragraph constitutes a heuristic justification for matched filtering since matched-filter processing is equivalent to correlation processing (see Chapter 2). In a manner similar to the examples given above, it is also possible to formally derive this result. In terms of the complex notation which was introduced in Section 5.6, the likelihood function is restated from (5-20):

$$L(\bar{x}; \theta_1, \theta_2) = \left[\frac{1}{2\pi N}\right]^{WT} \exp\left[-\frac{1}{2N}\right] \sum_{i=1}^{WT} |\gamma_i - \psi_i(\theta_1, \theta_2)|^2 \qquad (5\text{-}71)$$

With regard to (5-71), it is desired to determine the set of parameters θ_1 and θ_2, which represent the time and frequency position of the radar echo signal, that maximize $L(\bar{x}; \theta_1, \theta_2)$. As previously noted, insofar as the location of the maximum of $L(\bar{x}; \theta_1, \theta_2)$ with respect to θ_1 and θ_2 is concerned, $\log L$ is an equivalent form. Thus

$$\log L = -WT \log 2\pi N - \frac{1}{2N} \sum_{i=1}^{WT} |\gamma_i - \psi_i(\theta_1, \theta_2)|^2 \qquad (5\text{-}72)$$

The first term on the right side of (5-72) is independent of the parameters θ_1 and θ_2, and therefore it can be ignored. Moreover, the second term on the right side of (5-72) can be expanded and yields

$$\sum_{i=1}^{WT} |\gamma_i - \psi_i(\theta_1, \theta_2)|^2 = \frac{1}{2N} \sum_{i=1}^{WT} |\gamma_i|^2 + \frac{1}{2N} \sum_{i=1}^{WT} |\psi_i(\theta_1, \theta_2)|^2$$
$$- \frac{2\,\mathrm{Re}}{2N} \sum_{i=1}^{WT} \gamma_i \psi_i^*(\theta_1, \theta_2) \qquad (5\text{-}73)$$

The first two terms on the right side of (5-73) are independent of the parameters, and can also be ignored. Attention is therefore drawn to the last term. Clearly, when all signs in (5-73) are accounted for, the likekihood function will be maximized when this term is maximized with respect to the parameters θ_1 and θ_2. The mathematical operation indicated by this term can be recognized as a correlation process between the raw input data γ_i and the waveform ψ_i which is varied as a function of θ_1 and θ_2. Hence, the formal justification for matched filtering has been obtained.

This discussion has ignored the fact that the amplitude as well as the phase of the radar signal are usually not known at the receiver. Since these parameters furnish no useful information about the position of a target, but can influence the accuracy of the estimates of the important parameters, it is consequently necessary to account for their presence by including them in the initial formulation of the likelihood function. The inclusion of these "stray" parameters will naturally affect the processing that is derived. It has been shown that the effect on the processing previously discussed is to require that the output of the matched filter pass through a square law envelope detector [9, 10, 14–16]. The maximum of this output, whose form is given by

$$\left| \frac{1}{N} \sum_{i=1}^{WT} \gamma_i \psi_i^*(\theta_1, \theta_2) \right|^2 \tag{5-74}$$

is then used to locate the estimates of θ_1 and θ_2. Based on this method of estimation, a linear approximation of the estimation error variances has been derived which is valid for large signal-to-noise ratios [9, 10]. This result can be expressed in terms of the ambiguity function as follows:

$$E\{(\hat{\theta}_j - \theta_j)(\hat{\theta}_k - \theta_k)\} = M_{jk} \tag{5-75}$$

where M_{jk} represents the elements of the error moment matrix and these can be determined from the elements of the inverse moment matrix

$$[M^{-1}]_{jk} = -\frac{2E}{4\pi^2 N_0} \frac{\partial^2}{\partial \theta_j \, \partial \theta_k} |\chi(\tau, \phi)|^2_{\tau, \phi = 0} \tag{5-76}$$

where the index jk denotes the different combinations of τ and ϕ. A straightforward substitution for $\chi(\tau, \phi)$ which is given by

$$\chi(\tau, \phi) = \int_{-\infty}^{\infty} u(t) u^*(t + \tau) \exp[-j2\pi\phi t] \, dt \tag{5-77}$$

where $\psi(t) = u(t) \exp[j2\pi f_0 t]$ and $u(t) = a(t) \exp[j\theta(t)]$, will yield the same results listed at the end of the last section. It is readily seen that the terms in the inverse of the error moment matrix are the coefficients of the second order terms of a Taylor series expansion of $\chi(\tau, \phi)$ [14]. These were derived in Section 4.9.

REFERENCES

1. C. W. Helstrom, "Statistical Theory of Detection." Pergamon Press, Oxford, 1960.
2. D. Middleton, "Introduction to Statistical Communications." McGraw-Hill, New York, 1960.
3. P. M. Woodward and I. L. Davies, A theory of radar information, *Phil. Mag.* [7] **41**, 1001–1017 (1950).
4. P. M. Woodward, "Probability and Information Theory with Applications to Radar." Pergamon Press, Oxford, 1953.
5. H. Cramér, "Mathematical Methods of Statistics." Princeton Univ. Press, Princeton, New Jersey, 1946.
6. For a complete list of papers by R. A. Fisher, the reader should refer to the bibliography in Cramér [5].
7. C. A. Shannon, Communications in the presence of noise, *Proc. IRE* **37**, 10–21 (1949).
8. W. B. Davenport and W. L. Root, "Random Signals and Noise." McGraw-Hill, New York, 1948.
9. E. J. Kelly, I. S. Reed, and W. L. Root, The detection of radar echoes in noise, Parts I and II, *J. Soc. Indust. Appl. Math.* **8**, 309–341, 481–507 (1960).

10. E. J. Kelly, The radar measurement of range velocity and acceleration, *IRE Trans*. **MIL-5**, 51–57 (1961).
11. D. Gabor, Theory of communications, *J. Inst. Elec. Engrs.* (*London*) **93**, Part III, 429–457 (1946).
12. D. Slepian, Estimation of signal parameters in the presence of noise, *IRE Trans*. **IT-3**, 68–89 (1954).
13. J. Dugundji, Envelopes and pre-envelopes of real waveforms, *IRE Trans*, **IT-4**, 53–57 (1958).
14. R. Manasse, Range and velocity accuracy from radar measurements, Mass. Inst. Technol., Lincoln Lab., Lexington, Massachusetts, Group Rept. 312-26 (February, 1955).
15. P. Swerling, Parameter estimation accuracy formulas, *IRE Trans*, **IT-10**, 302–314 (1964).
16. P. Bello, Joint estimation of delay, Doppler, and Doppler rate, *IRE Trans*. **IT-6**, 330–341 (1960).

CHAPTER 6

The Linear FM Waveform and Matched Filter

6.1 Introduction

Chapter 1 presented the intuitive relationships between a linear frequency modulated waveform and the matched filter necessary to obtain the pulse-compression phenomenon. The principle of stationary phase was developed in Chapter 3 as a tool to extend the intuitive reasoning to cover a broader class of coded waveform functions described by

$$s(t) = a(t) \cos[\omega_0 t + \theta(t)] \qquad (6\text{-}1)$$

In this and the following chapter, detailed topics associated with the linear FM pulse-compression signal will be developed in the context of its application to radar systems. The purpose of this development is twofold:

(1) The linear FM signal is applicable to a majority of system applications, having a number of useful characteristics (including ease of implementation), and thus is important in itself.

(2) A thorough understanding of the property of this matched-filter signal provides a basis for developing and comparing the properties of more complex matched-filter waveforms.

Thus, the linear FM matched-filter waveform will be considered as a representative useful function of the general class of signals defined by Eq. (6-1).

6.2 Linear FM Matched-Filter Waveform

In the discussion of the matched-filter concept in Chapter 1 two basic aspects of the characteristics of a matched filter were given:

(i) If the signal spectrum is described by $S(\omega)$, then the frequency response function, $H(\omega)$, of a filter that results in a maximum signal-to-noise ratio at the filter output is the complex conjugate of the spectrum function, or $H(\omega) = S^*(\omega)$.

(ii) If the signal waveform is defined by $s(t)$, the impulse response of the filter meeting the condition of (i) is $h(t) = ks(-t)$, and the filter output waveform is found by performing the operation

$$g(t) = k \int_{-\infty}^{\infty} s(\tau)s(t - \tau)\, d\tau$$

The relation of (i) is useful in defining the properties of the matched filter, while (ii) is useful in deriving the output waveform that results from matched filtering.

With the application to radar in mind, it will be assumed that the envelope $a(t)$ is limited to a flat-topped rectangular function. Under this condition the signal formed at the radar transmitter is [1]

$$s(t) = \cos\left[\omega_0 t + \frac{\mu t^2}{2}\right], \qquad -\frac{T}{2} < t < \frac{T}{2} \tag{6-2}$$

The impulse response of the filter matched to this signal is

$$h(t) = k \cos\left[\omega_0 t - \frac{\mu t^2}{2}\right], \qquad -\frac{T}{2} < t < \frac{T}{2} \tag{6-3}$$

By imposing the condition that the matched filter should exhibit unity gain at ω_0 the matched-filter constant is

$$k = \sqrt{\frac{2\mu}{\pi}} \tag{6-4}$$

and the matched-filter output signal is

$$g(t) = \sqrt{\frac{2\mu}{\pi}} \int_{-T/2}^{T/2} \cos\left[\omega_0 \tau + \frac{\mu \tau^2}{2}\right] \cos\left[\omega_0(t - \tau) - \frac{\mu(t - \tau)^2}{2}\right] d\tau \tag{6-5}$$

The integral expression (6-5) represents the optimum output, in the signal-to-noise sense, from the matched filter. In many applications the input to a radar receiver will not be an exact replica of the transmitted signal, having undergone various types of distortion. The most common form of distortion results when the signal is reflected from an object that has a radial velocity component relative to the radar system. This results in a center frequency Doppler shift of the reflected signal that is expressed as

$$\omega_d = \frac{2v}{c}\omega_0 \qquad \text{or} \qquad \phi = \frac{2v}{c}f_0 \tag{6-6}$$

where v is the radial velocity, c the velocity of light, and $\phi = \omega_d/2\pi$. The assumption will be made that this representation of the Doppler shift is the only velocity distortion effect present in the linear FM signal. This ignores second order effects, but is a very good approximation for time-bandwidth products, or compression ratios, up to 10,000. Proceeding on this basis, the general output of the matched filter becomes

$$g(t, \omega_d) = \sqrt{\frac{2\mu}{\pi}} \int_a^b \cos\left[(\omega_0 + \omega_d)\tau + \frac{\mu\tau^2}{2}\right] \cos\left[\omega_0(t - \tau) - \frac{\mu(t - \tau)^2}{2}\right] d\tau \tag{6-7}$$

where

$$a = -\frac{T}{2} + t, \qquad b = \frac{T}{2}, \qquad t > 0$$

and

$$a = -\frac{T}{2}, \qquad b = \frac{T}{2} + t, \qquad t < 0$$

When $\omega_d = 0$, $s(\tau)$ and $h(\tau)$ are "matched" and $g(t)$ is the autocorrelation function of the input signal. When $\omega_d \neq 0$, then $g(t, \omega_d)$ is the crosscorrelation of the two functions, representing all possible matched-filter outputs for the moving object condition. By making use of the trigonometric expansion for $\cos(a)\cos(b)$, and noting that this results in a high frequency term $(2\omega_0)$ when applied to Eq. (6-7) that can be ignored for most cases of practical interest, the matched-filter output becomes

$$g(t, \omega_d) = \frac{1}{2}\sqrt{\frac{2\mu}{\pi}} \int_a^b \cos\left[\omega_0 t + \omega_d\tau + \mu\tau t - \frac{\mu t^2}{2}\right] d\tau \tag{6-8}$$

$$= \frac{1}{2}\sqrt{\frac{2\mu}{\pi}} \left[\frac{\sin\left[\omega_0 t + \omega_d\tau + \mu\tau t - \frac{\mu t^2}{2}\right]}{\omega_d + \mu t}\right]_{\tau=a}^{\tau=b} \tag{6-9}$$

Evaluating Eq. (6-9) for the case of $t > 0$ the following is obtained:

$$g(t, \omega_d) = \frac{1}{2}\sqrt{\frac{2\mu}{\pi}} \left[\frac{\sin\left[\omega_0 t + \frac{\omega_d T}{2} + \frac{\mu t}{2}(T - t)\right]}{\omega_d + \mu t}\right.$$

$$\left. - \frac{\sin\left[\omega_0 t + \omega_d t - \frac{\omega_d T}{2} - \frac{\mu t}{2}(T - t)\right]}{\omega_d + \mu t}\right] \tag{6-10}$$

If $\omega_d t/2$ is appropriately added and subtracted in the argument of the first term, then one obtains

$$g(t, \omega_d) = \frac{1}{2}\sqrt{\frac{2\mu}{\pi}}\left[\frac{\sin\left[\left(\omega_0 + \frac{\omega_d}{2}\right)t + \frac{\omega_d}{2}(T-t) + \frac{\mu t}{2}(T-t)\right]}{\omega_d + \mu t}\right.$$

$$\left. -\frac{\sin\left[\left(\omega_0 + \frac{\omega_d}{2}\right)t - \frac{\omega_d}{2}(T-t) - \frac{\mu t}{2}(T-t)\right]}{\omega_d + \mu t}\right] \quad (6\text{-}11)$$

This has the form

$$\sin(\alpha + \beta) - \sin(\alpha - \beta) = 2\cos\alpha\sin\beta$$

where

$$\alpha = \left(\omega_0 + \frac{\omega_d}{2}\right)t, \qquad \beta = \frac{\omega_d + \mu t}{2}(T-t)$$

A similar expression is obtained for $t < 0$, and when the combined results are considered the matched-filter output waveform is

$$g(t, \omega_d) = \sqrt{\frac{2\mu}{\pi}}\,\frac{\sin\left[\frac{\omega_d + \mu t}{2}(T - |t|)\right]}{\omega_d + \mu t}\cos\left(\omega_0 + \frac{\omega_d}{2}\right)t, \qquad -T < t < T \quad (6\text{-}12)$$

For the special case of $\omega_d = 0$ Eq. (6-12) reduces to the matched-filter auto-correlation function, or

$$g(t) = \sqrt{\frac{2\mu}{\pi}}\,\frac{\sin\frac{\mu t}{2}(T - |t|)}{\mu t}\cos\omega_0 t, \qquad -T < t < T \quad (6\text{-}13)$$

It is of interest to note that the frequency shift of the output function is $\omega_d/2$ for an input frequency shift of ω_d. This can be attributed to the essentially rectangular bandlimiting characteristic of the matched filter (as shown in the next section) that controls the centroid of the output spectrum as illustrated in Fig. 6.1. Figures 6.2 and 6.3 plot several of the matched-filter output time functions in which can be noted the effect of compression ratio $(T\,\Delta f)$ on the autocorrelation waveform and the distortion effects caused by the Doppler shift conditions. These are a time shift of the waveform, loss in peak amplitude and spreading of the pulse width. For $\omega_d < \Delta\omega/2$ a good approximation for the value of the time shift is

$$t_s = -\frac{\omega_d}{\mu} = -\frac{\omega_d}{\Delta\omega}T$$

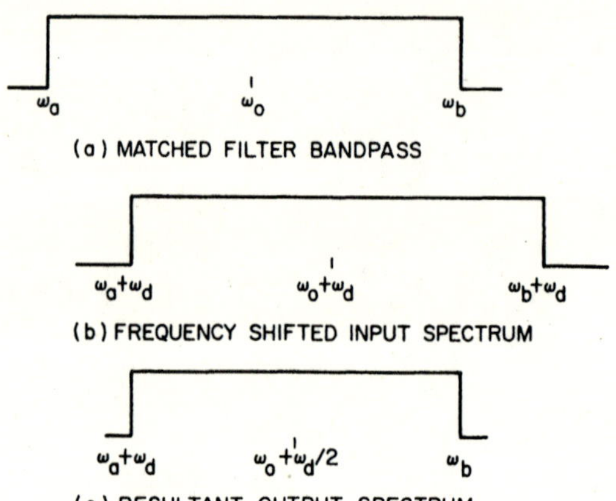

FIG. 6.1   Output spectrum modified by matched-filter bandpass.

Figure 6.4 combines the Doppler outputs in a three-dimensional model (discussed below) and Fig. 6.5 plots the time shift and amplitude fall-off as functions of $\omega_d/\Delta\omega$ or $\phi/\Delta f$.

It is observed that the Doppler-shifted matched-filter outputs are bounded by the triangle function having a baseline extent from $-T$ to T. This function is the autocorrelation function of the rectangular envelope of the matched-filter input, and the observed relationship is a property of the linear FM signal. It can be shown for a general envelope, $a(t)$, and linear frequency modulation, that the bound of the matched-filter outputs is the autocorrelation function of $a(t)$. For an envelope function truncated at $-T/2$ and $T/2$ the ideal matched-filter output has zero output

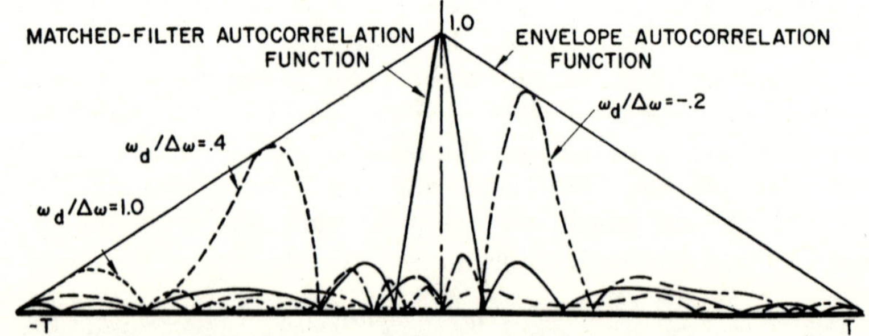

FIG. 6.2   Matched-filter output waveforms for frequency shifted inputs (from Cook [1]).

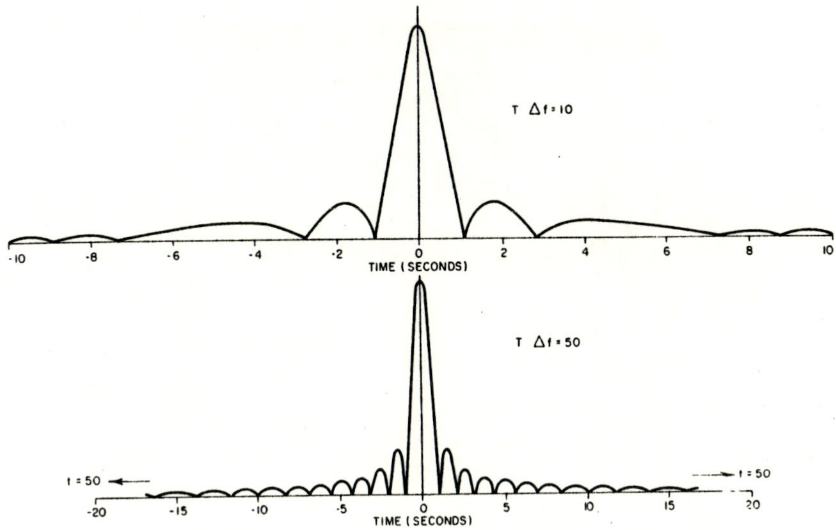

FIG. 6.3   Linear FM matched-filter autocorrelation waveforms for $T\,\Delta f = 10$ and $T\,\Delta f = 50$ (Δf normalized to unity).

for time values less than $-T$ and greater than T, that is $g(t, \omega_d) = 0$ for $|t| > T$.

In many instances the expected range of Doppler shifts is a very small percentage of the signal bandwidth. In this situation there is very little deterioration in the peak amplitude of the matched filter output. Thus, when there is a nominal range of Doppler shifts the detection capability of the linear FM matched-filter signal suffers negligible degradation. Equation (6-12) constitutes the so-called response function for the linear FM pulse-compression matched-filter waveform. The response function was introduced in Chapter 3 as describing the Doppler response characteristic of a matched filter. Figure 6.4 illustrates this function for the linear FM signal as a composite three-dimensional model of the Doppler-shifted output signals. The "ridge-line" characteristic of this function illustrates the "ambiguity" involved in determining, on a one pulse basis, target range and velocity without a prior knowledge of one or the other. Each slice of the response function model represents the matched-filter output, $g(t, \omega_d)$, for an incremental change of ω_d that is 10% of the radian signal bandwidth in magnitude, with the center being the autocorrelation function (i.e., $\omega_d = 0$). The deeper implications of the uncertainty or response function, which can be considered as *one* of several bases for applying a value judgement to a particular waveform, were examined in Chapter 4.

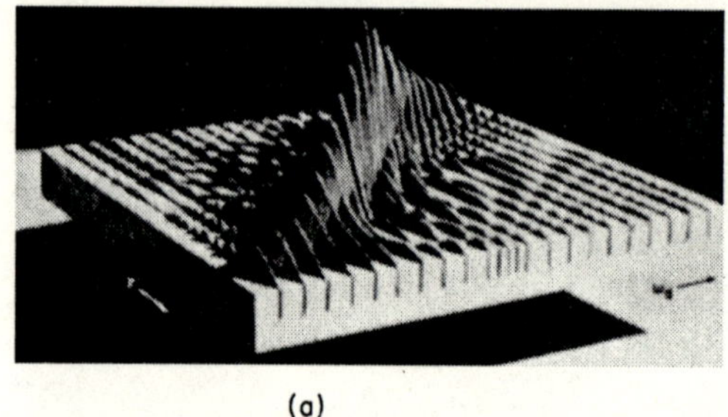

(a)

FIG. 6.4   Calculated response surface for a linear FM signal.

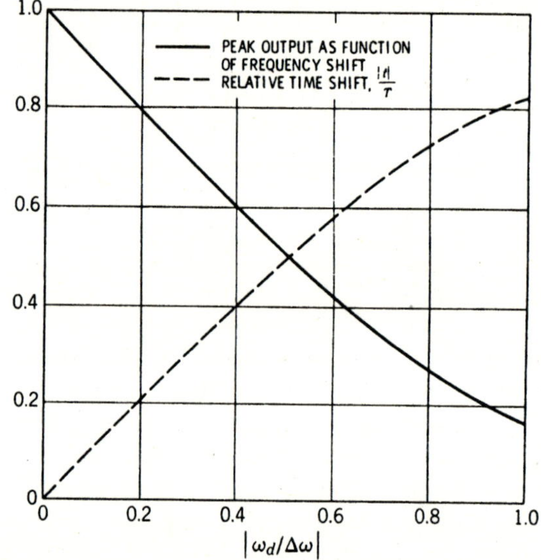

FIG. 6.5   Linear FM matched-filter output parameters. (From "Modern Radar—Analysis, Evaluation, and System Design" (R. S. Berkowitz, ed.). Wiley, New York, 1965.)

6.3   Linear FM Matched-Filter Characteristics

The analysis of the matched-filter condition for the linear FM signal leads directly to the time waveform relationship for the matched-filter input and output. This analysis tells little of a direct nature about the matched filter itself that has meaning for most engineers. For an understanding of the necessary filter characteristics it is more helpful to return to the basic relationship developed by North. Thus, the matched filter is related to the signal by nature of its transfer function, $H(\omega)$, being the complex conjugate of the signal spectrum. This tells us that once one knows all there is to know about the (ideal) signal spectrum then one will also know all there is to know about the (ideal) matched-filter amplitude and phase characteristics.

The spectrum of the linear FM signal is expressed by [2–4]

$$S(\omega) = \int_{-T/2}^{T/2} \cos\left[\omega_0 t + \frac{\mu t^2}{2}\right] \exp[-j\omega t]\, dt$$

$$= \frac{1}{2}\int_{-T/2}^{T/2} \exp\left[j\left\{(\omega_0 - \omega)t + \frac{\mu t^2}{2}\right\}\right] dt \qquad (6\text{-}14)$$

$$+ \frac{1}{2}\int_{-T/2}^{T/2} \exp\left[-j\left\{(\omega_0 + \omega)t + \frac{\mu t^2}{2}\right\}\right] dt$$

The second integral defines the spectrum at negative frequencies and has a negligible contribution at positive frequencies if the ratio of center frequency to bandwidth is large. The essential assumption here is that the tail of the negative frequency spectrum has an infinitesimally small value in the region of the positive spectrum. This is a realistic assumption for the ratio $\omega_0/\Delta\omega$ sufficiently large, as it would be in a normal receiver implementation. By completing the square of the bracketed term in Eq. (6-14) the following expression for the spectrum is obtained:

$$S(\omega) = \frac{1}{2}\exp\left[-j\left(\frac{(\omega - \omega_0)^2}{2\mu}\right)\right]\int_{-T/2}^{T/2} \exp\left[j\frac{\mu}{2}\left(t - \frac{\omega - \omega_0}{\mu}\right)^2\right] dt \quad (6\text{-}15)$$

The variable in Eq. (6-15) can be changed by letting

$$\sqrt{\mu}\left(t - \frac{\omega - \omega_0}{\mu}\right) = \sqrt{\pi}x \qquad (6\text{-}16)$$

so that

$$dt = \sqrt{\frac{\pi}{\mu}}\, dx$$

and the spectrum becomes

$$S(\omega) = \frac{1}{2}\sqrt{\frac{\pi}{\mu}}\, \exp\left[-j\left(\frac{(\omega - \omega_0)^2}{2\mu}\right)\right]\int_{-X_1}^{X_2} \exp\left[j\frac{\pi x^2}{2}\right] dx \qquad (6\text{-}17)$$

where

$$X_1 = \frac{\dfrac{\mu T}{2} + (\omega - \omega_0)}{\sqrt{\pi\mu}}, \quad X_2 = \frac{\dfrac{\mu T}{2} - (\omega - \omega_0)}{\sqrt{\pi\mu}}$$

This yields

$$S(\omega) = \frac{1}{2}\sqrt{\frac{\pi}{\mu}}\,\exp\left[-j\left(\frac{(\omega - \omega_0)^2}{2\mu}\right)\right][C(X_1) + jS(X_1) + C(X_2) + jS(X_2)]$$

$$(6\text{-}18)$$

where

$$C(X) = \int_0^X \cos\frac{\pi y^2}{2}\,dy \quad \text{and} \quad S(X) = \int_0^X \sin\frac{\pi y^2}{2}\,dy$$

are the Fresnel integrals, which have the property $C(-X) = -C(X)$ and $S(-X) = -S(X)$.[1]

It is helpful to consider the linear FM spectrum as having three major components, as given below.

Amplitude Term[2]

$$|S(\omega)| = \frac{1}{2}\sqrt{\frac{\pi}{\mu}}\left\{[C(X_1) + C(X_2)]^2 + [S(X_1) + S(X_2)]^2\right\}^{1/2} \qquad (6\text{-}19)$$

Square-Law Phase Term

$$-\Phi_1(\omega) = \frac{(\omega - \omega_0)^2}{2\mu} \qquad (6\text{-}20)$$

Residual Phase Term

$$\Phi_2(\omega) = \tan^{-1}\left[\frac{S(X_1) + S(X_2)}{C(X_1) + C(X_2)}\right] \qquad (6\text{-}21)$$

The residual phase term, Φ_2, is so designated because it is generally ignored in the design of a matched filter. For large values of $T\,\Delta f$ the ratio

$$\frac{S(X_1) + S(X_2)}{C(X_1) + C(X_2)} \doteq 1$$

---

[1] The Fresnel integrals can be found in tabulated form in several mathematical reference works [5, 6].

[2] At $\omega = \omega_0$ the combination of Fresnel integrals in (6–19) approximately equals $\sqrt{2}$, and $|S(\omega_0)| = \sqrt{\pi/2\mu}$. The unity gain factor of the matched filter given by (6–4) is obtained from this.

over the significant interval of the variables, and Φ_2 approximates a constant phase angle, $\pi/4$, in this region. Figure 6.6 plots functions for $|S(\omega)|$ and Φ_2 for several values of compression ratio. It is of interest to note that by making the substitutions

$$\mu = \frac{\Delta\omega}{T}, \qquad \Delta\omega = \frac{2\pi}{\tau}, \qquad \omega - \omega_0 = \frac{n\,\Delta\omega}{2}, \tag{6-22}$$

the argument of the Fresnel integrals becomes

$$\sqrt{\frac{T}{\tau}}\left(\frac{1 \pm n}{\sqrt{2}}\right) = \sqrt{T\,\Delta f}\left(\frac{1 \pm n}{\sqrt{2}}\right) \tag{6-23}$$

where n is now a normalized frequency parameter. The spectra are thus functions of the compression ratio, T/τ, independent of any particular values of center frequency and bandwidth.

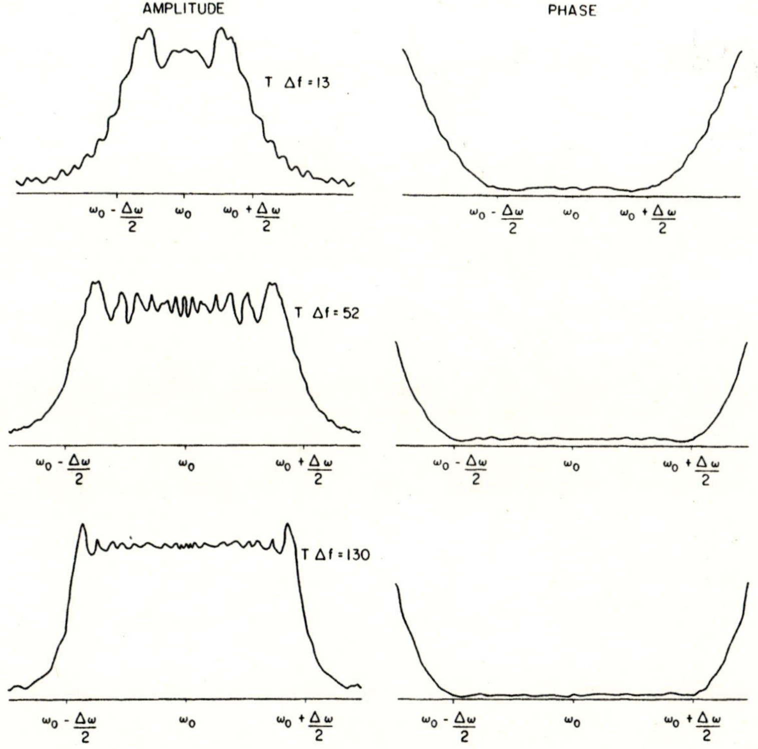

FIG. 6.6   Linear FM spectra after removal of square law phase term (from Cook [3]).

Although some broad assumptions can be made concerning the spectrum parameters, as outlined in the next section, the detailed nature of the spectrum becomes important in a number of practical applications to radar. Consideration of this is reserved to Chapter 7.

6.4   The Ideal vs the Practical Matched Filter

Ideally, the attempt to construct a matched filter for the rectangular envelope, linear FM signal should be based on the relationship previously described of

$$H(\omega) = S^*(\omega)$$

where the components of $S(\omega)$ are as given in Eqs. (6-19)–(6-21). It has already been stated that the phase term Φ_2 is neglected in the design problem. A further assumption is that $|S(\omega)|$ can be approximated by a rectangular distribution for large compression ratios (i.e., $T\Delta f > 30$). On this basis the design of a matched filter reduces to constructing an essentially rectangular bandpass (B.W. $= \Delta f$) and a conjugate phase function to

$$\Phi_1(\omega) = \frac{(\omega - \omega_0)^2}{2\mu} \tag{6-24}$$

The matched filter would then be required to have a square-law phase characteristic

$$-\Phi_1(\omega) = \beta_f(\omega) = -\frac{(\omega - \omega_0)^2}{2\mu} \tag{6-25}$$

or a time-delay function

$$T_d(\omega) = \frac{d\beta_f}{d\omega} = -\frac{(\omega - \omega_0)}{\mu} \tag{6-26}$$

In order to achieve a realizable delay a fixed constant is added to that of Eq. (6-26) so that positive time delay results over the major portion of the bandwidth. Thus

$$T_d(\omega) = -\frac{\omega - \omega_0}{\mu} + k \tag{6-27}$$

Figure 6.7 plots the delay function given by (6-27) and the type of delay curve that can be achieved in practice. (Time delay curves that vary as a function of frequency are designated as dispersive delay functions. This topic is treated in more detail in Chapters 12 and 13.) Since a realizable

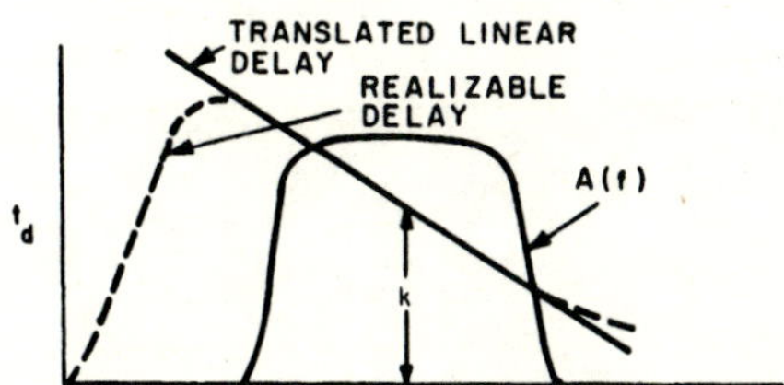

FIG. 6.7 Realizable linear FM matched-filter characteristics.

delay line must have zero group delay at zero frequency, the delay-line function that can be obtained in practice will be several steps removed from that of the ideal matched filter. However, this is not restrictive, as the waveforms of Fig. 6.8 show. The matched-filter output signal has basically the nearly $(\sin x)/x$ structure given by Eq. (6-12). Note also the effect of adding the constant delay, k, to the linear delay function. Figure 6.9 shows the effect of simulating the Doppler condition by shifting the center frequency of the linear FM input signal. The significant departure from the waveforms plotted from Eq. (6-12) are seen in the behavior of the range

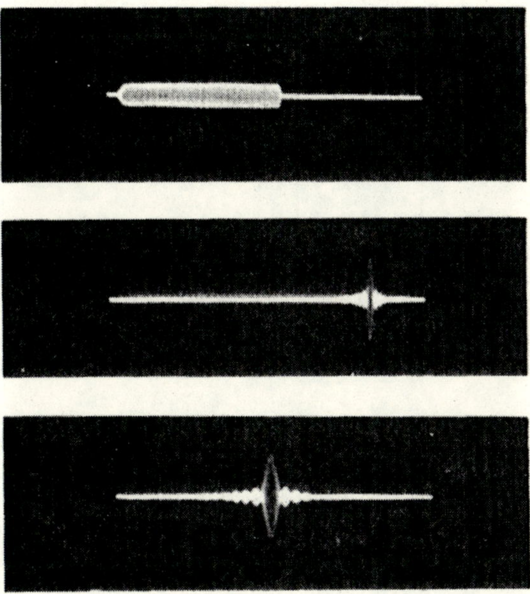

FIG. 6.8 Linear FM matched-filter waveforms. Top: Linear FM input; Middle: Matched-filter output; Bottom: Matched-filter output (Expanded scale).

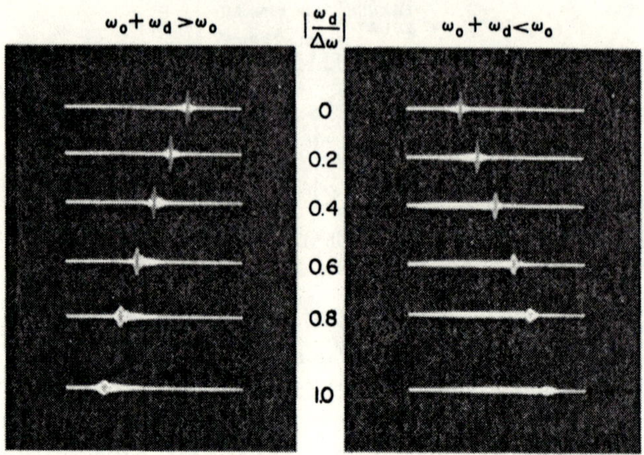

FIG. 6.9   Linear FM matched-filter outputs for frequency shifts, ω_d. (From "Modern Radar Analysis, Evaluation, and System Design" (R. S. Berkowitz, ed.). Wiley, New York, 1965.)

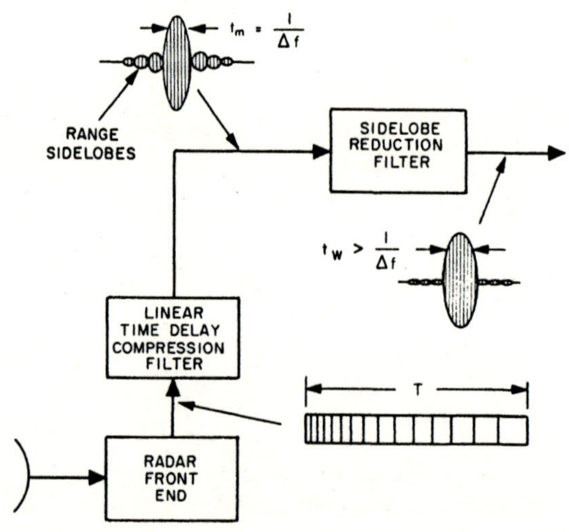

FIG. 6.10   Matched-filter sidelobe suppression.

sidelobes. From this type of waveform evidence the conclusion can be drawn that a good approximation to the linear FM pulse-compression matched filter can be achieved on the basis of the assumptions of a rectangular bandpass and a linear delay.

In many applications the presence of the large range sidelobes of the linear FM filter output limits the capability of a radar to examine a multiple

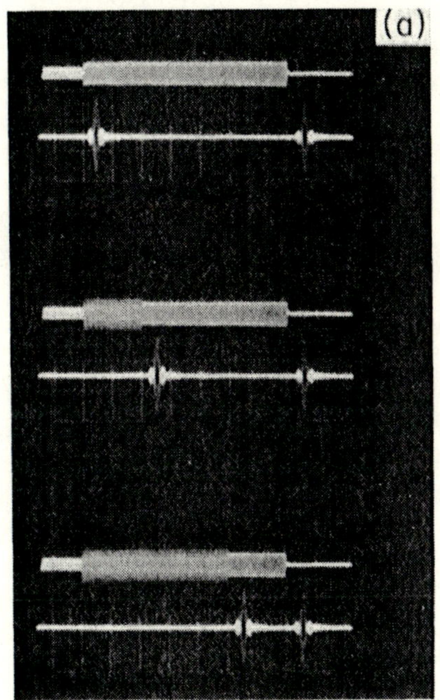
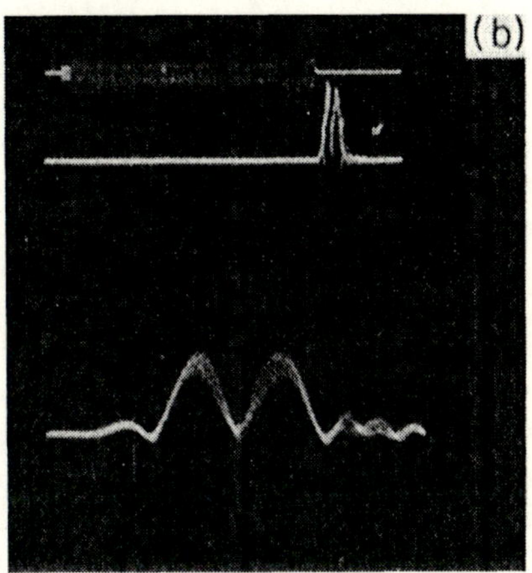

FIG. 6.11   Pulse-compression resolution capability. (a) Matched-filter output signals as overlap at input increases. (b) Minimum resolution of matched-filter waveform.

target environment over any extended dynamic range. In this situation a sidelobe suppression mismatch network is added to the matched filter, as shown in Fig. 6.10. Under this type of operation the assumption of a rectangular spectrum distribution is not justified in all cases, as is shown in the next chapter.

The application of the pulse-compression technique to radar systems depends on its functioning as a linear processing operation. Evidence of this is shown in Fig. 6.11 which shows that the pulse-compression matched filter provides a resolution among multiple signals on the basis of the compressed-pulse widths rather than the input-pulse widths. This demonstrates the superposition principle, which is one of the basic assumptions in a linear system.

6.5   Generating the Linear FM Matched-Filter Signal

There are two basic approaches to generation of the linear FM signal that is to be processed at the receiver with the linear delay matched filter. One approach is designated as active signal generation, since it is based on controlling the frequency of an oscillator with a voltage derived from an appropriate function generator. The alternate approach is designated as passive generation of the frequency modulated signal, since it is based on exciting a conjugate matched-filter network with an impulse. With this latter method the output signal of the conjugate filter can be considered as an impulse response function that is matched to the characteristic of the filter at the receiver. Several of these methods of signal formation will be described for the linear FM signal, although the techniques, with some modification of details, are applicable to the generation of arbitrary frequency modulated pulse-compression matched-filter signals.

Figure 6.12 shows two techniques for active generation of a linear FM signal. The frequency of the voltage controlled oscillator in Fig. 6.12(a) is linearly proportional to the voltage present on the control element of the oscillator, so that a linear ramp function will cause the necessary linear progression of frequency vs time. In this application the oscillator is free running, and gating circuits at the oscillator are required. If the linear FM is required to be free of any distortion, then the frequency excursion of the oscillator can be restricted to the most linear portion of the frequency-voltage characteristic. If a larger deviation is then required, frequency

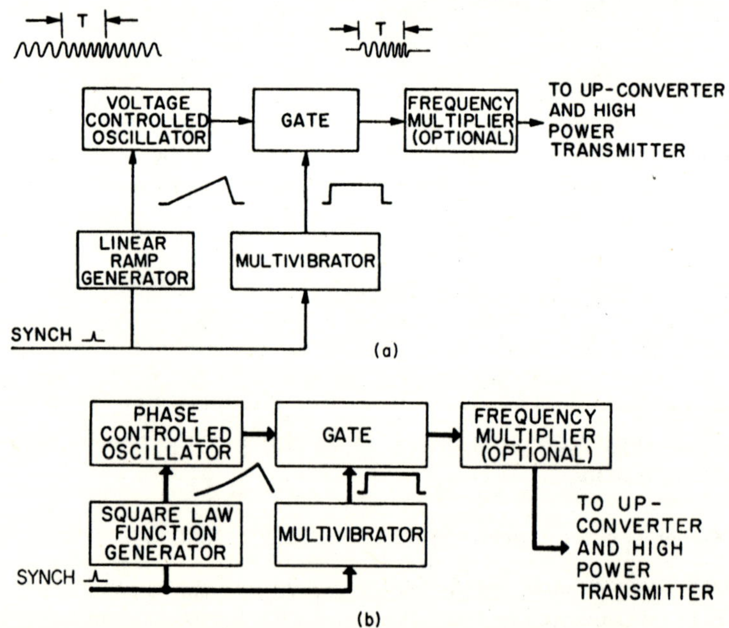

FIG. 6.12   Techniques for active generation of linear FM signal. (a) Frequency modulation. (b) Phase modulation. (From Bernfeld *et al.*, Ref. 4 of Chapter 1.)

multipliers can be used to build up the frequency deviation before transmission. Devices such as reactance tube modulators, klystrons, backward-wave oscillators, etc., lend themselves to this type of signal generation.

Figure 6.12(b) outlines an alternate active modulation technique in which the frequency of the controlled oscillator is proportional to the derivative of the voltage present at the control element [7]. Thus, in order to obtain a linear progression of frequency vs time a parabola or square-law voltage function is required. The use of gating circuits and frequency multiplication serve the same purpose as mentioned above. Figure 6.13 shows two methods

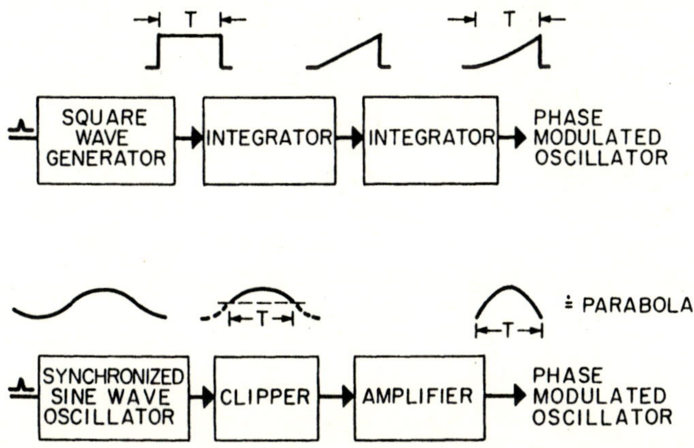

FIG. 6.13   Two methods of obtaining a square-law control voltage.

of obtaining the square-law oscillator control voltage for this application: (1) two stages of linear integration applied to a flat-topped square wave function, or (2) clipping of an undistorted sine wave near the peak of a cycle, where it fits a parabola quite closely, followed by linear amplification to the desired voltage level.

In either of these active modulation techniques the sense of the FM progression can be reversed by proper choice of a sideband output at a mixer stage. This can be done in the transmitter or in the receiver. The technique of active signal generation can be used readily in situations that call for control of the modulation characteristic of the signal independently of the matched filter at the receiver. Typical examples would be the injection of an additional modulation function to offset modulation distortion created elsewhere in the transmitter, or the generation of the modified-linear FM function described in Chapter 7. If a nonlinear FM matched-filter signal is required, then different types of voltage function generators are required to control the variable frequency oscillators.

Figure 6.14 blocks out the basic elements of passive generation of a pulse-compression matched-filter signal. The expansion filter has a dispersive delay characteristic, so that when driven by an impulse a stretched pulse waveform results at the filter output. Since the stretched pulse must contain essentially the same bandwidth information as the driving signal, this information must exist in the form of frequency, or phase, modulation since

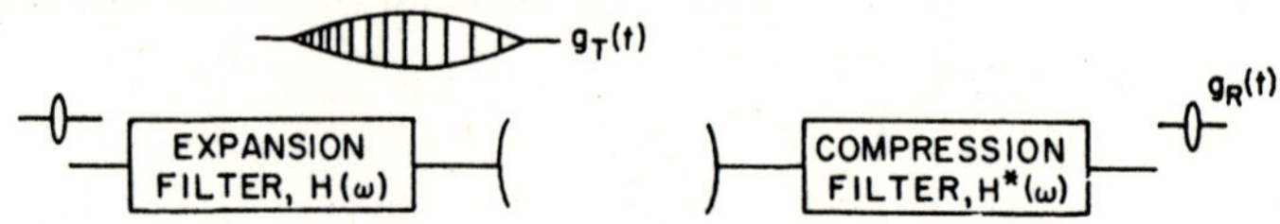

FIG. 6.14   Conjugate filter technique (from Bernfeld *et al.*, Ref. 4 of Chapter 1).

it is not contained in the stretched envelope function. Denoting the driving pulse spectrum by $S_i(\omega)$, the result of passing the signal through both the transmitter and receiver filters (assuming negligible distortion in the intervening transmission medium) can be described by

$$g_R(t) = \frac{1}{2\pi} \int_{-\infty}^{\infty} S_i(\omega)H(\omega)H^*(\omega) \exp[j\omega t]\, d\omega$$

$$= \frac{1}{2\pi} \int_{-\infty}^{\infty} S_i(\omega)|H(\omega)|^2 \exp[j\omega t]\, d\omega \tag{6-28}$$

If the spectrum of $g_T(t)$ is essentially rectangular and somewhat wider than the passband of $H(\omega)$, then the receiver output signal is

$$g_R(t) = \frac{1}{2\pi} \int_{-\infty}^{\infty} |H(\omega)|^2 \exp[j\omega t]\, d\omega \tag{6-29}$$

and a basis for a pulse expansion/compression system has been established.

For practical reasons the choice of the impulsing function should be one that restricts the major portion of the pulse expansion output to a finite time interval. Because of the nonideal delay characteristic outside the frequency band Δf (see Fig. 6.7) it is also desirable to restrict the frequency spectrum to prevent foldback of energy outside the band into the time region of the required linear FM characteristic. Based on these considerations, an impulsing function (approximating Eq. (6-12)) that meets these conditions is

$$s_i(t) = \frac{\sin \dfrac{\mu T t}{2}}{\dfrac{\mu T t}{2}} \cos \omega_0 t \tag{6-30}$$

Using this signal to drive the linear delay filter, the time response at the output of the expansion filter is

$$g_T(t) = \frac{1}{2\pi} \int_{-\infty}^{\infty} H(\omega) S_i(\omega) \exp[j\omega t] \, d\omega \tag{6-31}$$

Taking $H(\omega)$ from (6-25)[1], and using the exponential representation, yields

$$g_T(t) = \frac{1}{\mu T} \int_{\omega_0 - \mu T/2}^{\omega_0 + \mu T/2} \exp\left[j\left(\frac{(\omega - \omega_0)^2}{2\mu} + \omega t\right)\right] d\omega \tag{6-32}$$

The limits of the integral are derived from the rectangular spectrum associated with Eq. (6-30), which has a bandwidth μT. By means of the same analytical approach upon which the spectrum derivation of Section 6.3 was based, Eq. (6-32) transforms to

$$g_T(t) = \frac{1}{T}\sqrt{\frac{\pi}{\mu}} \exp\left[j\left(\omega_0 t - \frac{\mu t^2}{2}\right)\right] \int_{-X_1}^{X_2} \exp\left[j\frac{\pi y^2}{2}\right] dy \tag{6-33}$$

where

$$X_{1,2} = \frac{\dfrac{\mu T}{2}\left(1 \mp \dfrac{2t}{T}\right)}{\sqrt{\pi \mu}} = \sqrt{\frac{\mu T^2}{2\pi}}\left(\frac{1 \mp \dfrac{2t}{T}}{\sqrt{2}}\right)$$

The time function $g_T(t)$ can then be expressed as

$$g_T(t) = a(t) \exp\left[j\left\{\left(\omega_0 t - \frac{\mu t^2}{2}\right) + \theta_2(t)\right\}\right] \tag{6-34}$$

The important components of $g_T(t)$ can be identified as:

Envelope Term

$$a(t) = \frac{1}{T}\sqrt{\frac{\pi}{\mu}}\left\{[C(X_1) + C(X_2)]^2 + [S(X_1) + S(X_2)]^2\right\}^{1/2} \tag{6-35}$$

---

[1] Note that the convention $H(\omega) = \exp[-j\beta_f(\omega)]$ is used, with time delay being expressed as $T_d = d\beta/d\omega$.

Linear Frequency Modulation Term[1]

$$\frac{d}{dt}\left(\omega_0 t - \frac{\mu t^2}{2}\right) = \omega_0 - \mu t \tag{6-36}$$

Residual Phase Modulation Term

$$\theta_2(t) = -\tan^{-1}\left[\frac{S(X_1) + S(X_2)}{C(X_1) + C(X_2)}\right] \tag{6-37}$$

For large values of $T\,\Delta f$ the envelope function approximates a rectangular time envelope, and the residual modulation term approximates a constant phase angle, $\pi/4$. The choice of a $(\sin x)/x$ impulse function shape produces an expanded pulse having essentially the correct amplitude and phase properties for a linear FM pulse-compression radar signal. In a practical application the expanded pulse would be further "squared up" by gating and limiting, and as a result the signal spectrum observed at the transmitter output would be nearly indistinguishable from that produced by an active modulation technique.

The required $(\sin x)/x$ impulsing function can be obtained by passing a narrow pulse (bandwidth $> \Delta f$) through a rectangular bandpass filter (bandwidth $= \Delta f$) that has a linear phase characteristic. In essence, this reduces the pulse expansion operation to conjugate matched filtering, from which the approximate output signal characteristics could have been obtained by the principle of stationary phase. However, Eq. (6-35) points out that the expanded pulse envelope has the Fresnel ripple component. This can be an important consideration, since the limiting action that occurs in a high power radar transmitter will cause time function ripples to convert into frequency spectrum ripples. How this affects certain aspects of pulse-compression radar operation is examined in Chapter 7.

Figure 6.15 gives an illustrative example of a practical conjugate matched-filter radar implementation. This system is based on converting a signal having an odd symmetrical frequency modulation function into its time inverse. This is done by passing the signal through a frequency mixing device in which the output circuit is tuned to the difference sideband rather than the sum sideband. Symmetrical modulation inversion takes place if the mixer local oscillator frequency is higher than any frequency component of the input signal. Since this technique creates a conjugate time function at the receiver, the expansion and compression filters can both have identical characteristics, eliminating the need to design two different filters. Figure 6.16 shows how the sideband inversion system can be further instrumented

---

[1] The sense of the frequency modulation is the reverse of that of Eq. (6-2). This illustrates the fact that the compression filter matched to a positive slope linear FM signal will yield a negative slope linear FM signal when used as a pulse-expansion filter.

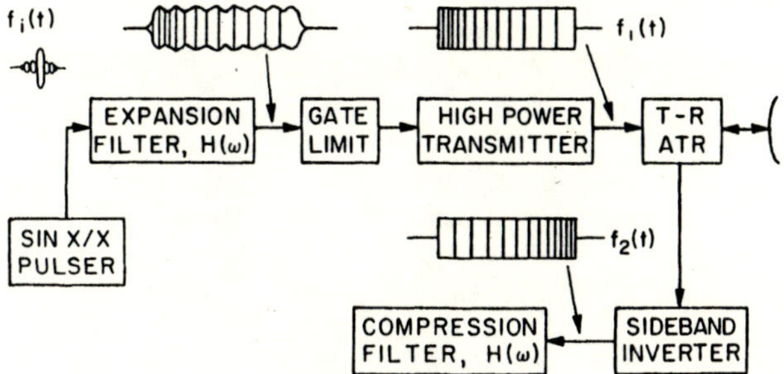

FIG. 6.15   Matched-filter radar using sideband inversion (from Bernfeld *et al.*, Ref. 4 of Chapter 1).

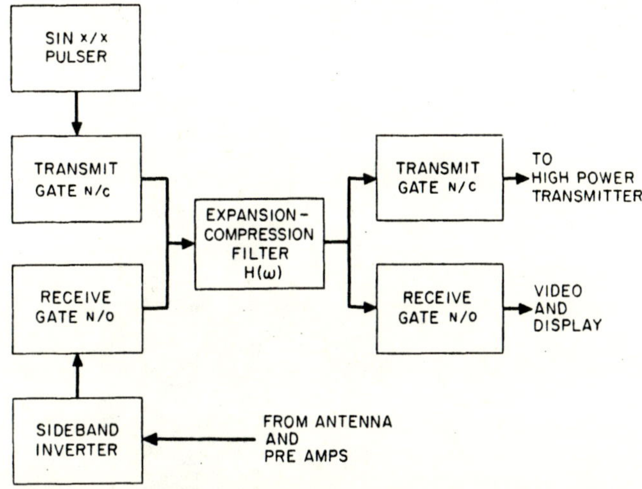

FIG. 6.16   A conjugate matched-filter system using a single filter.

to allow a single expansion/compression filter to be shared between the transmitter and receiver. These techniques have been described by Cook and Chin [8], Cook and Brockner [9], and Ramp and Wingrove [10], and are applicable to any pulse-compression matched-filter system that utilizes an odd symmetrical frequency modulation function.

Some early experimental pulse-compression radar waveforms obtained at the Sperry Gyroscope Company are shown in Figs. 6.17 and 6.18. The A-scope presentation shown in Fig. 6.17 was obtained with a conjugate filter (passive generation), X-band radar setup on the roof of the Sperry building at Great Neck in 1955. The A-scope trace shows a number of large clutter targets near the building and at a greater range a relatively large

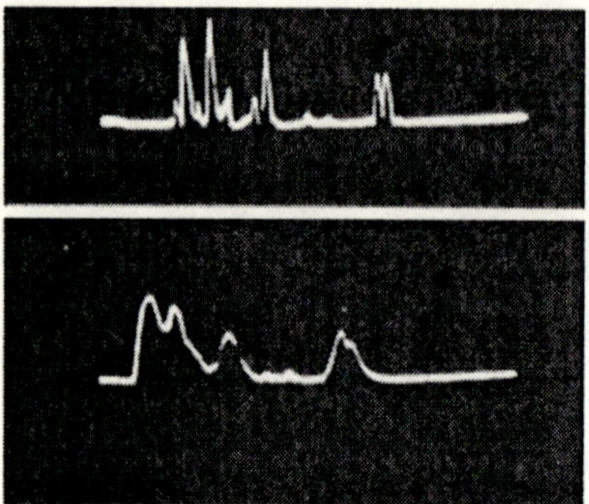

FIG. 6.17   A-scope trace, conjugate matched-filter system. Top: Compressed signals; Bottom: Uncompressed signals. (From Bernfeld *et al.*, Ref. 4 of Chapter 1).

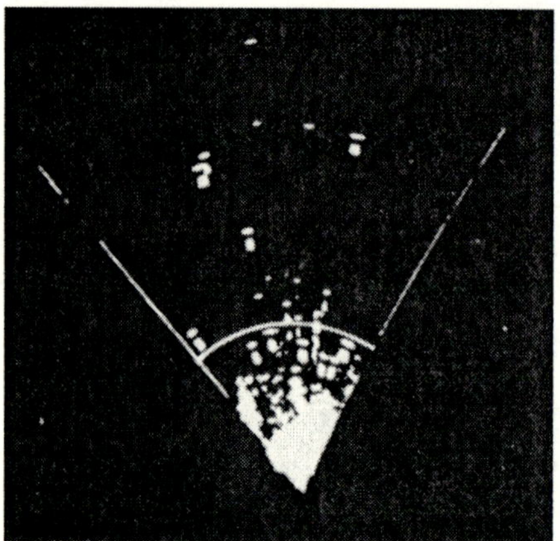

FIG. 6.18   Superimposed compressed and uncompressed signals on PPI presentation, (courtesy of A. E. Hylas, Sperry Gyroscope Company, Great Neck, New York).

return from corner reflectors. These were mounted on the top of automobiles stationed on a road near the Glen Oaks Golf Course. The two smaller signals midway between the clutter and corner reflectors are returns from automobiles or trucks travelling on a nearby road. The clutter breakup and enhanced resolution as a result of pulse compression are clearly illustrated. Figure 6.18 shows a PPI display in which detected uncompressed and compressed signals are superimposed. The signals at far ranges are aircraft in approach patterns for the local commercial airports. In this picture the

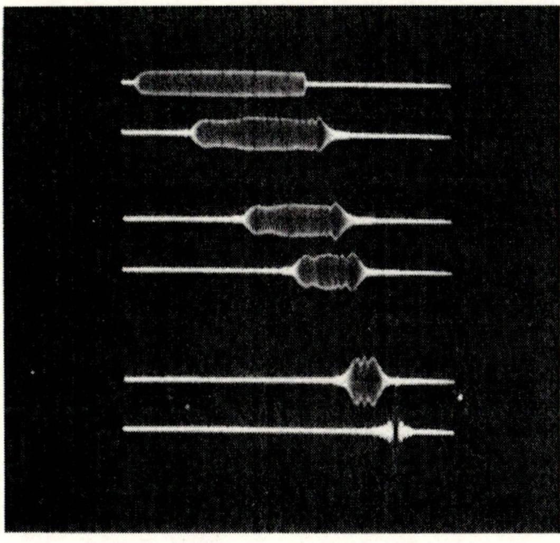

FIG. 6.19 Progression of linear FM signal through pulse-compression delay line, $T \, \Delta f = 70$ (from Bernfeld *et al.*, Ref. 4 of Chapter 1).

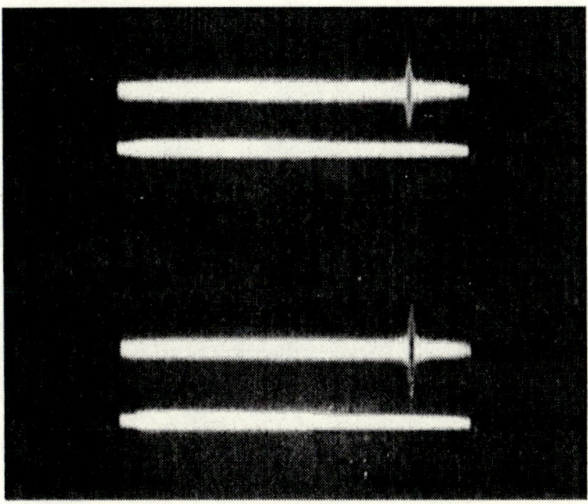

FIG. 6.20 Matched-filter signals immersed in noise.

greater effective peak power of the matched-filter output signal is demonstrated. Figure 6.19 illustrates the progressive compression of the linear FM pulse as it propagates along the compression line. The top waveform is the input signal and the bottom one the output compressed pulse. The intermediate waveforms show how the dispersive delay of the compression filter acts on the signal at various interior points along the delay line. Figure 6.20 illustrates the peak amplitude buildup on an A-scope trace when the uncompressed pulse is immersed in noise.

6.6 Effect of Linear Delay Mismatch on the Compressed-Pulse Signal

When the pulse-compression matched-filter phase response is not the exact conjugate of the transmitted spectrum phase a mismatch condition exists. A common type of mismatch condition occurs for the linear FM signal when the linear delay function at the receiver does not have the correct slope to match the linear FM function that is transmitted. This condition is referred to as linear delay or linear FM mismatch, and is

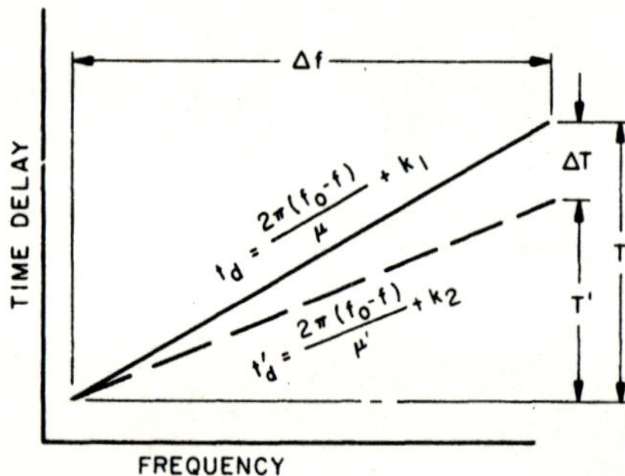

FIG. 6.21   Matched and mismatched linear delay functions.

illustrated by the linear delay functions shown in Fig. 6.21. The correct linear delay function for the linear FM signal of Eq. (6-2) is

$$t_d = -\frac{2\pi(f - f_0)}{\mu} + k_1 \qquad (6\text{-}38)$$

The delay difference of the matched filter is $T = 2\pi\,\Delta f/\mu$. The linear delay of the mismatched compression filter is given by

$$t'_d = -\frac{2\pi(f - f_0)}{\mu'} + k_2 \qquad (6\text{-}39)$$

where μ' can be greater or smaller than μ. The delay difference associated with (6-39) is $T' = 2\pi\,\Delta f/\mu'$. The effect of this mismatched linear delay function on the compressed pulse can be analyzed on the basis of an ideal matched filter that is followed by a linear delay mismatch filter, as shown in Fig. 6.22.

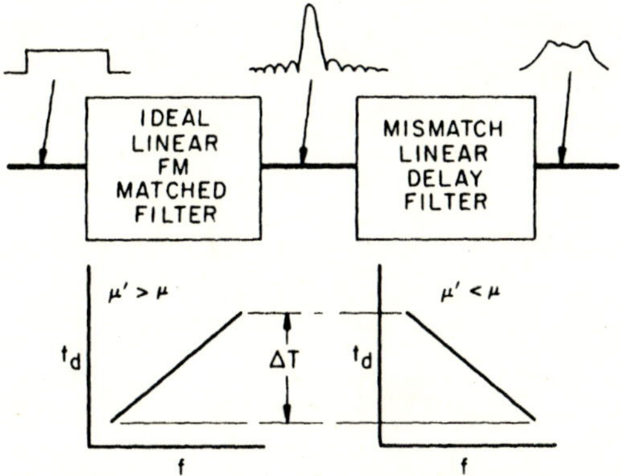

FIG. 6.22 Characteristics of mismatch linear delay filter.

The linear delay difference of the mismatch filter is given by

$$\Delta T = 2\pi \,\Delta f \left(\frac{1}{\mu} - \frac{1}{\mu'} \right) = \frac{2\pi \,\Delta f}{\mu_{\text{eq}}} \tag{6-40}$$

where μ_{eq} defines the delay slope of the mismatch filter, and is

$$\mu_{\text{eq}} = \frac{\mu\mu'}{\mu' - \mu}$$

A mismatch factor $\gamma = (\mu' - \mu)/\mu'$ can be defined so that $\mu_{\text{eq}} = \mu/\gamma$, $\Delta T = \gamma T$, and the time-bandwidth factor of the mismatch filter is $|\gamma| T \,\Delta f$. For compression ratios of 20 or greater the output of the ideal matched filter in Fig. 6.22 can be approximated by

$$g(t) = T \sqrt{\frac{\mu}{2\pi}} \frac{\sin \mu T t/2}{\mu T t/2} \cos \omega_0 t \tag{6-41}$$

From Eqs. (6-30)–(6-35) of the previous section, the output response of the linear delay mismatch filter to the input given by (6-41) is

$$g_d(t) = \frac{T}{\Delta T} \sqrt{\frac{\mu}{2\mu_{\text{eq}}}} \left\{ [C(X_1) + C(X_2)]^2 + [S(X_1) + S(X_2)]^2 \right\}^{1/2}$$

$$\times \exp[\,j\{\omega_0 t + \tfrac{1}{2}\mu_{\text{eq}}t^2 + \theta_2(t)\}\,] \tag{6-42}$$

where $\theta_2(t)$ is given by (6-37) and

$$X_{1,2} = \frac{\mu_{\text{eq}} \Delta T[1 \mp (2t/\Delta T)]}{2\sqrt{\pi\mu_{\text{eq}}}}$$

Applying the mismatch filter parameters defined above, the Fresnel integral variables in (6-42) become

$$X_{1,2} = \frac{\gamma\mu T[1 \mp n]}{2\sqrt{\gamma\pi\mu}} \tag{6-43}$$

where $n = 2t/\gamma T$, and the magnitude of the distorted output can be expressed as

$$|g_d(t)| = \sqrt{\frac{\mu'}{2|\mu' - \mu|}}\left\{[C(X_1) + C(X_2)]^2 + [S(X_1) + S(X_2)]^2\right\}^{1/2} \tag{6-44}$$

The approximate pulse width of this distorted signal is $\tau = \gamma T$, and the average amplitude (for $\gamma T \Delta f > 5$) $\approx (\mu'/|\mu' - \mu|)^{1/2} = 1/\sqrt{|\gamma|}$. The ratio of the amplitude of the distorted mismatched-filter output to that of the ideal matched filter is[1]

$$\frac{1/\sqrt{|\gamma|}}{\sqrt{T\,\Delta f}} = \frac{1}{\sqrt{|\gamma|T\,\Delta f}} \tag{6-45}$$

The distorted time functions described by (6-44) resemble linear FM spectra for low time-bandwidth products. Examples of these distorted waveforms are shown in Figs. 6.23–6.26. The normalized values of time and amplitude shown in these figures can be converted to real time and amplitude factors for specific values of γ, T, and $T\,\Delta f$. Thus, for $T\,\Delta f = 50$, $\gamma = 0.1$ and

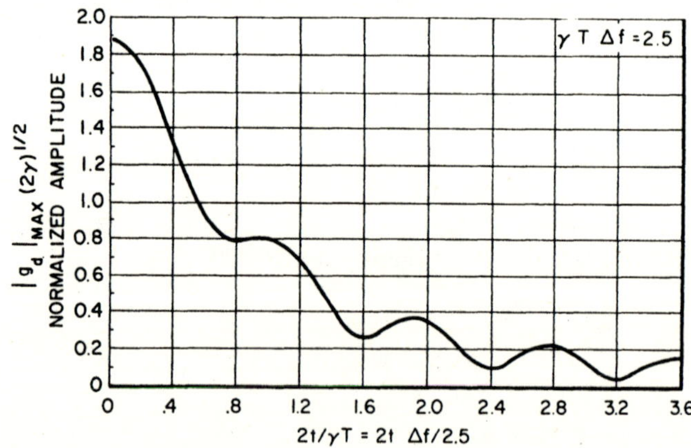

FIG. 6.23 Normalized distorted linear FM compressed pulse for delay mismatch: $\gamma T \Delta f = 2.5$. (Figs. 6.23–6.26 courtesy of J. Paolillo, Sperry Gyroscope Company, Great Neck, New York.)

---

[1] If the peak is used this ratio will be smaller, and can be calculated from Figs. 6.24–6.26.

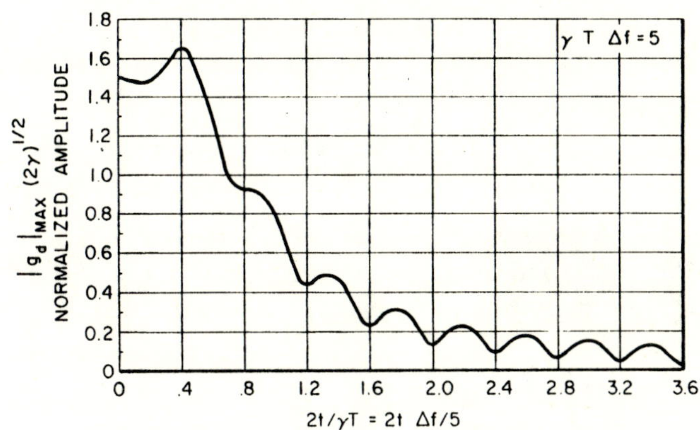

FIG. 6.24 Normalized distorted linear FM compressed pulse for delay mismatch: $\gamma T \Delta f = 5$.

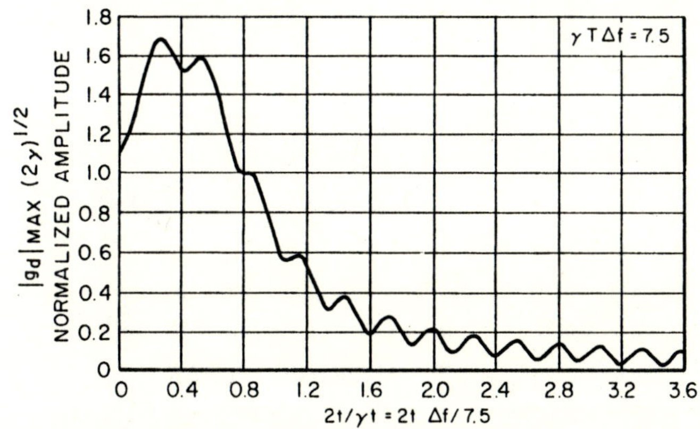

FIG. 6.25 Normalized distorted linear FM compressed pulse for delay mismatch: $\gamma T \Delta f = 7.5$.

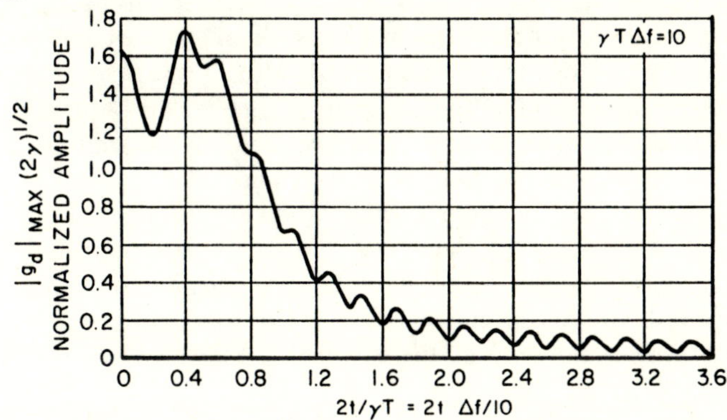

FIG. 6.26 Normalized distorted linear FM compressed pulse for delay mismatch: $\gamma T \Delta f = 10$.

$\gamma T \Delta f = 5$, the distorted waveform will resemble Fig. 6.24. The 3 db undistorted pulse width is $0.9/\Delta f$. From Fig. 6.24 the 3 db distorted pulse width is given by $2t/\gamma T = \pm 0.67$, or

$$t = \pm \frac{0.67\gamma T}{2} = \pm \frac{0.67\gamma T \Delta f}{2\,\Delta f} = \pm \frac{0.67 \times 5}{2\,\Delta f}$$

This yields a 3 db pulse width ratio of the distorted to undistorted signal of $0.67 \times 5/0.9 = 3.72$ and an amplitude ratio, from (6-45), of $\sqrt{\gamma T \Delta f} = \sqrt{5}$. Figure 6.27 plots the 3 db pulse widening ratio as a function of the general

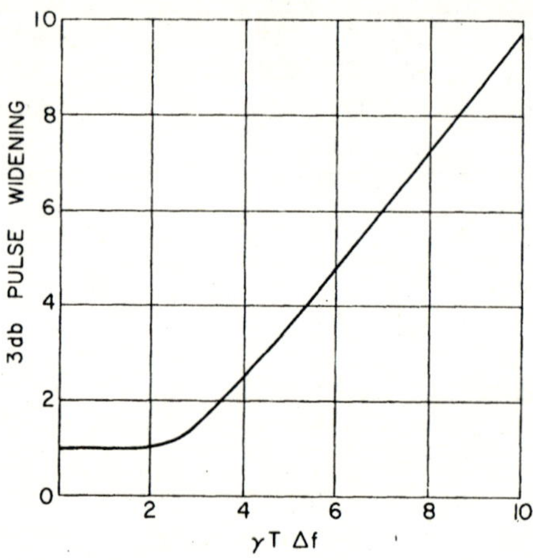

Fig. 6.27   The effect of linear delay mismatch (quadratic phase error) on 3 db pulse width as a function of the compression ratio-mismatch factor product, $\gamma T \Delta f$.

parameter $\gamma T \Delta f$. As this figure indicates a smaller mismatch factor γ is required to yield the same amount of pulse widening as the compression ratio becomes larger. If the transmitted bandwidth is swept over an interval $\Delta f' > \Delta f$, the filter bandwidth, the analysis above can be applied by first assuming that the ideal matched filter performs a truncation upon both the time duration and bandwidth of the transmitted signal by the factor $\Delta f/\Delta f'$. This results in an effective time-bandwidth product of the matched-filter input of $T \Delta f' (\Delta f/\Delta f')^2 = T(\Delta f)^2/\Delta f'$. The mismatched FM sweep rate for this case is $\mu' = 2\pi\,\Delta f'/T$. This approximate analysis can be extended to other combinations of mismatched input signal duration and sweep rate to obtain an estimate of the form of the mismatched signal at the compression-filter output.

When a frequency weighting filter $W(\omega)$, as described in the next chapter, follows the mismatched linear delay filter the distorted output pulse is expressed in exponential form as

$$g_d(t) = \frac{1}{\sqrt{2\pi\mu}} \int_{\omega_0 - \Delta\omega/2}^{\omega_0 + \Delta\omega/2} W(\omega) \exp\left[j\left\{ \frac{(\omega - \omega_0)^2}{2\mu_{eq}} + \omega t \right\} \right] d\omega \qquad (6\text{-}46)$$

The effect of the weighting function on the distorted compressed pulse can be illustrated with the general $(\text{cosine})^2$-on-pedestal function given by

$$W(\omega) = k + (1 - k)\cos^2 \frac{\pi(\omega - \omega_0)}{\Delta\omega}$$

$$= \frac{k + 1}{2} + \frac{1 - k}{4}\left(\exp\left[j\frac{2\pi(\omega - \omega_0)}{\Delta\omega} \right] + \exp\left[-j\frac{2\pi(\omega - \omega_0)}{\Delta\omega} \right] \right) \qquad (6\text{-}47)$$

The terms in parentheses can be associated with paired echo signals, the theory of which is developed in Chapter 11. Insert (6-47) into (6-46) for $W(\omega)$, and use the development for the unweighted distorted pulse given above; this yields

$$g_d(t) = \frac{k + 1}{2\sqrt{2\gamma}}\left\{ [C(X_1) + C(X_2)]^2 + [S(X_1) + S(X_2)]^2 \right\}^{1/2}$$

$$\times \exp\left[j\left\{ \omega_0 t + \frac{\mu}{2\gamma} t^2 + \theta_2(t) \right\} \right]$$

$$+ \frac{1 - k}{4\sqrt{2\gamma}}\left\{ [C(X'_1) + C(X'_2)]^2 + [S(X'_1) + S(X'_2)]^2 \right\}^{1/2}$$

$$\times \exp\left[j\left\{ \omega_0 t + \frac{\mu}{2\gamma}\left(t + \frac{1}{\Delta f} \right)^2 + \theta_2\left(t + \frac{1}{\Delta f} \right) \right\} \right]$$

$$+ \frac{1 - k}{4\sqrt{2\gamma}}\left\{ [C(X''_1) + C(X''_2)]^2 + [S(X''_1) + S(X''_2)]^2 \right\}^{1/2}$$

$$\times \exp\left[j\left\{ \omega_0 t + \frac{\mu}{2\gamma}\left(t - \frac{1}{\Delta f} \right)^2 + \theta_2\left(t - \frac{1}{\Delta f} \right) \right\} \right] \qquad (6\text{-}48)$$

where X_1 and X_2 are given in (6-43), and

$$X'_{1,2} = \frac{\gamma\mu T[1 \mp 2\{t + (1/\Delta f)\}/\gamma T]}{2\sqrt{\gamma\pi\mu}}$$

$$X''_{1,2} = \frac{\gamma\mu T[1 \mp 2\{t - (1/\Delta f)\}/\gamma T]}{\sqrt{\gamma\pi\mu}}$$

In regions where the Fresnel magnitude functions are relatively constant the last two terms of (6-48) combine to yield a linear frequency modulation equivalent to that of the first term, plus FM ripple terms. Klauder *et al.* [4] have calculated a number of cases for $k = 0.088$ (-40 db weighted range sidelobes). Figure 6.28 compares the undistorted weighted pulse to that

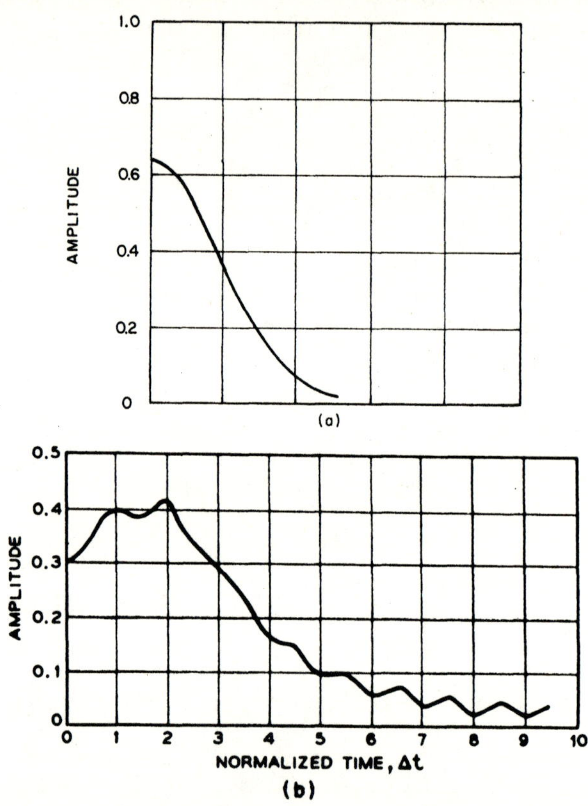

FIG. 6.28   Quadratic phase distortion effects on weighted pulse: $\gamma T \, \Delta f = 8$. (a) Weighted. (b) Unweighted. (Courtesy of A. C. Price, Bell Telephone Laboratories, Whippany, New Jersey.)

of the distorted pulse for $\gamma T \Delta f = 8$. For this amount of mismatch distortion the sidelobes increased to only -36 db, as compared to the much higher level of distortion lobes for the unweighted case. Figure 6.29 gives the pulse amplitude and pulse widening degradation for this case, as derived from the data of Klauder *et al.*, who conclude that for -40 db weighted compressed-pulse sidelobes a value of $\gamma T \Delta f = 4$ is tolerable. Figure 6.30 shows actual compressed-pulse waveforms for both the weighted and unweighted case and different factors of $\gamma T \Delta f$. The linear delay mismatch can also be expressed as a quadratic phase error, defined by the band-edge deviation

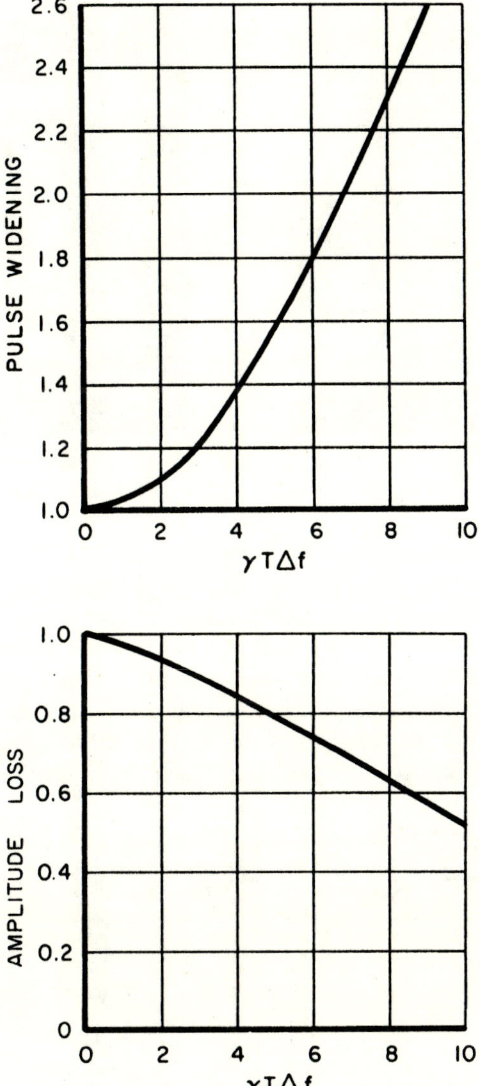

FIG. 6.29   Pulse widening and amplitude degradation for weighted, linear delay mismatched pulse.

from the ideal phase characteristic. In terms of the mismatch factor γ, and time-bandwidth product of the signal, this band-edge phase mismatch is given by $\Phi = \pi\gamma T\Delta f/4$.

Calculations based on (6-46) can become quite complicated for the general weighting function. If the mismatch time-bandwidth factor $\gamma T\Delta f$ is

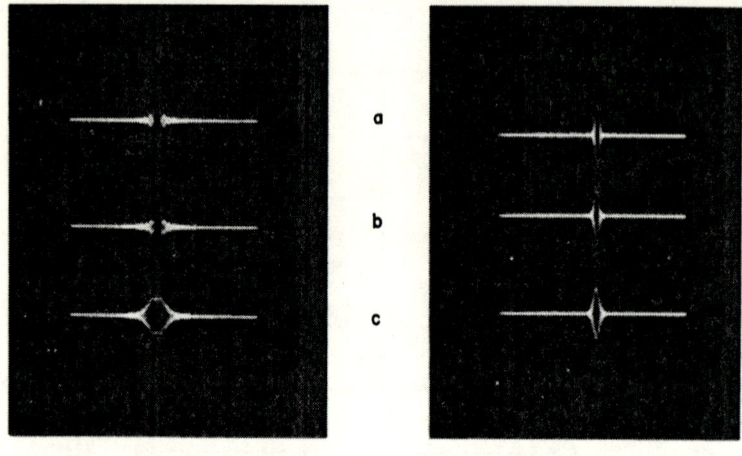

FIG. 6.30 Compressed-pulse distortion caused by linear FM mismatch. (a) $\gamma T \,\Delta f = 0$. (b) $\gamma T \,\Delta f = 2$. (c) $\gamma T \,\Delta f = 8$. (From Cook and Heiss [21].)

relatively large (>4) the distorted output time response is approximately described by the same function as the frequency weighting response, as shown in Fig. 6.31, where the time interval ΔT replaces the frequency interval Δf. This approximation ignores the sidelobe levels of the distorted pulse, but is useful for calculating the pulse width degradation caused by the linear FM mismatch or quadratic phase error on the weighted linear FM signal.

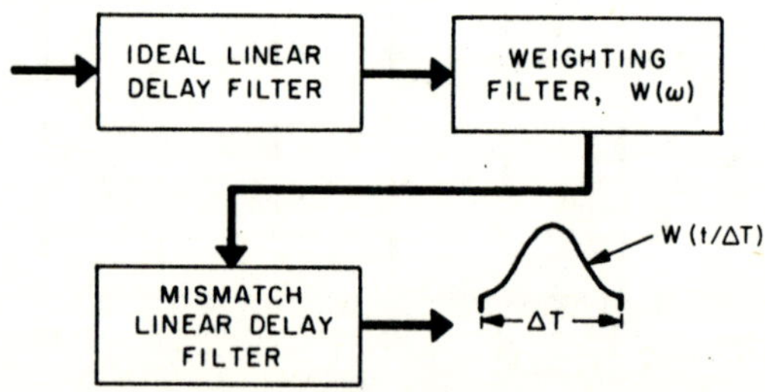

FIG. 6.31 Effect of weighting function on compressed pulse at mismatch linear delay filter output.

6.7 Large Time-Bandwidth Techniques

When large time-bandwidth signals are required the design of the receiver matched filter, as outlined in Sections 6.2–6.4, may become complicated through factors such as the build-up of distortion in a long delay line, insertion losses and bandwidth limitations. Two techniques for the design of large time-bandwidth linear FM matched-filter systems that have been developed to cope with these problems are discussed in this section.

(a) Parallel Channel Design

The basic parallel channel matched-filter design for the linear FM signal is shown in Fig. 6.32. Each channel has a dispersive filter that has a compression ratio $T_c \Delta f_c$, where $T_c = T/N$ and $\Delta f_c = \Delta f/N$, T being the total

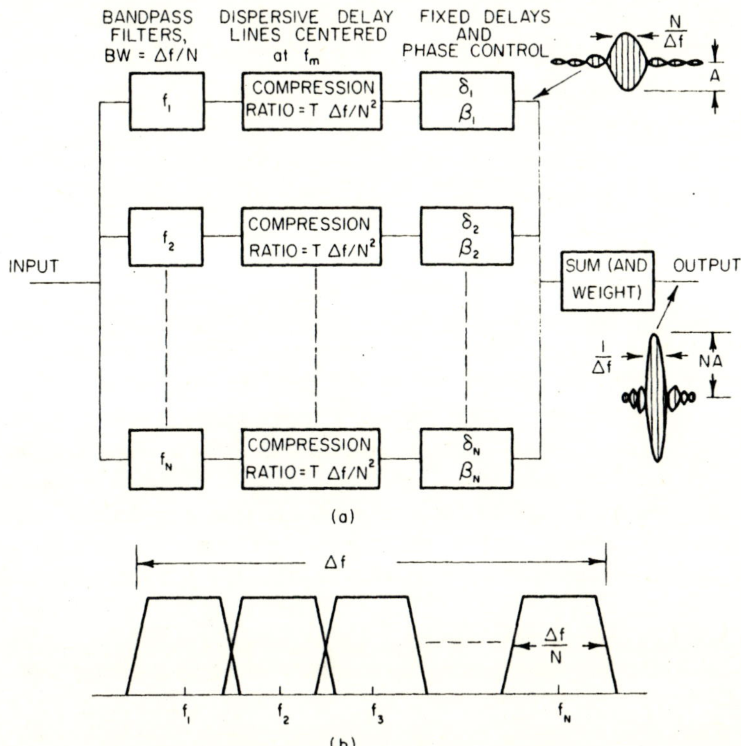

FIG. 6.32 Parallel channel pulse compression processor for large time-bandwidth linear FM signals. (a) Parallel channel block diagram. (b) Bandpass filter frequency responses.

signal duration and Δf the total swept bandwidth. The channel center frequencies are spaced apart by $\Delta f/N$, and are given by

$$f_m = f_0 - \frac{\Delta f}{2}\left(\frac{N - 2m + 1}{N}\right), \qquad 1 < m < N \qquad (6\text{-}49)$$

If the input to each channel is considered to be a rectangular envelope linear FM signal of duration T/N and swept bandwidth $\Delta f/N$, then the output from each channel is described by [11]

$$g_m(t) = \left(\frac{T\Delta f}{N^2}\right)^{1/2} \frac{\sin \pi \Delta f t/N}{\pi \Delta f t/N} \cos\left[2\pi f_m t - \frac{\pi}{4}\right] \qquad (6\text{-}50)$$

The -4 db compressed-pulse width at each channel output is $\tau_c = N/\Delta f$. The fixed delay lines in each channel bring their respective outputs into time coincidence, and the phase controls are adjusted to cause the carrier frequencies of the N channel signals to be in phase (in the absence of Doppler shift) at their coinciding peaks. The summation of the N channel outputs, as shown in Fig. 6.32, yields a composite output signal given by

$$
\begin{aligned}
g_N(t) &= \sum_{m=1}^{N} g_m(t) \\
&= \left(\frac{T\Delta f}{N^2}\right)^{1/2} \frac{\sin \pi \Delta f t/N}{\pi \Delta f t/N} \frac{\sin \pi \Delta f t}{\sin \pi \Delta f t/N} \cos\left(2\pi f_0 t - \frac{\pi}{4}\right) \\
&= (T\Delta f)^{1/2} \frac{\sin \pi \Delta f t}{\pi \Delta f t} \cos\left(2\pi f_0 t - \frac{\pi}{4}\right) \qquad (6\text{-}51)
\end{aligned}
$$

Equation (6-51) shows that the overall time-bandwidth product is N^2 times that of a single channel; or, that if the time-bandwidth is $T\Delta f$, decomposition of the matched filter into N similar channels results in a time-bandwidth product per channel of $T\Delta f/N^2$. This leads to a marked simplification of the dispersive delay line design for each channel as compared to a single channel implementation. Since the size of a dispersive delay filter is directly proportional to its time-bandwidth product (see Chapters 12–14), it can be seen that the number or size of dispersive components required for the parallel channel implementation is approximately $1/N$ that of a single channel compression filter. Leakage between adjacent channels, resulting from their overlapping bandpass skirt characteristics, requires that the linear delay in each channel extend over a band greater than $\Delta f/N$. Thus, the reduction of dispersive delay size will be somewhat less than $1/N$. This savings in dispersive components is partially offset by the need for the fixed delay filters and phase controls, as well as the more complex input and output circuits. The basic advantage of the parallel channel technique is that the smaller time-bandwidth product in each channel permits greater control over the time-delay distortions in each dispersive filter. This type of distortion, discussed in greater detail in Chapter 11, becomes larger as the time-bandwidth product of the dispersive filter increases. The compression filters in the N channels can be any of the types described in Chapters 12–14. Sidelobe reduction can be performed with a weighing network following the channel recombination or by controlling the gain level in each channel to achieve a stepped weighting response. For the particular case of the narrower band devices the parallel channel approach is also a method of achieving increased bandwidth.

The design details of a ten channel 1000:1 compression ratio linear FM filter are described by Jacobus [12]. The compression ratio per channel is $1000/N^2$, or 10. This equipment was used as a pulse-expansion as well as a pulse-compression filter. The waveforms obtained with this matched filter are illustrated in Fig. 6.33. The 1000 microsecond expanded pulse is shown in Fig. 6.33(a). After sideband inversion to reverse the sense of the FM sweep the same signal was fed back into the parallel channel matched filter to obtain the 1 microsecond (-4 db pulse width) compressed pulse shown in Fig. 6.33(b). In this implementation input mixers and local oscillators were used in each channel so that the compression filters in each channel had the same center frequency, and thus identical designs. The same local

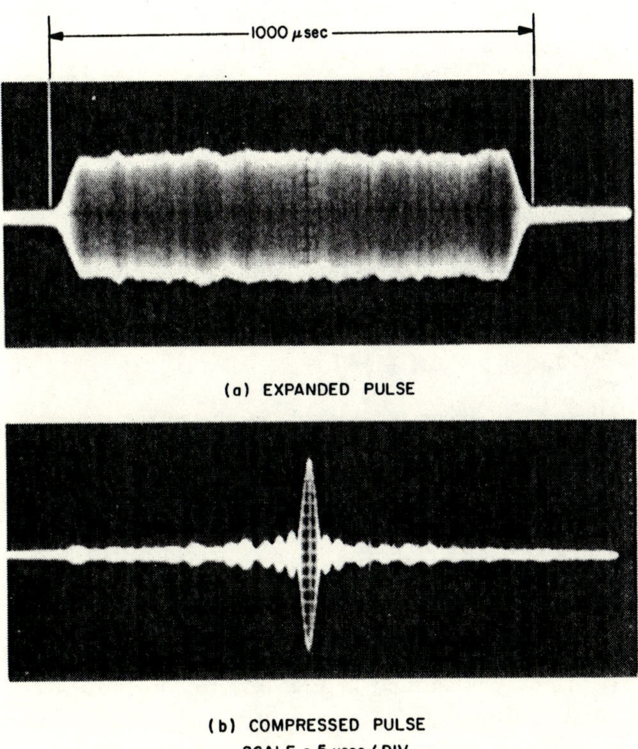

FIG. 6.33 Waveforms for 1000:1 parallel channel linear FM pulse-compression system, (courtesy of R. W. Jacobus and The MITRE Corporation, Bedford, Massachusetts).

oscillators and output mixers placed the channel subspectrum contributions in their proper frequency locations (see Fig. 6.32(b)) before summation of the component pulses from each channel. A related design technique based on a tapped delay line synthesis of a dispersive characteristic is described by Haggarty [13]. Another application of the parallel channel method is to the design of nonlinear FM signals. In this case the nonlinear FM function is approximated by a set of contiguous linear FM segments of different slopes. Each of these linear FM segments is processed in its own linear dispersive delay channel. An application of this type of nonlinear FM large time-bandwidth processor is described by Thor [14].

(b) Spectrum Analyzer Technique

The frequency vs time relationship of the linear FM signal permits the measurement of range by heterodyning, or comparing, the linear FM function of the received signal with a local oscillator signal that is a linear FM ramp function with the same FM sweep rate as the transmitted signal. In this type of implementation the start of the matching linear FM local oscillator signal is normally delayed until a time t_0 after transmission, but prior to the time of arrival of the received signal. The local oscillator signal is described by

$$\cos[\omega_l \tilde{t} + \tfrac{1}{2}\mu \tilde{t}^2], \qquad \tilde{t} > 0 \tag{6-52}$$

where $\tilde{t} = t - t_0$.

A received signal arriving Δt seconds after the start of the local oscillator swept signal has the form

$$a(\tilde{t} - \Delta t) \cos[\omega_l(\tilde{t} - \Delta t) + \tfrac{1}{2}\mu(\tilde{t} - \Delta t)], \qquad \Delta t < \tilde{t} < T + \Delta t \tag{6-53}$$

The multiplication of these two signals in the mixer shown in Fig. 6.34 yields

$$a(\tilde{t} - \Delta t) \cos[\mu \Delta t\, \tilde{t} - \theta(\Delta t)] + \text{a term in } 2\omega_0, \qquad \Delta t < \tilde{t} < T + \Delta t \tag{6-54}$$

where $\theta(\Delta t) = \omega_0 \Delta t + \tfrac{1}{2}\mu(\Delta t)^2$.

The first term in (6-54) is a constant frequency pulse of duration T in which the constant frequency is a function of the time delay, or range, of the received signal relative to the start of the local oscillator ramp, as shown by the FM functions illustrated in Fig. 6.35. In order to avoid the zero frequency condition for signals arriving at time t_0 an i.f. offset frequency is added to the local oscillator. The output signal from the mixer is then given by

$$a(\tilde{t} - \Delta t) \cos[\omega_{i.f.}\tilde{t} + \mu \Delta t\, \tilde{t}], \qquad \Delta t < \tilde{t} < T + \Delta t \tag{6-55}$$

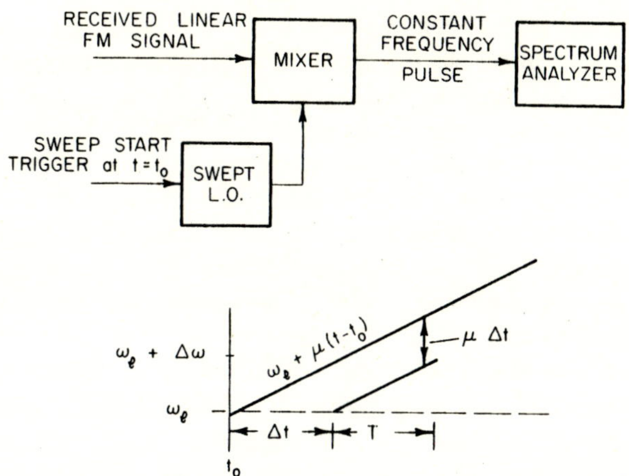

FIG. 6.34   Basic spectrum analyzer reception technique.

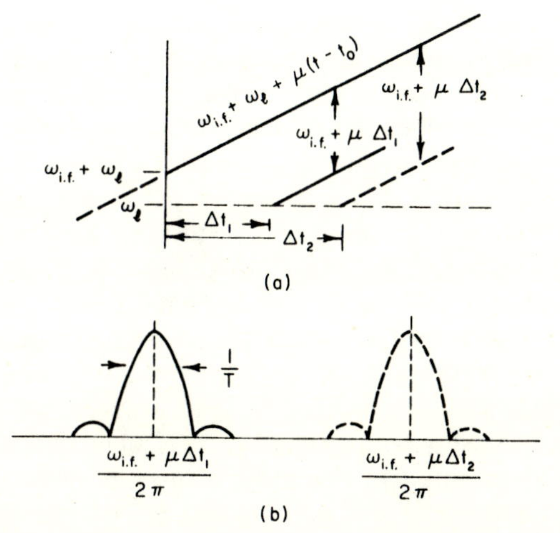

FIG. 6.35   Multiple signal resolution and range dependence properties of linear FM spectrum analyzer technique. (a) Offset swept local oscillator and two received signals. (b) Spectra at analyzer input.

For $a(t)$ a rectangular envelope the spectrum of this signal is

$$F(\omega) = \frac{\sin \pi T\left(f - \dfrac{\omega_{\text{i.f.}} - \mu\,\Delta t}{2\pi}\right)}{\pi T\left(f - \dfrac{\omega_{\text{i.f.}} - \mu\,\Delta t}{2\pi}\right)} \tag{6-56}$$

The width of this spectrum between its $-4\,\mathrm{db}$ points is $1/T$ and is the spectrum resolution cell size. Since $\mu = 2\pi\,\Delta f/T$, it can be seen from (6-55) that a second signal with a range delay $\Delta t + 1/\Delta f$, that is displaced by one normally compressed-pulse resolution cell, will shift the center frequency of the mixer output by $\mu/2\pi\,\Delta f$ or $1/T$. Thus, the process of heterodyning the received signals with the local oscillator ramp function converts the range difference information that would be observed at the output of a pulse-compression filter into frequency difference information that has the equivalent range resolution as that of the compressed pulses. In this case, range is measured as a frequency difference.

Implementation of a system based on this technique requires the application of the mixer output to a spectrum analysis device. This could be a bank of narrow band filters (bandwidth $= 1/T$) spaced $1/T$ apart in frequency. This approach can become rather cumbersome if many range cells are to be examined simultaneously. A more compact arrangement makes use of the coherent memory filter [15, 16], a recirculating delay line heterodyning device that is capable of displaying in real time at its output the spectra of multiple simultaneous inputs [17]. In general, the spectrum analyzer technique is confined to tracking or narrow range "window" applications in order to avoid the problems associated with controlling the linearity of the local oscillator ramp function over long regions of time and very wide swept frequency intervals. In the tracking application the frequency information at the mixer output is fed back to a circuit controlling the start of the ramp function to keep the mixer output at a constant center frequency. The change in timing of the ramp start necessary to achieve this constitutes the range tracking data. Temes *et al.* [18] describe an application of this type in which the transmitted pulse duration is 2000 μsec and the swept bandwidth is 4 MHz, for a time-bandwidth product of 8000. If several signals have practically identical arrival times the sidelobes of the larger spectral responses may obscure the smaller signals. The sidelobes can be reduced by weighting or tapering the envelope of the received signal before it enters the spectrum analyzer. This can be accomplished by controlling the frequency response of the receiver before the mixer of Fig. 6.37 or by applying a time amplitude modulation to the received pulse envelope. This latter approach was followed by Temes *et al.*, and requires synchronizing the receiver time weighting function with the time of arrival of the received signals. The Hamming weighting function, described in Chapter 7, lends itself to straightforward implementation with an amplitude modulator at the receiver [18]. However, time weighting at the receiver is not the optimum technique when several signals of interest have differential times of arrival commensurate with the transmitted signal duration. If the received signals are Doppler shifted the range measurements derived from the spectrum analyzer will be in error. The linear FM waveform lends itself to the automatic correction of this range error by the range prediction techniques discussed in Section 9.5.

The spectrum analyzer technique performs best when the FM functions of both the transmitted signal and the swept local oscillator are extremely linear. Temes *et al.* describe a push-pull frequency multiplication technique for generating a linear FM function with very small amounts of distortion. Peebles and Stevens [19] discuss an alternate technique based on generating an accurately controlled stair-case step FM function and filling in the steps with a sawtooth FM function to achieve phase modulation nonlinearities in the order of one degree.

6.8 Doppler Shift Distortion of Large Time-Bandwidth Linear FM Signals

In Section 6.2 it was assumed that the major Doppler shift effect on the linear FM signal was a uniform translation of its spectrum, with all other characteristics of the signal remaining unaffected. It has been noted that this approximation is not sufficiently accurate for very large time-bandwidth product signals [20]. For narrow band signals the Doppler effect is more accurately represented, as discussed in Chapter 4, by a general transformation on the signal $s(t)$ to yield $s_d(t)$ such that

$$s_d(t) = \sqrt{\alpha}\, s(\alpha t) \tag{6-57}$$

where $\alpha = 1 + 2v/c$ and v is the vector radial velocity, assumed positive for closing velocities.

If $s(t)$ is of finite duration then the received pulse length is changed by the factor $1/\alpha$. Applying (6-57) to the linear FM signal results in

$$s_d(t) = \sqrt{\alpha}\, \cos[\alpha\omega_0 t + \tfrac{1}{2}\mu(\alpha t)^2], \qquad -\frac{T}{2\alpha} < t < \frac{T}{2\alpha}$$

$$= \sqrt{1 + \frac{2v}{c}}\, \cos\left[\omega_0 t + \frac{2v}{c}\omega_0 t + \tfrac{1}{2}\mu\left(1 + \frac{2v}{c}\right)^2 t^2\right] \tag{6-58}$$

The term $2v\omega_0/c$ is the radian Doppler shift frequency considered in Section 6.2. Generally, the effect of the velocity factor on the received signal amplitude and duration is ignored. For this case Ramp and Wingrove [20] and Thor [14] assume that the received signal is processed through a linear delay conjugate filter (i.e., matched-filter bandlimiting effect neglected) that is optimum for the zero Doppler signal (see (6-25)). The received spectrum of (6-58) for large time-bandwidth signals is essentially rectangular over a bandwidth $\Delta\omega(1 + 2v/c)$ centered at $\omega_0 + 2v\omega_0/c$. The square-law phase of the Doppler distorted spectrum can be obtained from (6-20), letting $\omega_0 = (1 + 2v/c)\omega_0$ and $\mu = \mu(1 + 2v/c)^2$. After being processed through the linear delay filter described by (6-25) the spectrum at the filter output has a residual phase term that is approximated by

$$\Phi_d(\omega) = -\frac{\dfrac{2v}{c}\left(2 + \dfrac{2v}{c}\right)(\omega - \omega_0)^2}{2\mu\left(1 + \dfrac{2v}{c}\right)^2} - \frac{\dfrac{2v}{c}\omega\omega_0}{\mu\left(1 + \dfrac{2v}{c}\right)^2}$$

$$+ \frac{\dfrac{2v}{c}\left(2 + \dfrac{2v}{c}\right)\omega_0^2}{2\mu\left(1 + \dfrac{2v}{c}\right)^2} = \Phi_1 + \Phi_2 + \Phi_3 \tag{6-59}$$

The last term of (6-59) is a constant phase that has no effect on the output pulse shape. The middle term represents a time shift given by

$$\frac{d\Phi_2}{d\omega} \doteq -\frac{\dfrac{2v}{c}\left(1 - \dfrac{4v}{c}\right)\omega_0}{\mu} = -\frac{\omega_d}{\mu} + \frac{4v}{c}\frac{\omega_d}{\mu} \tag{6-60}$$

where $\omega_d = 2\pi\phi$. The term $-\omega_d/\mu = -(\phi/\Delta f)T$ is the time shift at the matched-filter output for the simple Doppler shift approximation. For most practical radar cases the second term of (6-60) is negligible since $2v/c$ is generally in the order of 10^{-4} or less. The first term of (6-59) is a square-law phase term that can be associated with an equivalent linear delay mismatch filter similar to that discussed in Section 6.6. The Doppler shift distortion resulting from this phase term is analogous to the case of sending the ideal $(\sin x)/x$ compressed pulse through a linear delay filter that has a time delay function given by

$$T_d(\omega) = \frac{d\Phi_1}{d\omega} = -\frac{\dfrac{2v}{c}\left(2 + \dfrac{2v}{c}\right)(\omega - \omega_0)}{\mu\left(1 + \dfrac{2v}{c}\right)^2} \doteq -\frac{4v}{c}T(\omega - \omega_0)/\Delta\omega \tag{6-61}$$

The total differential delay of this filter is approximately

$$|\Delta T| = \left|\frac{4v}{c}T\right| \tag{6-62}$$

In order to make use of the development of Section 6.6 one lets $|4v/c| = \gamma_v$, where γ_v is the effective linear delay mismatch associated with the Doppler effect. Figure 6.27 can then be used to determine the pulse widening caused by this mismatch by calculating $\gamma_v T \Delta f$. Thus, for $v = 7500$ meters/sec and $T\Delta f = 5 \times 10^5$

$$\gamma_v T \Delta f = \frac{4 \times 7.5 \times 10^3 \times 5 \times 10^4}{3 \times 10^8} = 5$$

and the 3 db pulse widening due to the Doppler effect is 3.72. The distorted pulse shape would be similar to Fig. 6.24. If the criterion is adopted that the pulse spreading factor should be less than 1.1, then a limit on the usable time-bandwidth product is given by

$$\left|\frac{4v}{c}\right| T\Delta f < 2 \qquad \text{or} \qquad \left|\frac{2v}{c}\right| T\Delta f < 1 \qquad (6\text{-}63)$$

The criterion of (6-63) agrees with that derived for the general matched filter case by a number of workers. If sidelobe reduction weighting is used at the receiver then the pulse width degradation resulting from the Doppler effect can be markedly reduced for a given value of $\gamma_v T \Delta f$. For an allowable 3 db pulse widening factor of 2 the maximum value of $\gamma_v T \Delta f$ from Fig. 6.29 is 6.8. This assumes that the magnitude of the center frequency shift $2v\omega_0/c$ is less than 10% of the signal bandwidth, or that the center frequency of the receiver matched-filter bandpass has been shifted to account for $2v\omega_0/c$ as suggested by Cook and Heiss [21].

From the discussion above it can be seen that linear FM time-bandwidth products between 10^4 and 10^5 represent an upper limit before special compensation techniques are required to offset the Doppler dispersion effect for radar applications [14, 20]. Thor [14] has suggested the use of an hyperbolic FM function for the very large time-bandwidth case, and has shown that the compressed-pulse shape, which closely resembles the $(\sin x)/x$ waveform, is invariant under Doppler shift conditions. Thor shows that the phase modulation function for this signal is logarithmic, being described by

$$\omega_0 t + \theta(t) = \frac{\omega_0}{\mu} \log\left(1 - \frac{\mu t}{\omega_0}\right) = \omega_0 t + \frac{1}{2}\mu t^2 + \frac{2\mu^2 t^2}{3\omega_0} + \cdots$$

$$\doteq \omega_0 t + \frac{1}{2}\mu t^2 \left(1 + \frac{4\mu}{3\omega_0}\right) \qquad (6\text{-}64)$$

This modulation function may be more useful in the sonar case where the maximum velocities are much greater relative to the signal velocity of propagation than for the radar case. This application of the hyperbolic FM signal is discussed by Rowlands [22], who describes the analysis of a sidelobe reduction weighting function for this signal based on a tapped delay line implementation.

If the factor $\gamma_v T \Delta f$ is large and the range of Doppler shifts is relatively narrow the dispersive delay of the receiver matched filter can be designed to compensate for the Φ_1 phase term of (6-59). If the maximum value of $\gamma_v T \Delta f$ is very large ($\gg 10$) and the expected velocities are uniformly spread over the interval $-v_{\max}$ to $+v_{\max}$, then the implementation suggested by Benjamin may be used [23]. The receiver compression filter for this case is shown by Fig. 6.36. In this arrangement incremental linear delay is

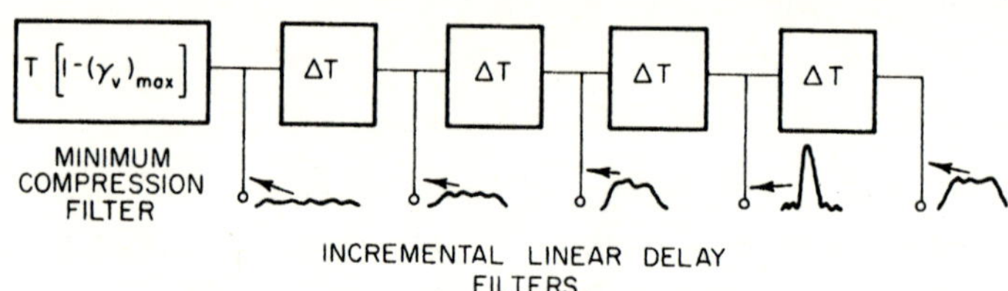

FIG. 6.36   Technique for velocity compensation reception for large time-bandwidth linear FM signals (after Benjamin [23]).

added between each of the output taps. A signal matched to the correct combination of linear delay will appear at that output tap as an optimally compressed pulse. The outputs from the other taps will be dispersed and reduced in peak amplitude if the time-bandwidth products of the incremental delay units can be made fairly large. This will provide a coarse indication of velocity if not too many signals are present to build up the interference resulting from the nonmatched outputs. A related velocity measuring technique is described by Cook and Chin [24].

The more exact effects of velocity on the general matched-filter output can be obtained by evaluating the matched filter response

$$\chi_d(\tau, \phi) = \sqrt{\alpha} \int u(\alpha t) u^*(t + \tau) \exp[-j2\pi\phi t] \, dt \tag{6-65}$$

where the limits on the integral are $-T/2\alpha < t < T/2\alpha$ for $2v/c > 0$; and $-T/2 < t < T/2$ for $2v/c < 0$. Rubin and DiFranco [25] have shown for this exact case that the time-bandwidth product of the signal $\sqrt{\alpha} u(\alpha t)$ is invariant under Doppler shift conditions. It can be assumed for the linear FM signal that the receiver filter is compensated for the center frequency translation but not for the linear FM mismatch. For this case the evaluation of (6-65) yields the following ratios of peak amplitude of the Doppler distorted compressed pulse to the optimum matched-filter output, for the condition $(2v/c)T\,\Delta f \leqslant 1$, or $\gamma_v T\,\Delta f \leqslant 2$

$$\frac{\left|\chi_{\max}\left(\dfrac{2v}{c}\right)\right|}{|\chi(0,0)|} = \frac{[1 - k_1(\gamma_v T\,\Delta f/4)^2]^{1/2}}{\left(1 + \dfrac{2v}{c}\right)^{1/2}}, \qquad \frac{2v}{c} > 0$$

$$= \left(1 - \frac{2v}{c}\right)^{1/2}[1 - k_2(\gamma_v T\,\Delta f/4)^2]^{1/2}, \qquad \frac{2v}{c} < 0 \qquad (6\text{-}66)$$

where

$$k_1 = \frac{\pi^2}{45}\left(\frac{2 + \dfrac{2v}{c}}{1 + \dfrac{2v}{c}}\right)^2, \qquad k_2 = \frac{\pi^2}{45}\left(2 - \frac{2v}{c}\right)^2$$

For the maximum value of $\gamma_v T\,\Delta f = 2$ the minimum values of (6-66) are approximately 0.88, resulting in slightly more than 1 db loss in signal-to-noise ratio at the compression-filter output from the Doppler dispersion effect. How this would be affected by the presence of a sidelobe reduction weighting filter at the receiver has not been evaluated.

Further distortion of the modulation function of the received signal would result if acceleration were a factor in addition to uniform radial velocity. This would add cubic and higher order phase terms to the Doppler distorted signal spectrum. A matrix of different combinations of velocity and acceleration compensation filters can be envisioned as a possible receiver implementation. Radar signal design applications of this complexity are beyond the introductory nature of the present volume and will not be treated. Suffice it to say that the linear FM signal, for most values of compression ratios that will be encountered in general systems applications, has the advantage of remaining essentially fully correlated at the matched filter output under Doppler shift conditions. Thus, there is very little loss in detection capability for Doppler shifted signals at the matched-filter output. The effect of the linear range displacement as a function of Doppler shift on range measurement accuracy is discussed in Section 9.5.

REFERENCES

1. C. E. Cook, A general matched-filter analysis of linear FM pulse compression, *Proc. IRE (Correspondence)* **49**, 831 (1961).
2. J. E. Chin and C. E. Cook, The mathematics of pulse compression—a problem in systems analysis, *Sperry Eng. Rev.* **12**, 11–16 (October, 1959).

3. C. E. Cook, Pulse compression—key to more efficient radar transmission, *Proc. IRE* **48**, 310–316 (1960).

4. J. R. Klauder, A. C. Price, S. Darlington, and W. J. Albersheim, The theory and design of chirp radars, *Bell System Tech. J.* **39**, 745–808 (1960).

5. E. Jahnke and F. Emde, "Tables of Functions." Dover, New York, 1945.

6. A. VanWijngaarden and W. L. Scheen, Tables of Fresnel integrals, *Computation Dept., Math. Center, Amsterdam*, Rept. R49 (1949).

7. L. B. Arguimbau, "Vacuum Tube Circuits." Wiley, New York, 1948.

8. C. E. Cook and J. E. Chin, A radar system employing complementary pulse expansion and compression filters, Unpublished research notes, 1956.

9. C. E. Cook and C. E. Brockner, Pulse Stretching and Compression Radar System, U.S. Patent 3,400,396, September 3, 1968 (filed November 14, 1955).

10. H. O. Ramp and E. R. Wingrove, Jr., Principles of pulse compression, *IRE Trans.* **MIL-5**, 109–116 (1961).

11. M. Bernfeld, Pulse-compression techniques, *Proc. IEEE* (*Correspondence*) **51**, 1261 (1963).

12. R. W. Jacobus, A linear FM 1000:1 pulse-compression system, The MITRE Corporation, Bedford, Massachusetts, Tech. Memo. TM-3529 (January, 1963).

13. R. D. Haggarty, Tapped delay line synthesis of large time-bandwidth signals, The MITRE Corporation, Bedford, Massachusetts, Tech. Memo. TM-3535 (January, 1963).

14. R. C. Thor, A large time-bandwidth product pulse-compression technique, *IRE Trans.* **MIL-6**, 169–173 (1962).

15. H. J. Bickel, Spectrum analysis with delay-line filters, *IRE WESCON Conv. Record* (Part 8), 59–67 (1959).

16. H. J. Bickel and R. I. Bernstein, Coherent Memory Filter, U.S. Patent No. 3,013,209.

17. J. Capon, High-speed Fourier analysis with recirculating delay-line-heterodyner feedback loops, *IRE Trans.* **I-10** (1961).

18. C. L. Temes, A. Citrin, M. A. Laviola, H. K. Boyce, and J. P. Biggs, Pulse-compression subsystem for a down-range tracker, *IEEE Intern. Conv. Record* (Part 8), 71–81 (1963).

19. P. Z. Peebles, Jr. and G. H. Stevens, A technique for the generation of highly linear FM pulse radar signals, *IEEE Trans.* **MIL-9**, 32–38 (1965).

20. H. O. Ramp and E. R. Wingrove, Jr., Performance degradation of linear FM pulse-compression systems due to the Doppler effect, *Proc. IRE* (*Correspondence*) **49**, 1693–1694 (1961).

21. C. E. Cook, W. H. Heiss, H. O. Ramp, and E. R. Wingrove, Jr., Linear FM Doppler distortion effects, *Proc. IRE* (*Correspondence*) **50**, 1535–1536 (1962).

22. R. O. Rowlands, Detection of a Doppler-invariant FM signal by means of a tapped delay line, *J. Acoust. Soc. Am.* **37**, 608–615 (1965).

23. R. Benjamin, Recent developments in radar modulation and processing techniques, *Proc. Inst. Elec. Engrs.* (*London*) **111**, 2002–2015 (1964).

24. C. E. Cook and J. E. Chin, Velocity Measuring Radar Apparatus for High Speed Vehicles, U.S. Patent No. 3,105,967, October, 1963.

25. W. L. Rubin and J. V. DiFranco, The effects of Doppler dispersion on matched-filter performance, *Proc. IRE* (*Correspondence*) **50**, 2127–2128 (1962).

Matched-Filter Waveform Considerations—Range Sidelobe Reduction

7.1 Introduction

In Chapter 6 it was shown that the matched-filter waveform associated with the linear FM pulse-compression signal has essentially a $(\sin x)/x$ shape, with time or range sidelobes extending on either side of the compressed pulse. The first, and largest, of these range sidelobes are only 13.2 db below the peak of the compressed pulse and the near sidelobes fall off at approximately 4 db per sidelobe interval, with the sidelobe null points being spaced $1/\Delta f$ seconds apart (Δf being the nominal spectrum bandwidth) for the ideal $(\sin x)/x$ waveform. When many objects, having a wide range of reflecting cross section areas, must be examined by the pulse-compression radar, the range sidelobes of the type described above represent sources of interference that can obscure weaker signals. This is illustrated in Fig. 7.1. The successful application of pulse-compression matched-filter techniques to radar systems often depends, quite critically in some instances, on producing at the output of the matched filter a compressed-pulse waveform having a low level of range sidelobes. For the general FM matched-filter signal, described earlier as

$$s(t) = a(t)\cos[\omega_0 t + \theta(t)] \tag{7-1}$$

the matched-filter output is given by either

$$g(t) = \int_{-\infty}^{\infty} s(\tau)s(\tau - t)\,d\tau \tag{7-2}$$

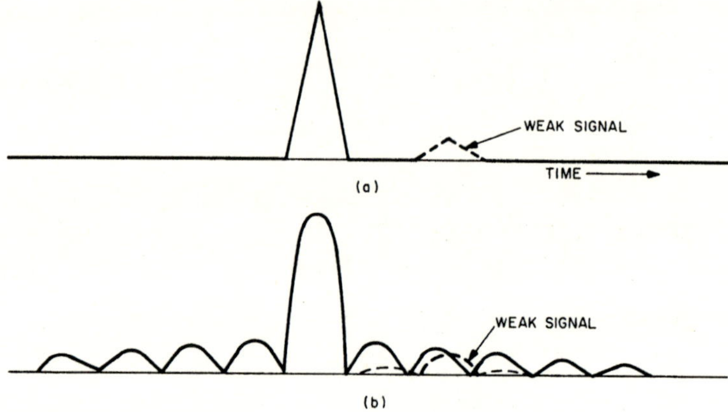

FIG. 7.1   Effect of range sidelobes on weak signal detection. (a) Normal pulse signals. (b) Matched-filter signals.

or

$$g(t) = \frac{1}{2\pi} \int_{-\infty}^{\infty} S(\omega)S^*(\omega) \exp[j\omega t]\, d\omega$$

$$= \frac{1}{2\pi} \int_{-\infty}^{\infty} |S(\omega)|^2 \exp[j\omega t]\, d\omega \tag{7-3}$$

The presence of the undesirable range sidelobes represents an integral part of the pulse-compression signal, being related to the modulus of the signal spectrum as shown by Eq. (7-3). To minimize the effects of these unwanted signals on system performance various approaches for controlling the spectrum modulus have been considered. These have fallen into the general areas of weighting, or shaping, the spectrum by frequency domain filtering or time domain envelope shaping; or by controlling the FM function, as outlined in the stationary phase discussion of Chapter 3, to directly produce a suitable matched-filter signal spectrum. Delay line cancellation techniques, referred to as transversal filtering, have also been employed to synthesize an effective frequency response function to yield a prescribed spectrum amplitude distribution at the matched-filter output. Regardless of the technique advocated, one of the important aspects of these general approaches has been the search for *optimum* spectrum modulus functions. Here, "optimum" is a broadly defined term that is usually taken to mean minimum loss in resolution and signal detectability for a maximum reduction of the interfering range sidelobes. However, the definition of optimum may also include the practical aspects of ease of implementation, sensitivity to changes in spectrum modulus parameters (design tolerances or errors), and the effect of changes in input signal center frequency on the range sidelobe level. The rate at which the range sidelobes fall off may also be an important consideration, in that the system designer may decide to accept somewhat larger sidelobes adjacent to the compressed pulse if lower side-lobes can be achieved further away, especially if in so doing there is less loss in the resolution and/or signal-to-noise degradation factors.

This chapter will treat in detail the various approaches to achieving reduced range sidelobes levels for the FM class of matched-filter signals. These are discussed separately from the digital, or phase coded, class of matched-filter signals, since the range sidelobe behavior of each class of functions exhibits different characteristics. The sidelobes of the FM matched-filter functions usually decrease as a function of time displacement from the center of the compressed pulse, and once the spectrum shape is fixed are relatively independent of the time-bandwidth product, or compression ratio. Thus it becomes important to emphasize the reduction of those range sidelobes adjacent to the desired signal. The level of the range sidelobes of the digital, or phase coded, class of functions are usually a function of the time-bandwidth product, and tend to continue at an almost constant level out to $\pm T$ seconds on either side of the compressed pulse, where T is the duration of the uncompressed signal. A comparison of these two types of range sidelobe behavior is shown in Fig. 7.2. The long

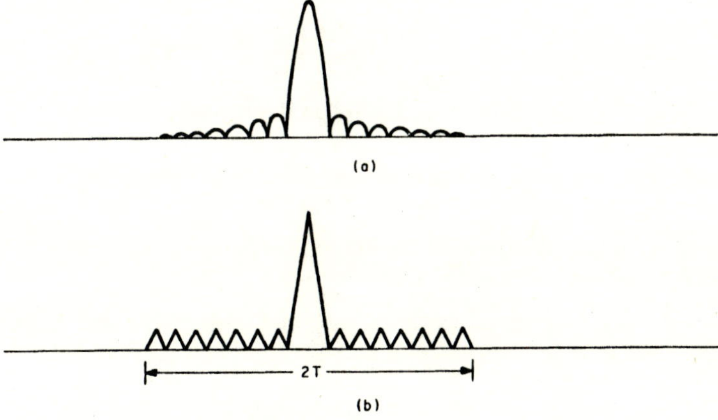

FIG. 7.2   Two types of matched-filter waveform range sidelobe structure. (a) FM matched-filter signal. (b) Phase coded matched-filter signal.

time duration of the range sidelobes of the phase coded signals can offer more serious problems in masking weaker signals than the range sidelobes of the FM type of signal. This becomes important in the dense target environment, discussed in Chapter 10. Treatment of the different approaches to phase coded matched-filter signals appears in Chapter 8.

One of the main conclusions that can be drawn from the topics covered in this chapter is that, for many applications, the linear FM signal with receiver frequency weighting to reduce sidelobes represents the best pulse-compression implementation. For this reason several special subjects related to this signal are included. Among these is the development of an FM pre-distortion function that can contribute to the efficient realization of low range sidelobes with the linear FM signal.

7.2 Spectrum Amplitude Functions for Desirable Matched-Filter Waveform Properties

In the context of the definition of the desired waveform resulting from an optimum spectrum amplitude distribution, attention has been naturally directed to the related topic of shaping a one-dimensional antenna illumination function. Such antenna illumination control is intended to yield reduced spatial sidelobes of the antenna beam in the far field pattern with minimum beam spreading and loss of directive power gain. This is analogous to achieving minimum pulse time spreading and signal-to-noise degradation for a maximum reduction of range sidelobes in the matched-filter waveform case. For a continuous finite dimension antenna of length d (where d is much larger than the radiated wavelength), in which the current distribution is given by $W(z)$, the far field electric field intensity as a function of direction angle ϕ, is given by the Fourier integral

$$E(\phi) = \int_{-d/2}^{d/2} W(z) \exp[j2\pi(z/\lambda)\sin\phi]\,dz \tag{7-4}$$

Equation (7-3) described the matched-filter output time function as

$$g(t) = \frac{1}{2\pi}\int_{-\infty}^{\infty} |S(\omega)|^2 \exp[j\omega t]\,d\omega$$

With a suitable change of variable, the equations above are seen to be equivalent, so that a spectrum modulus-squared function given by

$$|S(\omega)|^2 = W(z = \omega) \tag{7-5}$$

where $W(z)$ is a known antenna distribution function that yields a desirable radiation pattern, will result in a matched-filter output that has an amplitude vs time function that is the same as the radiation intensity vs $\sin\phi$ function. These relationships are summarized in Fig. 7.3. Many possible functions $W(z)$ have been studied in connection with antenna theory and a number of the more useful ones will be reviewed in this section. To present and compare data the idealized rectangular spectrum

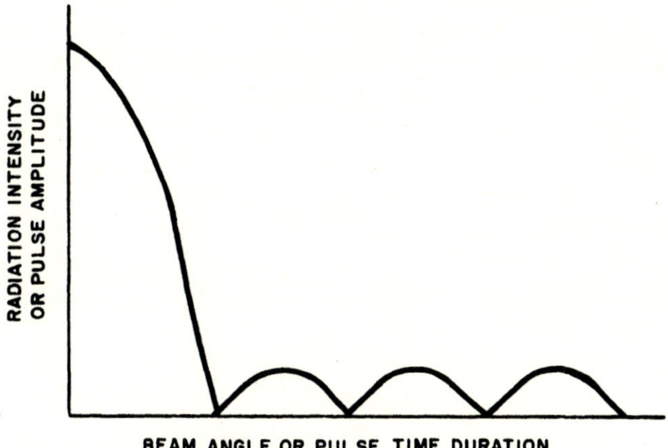

FIG. 7.3   Comparison of antenna and time waveform parameters.

and its $(\sin x)/x$ waveform will be used as a reference. It is assumed for many applications that the spectrum distribution functions obtainable by the FM class of pulse-compression signals are essentially smooth, ripple free, functions. The effect on actual matched-filter sidelobe behavior caused by not meeting this last condition in practice will be developed in a later section.

For the ideal matched-filter system Eq. (7-5) will hold, and the transmitted signal spectrum distribution is

$$|S(\omega)| = \sqrt{W(\omega)} \qquad (7\text{-}6)$$

This same relationship also describes the matched-filter receiver amplitude response, $|H(\omega)|$. For the case of

$$|S(\omega)| \neq |H(\omega)|$$

the relationship equivalent to Eq. (7-5) is

$$|S(\omega)| \cdot |H(\omega)| = W(\omega) \qquad (7\text{-}7)$$

This condition leads to a nonmatched amplitude response, and results in a signal-to-noise degradation at the matched-filter output compared to that obtained with the ideal matched filter. However, this may still be a useful approach to sidelobe reduction if the $|S(\omega)|$ implied by Eq. (7-6) is difficult to generate, or if this form of amplitude distribution implies a matched filter that has undesirable properties when the matched-filter signal parameters change. An example is the higher range sidelobes of certain nonlinear FM matched-filter outputs associated with the Doppler shifted return from a moving target. For a desired $W(\omega)$, the spectrum distribution $|S(\omega)|$ determines the receiver response, $|H(\omega)|$, and in the limiting case of the rectangular spectrum $|H(\omega)| = W(\omega)$. The general assumption will be that any intentional mismatching concerns only the spectrum and receiver amplitude distribution functions, and that the receiver transfer phase and the spectrum phase remain conjugate functions as described in Chapter 1.

As a general procedure, the function $W(\omega)$ required for any desired output time function $g(t)$ can be found by the Fourier relationship

$$W(\omega) = \int_{-\infty}^{\infty} g(t)\exp[-j\omega t]\,dt \tag{7-8}$$

Methods related to this approach have been developed for the synthesis of prescribed antenna field patterns. However, more specific methods are used where the criterion is narrowest beamwidth (or pulse width) for a desired sidelobe level.

Dolph–Chebyshev Distribution

The solution of the antenna problem of specifying the distribution function that results in the narrowest beamwidth for a desired sidelobe level was obtained by Dolph [1] for the case of equally spaced point sources fed in phase, and operated as a broadside array. By making use of Chebyshev polynomials to define the radiation pattern, Dolph showed that an optimum beam pattern was obtained for the criterion of minimum beamwidth for a given sidelobe level. In addition, the antenna pattern derived was characterized by all the sidelobes being of equal amplitude. As the number of elements in the Dolph–Chebyshev array increases the current density in the end elements becomes very large, becoming infinite for the case of the continuous antenna. An example of the Dolph–Chebyshev distribution function is given in Fig. 7.4. The expression for the distribution function for the limiting case of the continuous distribution has been derived by Van der Maas [2], and is

$$W(\omega)_D = \left\{ \frac{2\pi^2 A I_1\left[\pi A\left(1 - \dfrac{4\omega^2}{(\Delta\omega)^2}\right)\right]}{\Delta\omega\cosh\pi A\left(1 - \dfrac{4\omega^2}{(\Delta\omega)^2}\right)} \left[U\left(\omega + \frac{\Delta\omega}{2}\right) - U\left(\omega - \frac{\Delta\omega}{2}\right)\right] \right.$$

$$\left. + \left[\delta\left(\frac{2\omega}{\Delta\omega} - 1\right) + \delta\left(\frac{2\omega}{\Delta\omega} + 1\right)\right]\frac{2\pi}{\Delta\omega\cosh\pi A} \right\} \tag{7-9}$$

where $U(x)$ is the unit step function; $\delta(x)$ the dirac delta function; $I_1(x)$ the

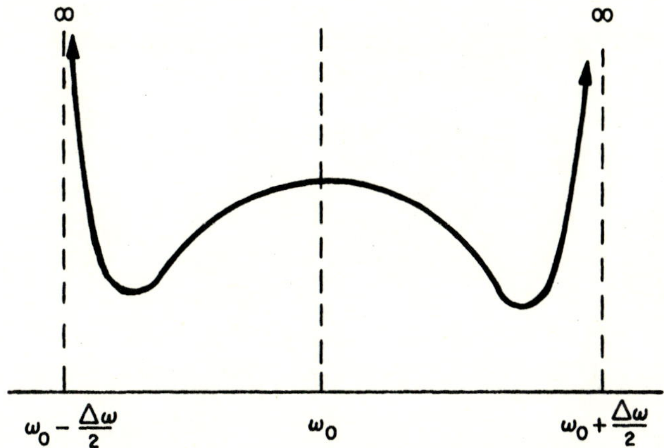

FIG. 7.4 Example of a continuous Dolph–Chebyshev distribution.

modified Bessel function, first kind, first order; and $\omega = 0$ is equivalent to the center of the i.f. spectrum.

Transformation of the Dolph–Chebyshev function to the frequency domain, as a matched-filter output spectrum distribution function, results in the time function.

$$g(t)_D = \frac{\cos \pi\sqrt{[(\Delta\omega/2\pi)t]^2 - A^2}}{\cosh \pi A}, \qquad -\infty < t < \infty \qquad (7\text{-}10)$$

The factor A is defined by the desired sidelobe level, such that

$$20 \log\left(\frac{1}{\cosh \pi A}\right) = \text{sidelobe level in db} \qquad (7\text{-}11)$$

The time waveform given by Eq. (7-10) has a constant level of sidelobes over the infinite time extent, as shown in Fig. 7.5. Klauder *et al.* [3] and Temes [4] have commented on the physical unrealizability of this time waveform sidelobe structure because of the infinite energy contained in $g(t)_D$. Essentially, this results from the fact that the continuous Dolph–Chebyshev function $W(\omega)_D$ is itself physically unrealizable because of the required infinite gain at the spectrum band edges. In practical applications to pulse-compression matched-filter systems, the characteristics of the Dolph–Chebyshev response are used as a standard against which to compare the results of other methods used to derive realizable approximations to the Dolph function.

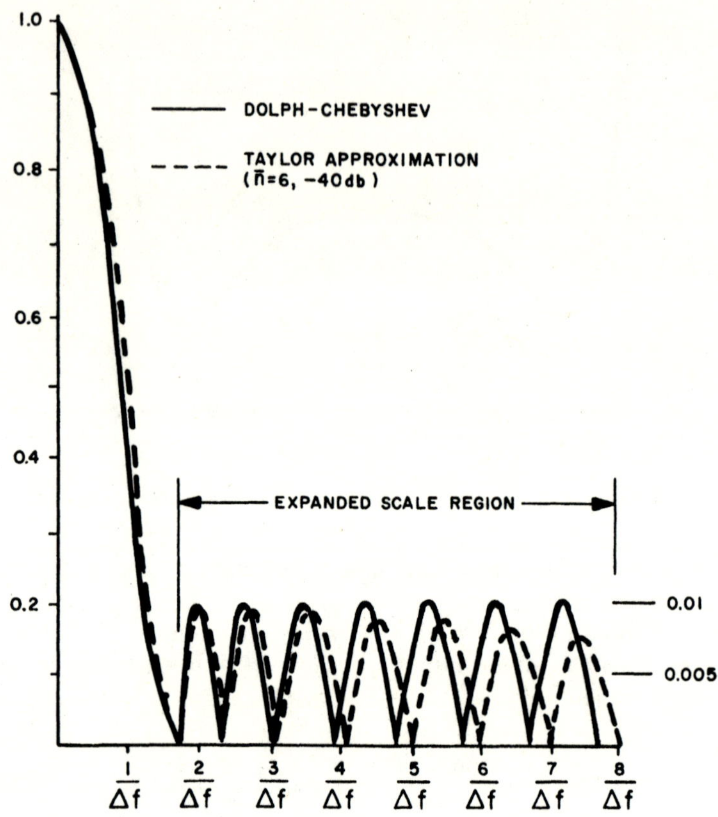

FIG. 7.5 Comparison of Dolph–Chebyshev and Taylor waveforms (from Klauder *et al.*
[3], courtesy of the Bell System Tech. Journal).

Taylor Distribution

The approximation to the Dolph–Chebyshev distribution, derived by
Taylor [5], is

$$W(\omega)_T = 1 + 2 \sum_{m=1}^{\bar{n}-1} F_m \cos \frac{2\pi m\omega}{\Delta\omega} \tag{7-12}$$

where

$$F_m = \frac{0.5(-1)^{m+1}}{\prod\limits_{p=1}^{\bar{n}-1}\left(1 - \dfrac{m^2}{p^2}\right)} \prod_{n=1}^{\bar{n}-1}\left[1 - \frac{\sigma_p^{-2}m^2}{A^2 + (n - \tfrac{1}{2})^2}\right]$$

$$\sigma_P = \frac{\text{Taylor 3 db pulse width}}{\text{Dolph–Chebyshev 3 db pulse width}}$$

The parameter A is defined in Eq. (7-11). The number of terms used in the Taylor approximation determines the closeness of the fit to the optimum, but unrealizable, response. When the desired sidelobe level has been specified, thus determining the factor A, and the number of terms to be used in Eq. (7-12) has been chosen, then the additional pulse widening over the Dolph–Chebyshev time response is given by

$$\sigma_P = \frac{\bar{n}}{[A^2 + (\bar{n} - \frac{1}{2})^2]^{1/2}} > 1 \tag{7-13}$$

Equation (7-13) states that the lower the sidelobe level to be achieved, the greater the number of terms in the expansion of (7-12) must be in order to keep the pulse widening σ_P near unity.

The time response of a Taylor weighted matched-filter output is

$$g(t)_T = \frac{\Delta\omega}{2\pi} \left[\frac{\sin(\Delta\omega\, t/2)}{\Delta\omega\, t/2} + \sum_{m=1}^{\bar{n}-1} F_m \frac{\sin[(\Delta\omega\, t/2) + m\pi]}{(\Delta\omega\, t/2) + m\pi} \right. $$
$$\left. + \sum_{m=1}^{\bar{n}-1} F_m \frac{\sin[(\Delta\omega\, t/2) - m\pi]}{(\Delta\omega\, t/2) - m\pi} \right] \tag{7-14}$$

Expression (7-14) represents the summation of time shifted $(\sin x)/x$ pulses weighted by the factors F_m. The representation of the signal in this form leads to the conclusion that the Taylor frequency response function, as well as other response functions, can be synthesized by tapped delay-line techniques. This approach is known as transversal filtering, and is described in Section 7.8. The Taylor time waveform is characterized by constant level sidelobes in the vicinity of the main lobe, or compressed pulse. The factor $\bar{n}$ governs the extent of the interval in which this property holds. Outside this region the sidelobes fall off. As seen from Eq. (7-13), $\bar{n}$ also controls the pulse widening factor σ_P. The larger $\bar{n}$ is, the greater is the extent of the constant sidelobe region and the smaller the pulse widening factor becomes. However, if $\bar{n}$ becomes too large the Taylor response function $W(\omega)_T$ develops some of the same features that are objectionable in the Dolph–Chebyshev response. Thus, for $\bar{n}$ large, the response function peaks up at the band edges as well as in the band center. For low values of $\bar{n}$, $W(\omega)_T$ has a maximum at the center frequency of the distribution and a minimum at the band edges, with decreasing monotonic behavior from center frequency to the band edges. This latter type of response function is preferable from the point of view of attempting to synthesize receiver bandpass functions to obtain a Taylor spectrum distribution. However, for any given level of sidelobes there exists a minimum value of $\bar{n}$ that must be used to achieve the results given by Taylor. As an example, if sidelobes 25 db below the peak signal are required the minimum value of $\bar{n}$ is 3, and if sidelobes 40 db down on are required a value of $\bar{n} = 6$ is needed.

Figure 7.5 compares a Taylor weighted time function with a Dolph–Chebyshev weighted time function for a -40 db sidelobe level and $\bar{n} = 6$. Figure 7.6 plots some typical Taylor distribution functions for various sidelobe levels and values of $\bar{n}$. The relative pulse widening factors, compared to the $(\sin x)/x$ pulse width (rectangular spectrum $= \Delta\omega$), for both the Dolph–Chebyshev and Taylor functions as a function of the sidelobe level specified are plotted in Fig. 7.15. More detailed data can be obtained from Taylor's original paper.

Modified Taylor Functions—Hamming Weighting

The Taylor functions shown in Fig. 7.5 are obtained from a weighted cosine series added to a constant, or pedestal. These functions may not be easily approximated by filter design or time function amplitude modulation. A somewhat simpler expression can be obtained by dropping some of the higher order terms of Eq. (7-12) that have very small values of the coefficient F_m. This can be done without too serious an effect on the time function of Eq. (7-14). As an example, if all but the first terms are dropped a truncated Taylor function is obtained

$$\overline{W(\omega)_T} = 1 + 2F_1 \cos \frac{2\pi\omega}{\Delta\omega} \tag{7-15}$$

For the case of $\bar{n} = 6$ and a -40 db design sidelobe level, the truncated Taylor function becomes

$$\overline{W(\omega)_T} = 1 + 0.84 \cos \frac{2\pi\omega}{\Delta\omega} \tag{7-16}$$

This particular response function results in range sidelobes below -40 db and a pulse width only slightly wider than the exact Taylor time function at the -3 db points.

Equation (7-16) represents a smoother function that is less difficult to approximate by either frequency or time weighting methods. Normalization of this equation to provide unit amplitude for $\omega = 0$, and application of a standard trigonometric identity, yields

$$\overline{W(\omega)_T} = 0.088 + 0.912 \cos^2 \frac{\pi\omega}{\Delta\omega} \tag{7-17}$$

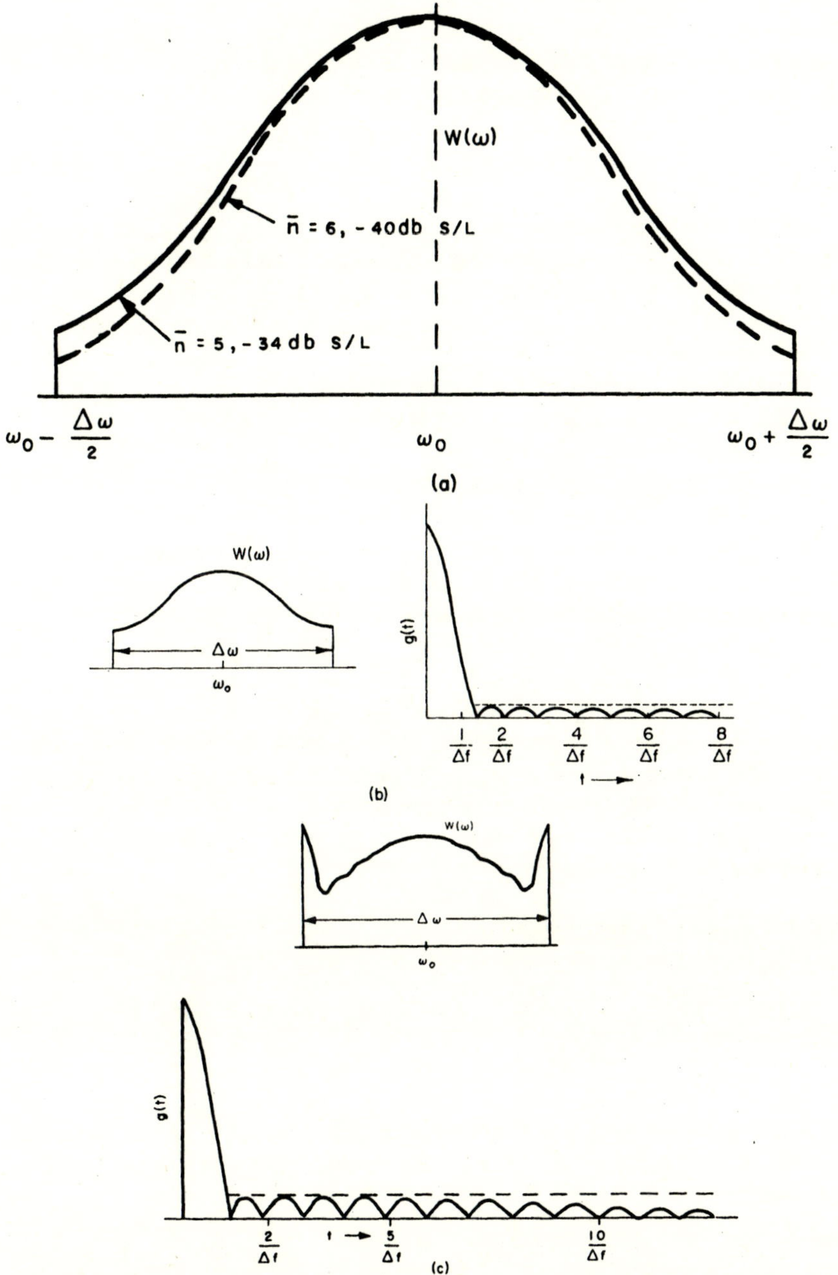

FIG. 7.6   Taylor weighting functions and waveforms. (a) Taylor frequency weighting functions. (b) Spectrum function and waveform for $\bar{n} = 5$, -25 db S/L. (c) Spectrum function and waveform for $\bar{n} = 10$, -20 db S/L. (From Bernfeld *et al.*, Ref. 4, Chapter 1 and Taylor [5]).

This is seen to be a weighted cosine-squared function on a baseline pedestal. A well-known weighting function that is close to this one is the Hamming function [4], which can be expressed by

$$W(\omega)_H = 0.08 + 0.92 \cos^2 \frac{\pi\omega}{\Delta\omega} \tag{7-18}$$

The time function characteristics associated with the Hamming weighting function yield a peak sidelobe level of -42.8 db and a 3 db pulse width of 1.47 compared to a 3 db Taylor pulse width (-40 db sidelobes) of 1.41 (pulse width associated with rectangular spectrum of width $\Delta\omega$ establishes a reference pulse width $= 1.0$).

Equations (7-17) and (7-18) can be treated as special cases of a more general weighting function that is given by

$$W(\omega) = k + (1 - k)\cos^2 \frac{\pi\omega}{\Delta\omega} \tag{7-19}$$

This represents a cosine-squared response weighted by the factor $(1 - k)$ sitting on a baseline pedestal of height k, where $k \leqslant 1$. The pulse width and range sidelobe levels associated with this weighted spectrum function vary as a function of the pedestal height. The Hamming weighting ($k = 0.08$) yields the lowest sidelobe levels for this class of functions. Comparative data for the weighting functions discussed in this section are given in Section 7.5 where it is shown that the value of the weighting function pedestal can have a critical effect on the matched-filter output time response, indicating that the specification of the Taylor or Hamming spectrum functions for -40 db range sidelobes requires a rigid control over the weighting function parameters.

General Cosine-Power Weighting

Equation (7-19) provides a basis for an even more general weighting function given by

$$W(\omega) = k + (1 - k)\cos^n \frac{\pi\omega}{\Delta\omega} \tag{7-20}$$

When n is an integer this function includes the cosine power weighting functions that have been studied in antenna applications [6]. The time response function for this class of spectrum response functions is given by

$$g(t) = \frac{\Delta\omega}{2\pi}\left[k\frac{\sin(\Delta\omega\, t/2)}{\Delta\omega\, t/2} + \frac{1 - k}{\pi}\int_0^\pi \cos^n u \cos\left(\frac{\Delta\omega}{\pi}ut\right)du \right] \tag{7-21}$$

Data has been calculated for noninteger values of n in the range from

$n = 1.8$ to $n = 2.2$, for various values of k, to evaluate the effect of errors in the Hamming response function. This data also provides approximate results on the effects of parameter errors for the -40 db Taylor weighting function.

7.3 Comparison of Time Weighting and Frequency Weighting for Linear FM Sidelobe Reduction

Methods for obtaining the spectrum function $W(\omega)$ that yields the desired range sidelobe structure can be based on amplitude weighting the envelope of the transmitted FM signal or on amplitude weighting the receiver matched filter frequency response. This section discusses the application of these two methods to the linear FM signal.

The linear FM signal that is time weighted by the function $w(t)$ can be represented by its complex form

$$u(t) = w(t) \exp\left[j\left(\omega_0 t + \frac{\mu t^2}{2}\right)\right], \qquad -\frac{T}{2} \leqslant t \leqslant \frac{T}{2} \qquad (7\text{-}22)$$

$$= 0 \qquad \text{elsewhere}$$

where $w(t)$ is a real, even function. Assume a receiver filter without band-limiting, and a phase characteristic given by (as shown in Fig. 7.7)

$$H(\omega) = \exp\left[-j\frac{(\omega - \omega_0)^2}{2\mu}\right] \qquad (7\text{-}23)$$

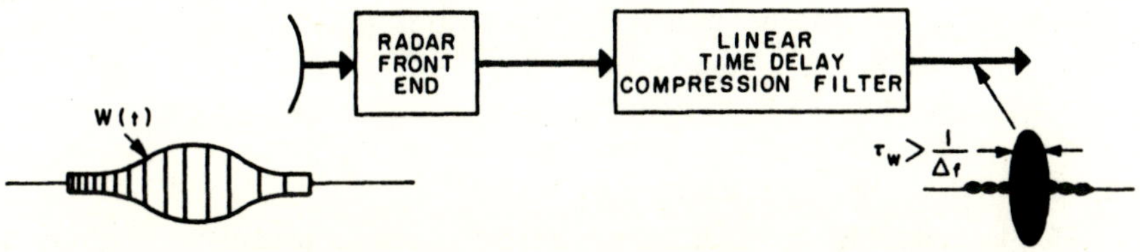

FIG. 7.7   Range sidelobe reduction by time weighting of transmitted signal.

The spectrum of the filter output is described by

$$G(\omega) = H(\omega)\int_{-T/2}^{T/2} w(\tau) \exp\left[j\left(\omega_0 \tau + \frac{\mu\tau^2}{2}\right)\right] \exp[-j\omega t]\, d\tau \qquad (7\text{-}24)$$

and the corresponding output time function is

$$g(t) = \frac{1}{2\pi}\int_{-\infty}^{\infty} H(\omega)\left\{\int_{-T/2}^{T/2} w(\tau) \exp\left[j\left(\omega_0 \tau + \frac{\mu\tau^2}{2}\right)\right] \exp[-j\omega\tau]\, d\tau\right\}$$

$$\times \exp[j\omega t]\, d\omega \qquad (7\text{-}25)$$

The order of integration of Eq. (7-25) can be rearranged, so that by making use of the integral

$$\int_{-\infty}^{\infty} H(\omega)\exp[\,j\omega(t-\tau)]\,d\omega = \sqrt{2\pi\mu}\,\exp\left[\,j\left(\omega_0(t-\tau)-\frac{\mu(t-\tau)^2}{2}+\frac{\pi}{4}\right)\right]$$

$$(7\text{-}26)$$

Equation (7-25) can be written

$$g(t)=\sqrt{\frac{\mu}{2\pi}}\,\exp\left[\,j\left(\omega_0 t-\frac{\mu t^2}{2}+\frac{\pi}{4}\right)\right]\int_{-T/2}^{T/2}w(\tau)\exp[\,j\mu\tau t]\,d\tau \qquad (7\text{-}27)$$

Since $w(t)$ was given as a real, even function, the magnitude of $g(t)$ is

$$|g(t)|=\sqrt{\frac{\mu}{2\pi}}\int_{-T/2}^{T/2}w(\tau)\exp[\,j\mu\tau t]\,d\tau \qquad (7\text{-}28)$$

Equation (7-28) establishes a Fourier pair relationship between the input and output time envelope functions for the linear FM signal processed through a linear-delay-only filter. When the time-bandwidth product of the linear FM waveform is large, then Eq. (7-28) is also a good representation of the waveform seen at the output of the filter that is matched to the rectangular envelope linear FM signal but has the input signal weighted by $w(t)$. This was shown in the previous chapter for $w(t)$ a rectangular function, for which the matched-filter output given by Eq. (6-13) is a close approximation to the $(\sin x)/x$ waveform obtained by Eq. (7-28). It can be seen that the development above leads to an expression for the compressed-pulse waveform that is the exact analog of the antenna pattern function given by Eq. (7-4).

If the linear FM spectrum is assumed to be a rectangular distribution, then functionally equivalent frequency response $W(\omega)$ and time weighting $w(t)$ will yield similar compressed-pulse waveform shapes. Figure 7.8

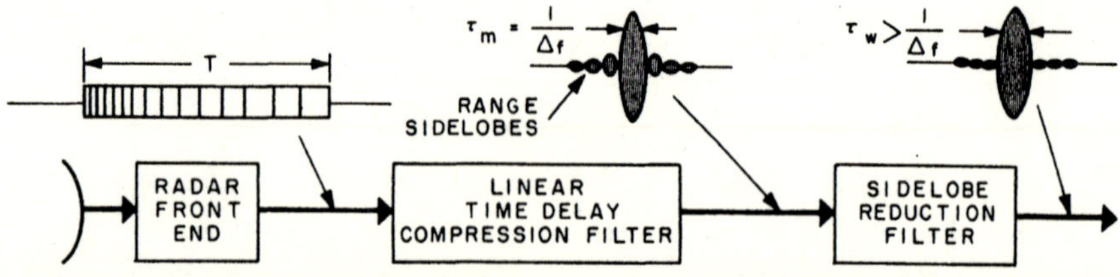

FIG. 7.8 Range sidelobe reduction by weighting of receiver frequency response.

illustrates one form of frequency response weighting for a linear FM signal. The receiver weighting function $W(\omega)$ results in a departure from the ideal matched-filter condition. This causes a degradation in the compression-filter output signal-to-noise ratio compared to the optimum condition, and is one of the costs associated with receiver frequency response weighting for sidelobe reduction. However, when the transmitter is peak power limited, this signal-to-noise loss is less severe than the loss in received energy for the case of the amplitude weighted transmitted signal and matched amplitude response receiver. This is discussed in detail in Section 7.4.

The equivalency of the weighting functions $w(t)$ and $W(\omega)$ cited above actually represents an approximation that improves as the time-bandwidth product of the signal increases. The exact frequency weighting function equivalent of $w(t)$ can be obtained by taking the ratio of the spectrum of (7-24) and the spectrum of the rectangular envelope linear FM signal $S(\omega)$, where $S(\omega)$ is described by Eqs. (6-19)–(6-21). Thus,

$$\frac{G(\omega)}{S(\omega)} = \frac{H(\omega)}{S(\omega)} \int_{-T/2}^{T/2} w(\tau) \exp\left[j\left(\omega_0 \tau + \frac{\mu \tau^2}{2}\right)\right] \exp[-j\omega\tau]\, d\tau \quad (7\text{-}29)$$

From (7-29) is obtained the desired relationship

$$W(\omega) = \frac{G(\omega)}{S(\omega)H(\omega)} \quad (7\text{-}30)$$

As an example, the time weighting function for the linear FM signal given by

$$w(t) = k + (1 - k)\cos^2 \frac{\pi t}{T}, \qquad -\frac{T}{2} \leqslant t \leqslant \frac{T}{2} \quad (7\text{-}31)$$

results in the spectrum

$$G(\omega) = H(\omega)\left[\frac{k+1}{2} S(\omega) + \frac{1-k}{4} S\left(\omega + \frac{2\pi}{T}\right) + \frac{1-k}{4} S\left(\omega - \frac{2\pi}{T}\right)\right] \quad (7\text{-}32)$$

Making use of Eq. (7-30), we obtain the frequency weighting function for this class of signals:

$$W(\omega) = \frac{k+1}{2} + \frac{1-k}{4}\left[\frac{S[\omega + (2\pi/T)] + S[\omega - (2\pi/T)]}{S(\omega)}\right] \quad (7\text{-}33)$$

The filter response function given by Eq. (7-33) will contain a phase term, since the $H(\omega)$ given by Eq. (7-23) is not an exact match for the spectrum phase term of a time truncated linear FM signal. For the large time-bandwidth product case the effect of this phase term is negligible, as has been shown by Cook [7] for the rectangular envelope linear FM signal. For this case an approximation for $W(\omega)$ can be obtained by noting that the desired compression-filter output spectrum should have the form of $w(t)$, substituting frequency for time. Thus,

$$W(\omega)_{\text{approx}} = \frac{w(t = \omega)}{|S(\omega)|} \tag{7-34}$$

so that for the time weighting functions described by (7-31) the equivalent frequency weighting function is given by

$$W(\omega)_{\text{approx}} = \frac{k + (1 - k)\cos^2[\pi(\omega_0 - \omega)/\Delta\omega]}{|S(\omega)|} \tag{7-35}$$

Figure 7.9 plots the equivalent $W(\omega)$ for the Hamming weighted ($k = 0.08$)

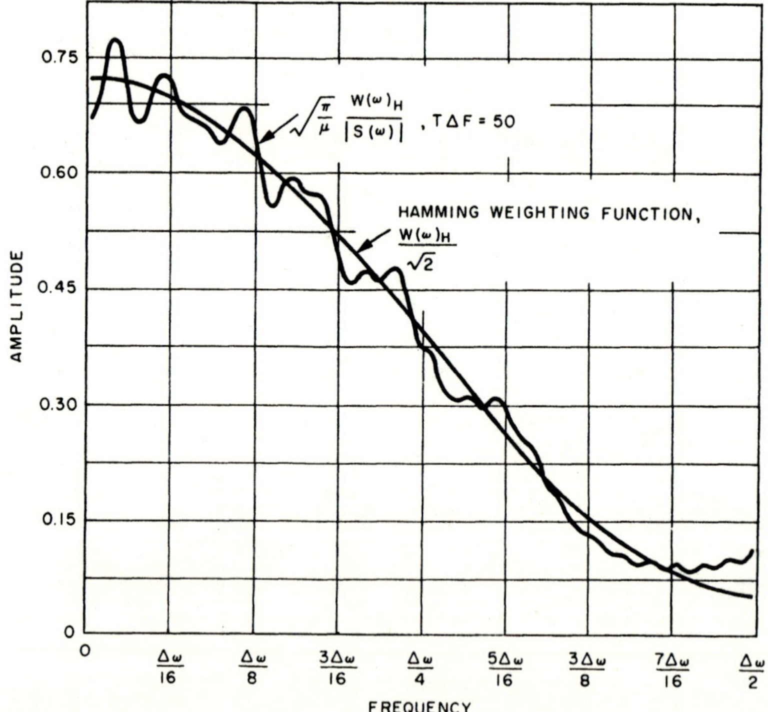

FIG. 7.9   Example of required frequency weighting function to yield a Hamming weighted output signal for a linear FM input (ripple structure is a function of compression ratio). (Courtesy of J. Paolillo, Sperry Gyroscope Company, Great Neck, New York.)

linear FM signal for the case of $T \Delta f = 50$. The rippling nature of this response function is difficult to synthesize, and in addition would yield a compressed pulse sidelobe structure that is sensitive to frequency shifts of the input signal. A more practical approach is to remove or reduce the ripples from the linear FM spectrum, as described in Section 7.7, and design a smooth frequency response function that has the same shape as $w(t)$.

Ordinarily, weighting of the envelope of the transmitted signal in a high power radar is not attempted, since the final amplifier stages inherently operate class C and are not subject to amplitude control. However, control of the linear FM time envelope in a low power laboratory instrumentation can provide a valuable means for examining the effect of various weighting function characteristics and determining the sidelobe behavior of a given compression-filter implementation. This is illustrated in Fig. 7.10 which

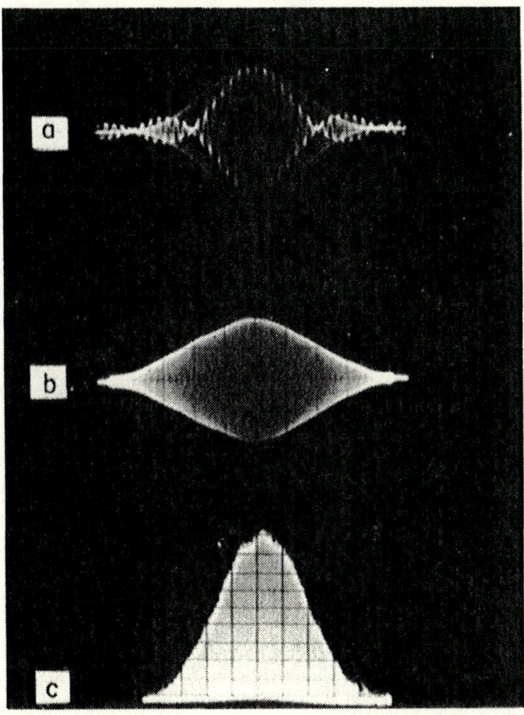

FIG. 7.10 Envelope weighting with odd harmonics (triangle approximation), $k = 0.08$. (a) Weighted compressed pulse overlapping $(\sin x)/x$ pulse. (b) Weighted linear FM signal. (c) Signal spectrum.

shows the effect of the weighting function on the linear FM spectrum, as well as on the sidelobe structure and pulse width of the compressed pulse. Another example is shown in Fig. 7.11, in which the time weighting function

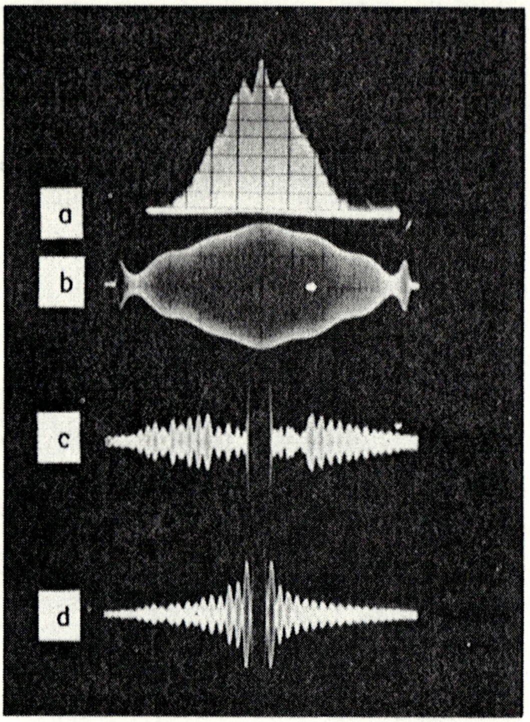

FIG. 7.11 Envelope weighting based on sidelobe cancellation by amplitude paired-echoes. (a) Signal spectrum. (b) Weighted linear FM signal. (c) Sidelobe detail showing cancellation effect. (d) $(\sin x)/x$ sidelobes of unweighted compressed pulse, 6 db attenuation relative to (c).

is composed of independently controlled cosine functions of different frequencies. Each cosine function can be adjusted to yield a maximum reduction of the sidelobes of the $(\sin x)/x$ waveform in a specific interval. If a sufficient number of cosine components are available, this technique can serve as an analog computer for arriving at the best combination of output tap gain levels for a transversal filter realization of a frequency weighting response that will also include compensation for the effects of amplitude distortion in the pulse-compression filter. This technique has been used to demonstrate that closer approximations to the Dolph–Chebychev results than given by the Taylor weighting technique can be achieved over a restricted range of specified sidelobe levels.

7.4 Effect of Weighting on Matched-Filter Output Signal-to-Noise Ratio

In the preceding section it was shown that either time weighting of the transmitted signal or frequency weighting of the receiver response function can be used to obtain identical pulse-compression output time functions. In addition, there is no loss of generality to this procedure if the overall weighting is shared between the transmitted signal and the pulse-compression receiver, as would be the case for a matched-filter implementation. For the linear FM signal the time weighting $w(t)$ and the frequency weighting $W(\omega)$ required to produce the same output time response are essentially identical.

If time weighting is employed, and the average energy in the transmitted signal can always be adjusted to equal that of the unweighted signal (Fig. 7.12(a)), then the receiver output signal-to-noise ratio is independent of whether transmitter or receiver weighting is used. However, the more realistic condition is that the transmitted signal, whether weighted or not, always has the same peak power limit (Fig. 7.12(b)). For this situation it is

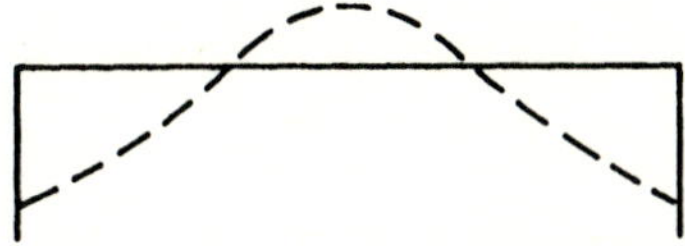

(a) EQUAL AVERAGE POWER   (b) PEAK POWER LIMITED

FIG. 7.12   Time weighted pulse envelope functions compared to rectangular envelope.

preferable to have all the weighting done at the receiver. This conclusion results from considering the four cases of time and/or frequency weighting illustrated in Fig. 7.13. It is assumed, for the development that follows, that the unweighted linear FM spectrum is rectangular and bandlimited over the interval $\Delta\omega$. This allows the use of functionally identical time and frequency weighting responses in the derivations below, which for cases *I–III* are based on obtaining the pulse-compression filter output signal-to-noise ratio for similar compressed-pulse waveforms. In all the cases considered the receiver input noise is assumed to be white, having a power density in the positive frequency region of N_0 watt/cycle/second.

Case I: Time Weighting.   Using the result of Eq. (7-28), the peak output signal voltage, which occurs at $t = 0$, is

$$g(0) = g(t)_{\max} = \sqrt{\frac{\mu}{2\pi}} \int_{-T/2}^{T/2} w(t)\, dt \qquad (7\text{-}36)$$

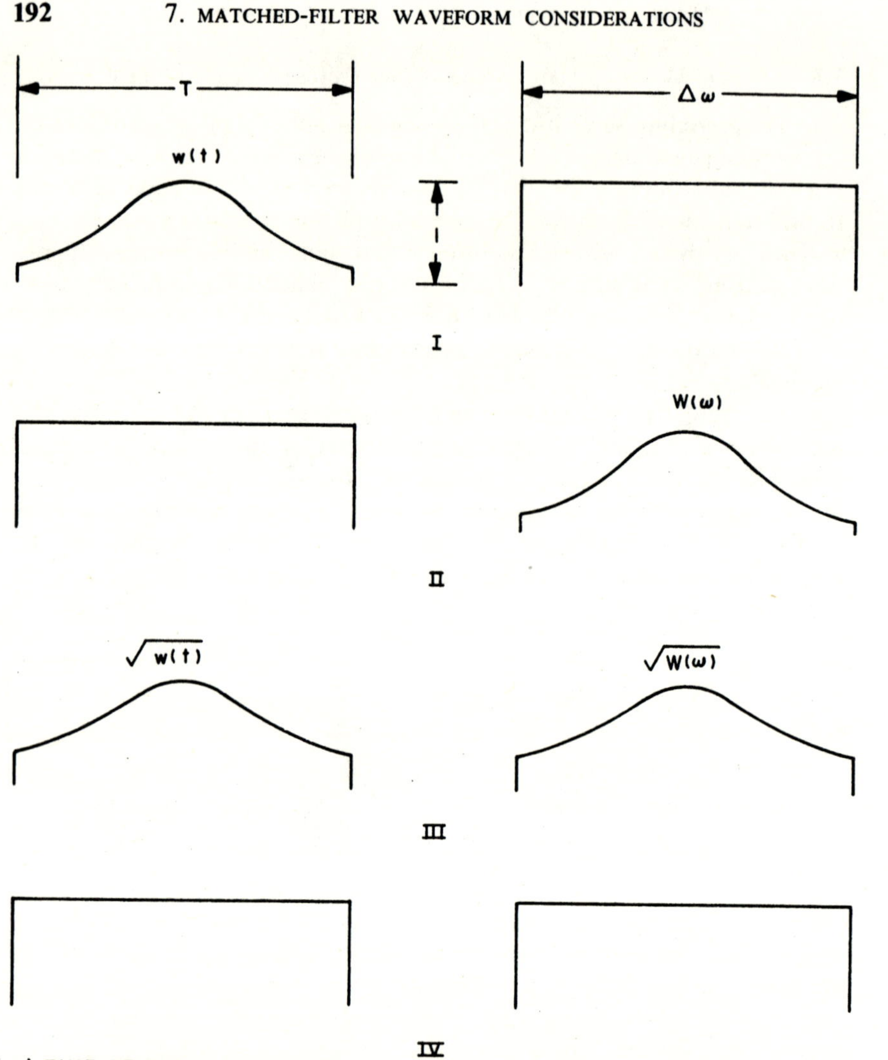

FIG. 7.13   Four basic cases for matched-filter output S/N considerations.

The average output noise power is given by

$$\sigma^2 = \frac{N_0}{2\pi}\int_{-\Delta\omega/2}^{\Delta\omega/2} W^2(\omega + \omega_0)\,d\omega \tag{7-37}$$

For case I (rectangular receiver bandpass), this is

$$\sigma_1^2 = N_0 \frac{\Delta\omega}{2\pi} \tag{7-38}$$

Thus, the output signal-to-noise ratio, defined by

$$\left(\frac{S}{N}\right)_1 = \frac{[g(t)_{\max}]^2}{\sigma_1^2}$$

is

$$\left[\sqrt{\frac{\mu}{2\pi}} \int_{-T/2}^{T/2} w(t)\, dt\right]^2 \Bigg/ N_0 \frac{\Delta\omega}{2\pi} \tag{7-39}$$

Case II: Frequency Weighting. The equivalent frequency weighting function for the time weighting of case I is

$$W(\omega) = w\left(\frac{\omega - \omega_0}{\mu}\right) \tag{7-40}$$

where the equivalence is obtained from the linear FM frequency-time relation

$$\omega = \omega_0 + \mu t$$

The amplitude of the assumed rectangular input spectrum is

$$|S(\omega)| = \sqrt{\frac{\pi}{2\mu}}$$

for the normalized unit amplitude received signal, so that the output voltage time function is given by

$$g(t) = \frac{\cos \omega_0 t}{\sqrt{2\pi\mu}} \int_{-\Delta\omega/2}^{\Delta\omega/2} w\left(\frac{\bar{\omega}}{\mu}\right) \exp[j\bar{\omega}t]\, d\bar{\omega} \tag{7-41}$$

where $\bar{\omega} = \omega - \omega_0$. By making a further change of variable

$$\bar{\omega}/\mu = x$$

and noting that

$$\Delta\omega/2 = \mu T/2$$

the peak output voltage is

$$g(t)_{\max} = \sqrt{\frac{\mu}{2\pi}} \int_{-T/2}^{T/2} w(x)\, dx \tag{7-42}$$

The expression above is identical to Eq. (7-36).

For this case the average output noise power is (letting $\omega_0 = 0$)

$$\sigma_{II}^2 = \frac{N_0}{2\pi} \int_{-\Delta\omega/2}^{\Delta\omega/2} W^2(\omega)\, d\omega \tag{7-43}$$

Thus, the output signal-to-noise ratio is

$$\left(\frac{S}{N}\right)_{II} = \frac{\left[\sqrt{\dfrac{\mu}{2\pi}} \displaystyle\int_{-T/2}^{T/2} w(x)\, dx\right]^2}{\dfrac{N_0}{2\pi} \displaystyle\int_{-\Delta\omega/2}^{\Delta\omega/2} W^2(\omega)\, d\omega} \tag{7-44}$$

Case III: Matched-Filter Implementation—Time and Frequency Weighting.   In a manner similar to that of case II it can be shown that the peak voltage is also identical to Eq. (7-42) (the details of this are left to the reader as an exercise). However, for this case the average output noise power is

$$\sigma_{III}^2 = \frac{N_0}{2\pi} \int_{-\Delta\omega/2}^{\Delta\omega/2} W(\omega)\, d\omega \tag{7-45}$$

and the output signal-to-noise ratio is

$$\left(\frac{S}{N}\right)_{III} = \frac{\left[\sqrt{\dfrac{\mu}{2\pi}} \displaystyle\int_{-T/2}^{T/2} w(x)\, dx\right]^2}{\dfrac{N_0}{2\pi} \displaystyle\int_{-\Delta\omega/2}^{\Delta\omega/2} W(\omega)\, d\omega} \tag{7-46}$$

By putting the integrals of Eqs. (7-39), (7-43), and (7-46) on a normalized basis, the respective signal-to-noise ratios for each of the cases above can be written (noting the interchangeability of $w(t)$ and $W(\omega)$):

$$\left(\frac{S}{N}\right)_{I} = \frac{T\,\Delta f \left[\displaystyle\int_{-1/2}^{1/2} W(y)\, dy\right]^2}{N_0\, \Delta f} \tag{7-47}$$

$$\left(\frac{S}{N}\right)_{II} = \frac{T\,\Delta f \left[\displaystyle\int_{-1/2}^{1/2} W(y)\, dy\right]^2}{N_0\, \Delta f \displaystyle\int_{-1/2}^{1/2} W^2(y)\, dy} \tag{7-48}$$

$$\left(\frac{S}{N}\right)_{III} = \frac{T\,\Delta f \left[\displaystyle\int_{-1/2}^{1/2} W(y)\, dy\right]^2}{N_0\, \Delta f \displaystyle\int_{-1/2}^{1/2} W(y)\, dy} \tag{7-49}$$

From the nature of the receiver frequency response functions for these cases it can be seen that

$$\int_{-1/2}^{1/2} W^2(y)\, dy \leqslant \int_{-1/2}^{1/2} W(y)\, dy \leqslant 1 \tag{7-50}$$

Thus,

$$(S/N)_{\mathrm{II}} \geqslant (S/N)_{\mathrm{III}} \geqslant (S/N)_{\mathrm{I}} \tag{7-51}$$

The equality signs in Eqs. (7-50) and (7-51) apply only if the weighting function is uniform (i.e., $w(y)$ or $W(y) = 1$). When the weighting is not uniform, which will be the case when reduced range sidelobes are desired, applying the weighting function at the receiver represents the best possible way to preserve detection capability despite the mismatch loss that results from not having the receiver response matched to the spectrum amplitude distribution.

The mismatch loss can be established by noting that the output signal-to-noise ratio for the rectangular envelope linear FM matched-filter system (case IV) is given by

$$\left(\frac{S}{N}\right)_{\mathrm{IV}} = \frac{T\,\Delta f}{N_0\,\Delta f} = \frac{T}{N_0} \tag{7-52}$$

The ratio of the results for cases II and IV gives the degradation, or mismatch loss, that results from receiver weighting. Thus,

$$\frac{(S/N)_{\mathrm{II}}}{(S/N)_{\mathrm{IV}}} = \frac{\dfrac{T}{N_0}\left[\displaystyle\int_{-1/2}^{1/2} W(y)\, dy\right]^2 \Big/ \displaystyle\int_{-1/2}^{1/2} W^2(y)\, dy}{\dfrac{T}{N_0}} = \frac{\left[\displaystyle\int_{-1/2}^{1/2} W(y)\, dy\right]^2}{\displaystyle\int_{-1/2}^{1/2} W^2(y)\, dy} \tag{7-53}$$

In making use of expression (7-53) to calculate the mismatch loss for various weighting functions it should be remembered that $W(y)$ represents the receiver bandpass function normalized to a unit bandwidth parameter.

For the rectangular envelope input signal of arbitrary peak amplitude A, the matched-filter peak output signal-to-noise ratio is, from Eq. (7-52)

$$(S/N) = \frac{A^2 T}{N_0} \tag{7-54}$$

Since the energy of this i.f. input signal is

$$E = A^2 T/2 \tag{7-55}$$

the matched-filter signal-to-noise ratio for the general case is

$$(S/N) = 2E/N_0 \tag{7-56}$$

Equation (7-56) is a relationship that applies to all matched-filter signals [8]. Equation (7-53) for the receiver weighting mismatch loss also applies to the general matched-filter signal if the receiver input time envelope is a rectangular function.

The conclusion that weighting at the receiver is a more efficient approach than weighting at the transmitter is of practical value, since control of the time envelope of a high power transmitter output is not feasible in most cases. An example of matched-filter receiver/transmitter weighting is the Gaussian envelope linear FM signal and Gaussian weighted receiver described by Fowle *et al.* [9]. This signal was shown to exhibit -40 db range sidelobes in a laboratory implementation. A comparison of the receiver signal-to-noise loss of this signal with that of a rectangular envelope linear FM signal that is frequency weighted at the receiver to yield -40 db range sidelobes is given below.

In comparing the Gaussian envelope linear FM filter signal the following assumptions are made:

1. A dispersive linear delay vs frequency compression filter is available having a differential delay of T_0 over a bandwidth Δf. This basically defines a time interval, T_0, within which either a Gaussian or rectangular envelope may be used for the transmitted signal.

2. The Gaussian signal is essentially truncated at the e^{-2} points without materially affecting the compressed-signal characteristics as described by Fowle *et al.* [9].

3. Each signal is limited to the same peak power level.

Starting with these assumptions, the Gaussian envelope can be compared to the rectangular envelope signal on a per pulse basis.

For the case of $T\Delta f \gg 1$, the normalized expression for the Gaussian signal, as shown by Fowle *et al.* [9], approximates

$$u(t)_N = \exp\left[-4\left(\frac{t}{T}\right)^2\right]\exp\left[j\left(2\pi f_0 t - \frac{\pi\,\Delta f\,t^2}{T} + \frac{\pi}{4}\right)\right] \qquad (7\text{-}57)$$

Letting this function become e^{-2} at $t = \pm T_0/2$, one obtains

$$T = T_0/\sqrt{2} \qquad (7\text{-}58)$$

Since the sweep rates of the rectangular and Gaussian signals are the same, then

$$\Delta f = \mu T/2\pi \qquad (7\text{-}59)$$

It can be shown, using the ambiguity function given by Fowle *et al.*, that the envelope of the Gaussian compressed pulse is

$$|\chi(\tau, 0)| = \frac{\Delta f}{2}\sqrt{\frac{\pi}{2}}\, \exp\left[-\frac{(\Delta f t)^2}{8}\right] \tag{7-60}$$

Comparing pulse widths at the 3 db points, one obtains

Gaussian envelope: compressed-pulse width $= 1.5/\Delta f$

Rectangular envelope
(no weighting):   compressed-pulse width $= 0.9/\Delta f$

(Hamming weighting):   compressed-pulse width $= 1.3/\Delta f$

On the basis of assumption (3), the energy in the Gaussian pulse is

$$E = \frac{1}{2}\int_{-\sqrt{2}T/2}^{\sqrt{2}T/2} \exp\left[-8\left(\frac{t}{T}\right)^2\right] dt = \frac{T\sqrt{2\pi}}{8} \tag{7-61}$$

and the energy in the rectangular pulse is

$$E = \frac{1}{2}\int_{-\sqrt{2}T/2}^{\sqrt{2}T/2} dt = \frac{\sqrt{2}T}{2} \tag{7-62}$$

and the comparison shown in the accompanying tabulation can be made.

	Relative signal energy (db)	Mismatch loss (db)	Relative signal efficiency (db)
Rectangular envelope	0	1.34 (Hamming weighting)	− 1.34
Gaussian envelope	− 3.53	0	− 3.53

Thus, under the conditions outlined above, the Gaussian envelope signal has a 2.2 db penalty in detection capability for the ideal − 40 sidelobe case when compared to the rectangular envelope signal.

7.5 Spectrum Weighting Data—Compressed-Pulse Waveform Characteristics

Data on the effect of matched-filter spectrum weighting on the pulse-compression output waveform characteristics is presented in this section. Figure 7.14 shows several of the waveforms that result from the weighting

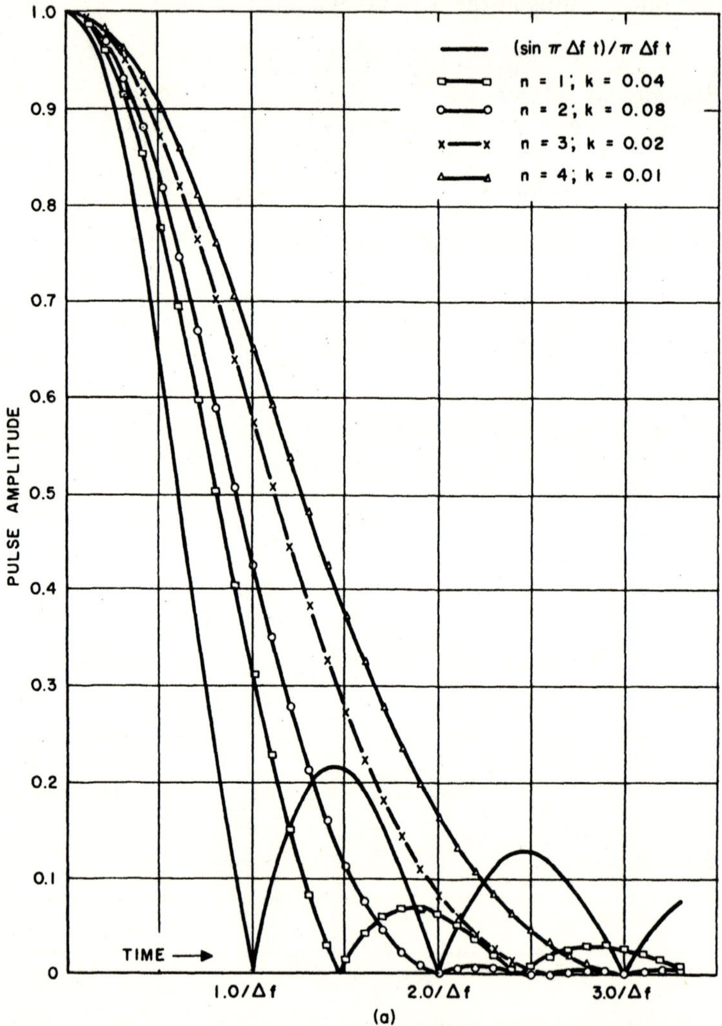

FIG. 7.14a Waveforms for $\cos^n$ weighting on pedestal of height k (from Bernfeld *et al.*, Ref. 4, Chapter 1).

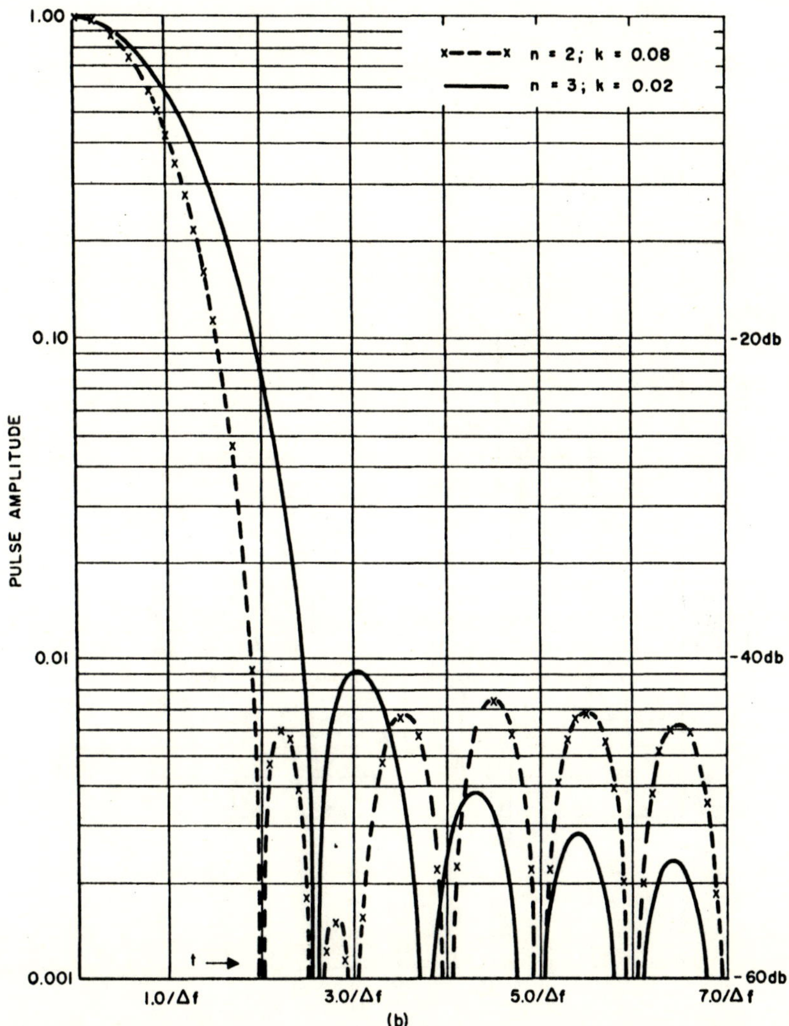

FIG. 7.14b   Sidelobe detail of two $|k + (1 - k)\cos^n|$ spectrum weighted time functions (from Bernfeld *et al.*, Ref. 4, Chapter 1).

functions considered in Section 7.2 (Taylor waveforms are shown in Figs. 7.5 and 7.6). All of the weighting functions discussed are truncated at the band edges, $\omega_0 \pm \Delta\omega/2$. The cosine-power (i.e., cos, $\cos^2$, $\cos^3$, $\cos^4$) weighted pulse signals exhibit a large amount of pulse widening for the sidelobe levels obtained. This can be offset by the addition of the pedestal to obtain the weighting function

$$W(\omega) = k + (1 - k)\cos^n \frac{\pi(\omega - \omega_0)}{\Delta\omega} \qquad (7\text{-}63)$$

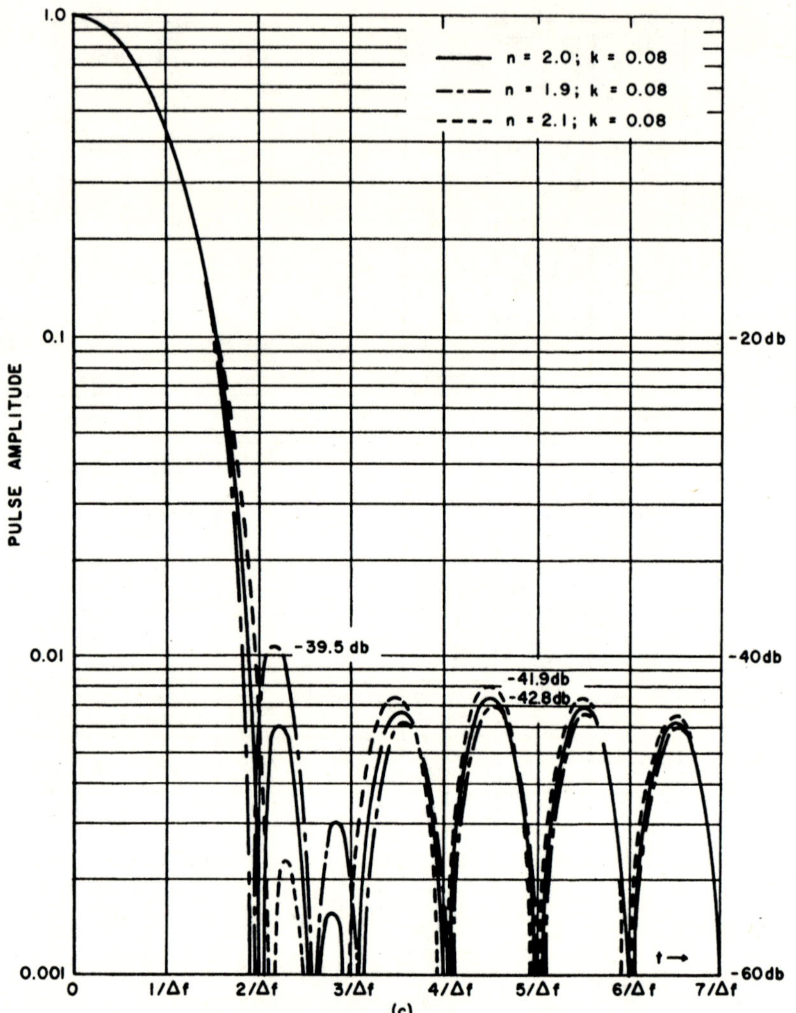

FIG. 7.14c   Effect of varying n on sidelobe structure of time function of a $[k + (1 - k)\cos^n]$ weighted spectrum (from Bernfeld *et al.*, Ref. 4, Chapter 1).

The pulse waveform associated with this function can be considered to be composed of two parts. One is the waveform that results from the $(1 - k)\cos^n$ part of $W(\omega)$. The other is a $(\sin x)/x$ waveform of width $\tau = 1/\Delta f$, weighted by the factor k. The effect of the addition of the weighted $(\sin x)/x$ waveform is to partially cancel existing sidelobes and to make the central time response narrower. Some waveform examples of this type of weighting function are shown in Fig. 7.14(a). Since a cancellation effect can be associated with the pedestal k, it is to be expected that sidelobe levels can vary considerably if the pedestal height changes. The Taylor weighting functions can be considered to belong to that class of functions for which the pedestal height is critical in determining the range sidelobe level. Figure 7.14(b) compares, on a logarithmic scale, the sidelobe structures of $\cos^2$-plus-pedestal and of $\cos^3$-plus-pedestal weighted time functions. The Hamming function ($k = 0.08$) represents the optimum for this family of weighted pulse shapes. Figure 7.14(c) shows the effect of changes in the weighting function curvature, made by varying the power of the cosine term of Eq. (7-63) about $n = 2$. The general conclusion drawn from these results is that the weighted time function is much more sensitive to the pedestal height than to variations in the function shape.

Figure 7.15 plots the Dolph–Chebyshev and Taylor pulse widening factors against range sidelobe level. Also shown are the experimental data

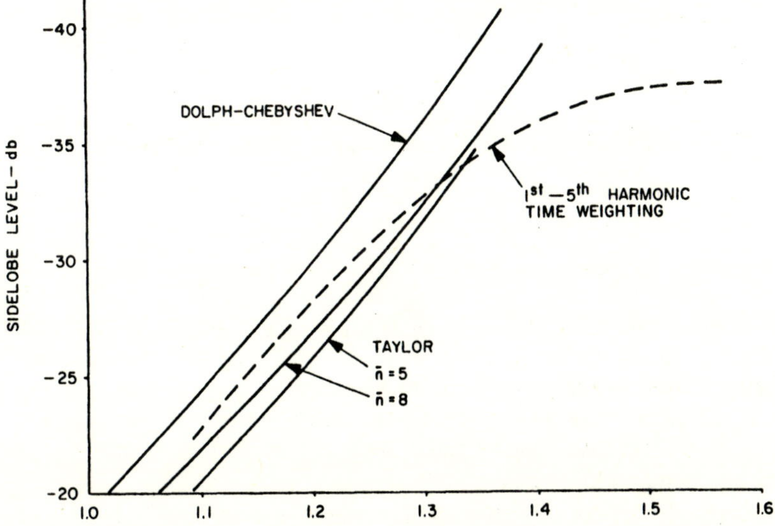

FIG. 7.15.   Pulse widening of weighted compressed pulse compared to $(\sin x)/x$.

obtained with the time amplitude weighting technique described in Section 7.3. It is interesting to note that, while the Taylor functions yield the best uniform approximation to the Dolph–Chebyshev characteristics over all ranges of sidelobe levels, it is possible with the same number of terms in the weighting function expansion (see Eq. (7-12)) to obtain better results over restricted sidelobe level/pulse widening regions. This is achieved by adjusting the values of the Taylor coefficients F_m, and is one of the possible techniques for improving sidelobe levels in a transversal filter implementation. Figure 7.16 plots the mismatch loss when Taylor weighting is used

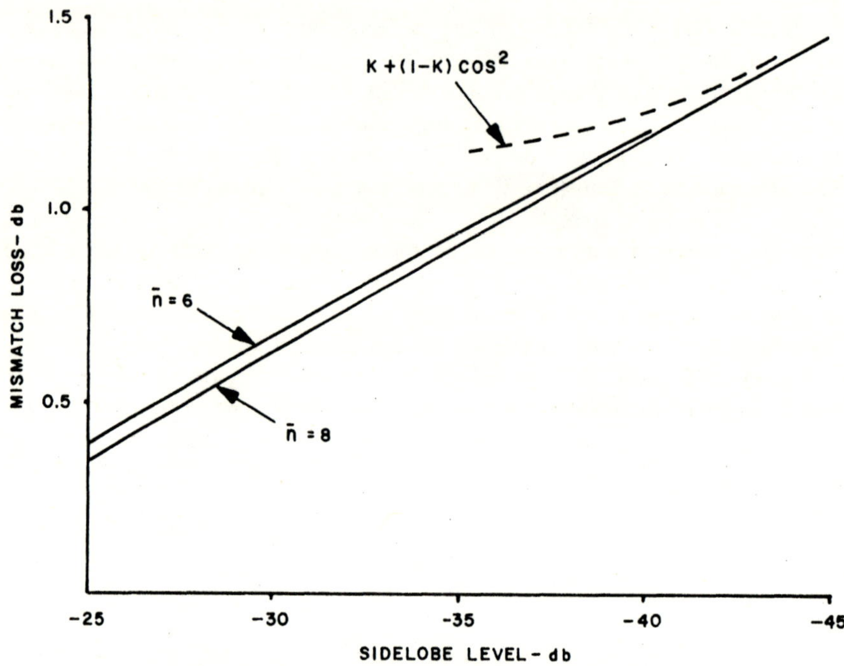

FIG. 7.16   Mismatch loss of receiver frequency weighting.

at the matched-filter receiver. Also shown in this figure is the mismatch loss resulting from the $k + (1 - k) \cos^2$ function. From the data in this figure and that in Fig. 7.17 it can be seen that the $k + (1 - k) \cos^2$ function yields sidelobe, pulse widening and mismatch loss results that are not too different from those theoretically obtained by Taylor weighting in the region of -40 db sidelobes.

Figures 7.18–7.20 plot the data associated with the $k + (1 - k) \cos^n$ weighting functions. Figure 7.17 shows the effect of the pedestal height on the sidelobe levels, and Fig. 7.18 the pulse widening factor, for different integer values of n. Figure 7.19 plots the mismatch loss resulting from receiver weighting with these functions as the pedestal k varies. Figure 7.20 shows the effect of changes in n on the mismatch loss for different pedestal height values. These curves can be used to estimate the effect on mismatch loss for different approximations to the desired weighting function. Thus, the $k = 0.08$ curve in the region about $n = 2$ shows the effect of not having an exact $\cos^2$ shape in the Hamming design. This particular curve also approximates the effect of this design curve factor on the -40 db Taylor weighting results.

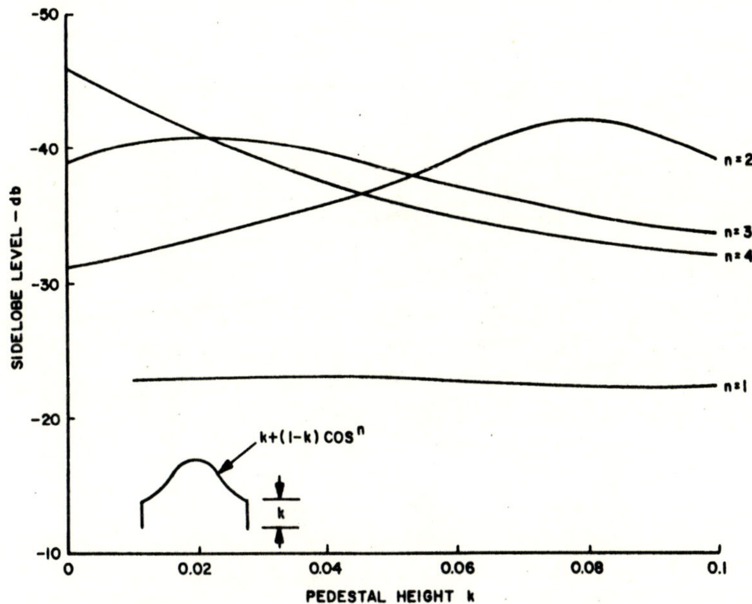

FIG. 7.17   Effect of pedestal height on sidelobe levels (from Bernfeld *et al.*, Ref. 4, Chapter 1).

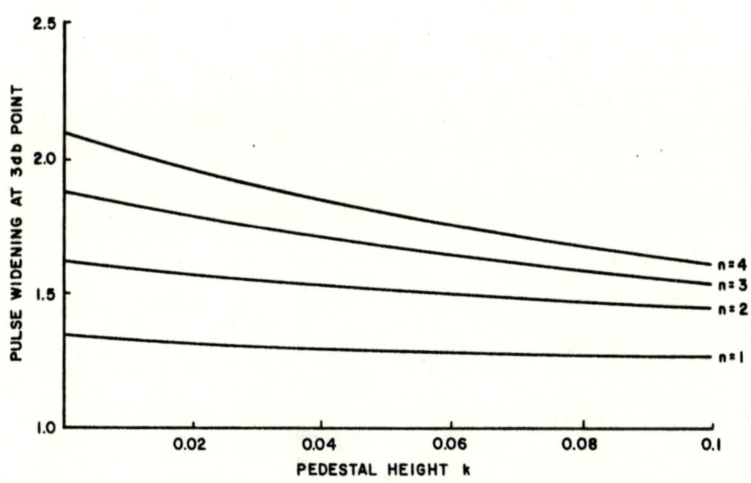

FIG. 7.18   Effect of pedestal height on pulse width (from Bernfeld *et al.*, Ref. 4, Chapter 1).

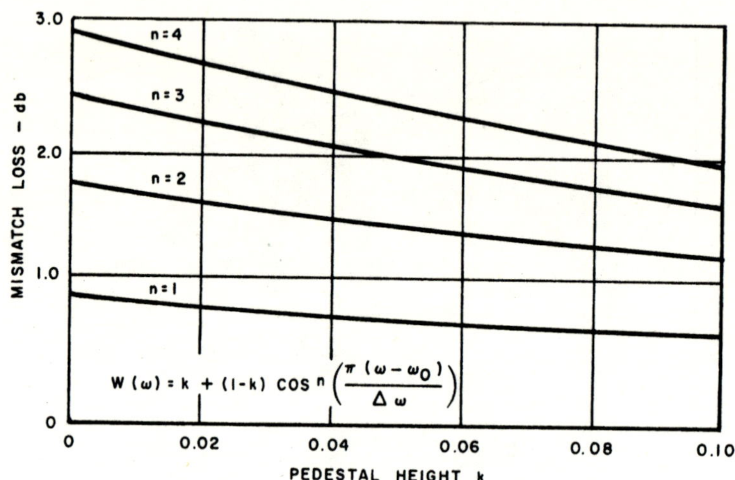

FIG. 7.19   Effect of k on mismatch loss.

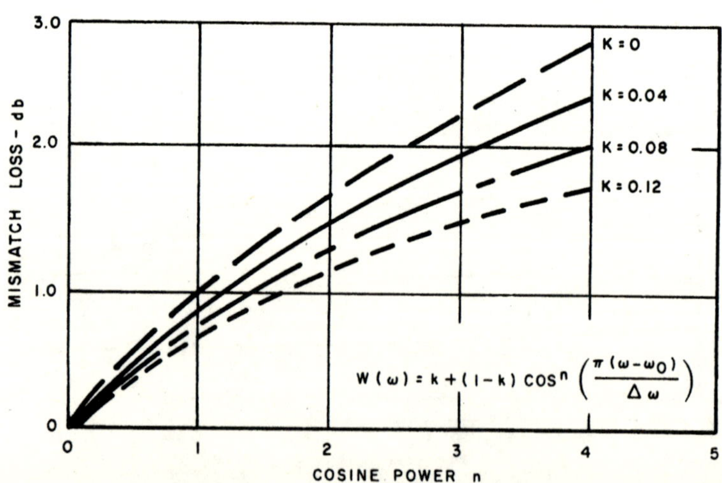

FIG. 7.20   Effect of n on mismatch loss.

Table 7-I gives comparative numerical data for several of the weighting functions discussed in this chapter, and Table 7-II presents data on the effect of noninteger values of n for pedestal heights about the optimum for $n = 2$.

TABLE 7-I

WEIGHTING FUNCTION DATA[a]

Weighting function	Peak sidelobe level (db)	Pulse widening	Mismatch loss (db)	Far sidelobe fall-off rate
Dolph–Chebyshev	−40	1.35	—	1
Taylor, $\bar{n} = 6$	−40	1.41	−1.2	$1/t$
$k + (1 - k)\cos^n$				
Hamming ($k = 0.08$, $n = 2$)	−42.8	1.47	−1.34	$1/t$
Cosine-squared ($k = 0$, $n = 2$)	−32.2	1.62	−1.76	$1/t^3$
Cosine-cubed ($k = 0$, $n = 3$)	−39.1	1.87	−2.38	$1/t^4$
$n = 1, k = 0.04$	−23	1.31	−0.82	$1/t$
$n = 2, k = 0.16$	−34	1.41	−1.01	$1/t$
$n = 3, k = 0.02$	−40.8	1.79	−2.23	$1/t$

[a] Note: All data are based on rectangular unweighted spectrum.

TABLE 7-II

PEAK-TO-SIDELOBE RATIOS IN db FOR (COSINE)^n WEIGHTING

	n				
k	1.8	1.9	2.0	2.1	2.2
0.07	35.3	38.0	41.8	43.7	43.0
0.08	35.9	39.5	42.8	42.1	41.5
0.09	36.6	40.6	41.5	40.8	40.0

The data presented in this section show that to obtain range sidelobes in the region of −40 db, with minimum effect on pulse widening, one must use a weighting function design that incorporates a pedestal. Since the sidelobe levels can be critically affected if the pedestal value is not exact, system implementation of these weighting functions may lead to results that are disappointing to the engineer who expects to obtain the results indicated by theory. It is perhaps more realistic to anticipate sidelobes at the −35 db level in an operating high power system unless frequent monitoring of the weighting function response is provided for. If sidelobes lower than this are a must in a system in which reliability is also an important factor, then the "safe" weighting functions that are less critical to parameter variations can be considered instead of the optimum functions. The penalty incurred by doing this is greater mismatch loss and pulse widening. Not much can be done about the former. In terms of absolute system resolution capability, the latter can be offset by designing for a narrower matched filter (i.e., unweighted) compressed-pulse width at the cost of an increased transmitted bandwidth.

As a final warning to the system designer, it should be pointed out that all of the data in this section is based on the assumption that the linear FM spectrum has a rectangular amplitude distribution. The amplitude ripple of the actual spectrum can be related to the existence of distortion sidelobes that can be considerably greater than the -40 db level. Thus, for many systems, it may be academic to talk about -40 db sidelobes. This is examined in more detail in the following sections, which also describe techniques for flattening out the spectrum amplitude ripple when low sidelobes are desired.

Another factor in obtaining low sidelobe levels is the effort that can be put into a system design to minimize amplitude and phase distortion throughout the system. Distortion analysis and correction is presented in Chapter 11. In the light of design procedures and component technology applied to present day high power radar systems, maintaining range sidelobes at -35 db, or even -30 db, under all conditions of field operation and maintenance represents a solid achievement.

7.6 Effect of Exact Linear FM Spectrum on the Weighted Compressed Pulse

Section 7.3 stated that in many cases it is often assumed that time and frequency weighting of the linear FM signal are completely analogous operations. This section will examine the effects of the ripple structure of the linear FM Fresnel spectrum on the range sidelobe behavior. It can be shown that the analogy cited above, with the associated inherent assumption of a rectangular spectrum, is not adequate when very low range sidelobes are desired [10].

The basic elements of a linear FM pulse-compression system were given in Fig. 7.8. The addition of the sidelobe reduction filter, $W(\omega)$ represents a departure from the optimum matched-filter condition. Assuming the simplified rectangular spectrum, Sections 7.2–7.5 have shown that frequency response functions for $W(\omega)$ can be specified that reduce the pulse sidelobes to 40 db or more below the peak output signal at the cost of a 40–50% increase in the pulse width and a signal-to-noise loss of between 1 and 2 db. Given the flat, rectangular spectrum model, the form of the idealized compressed pulse is

$$g(t) = \frac{1}{2\pi} \int_{\omega_0 - \Delta\omega/2}^{\omega_0 + \Delta\omega/2} W(\omega) \exp[j\omega t]\, d\omega \qquad (7\text{-}64)$$

To consider the effect of the actual spectrum in relation to the ideal sidelobe reduction characteristics cited above it is helpful to re-examine the amplitude component of the linear FM signal. This is:

Amplitude Term

$$|S(\omega)| = \frac{1}{2}\left(\frac{\pi}{\mu}\right)^{1/2} \{[C(X_1) + C(X_2)]^2 + [S(X_1) + S(X_2)]^2\}^{1/2} \qquad (7\text{-}65)$$

The Fresnel integrals $C(X)$ and $S(X)$, and the parameters X_1 and X_2 are as defined previously in Chapter 6. It is normally assumed that $|F(\omega)|$ is rectangular for compression ratios in excess of 30. This approximation is valid for a restricted range of system designs. The following discussion considers the effect of the simplifying approximations that have been made when higher performance characteristics are required.

The Fresnel integrals are quasi oscillatory in nature, and the actual linear FM spectrum contains amplitude ripples, the number and amplitude level of which are a function of $T\,\Delta f$, or compression ratio. Thus, a closer approximation to the weighted pulse-compression output signal would be, neglecting the residual phase term described by Eq. (6-21),

$$g(t) = \frac{1}{2\pi} \int_{\omega_1}^{\omega_2} |S(\omega)| W(\omega) \exp[j\omega t]\, d\omega \qquad (7\text{-}66)$$

The effects that must be considered for the exact case arise from a rippling variation of spectrum amplitude about a constant value. By approximating the amplitude variation by a combination of sinusoids, an analysis that yields useful results can be performed by means of paired-echo theory, thus avoiding tedious calculations based on the nonclosed form equation (7-66).

Figure 7.21 illustrates the factors that form the basis of the paired-echo analysis. The spectrum for $T\,\Delta f = 50$, as an example, is shown in Fig. 7.21(a). In Fig. 7.21(b) the amplitude ripple detail is assumed to be a cosinusoidal train, modulated by a more slowly varying cosine function. Using the modulated-cosine function model, the spectrum can be interpreted as being flat, with an associated error function. Thus, one may write, as a general formulation:

$$|S(\omega)| = [1 + (a_1 \cos C_1\omega)(1 + a_2 \cos C_2\omega)], \qquad \omega_1 \leqslant \omega \leqslant \omega_2 \qquad (7\text{-}67)$$

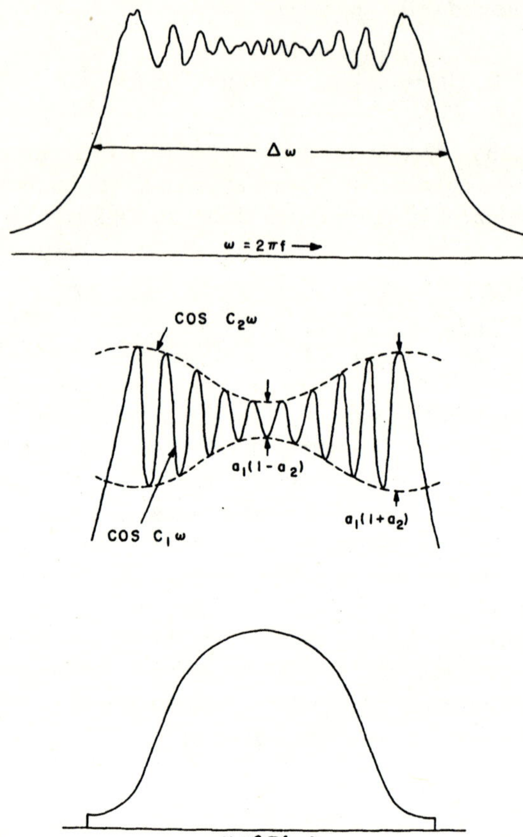

FIG. 7.21   Linear FM Fresnel spectrum and cosine model of spectrum ripple for range sidelobe calculation. Top: Linear FM spectrum $-T\Delta\omega/2\pi = 50$. Middle: Modulated-cosine model of spectrum ripple detail. Bottom: Frequency weighting function. (From Cook and Paolillo [10].)

where a_1 and a_2 may be positive or negative depending on the exact nature of the Fresnel integrals being approximated. Using the above, the weighted compressed time function of Eq. (7-66) becomes

$$g(t) = \frac{1}{2\pi}\int_{\omega_1}^{\omega_2} [1 + (a_1 \cos C_1\omega)(1 + a_2 \cos C_2\omega)]W(\omega)\exp[j\omega t]\,d\omega \qquad (7\text{-}68)$$

The factor

$$(a_1 \cos C_1\omega)(1 + a_2 \cos C_2\omega) \qquad (7\text{-}69)$$

can be expanded in its exponential form, permitting the weighted compressed-pulse function to be written as

$$
\begin{aligned}
g(t) = \frac{1}{2}\Bigg[& \int_{\omega_1}^{\omega_2} W(\omega)\exp[j\omega t]\,d\omega \\
& + \frac{a_1}{2}\left\{ \int_{\omega_1}^{\omega_2} W(\omega)\exp[j(t+C_1)\omega]\,d\omega + \int_{\omega_1}^{\omega_2} W(\omega)\exp[j(t-C_1)\omega]\,d\omega \right\} \\
& + \frac{a_1 a_2}{4}\left\{ \int_{\omega_1}^{\omega_2} W(\omega)\exp[j(t+C_1+C_2)\omega]\,d\omega \right. \\
& \qquad + \int_{\omega_1}^{\omega_2} W(\omega)\exp[j(t-C_1-C_2)\omega]\,d\omega \\
& \qquad + \int_{\omega_1}^{\omega_2} W(\omega)\exp[j(t+C_1-C_2)\omega]\,d\omega \\
& \qquad \left. + \int_{\omega_1}^{\omega_2} W(\omega)\exp[j(t-C_1+C_2)\omega]\,d\omega \right\} \Bigg]
\end{aligned}
\tag{7-70}
$$

The last six terms of this expression represent time displaced paired-echo
distortions. The use of paired-echo distortion analysis is a method of
establishing distortion criteria for signal processing and pulse-compression
systems, and is described in greater detail in Chapter 11.

Assuming that the proper choice of C_1 is governed by the rate of amplitude variation in the central region of the spectrum, the following
estimates of the constants of Eq. (7-67) are made for the range of compression ratios between $30:1$ and $250:1$

$$
C_1 = \frac{T\Delta f}{2}\,\text{cycles}/\Delta f
$$

$$
C_2 = 1\ \text{cycle}/\Delta f
$$

$$
a_1 = 0.06 \qquad a_2 = 0.6\text{--}0.7
$$

From these, one obtains the values shown in the accompanying tabulation.
The result of this triplet of paired echoes in adding to the idealized weighted
sidelobe level of the flat spectrum signal is illustrated in Fig. 7.22. The
spectrum ripple paired-echo terms, predicted on the basis of Eq. (7-67),
overlap and add vectorially to the calculated ideal sidelobe structure to
produce a composite formation well above the level derived on the basis
of the flat spectrum.

Paired-echo location relative to center of compressed pulse	Approximate paired-echo amplitude (db)
$\pm\left(\dfrac{T}{2} - \dfrac{1}{\Delta f}\right)$	-40
$\pm\dfrac{T}{2}$	-30
$\pm\left(\dfrac{T}{2} + \dfrac{1}{\Delta f}\right)$	-40

Figures 7.23 and 7.24 illustrate the phenomenon of the spectrum ripple paired-echo range sidelobes, and one method of reducing their amplitude. The left column of Fig. 7.23 shows a linear FM signal with various degrees of envelope rise and fall times. The middle column illustrates the response of a -40 db weighting network to the linear FM signal. The right column of waveforms shows in greatly expanded detail the range sidelobes at the

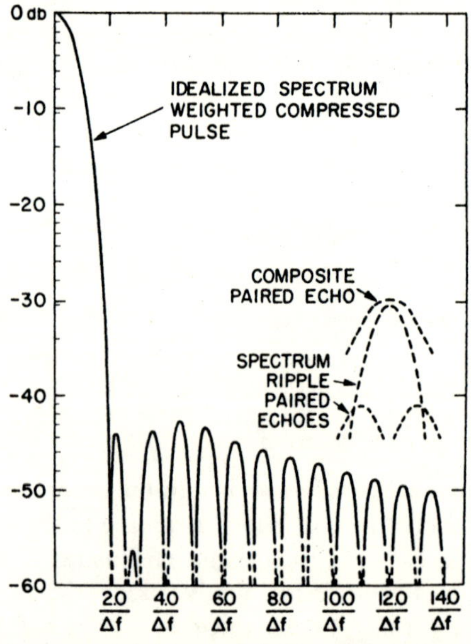

FIG. 7.22   Calculated effect of spectrum ripple on sidelobe level. Idealized pulse is based on Hamming weighting of a rectangular input spectrum (from Cook and Paolillo [10]).

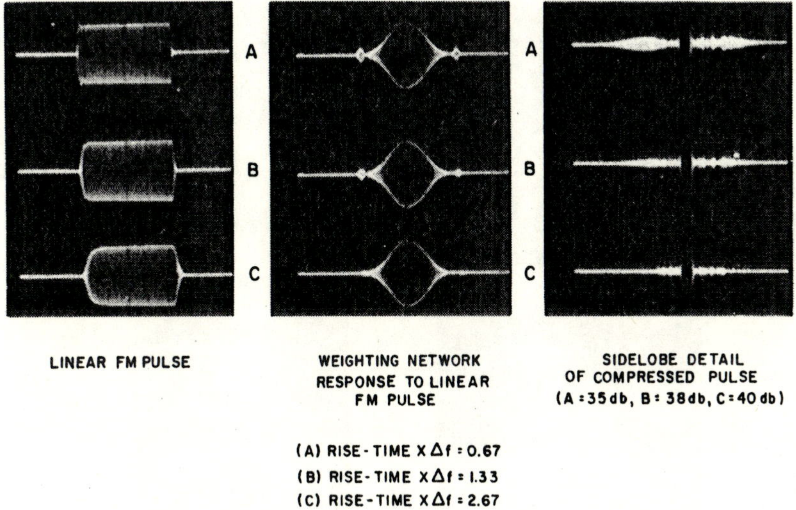

FIG. 7.23 Effect of linear FM pulse rise-time and fall-time on weighting network time response and compressed-pulse sidelobe level (from Cook and Paolillo [10]).

output of the combination of the weighting network and pulse-compression filter. Correlated with these waveforms are the signal spectra shown in Fig. 7.24. It is seen that the spectrum ripple decreases as the matched-filter input envelope function rise time increases. As the spectrum amplitude ripple decreases, so does the high level range sidelobes formation shown in Fig. 7.23(a, right), as well as the impulses on either end of the weighting network output. When the spectrum is very nearly a smooth distribution, the range sidelobes of the compressed pulse are at the desired -40 db level. The effect of the matched-filter input envelope rise time function on the linear FM Fresnel spectrum amplitude ripple is discussed by Cook and Paolillo [10] and Yost [11]. The causal relationship between the spectrum amplitude ripple and the range sidelobes predicted by Eq. (7-70) are indicated by the waveforms and spectra shown in Figs. 7.23 and 7.24. Paired-echo range sidelobes within 1 db of those predicted by the Fresnel spectrum model given by (7-67) have been observed experimentally. The following section describes a technique of FM predistortion to control the magnitude of the linear FM spectrum ripple, and thus the paired-echo range sidelobes associated with this ripple factor. This technique has application to high power systems, in which it is not usually feasible or desirable to control the transmitted pulse rise time because of the implied loss of power output when this is attempted.

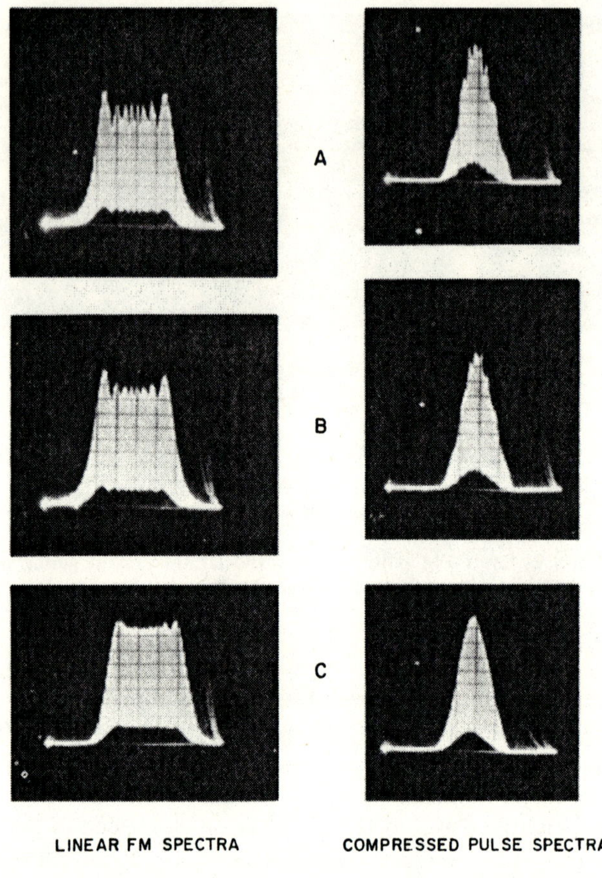

FIG. 7.24   Rise-time effect on linear FM and compressed-pulse spectra (from Cook and Paolillo [10]).

7.7   FM Predistortion

An approach other than modifying the transmitted pulse envelope is desirable for high power radar applications in order to control the spectrum amplitude ripples. This can be achieved by predistorting the phase or frequency modulation of the signal so that the transmitted pulse-compression waveform is

$$s_p(t) = \cos\left[\omega_0 t + \frac{\mu t^2}{2} + \theta_p(t)\right] \qquad (7\text{-}71)$$

where $\theta_p(t)$ is the phase predistortion function. Cook and Paolillo [10] discuss such a predistortion function, shown in Fig. 7.25. This function was derived on the basis of the equivalence, for small distortions, of the effects of envelope distortion and phase distortion that have functional similarity. The rise-time factor of Fig. 7.25(a) is equated to the phase modulation distortion shown in Fig. 7.25(b). Square-law phase modulation was assumed in order to obtain, for analytical simplicity, the linear FM distortion segments shown in Fig. 7.25(c) ($\Delta f_p = [d(\Delta\theta_p)/dt]_{\max}$). These yielded the monotonic frequency modulation function of Fig. 7.25(c).

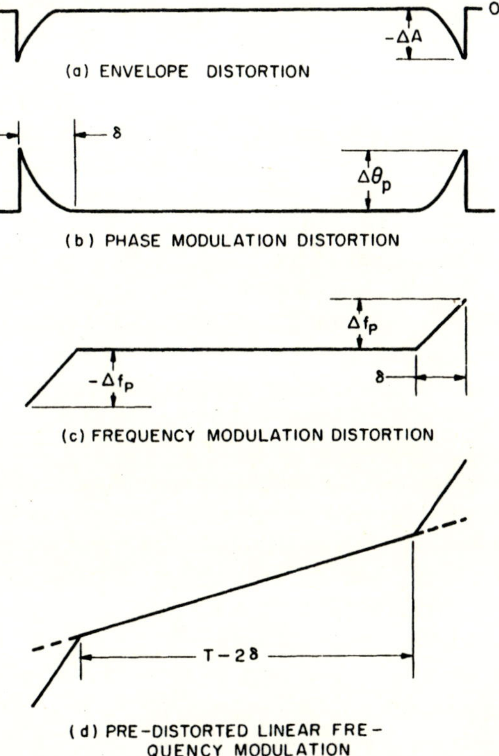

FIG. 7.25 Development of the predistorted linear FM function on the basis of replacing the distortion effect (i.e., spectrum ripple reduction) caused by the amplitude distortion term (a), with an equivalent phase (b), or frequency (c), distortion that produces the monotonic FM function shown in (d). (From Cook and Paolillo [10].)

The resulting predistorted spectrum is given by

$$S_p(\omega) = |S_1(\omega)| \exp[\,j\psi_1(\omega)] + |S_2(\omega)| \exp[\,j\psi_2(\omega)] + |S_3(\omega)| \exp[\,j\psi_3(\omega)]$$

$$(7\text{-}72)$$

$S_1(\omega)$ and $S_3(\omega)$ are the spectrum contributions of the linear FM end segments, and $S_2(\omega)$ is the spectrum of the matched linear FM portion of the signal. Each component function is a Fresnel spectrum with magnitudes given by Eq. (6-19) and phases ψ_1 to ψ_3 given by Eqs. (6-20) and (6-21). Equation (7-72) represents the vectorial combination of complex spectra that cannot be evaluated by a direct method. Equation (7-72) can be calculated for specific predistortion parameters δ and Δf_p. Any such calculation must satisfy the condition of phase continuity across the boundaries of the respective linear FM segments.

Values of predistortion parameters derived by Cook and Paolillo are $\Delta f_p = 0.75\, \Delta f$ and $\delta = 1/\Delta f$, Δf being the swept bandwidth of the undistorted linear FM signal, and $1/\Delta f$ the nominal unweighted compressed-pulse width. Experimental values of Δf_p and δ within 20% of these values for compression ratios of 40 and 80 resulted in sufficient reduction of the Fresnel spectrum ripple to greatly reduce the paired-echo range sidelobes described in Section 7.6. An associated effect was a slight raising of the spectrum skirt levels, as might be anticipated from the increased bandwidth swept out as a result of the FM predistortion function.

Figure 7.26 shows compression-filter input signals for both the linear FM and the predistorted linear FM functions. The effect of the predistortion function on the Fresnel spectrum for compression ratios of 40 and 80 are shown in Fig. 7.27. How the indicated spectrum ripple reduction carries over into the weighted compressed-pulse time function is shown in Fig. 7.28. Here, the spectrum paired-echo range sidelobes and their reduction as a result of the predistortion function are clearly illustrated.

Figures 7.29 and 7.30 show the effect of the FM predistortion function in maintaining low range sidelobes when the input signal has been frequency shifted (i.e., Doppler shift approximation). Figure 7.29 indicates a reduced level of distortion on the uncompressed pulse output of a weighting network when the input is a frequency shifted signal that has been predistorted as described above. Figure 7.30 illustrates the effect of the frequency shift on the predistorted signal after it has passed through the compression filter. The waveforms on the left show little net effect on the main pulse shape other than the expected shift in pulse position. The compressed-pulse sidelobes are shown in expanded detail on the right hand side of Fig. 7.30. A frequency shift of $\pm 7.5\%$ of the signal bandwidth caused no more than a 3 db variation in the -40 db sidelobe level obtained

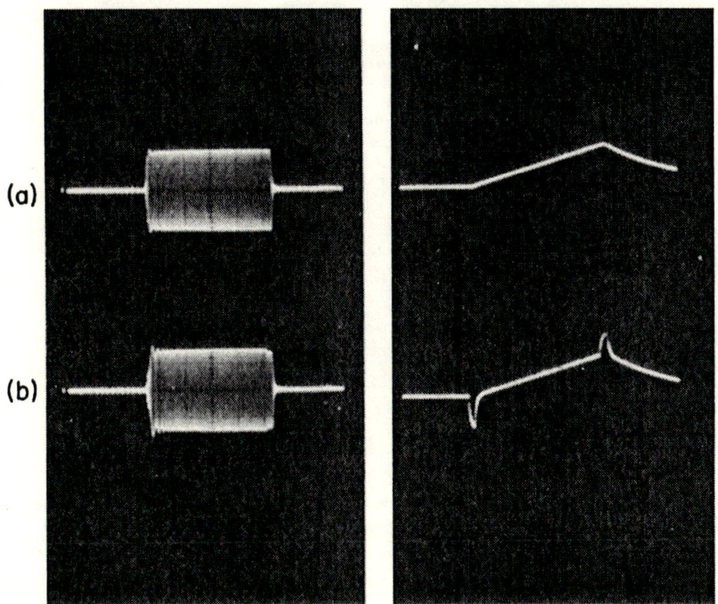

FIG. 7.26 (a) Linear FM pulse and FM sweep function. (b) Modified linear FM pulse and FM sweep function. (From Cook and Paolillo [10].)

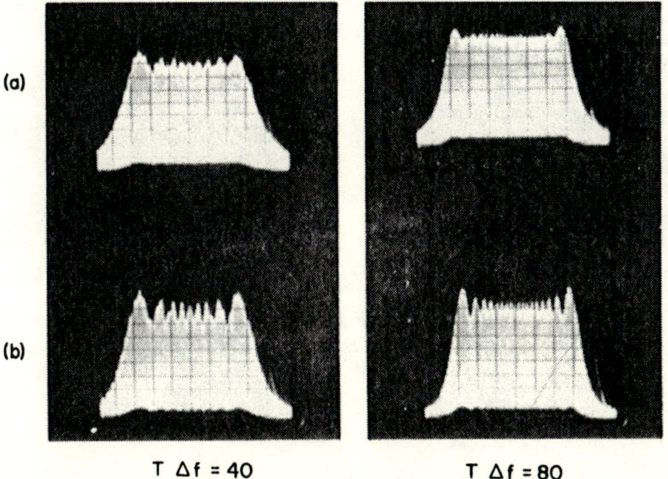

FIG. 7.27 (Top) Modified linear FM spectra. (Bottom) Linear FM spectra. (From Cook and Paolillo [10].)

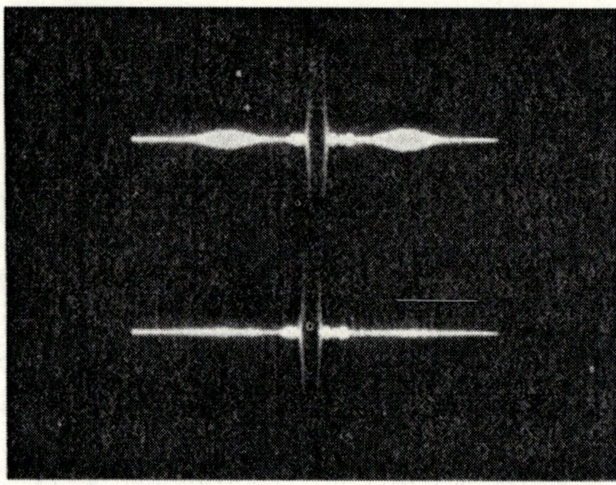

FIG. 7.28   Fresnel paired-echo sidelobe reduction for 80:1 compression ratio. (From Cook and Paolillo [10].)

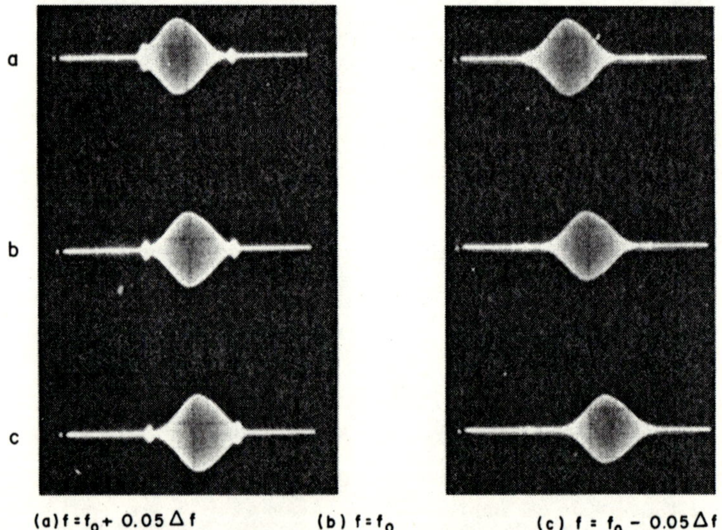

FIG. 7.29   Reduction of distortion in weighted uncompressed FM signal for input signal center frequency shift. Waveforms on right have FM predistortion. (From Cook and Paolillo [10].)

at zero frequency shift condition. In terms of Eq. (7-66), if the factor $|S(\omega)|W(\omega)$ stays relatively constant over the limits of the integral as a function of frequency shift there will be minimal effect on the weighted compressed pulse, provided that phase distortions are avoided. The linear delay characteristic of the compression filter, extending over sufficiently

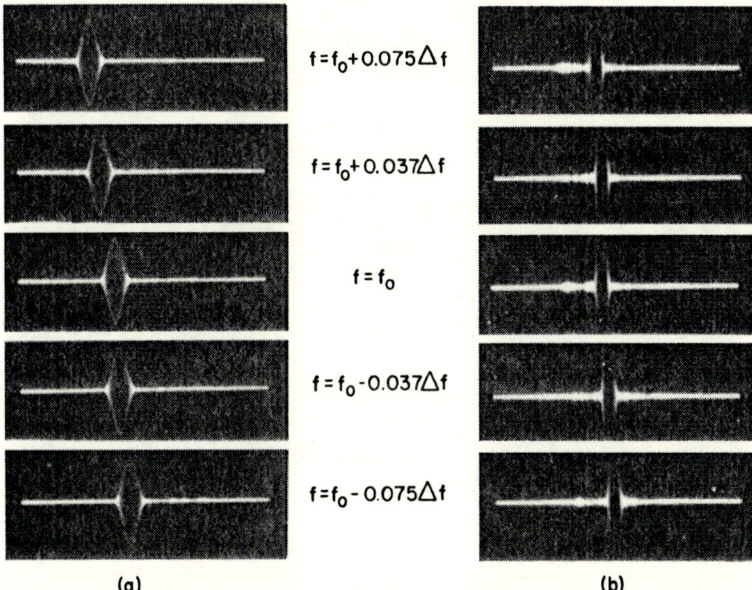

FIG. 7.30 (a) Modified linear FM compressed-pulse waveforms for frequency shifted inputs. (b) Sidelobe detail, -40 db nominal. (From Cook and Paolillo [10].)

wide band to include all possible Doppler shifts, is one of the conditions for minimizing additional phase distortion when Doppler shifted signals are expected. Predistortion functions other than that shown in Fig. 7.25 are possible, although the use of linear FM segments allows certain analytical simplifications in programming the predistorted spectrum function for computer calculations. The concept of predistortion at the transmitter can also be extended to compensate for distortion sidelobes associated with other factors besides the spectrum ripple.

The magnitude of the Fresnel spectrum ripple within the weighting network bandpass can also be reduced by deliberately widening the linear FM spectrum at the input to the weighting network, or conversely by narrowing the response of the weighting network. The larger time-bandwidth product signal resulting from the former approach results in a lower level of spectrum ripple in the frequency region covered by the weighting response. This represents a basic inefficiency in the use of the signal energy and bandwidth when compared to the case of the idealized spectrum function. This inefficiency can be expressed as a S/N loss that is a function of the bandwidth ratio of the widened spectrum to the ideal spectrum. The data given in the accompanying tabulation are indicative of the penalty associated with this approach. These data were obtained experimentally for compression ratios of 40 and 80, and the weighting network designed to obtain range sidelobes (ideal rectangular spectrum case) 40 db below the peak of the compressed pulse.

Spectrum widening ratio	Additional S/N loss (db)	Observed paired echo level (db)
1.00	0	-31
1.25	0.96	-34
1.50	1.76	-35
1.75	2.44	-37
2.00	3.00	-39

The design criteria for a precision radar (i.e., low distortion) are discussed in Chapter 11, where it is shown that the phase and amplitude fidelity of the system components must be rigidly controlled. The data in this and the preceding section show that the detailed structure of the signal spectrum can also be an important factor in optimizing system performance.

7.8 Realization of Weighting Function Responses by Transversal Filtering

Equation (7-14) in Section 7.2 showed that the Taylor weighted time response is expressed as the sum of weighted time-shifted $(\sin x)/x$ pulses. If, as in the case of the linear FM matched-filter output, the signal being fed into the Taylor weighting sidelobe reduction network is also a $(\sin x)/x$ waveform (or nearly so for large time-bandwidth products), then the output function represents the summation of delayed replicas of the original input signal. On this basis, an alternate synthesis of the Taylor weighted output function can be obtained from a tapped delay line. This form of weighting function realization is shown in Fig. 7.31. Since linear operation on an

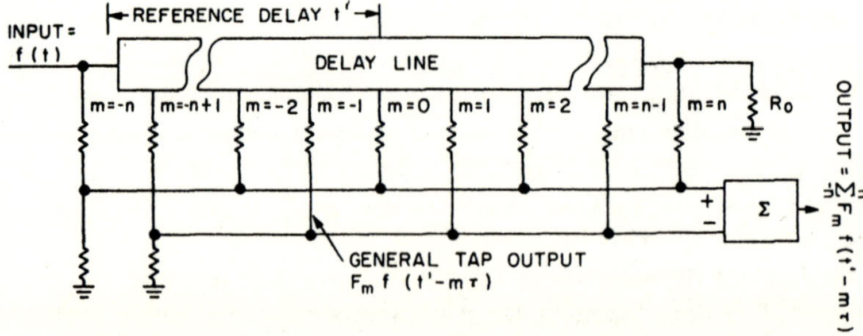

FIG. 7.31   Basic representation of a transversal filter for Taylor bandpass weighting.

ensemble of signals that overlap in time requires coherent addition of the delayed signals, this type of filter must be synthesized at an intermediate, rather than video, band of frequencies. Because of this, the transversal filter output is sensitive to changes in the center frequency of the input signal. A moderate frequency shift of narrow band signals can cause the delayed signal and the sidelobe at the same time interval to add instead of cancel.

The time delay between the taps on the transversal filter is the 4 db pulse width, or $1/\Delta f$, of the $(\sin x)/x$ input, and the overall frequency response is periodic in the frequency domain at a rate that is the reciprocal of the 4 db pulse width, as shown in Fig. 7.32. The definition of the weighting coefficients just following Eq. (7-12) indicates that the odd coefficients are

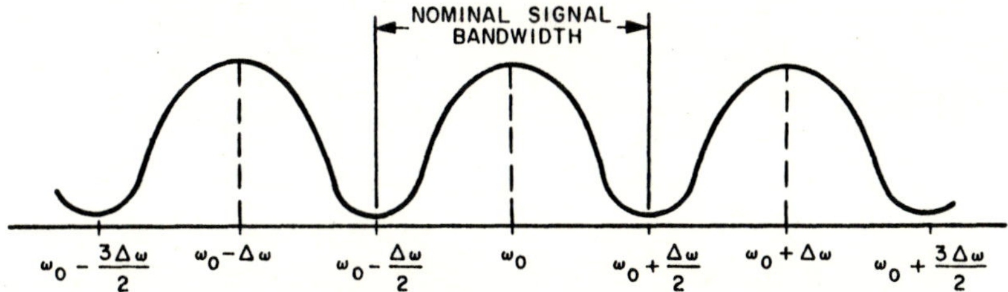

FIG. 7.32   Frequency response of a transversal filter.

positive and the even ones are negative. In the transversal filter the negative coefficients can be realized by summing the weighted outputs from the even numbered taps, and then inverting, or phase shifting the carrier frequency by 180 degrees, by means of an amplifier or transformer. Since the weighted tapped outputs are equally spaced about the input signal, the physically realizable transversal filter must incorporate sufficient fixed delay so that no negative delays are required. The amount of this fixed delay, and therefore the physical size of the transversal filter, depends on the number of Taylor coefficients used. For this reason it is desirable to use as small an $\bar{n}$ as possible in the Taylor design, or to drop the higher order terms in the synthesis procedure, as discussed in Section 7.2 under modified Taylor or Hamming weighting. If this is done there will be only minor degradations in sidelobe level and pulse width from the desired values.

In a practical implementation of a transversal filter delay line care must be taken to insure that there is a minimum amount of cross feed between the taps of the delay line. To meet this requirement will normally require the use of isolation devices at the tap outputs, such as amplifiers or directional couplers. The design of the line itself should be such that reflections up and down the line, and those set up by impedance mismatches at the tap points, are less than the desired sidelobe level at the final summed output.

Another important consideration is provision for control of the carrier phase at the tap points, since the carrier phase of the range sidelobes that are to be cancelled may not follow that based on the assumption of an ideal $(\sin x)/x$ input signal to the transversal filter. This will be particularly true if the matched-filter range sidelobes are combined with distortion paired echoes. A technique to provide for this necessary tap point phase control, that has been used with success, is shown in Fig. 7.33. The phase of the tap point signal is adjusted by adding an in-phase and a quadrature signal at each tap output. Variation of the amplitudes and signs of the two components will determine the phase of the combined tap output, which

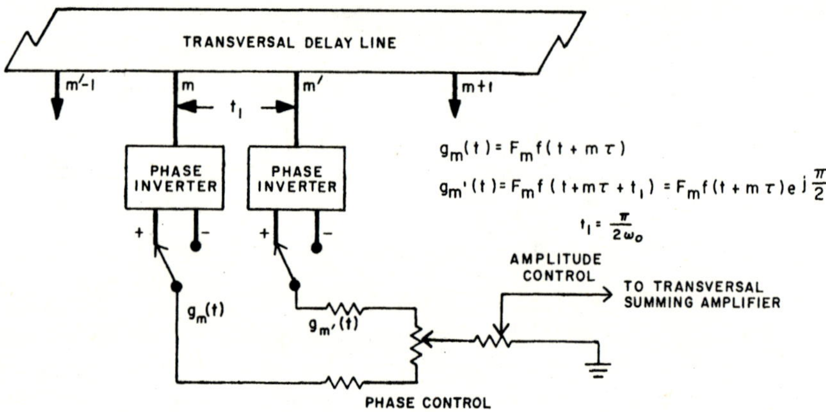

FIG. 7.33   Generation of quadrature signal for phase control of transversal filter tap output signal.

can vary over the complete range of 360 degrees. In the diagram of Fig. 7.33 the quadrature component is obtained by providing a second tap point for which the additional time delay is equivalent to a 90 degree phase shift at the carrier frequency. An alternate method is to use a 90 degree phase shifter circuit, and to split the signal at the tap point to produce the in-phase and quadrature components, which are then recombined as outlined above. It can be seen that a large number of independent controls are necessary in a practical transversal filter design. For this reason, and also because of the transversal filter sensitivity to frequency shifts, the use of this approach to range sidelobe reduction may not be feasible in those systems where weight and/or complexity must be kept to a minimum. Figure 7.34 illustrates the type of range sidelobe reduction that can be obtained by use of a transversal filter. The input waveform contains paired echo distortion signals, as well as the normal compressed-pulse sidelobes.

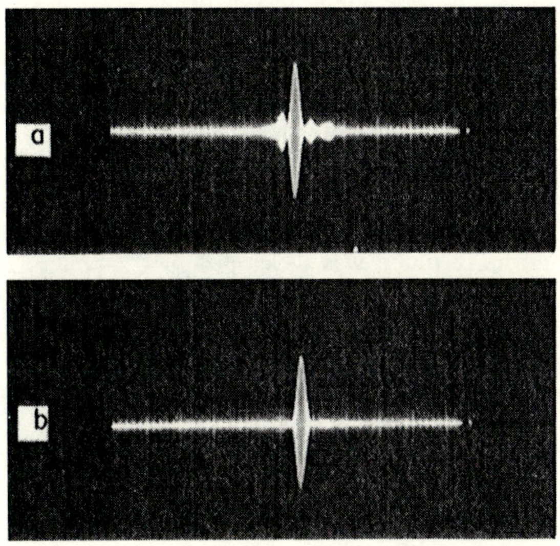

FIG. 7.34 Transversal filter waveforms. (a) Transversal filter input from compression filter (distortion sidelobes are also present). (b) Weighted transversal filter output (S/L approx. − 30 db). (Courtesy C. R. Wood, Sperry Gyroscope Company, Great Neck, New York.)

The transversal filter output has range sidelobes that are reduced to the vicinity of − 30 db with a pulse widening factor of 1.25.

The theory of Taylor weighting is based on the use of $(\sin x)/x$ functions that have regularly spaced zeros, and sidelobes in which the carrier frequency is successively shifted by 180 degrees. The linear FM matched-filter output is only a good approximation to a $(\sin x)/x$ function for relatively large compression ratios. The actual matched-filter output signal given by Eq. (6-13) has irregularly spaced sidelobes for low compression ratios, as shown by Fig. 6.3. Schreitmueller [12] has calculated for this 10 : 1 compression-ratio signal that the values of the Taylor weighting coefficients can be modified to obtain reduced sidelobes. For a single pair of taps spaced about the F_0 output, a delay interval of $2/\Delta f$, rather than $1/\Delta f$, and a modified value of $F_1 = 0.15$ are cited for − 23 db sidelobes. The tapped delay line synthesis can be applied to arbitrary functions, and in operation the coefficient values can be adjusted to compensate for irregularities in the system.

Transversal filtering can also serve as a supplement to the usual bandpass weighting filter. In this arrangement the basic sidelobe reduction is accomplished by the bandpass weighting. The function of the transversal filter is to cancel additional sidelobes caused by amplitude or phase distortion in the system. Pratt [13] gives a thorough discussion of this aspect of transversal filtering, as well as of some of the important design considerations. The basic theory of transversal filtering can be found in Kallmann [14].

7.9 Nonlinear FM Matched Filters for Sidelobe Reduction

The topic of the design of FM matched-filter signals based on the principle of stationary phase was developed in Chapter 3. It was shown that nonlinear FM pulse-compression signals could be designed so that the compressed-pulse waveform had very low level sidelobes. Since these represented matched-filter implementations, they did not suffer from the mismatch loss described in Section 7.4. The governing relationship between the signal spectrum and the FM modulation function was shown to be

$$|S(\omega)|^2 \doteq \frac{2\pi a^2(t)}{|\theta''(t)|} \tag{7-73}$$

where $\theta'(t)$ is the instantaneous frequency vs time characteristic of the transmitted nonlinear FM signal. Since the desirable frequency weighting functions are even symmetrical functions, it follows from (7-73) that $\theta''(t)$ must also be an even symmetrical function, and thus $\theta'(t)$ must possess odd symmetry. Several FM pulse-compression waveforms of this type were described in Chapter 3. Figure 3.4 illustrated the possible implementations for a nonlinear FM matched-filter system. The matched-filter output signal would be given by

$$g(t) = \frac{1}{2\pi} \int_{-\infty}^{\infty} |S(\omega)|^2 \exp[\,j\omega t\,]\, d\omega \tag{7-74}$$

where $|S(\omega)|^2 = W(\omega)$, $W(\omega)$ being the desired frequency weighting function. The design of a nonlinear dispersive delay line introduces problems that have not been solved as effectively as those of the linear delay design, although the flexibility offered by ultrasonic and optical filtering techniques may provide the effective design means for nonlinear delay lines (see Chapters 13 and 14).

The use of nonlinear FM pulse compression to control the compressed-pulse waveform is not without its disadvantages. Mainly, these occur when the received signal has been frequency shifted. This causes an increase in the range sidelobes as a result of a phase mismatch condition. For the general case, the received spectrum can be designated as

$$S_R(\omega) = A(\omega + \omega_d)\exp[\,j\Phi(\omega + \omega_d)\,] \tag{7-75}$$

where $\omega_d = 2\pi\phi$ and the matched-filter transfer function is

$$H(\omega) = A(\omega)\exp[-j\Phi(\omega)] \tag{7-76}$$

The spectrum function at the matched-filter output is given by

$$G(\omega) = S_R(\omega)H(\omega) = A(\omega + \omega_d)A(\omega)\exp[j\Phi_d(\omega)] \tag{7-77}$$

where $\Phi_d(\omega) = \Phi(\omega + \omega_d) - \Phi(\omega)$. For the linear FM signal $\Phi_d(\omega)$ is primarily the difference of two square-law phase functions, and thus is a linear function of frequency, having a slope proportional to the sign and magnitude of ω_d. This can be equated to a linear time shift of the matched-filter output signal as a function of ω_d, and is one of the properties, for small Doppler shifts, of the linear FM matched-filter signal discussed in Chapter 6. When the FM function is nonlinear, then $\Phi_d(\omega)$ can exhibit nonlinearities that create the equivalent of distortion sidelobes at the matched-filter output. An illustrative example is the nonlinear FM signal that has $|S(\omega)|^2 = \cos^2(2\pi[\omega_0 - \omega]/\Delta\omega)$ over the frequency interval $\Delta\omega$. The dispersive delay function for this matched-filter signal, given in Chapter 3, is

$$T(\omega) = T\left[\frac{\omega_0 - \omega}{\Delta\omega} + \frac{1}{2\pi}\sin\frac{2\pi(\omega_0 - \omega)}{\Delta\omega}\right] \tag{7-78}$$

The matched-filter output spectrum modulus and delay function for this signal are plotted in Figs. 3.10 and 3.11. Let $\bar{\omega} = \omega_0 - \omega$; then the phase function associated with (7-78) is given by

$$\beta_f(\bar{\omega}) = \frac{T}{2\,\Delta\omega}\bar{\omega}^2 - \frac{T\,\Delta\omega}{4\pi^2}\cos 2\pi\frac{\bar{\omega}}{\Delta\omega} \tag{7-79}$$

Under matched conditions the spectrum phase function $\Phi_s(\bar{\omega})$ is the negative of (7-79), yielding $\beta_f(\bar{\omega}) + \Phi_s(\bar{\omega}) = 0$. If the spectrum is shifted by an amount ω_d, the matched-filter output spectrum phase becomes, for small ω_d,

$$\beta_f(\bar{\omega}) - \beta_f(\bar{\omega} + \omega_d) = \frac{T}{2\,\Delta\omega}[-2\bar{\omega}\omega_d - \omega_d^2]$$

$$+ \frac{T\,\Delta\omega}{2\pi^2}\sin\frac{\pi\omega_d}{\Delta\omega}\sin 2\pi\left(\frac{\bar{\omega} + \frac{1}{2}\omega_d}{\Delta\omega}\right) \tag{7-80}$$

The first term of (7-80) is associated with a linear shift in time of the matched-filter output signal. The second term represents a sinusoidal spectrum phase error of approximately one cycle over the band $\Delta\omega$. This will cause the equivalent of paired-echo time distortions (see Chapter 11) at the matched-filter output. The magnitude of this phase error is

$$b_1 = \frac{T\,\Delta\omega}{2\pi^2}\sin\frac{\pi\omega_d}{\Delta\omega} \doteq \frac{T\omega_d}{2\pi} \qquad \text{for} \quad \omega_d \ll \Delta\omega \qquad (7\text{-}81)$$

and the magnitude of the associated paired-echo signal is

$$J_0(b_1) \doteq \frac{b_1}{2} = \frac{T\omega_d}{4\pi} \qquad (7\text{-}82)$$

If these paired-echo distortion sidelobes must be kept 30 db below the peak of the compressed pulse the restriction on ω_d is

$$\omega_d \leqslant \frac{0.12\pi}{T} \qquad \text{or} \qquad \phi \leqslant \frac{0.06}{T} \qquad (7\text{-}83)$$

For -20 db sidelobes the limitation on ω_d relaxes to

$$\omega_d \leqslant \frac{0.4\pi}{T} \qquad \text{or} \qquad \phi \leqslant \frac{0.2}{T} \qquad (7\text{-}84)$$

Equations (7-83) and (7-84) show that even a moderate shift in the frequency of the received signal can offset the advantages inherent in the nonlinear FM pulse-compression signal when there is no frequency shift (i.e., low range sidelobe autocorrelation function). These values of ω_d indicate the approximate frequency stability and/or restricted range of Doppler shifts permitted for this signal if the low level of range sidelobes is to be maintained.

This type of range sidelobe behavior is supported by the detailed analysis of the inverse-tangent delay matched-filter signal [15]. Figure 7.35 shows

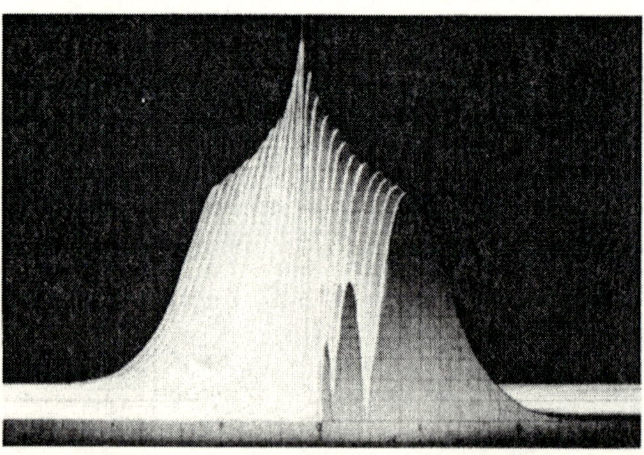

FIG. 7.35 Ambiguity function of tangent FM matched-filter signal for 100:1 compression ratio (the interval between each frequency shifted output is $1/4T$, $T = $ Transmitted pulse width) (Courtesy E. N. Fowle, The MITRE Corporation, Bedford, Massachusetts).

the model of the ambiguity function of the inverse-tangent (tangent function FM) signal. The gradual increase of the range sidelobes on one side of the compressed pulse is illustrated as the frequency shift at the matched-filter input increases. If the sign of the frequency shift is reversed, the enhanced sidelobe structure appears on the opposite side of the compressed pulse.

In the discussion above it has been assumed that the matched-filter application is in a search or tracking system in which it is necessary for the peak matched-filter output signal/range sidelobe ratio to remain relatively constant over a band of Doppler shifted inputs. There are other applications in which this type of behavior of the matched-filter output is not desired, and for which certain types of nonlinear FM signals can be used. Examples of these are discussed in Chapter 9, which is concerned with the application of pulse-compression matched-filter signals to the measurement, or estimation, of signal parameters.

REFERENCES

1. C. L. Dolph, A current distribution for broadside arrays which optimizes the relationship between beamwidth and sidelobe level, *Proc. IRE* **34**, 335–348 (1946).
2. G. J. Van der Maas, A simplified calculation for Dolph–Tchebycheff arrays, *J. Appl. Phys.* **25**, 121–124 (1954).
3. J. R. Klauder *et al.*, The theory and design of chirp radars, *Bell Syst. Tech. J.* **39**, 745–808 (1960).
4. C. L. Temes, Sidelobe suppression in a range-channel pulse-compression radar, *IRE Trans.* **MIL-6**, 162–169 (1962).
5. T. T. Taylor, Design of line-source antennas for narrow beamwidths and low sidelobes, *IRE Trans.* **AP-3** 16–28 (1955).
6. M. I. Skolnik, "Introduction to Radar Systems." McGraw-Hill, New York, 1962.
7. C. E. Cook, Modification of pulse-compression waveforms, *Proc. Nat. Electron. Conf.* **14**, 1058–1067 (1958).
8. P. M. Woodward, "Probability and Information Theory with Applications to Radar." Pergamon Press, Oxford, 1953.
9. E. N. Fowle, D. R. Carey, R. E. Vander Schuur, and R. C. Yost, A pulse-compression system employing a linear FM Gaussian signal, *Proc. IEEE* **51**, 304–312 (1963).
10. C. E. Cook and J. Paolillo, A pulse compression predistortion function for efficient sidelobe reduction in a high power radar, *Proc. IEEE* **52**, 377–389 (1964).
11. R. C. Yost, A linear FM radar pulse-compression system employing thickness-tapered dispersive delay lines, Mass. Inst. Technol., Lincoln Lab., Lexington, Massachusetts, Tech. Rept. 321 (July, 1963).
12. R. F. Schreitmueller, Notes on Taylor weighting. Unpublished memorandum (1959).
13. W. R. Pratt, Transverse equalizers for suppressing distortion echoes in radar systems, Proc. Pulse Compression Symp., pp. 119–128, Rome Air Development Center, Griffiss A. F. Base, New York, Tech. Document. Rept. *RADC-TDR-62-580.* (April, 1963).
14. H. E. Kallmann, Transversal filters, *Proc. IRE.* **28**, 302–310 (1940).
15. E. L. Key, E. N. Fowle, and R. D. Haggarty, A method of designing signals of large time-bandwidth product, *IRE Intern. Conv. Record Pt.* 4, 146–155 (1961).

CHAPTER 8

Discrete Coded Waveforms

8.1 Introduction

The radar waveform examples previously considered in this volume primarily have involved the utilization of analog codes employing continuous amplitude and frequency (or phase) modulations. The rectangular envelope, linear FM pulse is one of the better known representatives of the analog coded waveforms. In this chapter some additional waveform examples are considered which are derived from discrete codes. In general, these waveforms embody ordered sequences[1] that are impressed, at fixed time intervals,[2] on the amplitude, frequency and phase of a coherent cw carrier. The general description for these waveforms is given by[3]

$$\psi(t) = \sum_{n=1}^{N} a_n P_n(t) \exp[j\{(\omega_0 + \omega_n)t + \theta_n\}], \qquad 0 \leqslant t \leqslant N\delta$$

$$= 0, \qquad \text{elsewhere} \tag{8-1}$$

where $n = 1, 2, \ldots, N$, and $P_n(t)$ is a unit height pulse of fixed duration δ such that

$$P_n(t) = P(t - [n-1]\delta) = 1, \qquad (n-1)\delta \leqslant t \leqslant n\delta$$

$$= 0, \qquad \text{elsewhere} \tag{8-2}$$

---

[1] The general notation of an ordered sequence is $\{x_n\}$; for amplitude, frequency, and phase this becomes, respectively, $\{a_n\}$, $\{\omega_n\}$, and $\{\theta_n\}$.

[2] The size of this interval must be compatible with the requirements established for radar system performance and with practical restrictions that limit implementations.

[3] As presented here, the waveform description is not normalized; i.e. $\int_{-\infty}^{\infty} |\psi(t)|^2 \, dt \neq 1$. In order to accomplish this normalization it is necessary to multiply (8-1) by the factor $(1/\delta \sum_{n=1}^{N} a_n^2)^{1/2}$. Without affecting the general discussions to follow, this factor has been omitted for convenience.

By inserting (8-1) into Eq. (4-2a), it is seen that the general response function description for the discrete coded waveforms is given by

$$\chi(\tau, \phi) = \sum_{n=1}^{N} \sum_{m=1}^{N} \int_{-\infty}^{\infty} u_n(t) u_m^*(t + \tau) \exp[-j2\pi\phi t]\, dt \tag{8-3}$$

where

$$\psi(t) \exp[-j\omega_0 t] = \sum_{n=1}^{N} u_n(t) \tag{8-4}$$

The expression for $\chi(\tau, \phi)$ given in (8-3) will be reduced in the subsequent discussions concerning particular discrete coded waveforms.

The discrete coded waveform examples presented in this chapter are divided into three basic groups that are described as follows:

Group I: $\{\theta_n\} = \{\omega_n\} = 0;$ $\{a_n = 1, 0\}$
Group II: $\{\omega_n\} = 0$
Group III: $\{a_n = 1, 0\};$ $\{\theta_n\} = 0$

By making this division, it was possible to restrict the examples presented in this chapter to fundamental waveforms. These can serve as the basis for the design of more sophisticated composite waveforms that will provide the best performance for a particular radar application. Much of the information about the characteristics of these basic waveforms is presented in the experimental and computed data contained in the figures of this chapter.[1]

8.2. Constant Carrier Pulse Trains (Group I)

The coded waveform examples that are derived under Group I by restricting the sequence $\{a_n\}$ to values of 1 and 0 are called cw pulse trains. The description for these waveforms is obtained by deleting ω_n and θ_n from Eq. (8-1). This results in

$$\psi(t) = \sum_{n=1}^{N} a_n P_n(t) \exp[j\omega_0 t], \qquad 0 \leqslant t \leqslant N\delta$$

$$= 0, \quad \text{elsewhere} \tag{8-5}$$

The simplest pulse train structure is obtained when the sequence $\{a_n\}$ is defined by

$$a_n = 1, \qquad n - 1 = 0 \qquad \mathrm{mod}\ p$$

$$a_n = 0, \qquad n - 1 \neq 0 \qquad \mathrm{mod}\ p \tag{8-6}$$

---

[1] The experimental waveforms for the basic pulse train signals have been obtained with an analog correlation computer, with outputs plotted on an oscilloscope display.

where $p = 1, 2, \ldots$ and must divide evenly into N-1 to yield

$$M - 1 = \frac{N - 1}{p} \tag{8-7}$$

where M is the number of 1's in the sequence. It is assumed that the sequence always starts and ends with a 1. To illustrate, suppose, for example, $N = 17$ and $p = 4$. Then $M = 5$ and

$$\{a_n\} = 10001000100010001$$

The waveforms that are derived from such a sequence are called uniform pulse trains. Their utilization is essential to the basic radar process, and their history dates back to the beginning of radar. However, the earliest consideration of uniform pulse trains in the context of radar waveform coding appeared in Woodward [1] in 1953. Subsequent studies of these and other pulse train waveforms have appeared in the published works of Siebert [2], Fowle *et al.* [3], Resnick [4] and Rihaczek [5]. A more convenient form for $\psi(t)$ for the uniform pulse train is given by

$$\psi(t) = \sum_{m=0}^{M-1} P\left\{ t - \left(\frac{m}{M-1} \right) T \right\} \exp[\, j\omega_0 t], \qquad 0 \leqslant t \leqslant T + \delta$$

$$= 0, \quad \text{elsewhere} \tag{8-8}$$

where T is the duration of the train between the front edges of the first and last pulses. The envelope structure of this waveform is illustrated in Fig. 8-1. In the description of $\psi(t)$ given by (8-8), it is not essential that δ, the duration of a single pulse, divide evenly into the pulse period. Ordinarily, however, $T/(M - 1) \geqslant 2\delta$; in which case

$$|\chi(\tau, \phi)| = \sum_{k=-M}^{M} \left| \frac{\sin \pi\phi \left[[M - |k|] \dfrac{T}{M-1} \right]}{\sin \dfrac{\pi\phi T}{M-1}} \right| \cdot \left| \frac{\sin \pi\phi \left[\delta - \left| \tau - \dfrac{kT}{M-1} \right| \right]}{\pi\phi} \right| \tag{8-9}$$

This expression is derived by inserting $u(t) = \psi(t) \exp[\, -j\omega_0 t]$ into the general expression given by Eq. (8-1).

It is seen that this expression consists of two factors. The second factor is periodic along τ at intervals given by $T/(M - 1)$. The first factor is independent of τ but periodic along ϕ at intervals of $(M - 1)/T$. The spectrum for the uniform pulse train is obtained by setting $\tau = k = 0$. Thus,

$$\Psi(f + f_0) = |\chi(0, f)| = \left| \frac{\sin \pi f \delta}{\pi f} \right| \cdot \left| \frac{\sin \pi M f \dfrac{T}{M-1}}{\sin \pi f \dfrac{T}{M-1}} \right| \tag{8-10}$$

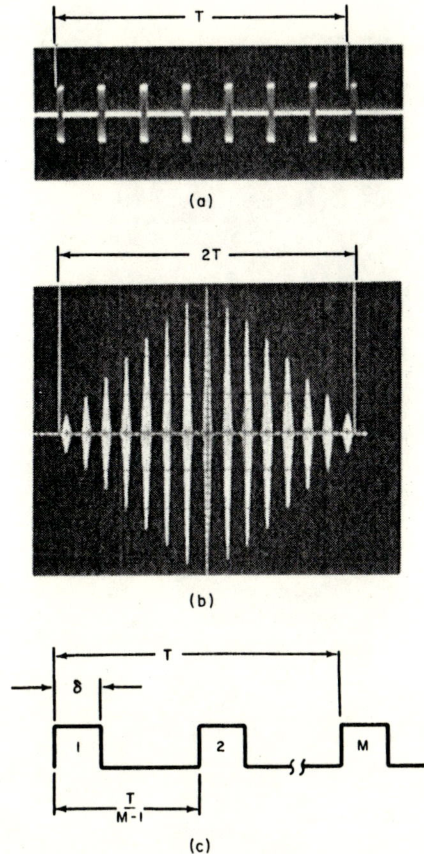

FIG. 8.1   The uniformly spaced train of identical pulses, $T/(M-1) = 5\,\delta$ (pulse positions: 1, 6, 11, 16, 21, 26, 31, 36). (a) Pulse train signal. (b) Matched-filter output (autocorrelation function). (c) Pulse and spacing parameters.

The matched-filter response (i.e., $\phi = 0$) for the uniform pulse train is

$$|\chi(\tau,0)| = M\delta \sum_{k=-M}^{M} \left[1 - \frac{|k|}{M}\right]\left[1 - \frac{\left|\tau - \dfrac{kT}{M-1}\right|}{\delta}\right] \qquad (8\text{-}11)$$

It is seen that this consists of a periodic train of triangles, 2δ wide at their base (see Fig. 8.2) which are weighted by $[1 - |k|/M]$. Figure 8.1 illustrates this matched-filter response, and Fig 8.2 illustrates some vertical profiles $[M\phi T/(M-1) = 1, 2, 3, \ldots, M/2, \ldots, M-2, M-1, M]$ of $|\chi(\tau,\phi)|$. An expanded view of one of the off-Doppler profiles is presented in Fig. 8.3.

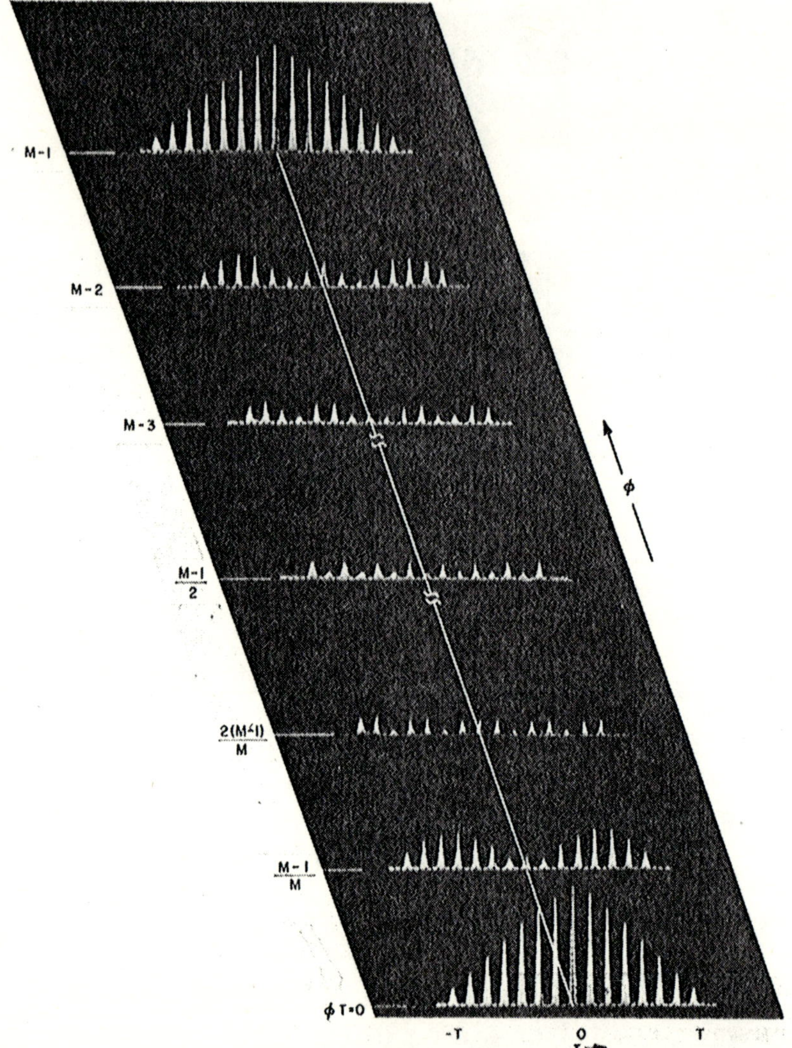

FIG. 8.2   Uniform pulse train response function profiles to first Doppler ambiguity, $M = 8$.

Figure 8.4 shows $|\chi(\tau, \phi)|$ contours at a constant level (say 3 to 4 db) down from $|\chi(0, 0)|$.

The importance of the uniform pulse train is furnished by two of its features. One feature is derived from the dead time between pulses. This allows the radar receiver to be activated for processing before the entire pulse train has left the transmitter. The second feature involves the volume

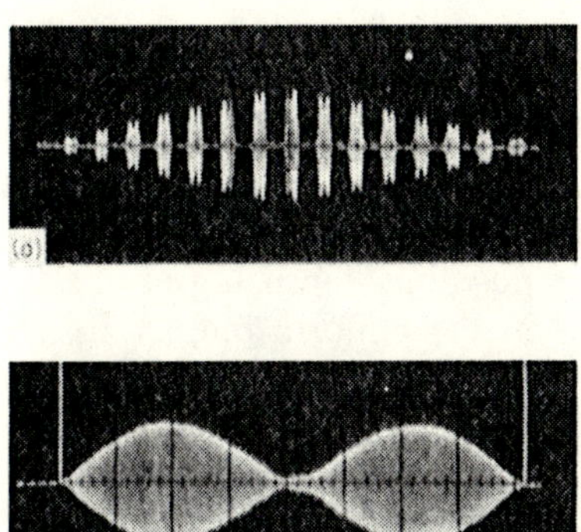

FIG. 8.3 Matched-filter Doppler shifted output of eight pulse uniform train for $\phi = 5(M - 1)/T$ (5th ambiguity lobe). (a) Matched-filter response. (b) Detail of central response.

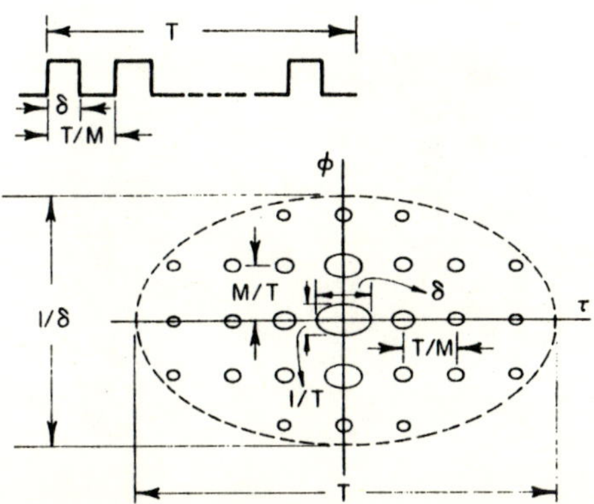

FIG. 8.4 Uniform pulse train response function level contours. (M here is equivalent to $M - 1$ of the analysis in this section).

free area (see Section 4.7) that surrounds the central matched-filter response. For many radar applications, it is possible to assign values to T and δ so that the *a priori* parameter estimation uncertainties and potentially ambiguous targets (or clutter) are confined to a subarea of the theoretical maximum volume free area surrounding the central response. There are situations, however, when this cannot be done. For these cases, the periodic ambiguity "pop-ups," which are inherent in the ambiguity structure associated with the uniform pulse train waveforms, need to be carefully weighed against the ambiguity structure of other signals.

A waveform that may nominally be well suited for many radar applications is derived by destroying the periodic structure of the pulse trains by spacing the pulses at nonuniform intervals. The ambiguity characteristics for three staggered pulse train examples are illustrated in Fig. 8.5 through 8.11[1]. Figures 8.5 and 8.6 result when the 1's in the sequence $\{a_n\}$ are restricted to three positions consisting of -1, 0, or $+1$ relative to the positions of the ones in the uniform pulse train sequence. This example

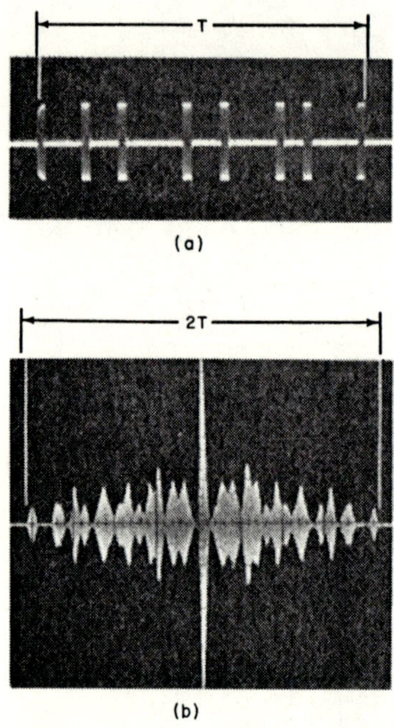

FIG. 8.5   Plus/minus position staggering of eight pulse train waveform (pulse positions: 1, 6, 10, 17, 21, 27, 30, 36). (a) Pulse train signal. (b) Matched-filter output (autocorrelation function).

is a special case of a more general pulse position stagger discussed by Rihaczek [5]. The particular sequence represented by these figures contains eight nonzero elements whose positions are given by

$$\{a_n\} = 1, \qquad n = 1, 6, 10, 17, 21, 27, 30, 36$$

$$= 0, \qquad \text{elsewhere}$$

---

[1] In order to retain the general feature of pulse trains that allows the receiver to be activated before the entire waveform is transmitted, it is necessary to insert an additional fixed interval between the pulses which can never be occupied. These intervals, however, can be omitted from the sequence $\{a_n\}$ when the basic ambiguity properties are studied.

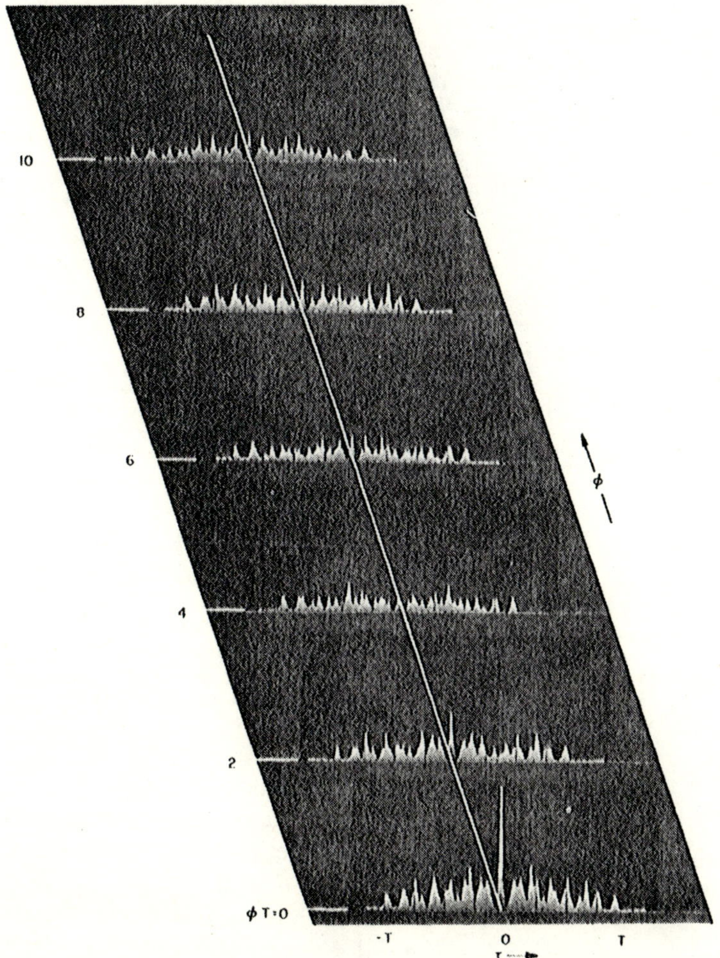

FIG. 8.6   Response function profiles for eight pulse plus/minus staggered pulse train.

The peak to highest sidelobe response, for no Doppler shift, is seen to be 8 to 3. These figures should be compared to the ambiguity properties of the uniformly spaced train shown in Figs. 8.1 and 8.2.

Figures 8.7 and 8.8 illustrate the ambiguity properties for a staggered pulse train having ten pulses, each positioned randomly in any one of 50 possible positions. For this case, the particular sequence is given by

$$\{a_n\} = 1, \qquad n = 1, 8, 11, 17, 27, 29, 32, 37, 45, 49$$

$$= 0, \qquad \text{elsewhere}$$

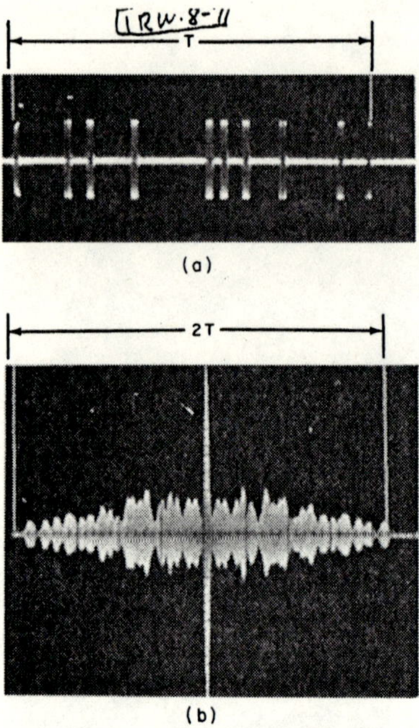

FIG. 8.7   Random stagger ten pulse train, $T = 50\,\delta$ (pulse positions: 1, 8, 11, 17, 27, 29, 32, 37, 45, 49). (a) Pulse train signal. (b) Matched-filter output (autocorrelation function).

The last case which is illustrated in Figs. 8.9 and 8.10 consists of a staggered pulse train where the positions of the nonzero elements were selected so that the sidelobe level would not exceed unity [6]. The sequence, for this case, contains seven ones and is given by

$$\{a_n\} = 1, \qquad n = 1, 4, 11, 17, 25, 29, 34$$

$$= 0, \qquad \text{elsewhere}$$

Figure 8.11 compares the Doppler axis responses (i.e., $\tau = 0$) of the preceding staggered pulse train examples. As in the case of the uniform pulse train, the Doppler axis responses $\chi(0, \phi)$ have the same structure as the signal spectrum. This property is clearly demonstrated in Fig. 8.12 which compares the Doppler axis response for the seven pulse example given above and its spectrum analyzer response. It can be noted that staggering the pulse positions destroys the periodicity of the Doppler axis response and spectrum, as well as of the autocorrelation response.

The general ambiguity function description for the staggered pulse trains,

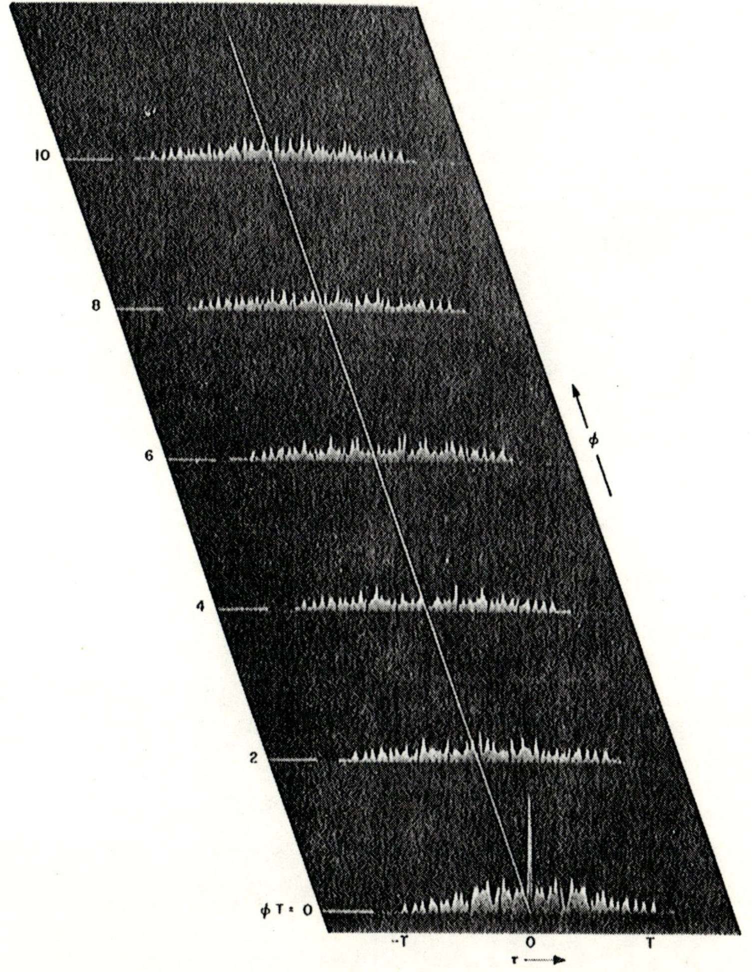

FIG. 8.8   Response function profiles for ten pulse random stagger pulse train.

obtained by direct substitution in (8-3) is

$$\chi(\tau, \phi) = \chi_p(\tau', \phi) \sum_{n=1}^{N-k} a_n a_{n+k} \exp[-j2\pi\phi(n-1)\delta]$$

$$+ \chi_p(\delta - \tau', \phi) \sum_{n=1}^{N-(k+1)} a_n a_{n+k+1} \exp[-j2\pi\phi n\delta], \quad (8\text{-}12)$$

$$0 < \tau < N\delta, \quad -\infty < \phi < \infty$$

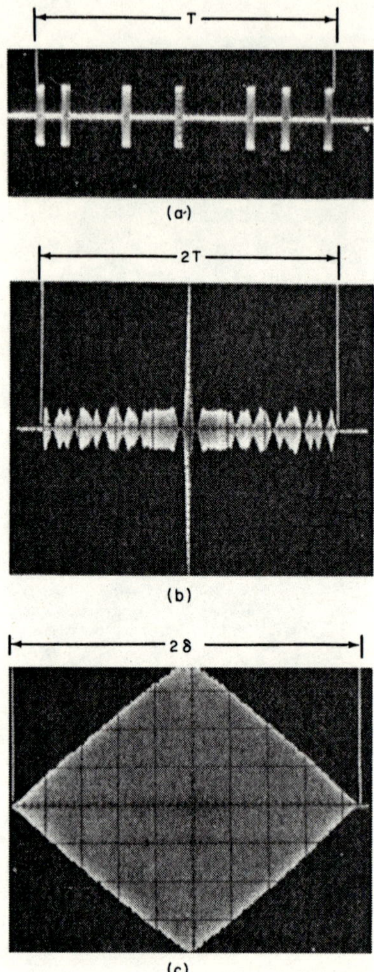

FIG. 8.9 Suboptimum seven pulse staggered pulse train (pulse positions: 1, 4, 11, 17, 25, 29, 34). (a) Pulse train signal. (b) Matched-filter output (autocorrelation function). (c) Central impulse of autocorrelation function. (After Rubin and Kaiteris [6].)

where τ' and k are defined by

$$\tau = k\delta + \tau', \quad \begin{cases} 0 < \tau' < \delta \\ k = 0, 1, 2, \ldots, N \end{cases} \tag{8-13}$$

and $\chi_p(\tau', \phi)$ is defined by

$$\chi_p(\tau', \phi) = \int_0^{\delta - \tau'} \exp[-j2\pi\phi t]\, dt, \qquad 0 < \tau' < \delta \tag{8-14}$$

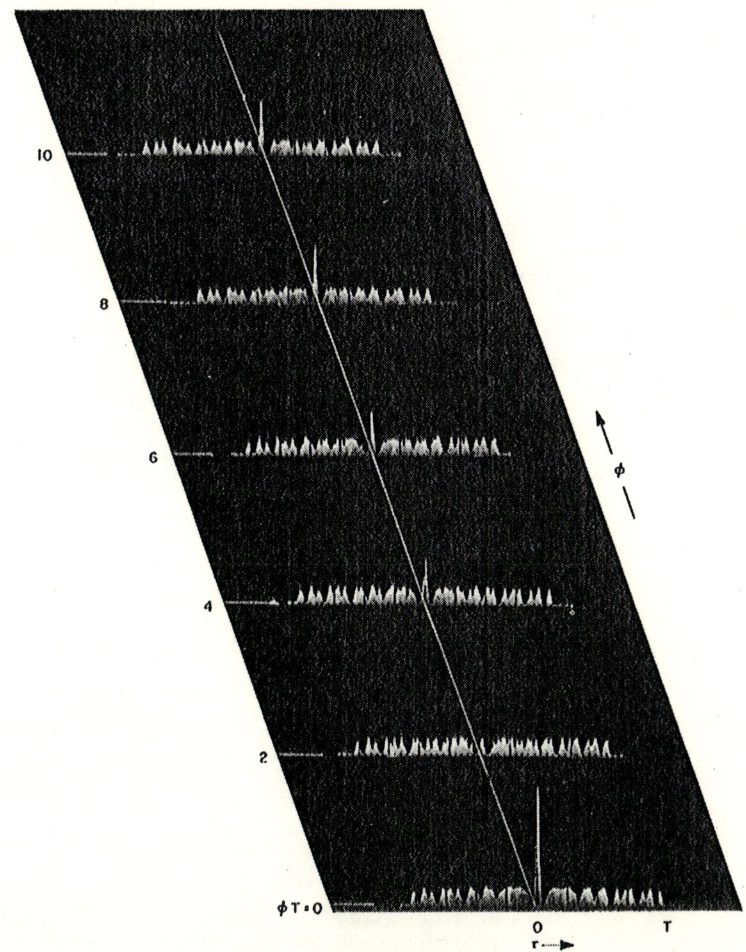

FIG. 8.10   Response function profiles for seven pulse suboptimum staggered pulse train.

The form of $\chi_p(\tau', \phi)$ is equivalent to the one-sided (i.e., $0 < \tau < \infty$) ambiguity characteristic derived for a single uncoded pulse waveform that is defined in the interval $0 \leqslant t \leqslant \delta$. Although Eq. (8-12) is valid for only $\tau \geqslant 0$, it is possible to extend the description of $\chi(\tau, \phi)$ to include the entire $\tau - \phi$ plane by applying an intrinsic property of the staggered pulse trains (and also of the binary phase codes which will be discussed in the next section). This property is given by

$$\chi(\tau, \phi) = \chi(-\tau, \phi) \exp[-j2\pi\phi\tau] \qquad (8\text{-}15)$$

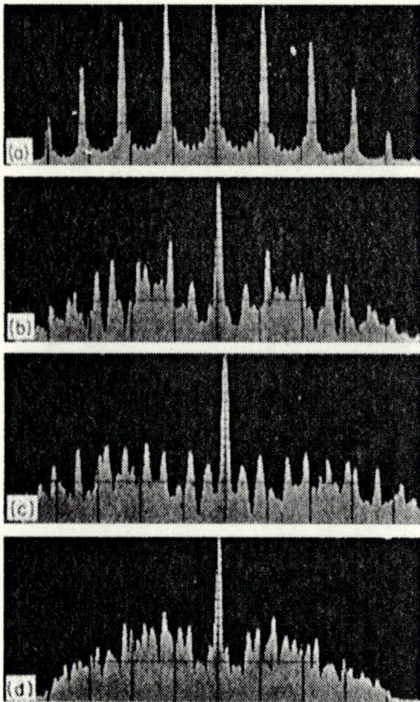

FIG. 8.11   Response function profiles for $\tau = 0$ (i.e., Doppler axis response). (a) Uniform pulse train, lobe spacing $= (M - 1)/T$. (b) Plus/minus stagger. (c) Ten pulse random stagger. (d) Seven pulse suboptimum stagger.

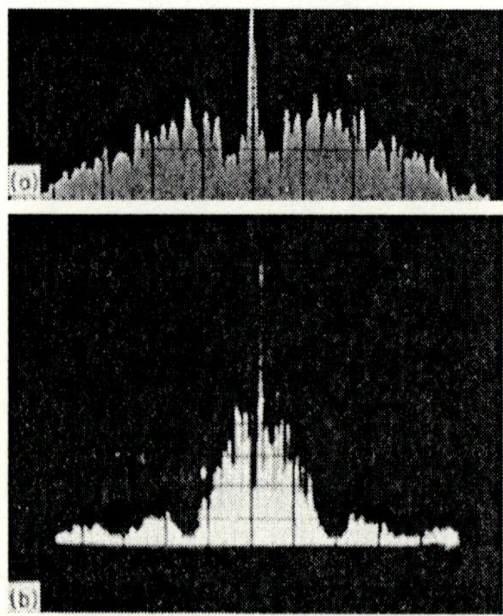

FIG. 8.12   Comparison of seven pulse suboptimum stagger spectrum and $\tau = 0$ ambiguity profile. (a) $\tau = 0$ ambiguity profile. (b) Pulse train spectrum. (Note scale difference; width of (a) represents central region of (b)).

The implication of (8-15) is that $|\chi(\tau, \phi)|$ is symmetric about the ϕ-axis. It may be recalled from the discussion in Chapter 4 of the ambiguity function properties that, in general

$$\chi(\tau, \phi) = \chi^*(-\tau, -\phi)\exp[j2\pi\phi\tau] \tag{8-16}$$

From Eqs. (8-15) and (8-16), it can be shown that

$$\chi(\tau, \phi) = \chi^*(\tau, -\phi) \tag{8-17}$$

Clearly, this relationship implies that $|\chi(\tau, \phi)|$ is also symmetric about the τ-axis. The combination of these properties make the staggered pulse trains very attractive when thumbtack ambiguity characteristics[1] are a desired waveform objective.

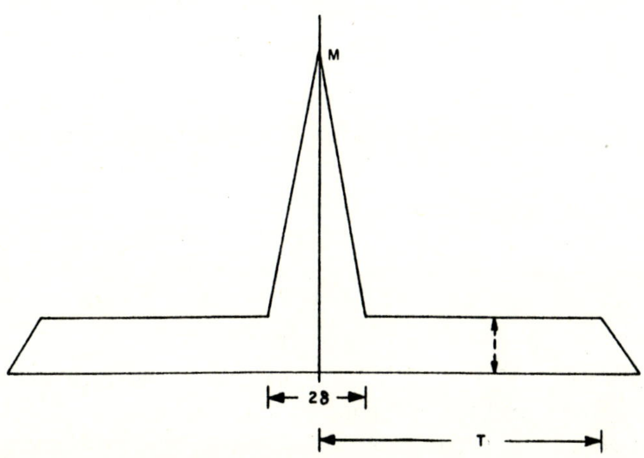

FIG. 8.13   Autocorrelation function of optimum stagger pulse train.

Resnick [4] has defined an "optimum" staggered pulse train as a sequence of nonuniformly spaced pulses for which the ambiguity characteristic $|\chi(\tau, 0)|$ has an absolutely uniform sidelobe level, equal to unity, in the region $\delta \leqslant |\tau| < T$ (see, for example, Fig. 8.13).[2] These staggered pulse trains

---

[1] The "thumbtack" ambiguity function was discussed in Chapter 4 and will also be discussed in Chapter 10.

[2] Although the last of the staggered pulse train examples previously given (see Figs. 8.9 and 8.10) also furnishes the ambiguity characteristic $|\chi(\tau, \phi)|$ with a sidelobe level that never exceeds unity, it actually falls under a broader heading of suboptimum [4] staggers which also provide unity sidelobes but do not have the most compact sequence structures.

provide very close realizations of the "thumbtack" ambiguity characteristic. In order to obtain this characteristic it is necessary to position the
nonzero elements in the stagger sequence $\{a_n\}$ so that the set of distances
between all possible nonzero element pairs are distinct and complete; that
is, every distance, starting with the unit distance and ending with the largest
possible distance of $N - 1$, must appear once and only once. These
sequences will then contain the maximum number of nonzero elements
for a given sequence length of N. Since every nonzero element must pair-up
with every other nonzero element, it is seen that a necessary condition for
an optimum stagger is that

$$(N - 1)\delta = \binom{M}{2}\delta = T \tag{8-18}$$

From Eq. (8-18), it can be shown that when $M \gg 1$

$$M = \sqrt{2T\,\Delta f} \tag{8-19}$$

where $\Delta f = 1/\delta$ is defined as the effective waveform bandwidth. Hence,
the peak to sidelobe ratio for these waveforms is uniform and equal to
$\sqrt{2T\,\Delta f}$.

The optimum stagger sequences for $M = 2, 3, 4$ are, respectively:

1.   11
2.   1101
3.   1100101

There are no others. That is, it is impossible, as Resnick has shown, to
find an optimum stagger sequence for $M > 4$. The proof follows from
Eq. (8-18). It is seen from this that the adjacent nonzero element distances

of the optimum stagger must add up to $N - 1 = \binom{M}{2}$. This result can only

be obtained if the adjacent distances between nonzero elements of the
sequence consist of all integers between 1 and $M - 1$. It follows from this
that each of the one and two unit distances can only be adjacent to the
$M - 1$ unit distance. Hence, the one and two unit distances must be
adjacent (e.g., 1101) or at most separated by the three unit distance
(e.g., 1100101). No other arrangement is possible except, of course, $\{a_n\} = 11$
where only the unit distance is present. However, when the restriction is
relaxed that the sidelobe structure be uniform the larger class of suboptimum staggers can be used. These yield autocorrelation responses
$\chi(\tau, 0)$ for which the sidelobes never exceed unity, but which also have
null values [6], as can be seen from Fig. 8.9.

8.3 Binary Phase Codes (Group II)

The general description obtained from Eq. (8-1) for binary phase codes by setting $a_n = 1$ and deleting ω_n is

$$\psi(t) = \sum_{n=1}^{N} P_n(t) \exp[j(\omega_0 t + \theta_n)], \qquad 0 \leqslant t \leqslant N\delta$$
$$= 0, \qquad \text{elsewhere}$$

$$(8\text{-}20)$$

where $\theta_n = 0, \pi$. The coefficient c_n is introduced for convenience such that

$$c_n = \exp[j\theta_n] \qquad (8\text{-}21)$$

Then, since $\theta_n = 0, \pi$, it is seen that

$$c_n = \pm 1 \qquad (8\text{-}22)$$

The binary sequences $\{\theta_n = 0, \pi\}$ and $\{c_n = \pm 1\}$ are therefore interchangeable. There are occasions, especially when deriving the properties of these sequences, when it is more convenient to represent the binary phase code sequence by:

$$d_n = 0, 1 \qquad (8\text{-}23)$$

Thus, there are three different element pairs by which it is possible to present a binary phase sequence. These are presented in the accompanying tabulations[1]:

θ_n	c_n	d_n
0	+	0
π	−	1

$$(8\text{-}24)$$

Associated with this table of binary elements are two kinds of arithmetic. One of these is used on the elements under c_n and is characterized by the following multiplication table:

	+	−
+	+	−
−	−	+

$$(8\text{-}25)$$

---

[1] The + and − under c_n imply, respectively, $+1$ and -1. The one is usually omitted for convenience.

The other arithmetic is used on the elements under d_n and is characterized by the following (mod 2) addition table:

	0	1
0	0	1
1	1	0

$$(8\text{-}26)$$

A typical binary phase sequence $\{\theta_n\}$ given, for example, in terms of $\{c_n\}$ is

$$\{c_n\} = + + + - + - - + \tag{8-27}$$

Figures 8.14(a) and 8.14(b) illustrate, respectively, the video and cw forms corresponding to the sequence given by (8-27). Note that the carrier in Fig. 8.14(b), is shown to be continuous at the points of phase inversion.

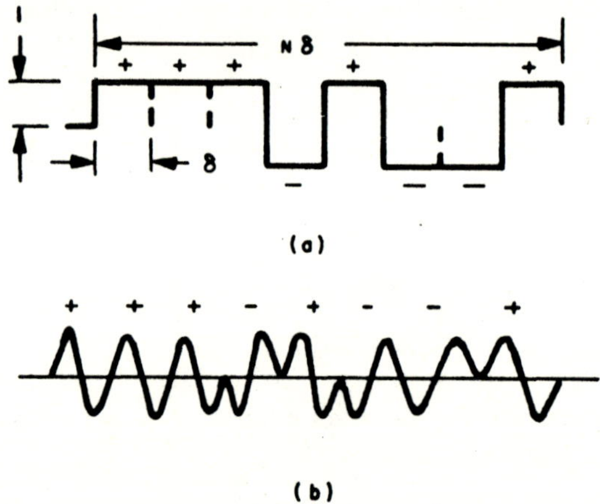

FIG. 8.14   Binary coded signals. (a) Video amplitude modulation. (b) Phase reversal cw coded signal.

However, this continuity occurs only as a result of the reciprocal relationship $\delta = 1/f_0$ which is employed in this illustration. This relationship was chosen only for convenience. As a rule $\delta \neq n/f_0$, where n implies integer multiples of $1/f_0$. Therefore, since the carrier is always coherent, the fine structure will generally be discontinuous at the phase inversion points.

The general response function description for the binary phase codes, obtained from Eq. (8-3), is

$$\chi(\tau, \phi) = \chi_p(\tau', \phi) \sum_{n=1}^{N-k} c_n c_{n+k} \exp[-j2\pi\phi(n-1)\delta]$$

$$+ \chi_p(\delta - \tau', \phi) \sum_{n=1}^{N-(k+1)} c_n c_{n+k+1} \exp[-j2\pi\phi n\delta], \quad (8\text{-}28)$$

$$0 < \tau < N\delta, \quad -\infty < \phi < \infty$$

This expression, with but one minor change from the coefficient a_n to c_n, was discussed in regard to the staggered pulse train in the previous section. It is seen for $\phi = 0$, that

$$\chi(\tau, 0) = \delta\left[1 - \frac{|\tau'|}{\delta}\right] \sum_{n=1}^{N-|k|} c_n c_{n+k} + |\tau'| \sum_{n=1}^{N-|k+1|} c_n c_{n+k+1} \quad (8\text{-}29)$$

and when $\phi = \tau' = 0$, then

$$\chi(k\delta, 0) = \delta \sum_{n=1}^{N-|k|} c_n c_{n+k} \quad (8\text{-}30)$$

Equation (8-30) can be evaluated conveniently by means of Bernfeld's Algorithm [7, 8]. This gives a construction technique that utilizes arithmetic consisting of noncarry multiplication involving $\{c_n\}$ and its reverse (or mirror image) $\{c_{-n}\}$. For example, taking the $\{c_n\}$ sequence given by (8-27), and letting $\delta = 1$ for convenience, $\chi(k, 0)$ is obtained as follows:

```
{c_n}  =  +  +  +  -  +  -  -  +
{c_-n} =  +  -  -  +  -  +  +  +
          ─────────────────────────
          +  +  +  -  +  -  -  +
             -  -  -  +  -  +  +  -
                -  -  -  +  -  +  +  -
                   +  +  +  -  +  -  -  +
                      -  -  -  +  -  +  +  -
                         +  +  +  -  +  -  -  +
                            +  +  +  -  +  -  -  +
                               +  +  +  -  +  -  -  +
```

$$\chi(k, 0) = \boxed{\ 1\ |\ 0\ |-1|-2|\ 1\ |\ 0\ |-1|\ 8\ |-1|\ 0\ |\ 1\ |-2|-1|\ 0\ |\ 1\ }$$

This technique is not only useful for binary sequences, but also for the more general case when c_n is a complex number. Clearly, $\chi(\tau, 0)$ is determined entirely from $\chi(k\delta, 0)$. Equation (8-30) expressed in terms of d_n, again letting $\delta = 1$, can be shown to be

$$\chi(k, 0) = [N - |k|] - 2 \sum_{n=1}^{N-|k|} d_n \oplus d_{n+k} \tag{8-31}$$

where the symbol $\oplus$ implies mod 2 addition; the proof is left as an exercise for the reader. A general property of $\chi(k, 0)$ can be obtained by decomposing the $\{c_n\}$ sequences as follows [8]:

$$\{c_n\} = \{p_n\} + \{q_n\} \tag{8-32}$$

where

$$p_n = \frac{c_n + 1}{2}, \qquad q_n = \frac{c_n - 1}{2}$$

For example, suppose $\{c_n\} = + + - +$, then $\{p_n\} = + + 0 +$ and $\{q_n\} = 0\,0 - 0$. When $\{c_n\}$ in Eq. (8-30) is replaced by $\{p_n\} + \{q_n\}$ the result is

$$\chi(k, 0) = \chi_+(k, 0) + \chi_-(k, 0) \tag{8-33}$$

where

$$\chi_+(k, 0) = \sum_{n=1}^{N-|k|} [p_n p_{n+k} + q_n q_{n+k}] \tag{8-34}$$

and

$$\chi_-(k, 0) = \sum_{n=1}^{N-|k|} [p_n q_{n+k} + q_n p_{n+k}] \tag{8-35}$$

It is seen that $\chi_+(k, 0)$ is always positive and $\chi_-(k, 0)$ is always negative. Next, form the summation obtained from (8-30) by replacing $\{c_n\}$ by $\{p_n\} - \{q_n\}$. This yields the following property:

$$N - |k| = \chi_+(k, 0) - \chi_-(k, 0) \tag{8-36}$$

or

$$\chi_+(k, 0) = \tfrac{1}{2}[\chi_{\max}(k, 0) + \chi(k, 0)] \tag{8-37}$$

and

$$\chi_-(k, 0) = \tfrac{1}{2}[\chi(k, 0) - \chi_{\max}(k, 0)] \tag{8-38}$$

where $\chi_{\max}(k, 0) = N - |k|$. The function $\chi_{\max}(k, 0)$ is obtained when there are no sign inversions in the sequence $\{c_n\}$.

The Barker Sequences

This family of binary sequences is characterized by

$$\chi(k, 0) = \begin{cases} N, & k = 0 \\ \pm 1, 0, & k \neq 0 \end{cases} \qquad (8\text{-}39)$$

The sequences of this group share, in common with the optimum stagger sequences, the property $|\chi(k, 0)| \leq 1$, $k \neq 0$. Thus, they are also called optimum sequences. The size of this family, however, is equally as restricted as the family of optimum stagger sequences. In fact, there exist no more than nine Barker sequences [9]. These are given in Table 8-I. Storer and

TABLE 8-I

BARKER SEQUENCES

N	$\{c_n\}$	$\chi(k, 0)$, $k = 0, 1, \ldots, (N - 1)$
2	+ +	2 +
2	− +	2 −
3	+ + −	3 0 −
4	+ + − +	4 − 0 +
4	+ + + −	4 + 0 −
5	+ + + − +	5 0 + 0 +
7	+ + + − − + −	7 0 − 0 − 0 −
11	+ + + − − − + − − + −	11 0 − 0 − 0 − 0 − 0 −
13	+ + + + + − − + + − + − +	13 0 + 0 + 0 + 0 + 0 + 0 +

Turyn [10] have shown that there can exist no odd Barker sequence of length greater than 13; however, the problem remains unsolved for the even length codes. It has been conjectured that there exist no even length Barker sequence beyond $N = 4$ [11]. The conjecture is supported by the fact that none have been found for $4 < N < 6084$. In the event that an even length sequence does exist, it must be a perfect square. This condition is obtained from the identity [11]

$$\left[\sum_{n=1}^{N} c_n \right]^2 = \chi(0, 0) + 2 \sum_{k=1}^{N-1} \chi(k, 0) = \chi(0, 0) + \sum_{k=1}^{N-1} [\chi(k, 0) + \chi(N - |k|, 0)]$$

$$(8\text{-}40)$$

and from the fact that for the even length codes ($N > 2$), $\chi(k, 0) + \chi(N - |k|, 0) = 0$. It is also necessary to recall that $\chi(0, 0) = N$. In addition to the identity (8-40), it may also be shown that when

(1) N is odd and $(N - 1)/2$ is even, then the sidelobes of $\chi(k, 0)$ are always positive ($N = 5, 13$);

(2) N is odd and $(N - 1)/2$ is odd, then the sidelobes of $\chi(k, 0)$ are always negative ($N = 3, 7, 11$).

These properties are illustrated in the table of Barker sequences. The following recurrence formula also exists between the elements of the odd length Barker sequences:

$$c_n = c_{N-n-1}(-1)^{n + (N-1)/2} \qquad (8\text{-}41)$$

In addition, it has also been shown that for N odd

$$\frac{c_n}{c_{n-1}} = \frac{c_{2n}}{c_{2n-1}} \qquad (8\text{-}42)$$

Finally, it has been observed that when N is odd and $(N - 1)/2$ is even the difference between the number of plus and minus signs is $\sqrt{2N - 1}$. When N is odd and $(N - 1)/2$ is odd this difference is 1.

Figures 8.15 through 8.17 illustrate the entire ambiguity function characteristics for the Barker $N = 11$ and $N = 13$ sequences as shown from

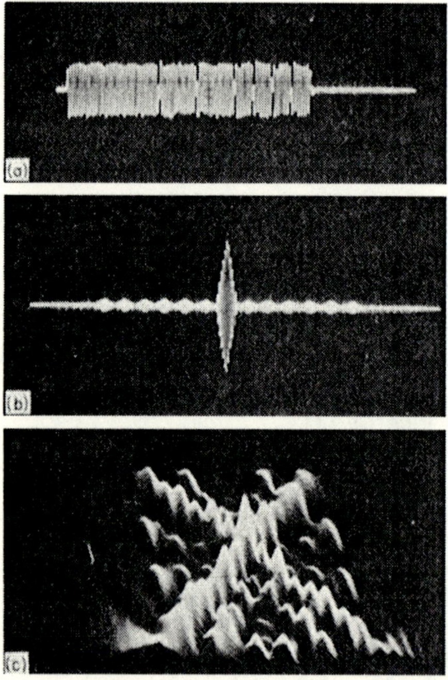

FIG. 8.15   Barker code (length = 13) waveforms and response surface. (a) Matched-filter input signal. (b) Matched-filter autocorrelation response. (c) Composite response surface. (Courtesy of C. E. Jagger and R. H. McLaughlin [25].)

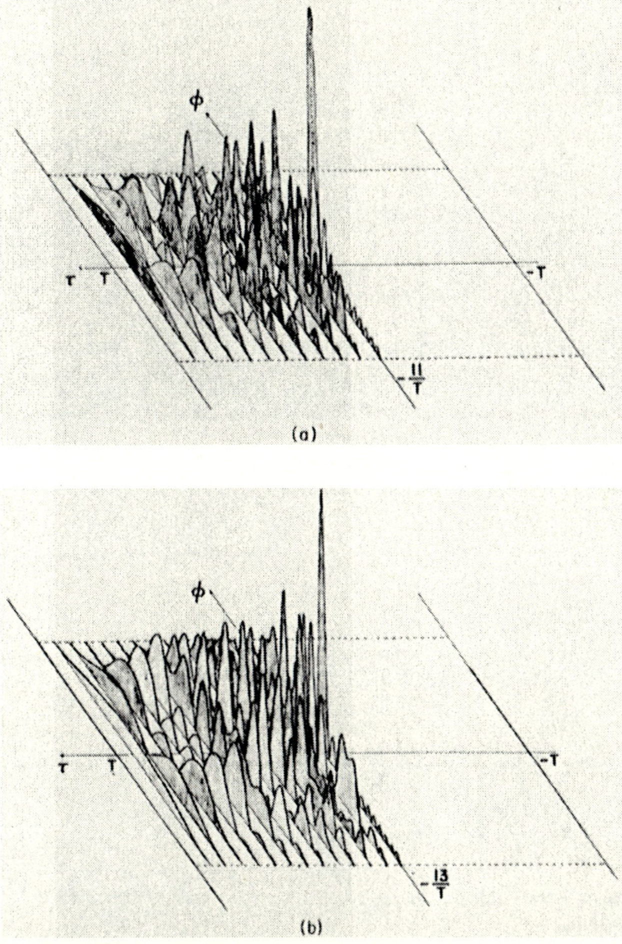

FIG. 8.16   Matched-filter response functions $|\chi(\tau, \phi)|$ for Barker coded signals. (a) Response function for Barker binary phase code, $N = 11$. (b) Response function for Barker binary phase code, $N = 13$. (Courtesy of T. Sakamoto *et al.* [14].)

different view points. The spectrum modulus of the Barker code of length 13 is also shown in Fig. 8.17. It can be seen that the ambiguity function for $N = 13$ exhibits some similarities to that of the V-FM signal with its crossing ridge line structure. This has also been noted for the $N = 11$ case.

Maximum Length Sequences

The maximum length sequences (or *M*-sequences) form another important class of binary sequences suitable for phase coding [2, 12–14]. These sequences are derived from recurrence formulas that are designed to provide

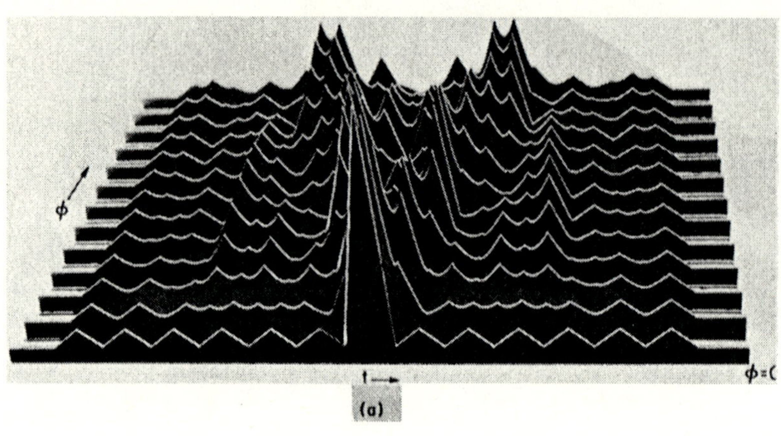

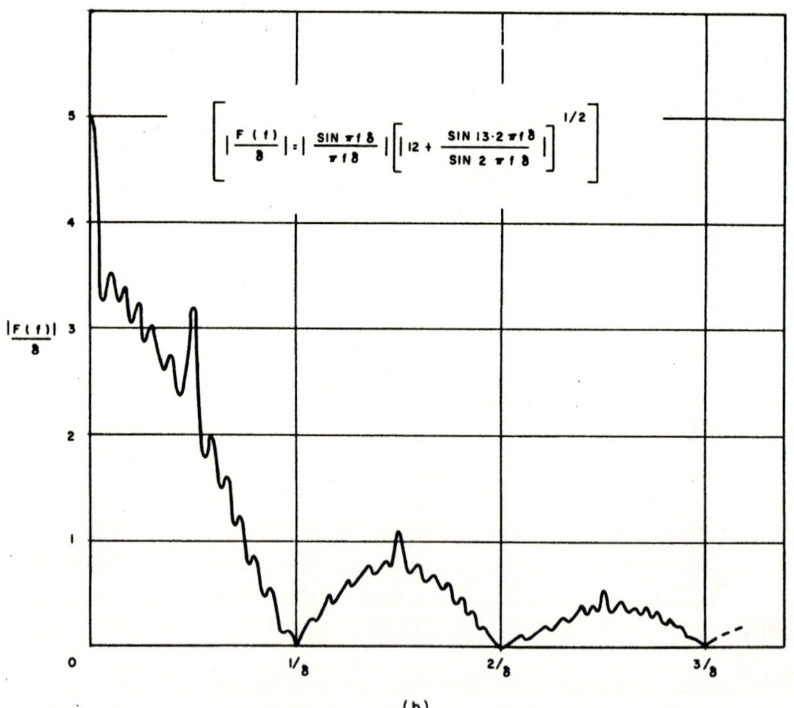

FIG. 8.17   Response function and spectrum modulus of Barker code of length 13. (a) Response function. (b) Spectrum modulus. (Courtesy of E. N. Fowle, The MITRE Corporation, Bedford, Massachusetts.)

a maximum number of sequence elements equal to $2^k - 1$ before recycling. In general, when these sequences are expressed in terms of $d_n = 0, 1$, then for $n > k$, each d_{i+1} is the modulo 2 sum of certain elements selected from the preceding k elements. This formula is given by

$$d_{i+1} = \alpha_1 d_i \oplus \alpha_2 d_{i-1} \oplus \cdots \oplus \alpha_k d_{i-k} \tag{8-43}$$

where the coefficients α are either 1 or 0. An alternate form leading to an equivalent sequence expressed in terms of c_n is given by

$$c_{i+1} = \prod_{j=1}^{k} c_{i-j+1}^{\alpha_j} \tag{8-44}$$

where the exponents α_j are the same as the coefficients α_j of (8-43).

The first k elements of these sequences, excluding all zeros (or plus signs), are arbitrarily selected. These k elements form the basis for one of $2^k - 1$ phases that can be derived from a given α sequence. Not all α sequences, however, lead to a maximum length sequence in d_n (or c_n). The conditions for these sequences can be found in a number of papers [15–17] which are recommended for further study. An extensive list of irreducible and primitive polynomials, equivalent to α sequences is furnished by Peterson [17].

The recurrence formula (8-43) is particularly suited for shift register implementations. A canonic configuration is illustrated in Fig. 8.18. It

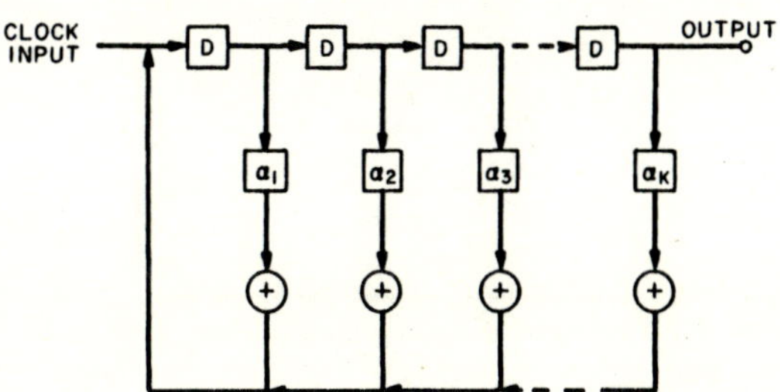

FIG. 8.18   Canonic shift register configuration.

consists of three basic elements: (1) a tandem array of single input, single output unit delays denoted by D (an operator introduced by Huffman [18]); (2) feedback switches denoted by α_i (elements of the α sequence); and (3) modulo 2 adders. These implementations are usually synchronized with clocked (uniformly spaced) pulses. For each successive pulse, the stored elements shift over one delay unit to the right and the computed element is stored at the first delay unit. The mathematical description of shift register implementations is usually expressed as a polynomial $P(D)$ in the unit delay operator D. In order to obtain a maximum sequence length, before the shift register recycles, this polynomial must be irreducible (unfactorable) and primitive (divide evenly into $D^m + 1$, where $m > 2^k - 1$). An example of such a polynomial is given by

$$P(D) = I \oplus D \oplus D^3 \tag{8-45}$$

where D^0 is denoted by I. The elements of the α sequence equivalent to (8-45) are $\alpha_1 = 1$, $\alpha_2 = 0$, $\alpha_3 = 1$. By choosing the initial state of the shift register to be 010, it is easily determined that the maximum length sequence derived from a shift register implemented according to (8-45) is (see Fig. 8.19)

$$\{d_n\} = \cdots 111, 0100111, 0100111, 010 \cdots \tag{8-46}$$

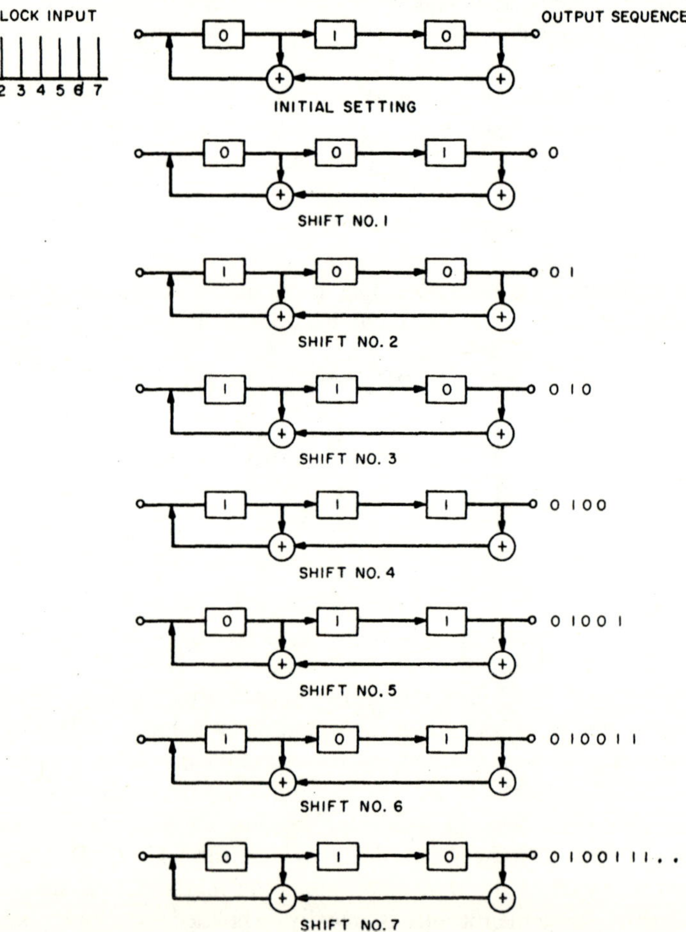

FIG. 8.19   Maximum length shift register states for $N = 7$.

where $N = 7$. Other phases of the sequence given by (8-46) can be derived by starting with a different set of three initial elements (excluding all zeros). There are six such remaining sets; i.e., 001, 011, 111, 100, 101, 110.

Periodic structures such as (8-46) and their associated polynomials have been extensively studied [15–17]. One of the important results of these studies is the property that

$$\chi(k, 0) = \begin{cases} 2^p - 1, & k = 0 \mod 2^p - 1 \\ -1, & k \neq 0 \mod 2^p - 1 \end{cases} \tag{8-47}$$

where p has been used here to denote the highest degree of the polynomial $P(D)$. This property is similar to the result obtained for the Barker sequences. However, in contrast with the Barker sequences, which consist of a finite number of elements, the structure of the maximum length sequence must be periodic to obtain (8-47). An experimental evaluation of $\chi(k, 0)$ for $N = 2^p - 1 = 255$ is illustrated in Fig. 8.20.

Periodic maximum length sequences have been employed in high resolution radar implementations [12–14]. This type of operation is equivalent to a cw radar system and thus has limited range unless the transmitter and receiver components are sufficiently isolated. For many radar applications the finite length sequence is preferred since it avoids the problem of cw operation, and also does not tie the maximum range of the radar to $(2^k - 1)\delta$, the unambiguous time interval of the periodic maximum length sequences. Examples of $\chi(\tau, 0)$ for maximum length sequences that are truncated beyond the first period are shown in Figs. 8.21 and 8.22 for $N = 7$, 15, and 31. These figures illustrate how the sidelobe structure of $\chi(\tau, 0)$ depends on the phase of these truncated maximum length sequences. In addition, Fig. 8.23 shows the response functions $\chi(\tau, \phi)$ corresponding to $N = 15$ and 31. For N large the peak-to-sidelobe ratio of the truncated maximum length sequences approaches $\sqrt{N}$.

8.4 Polyphase Codes (Group II)

The general description for the polyphase codes (except as noted in the footnote[1]) is also derived from Eq. (8-1) by setting $a_n = 1$, and deleting ω_n. The expression obtained for $\psi(t)$ is given by (8-20); however, θ_n is not restricted to values of 0 and π. In general, the $\{\theta_n\}$ sequences will be multivalued. Nevertheless, it is still convenient to replace the sequences in θ_n with sequences in c_n, where c_n is defined by (8-21). However, c_n will now represent a complex coefficient consisting of a real and imaginary part. The general ambiguity function description for the polyphase codes obtained from Eq. (8-3) is given by (8-28) except for the fact that the terms c_{n+k} and c_{n+k+1} in this expression are conjugated.

---

[1] The resultant expression does not describe the class of polyphase codes that are known as Huffman codes. The Huffman codes will be considered later in this section.

8. DISCRETE CODED WAVEFORMS

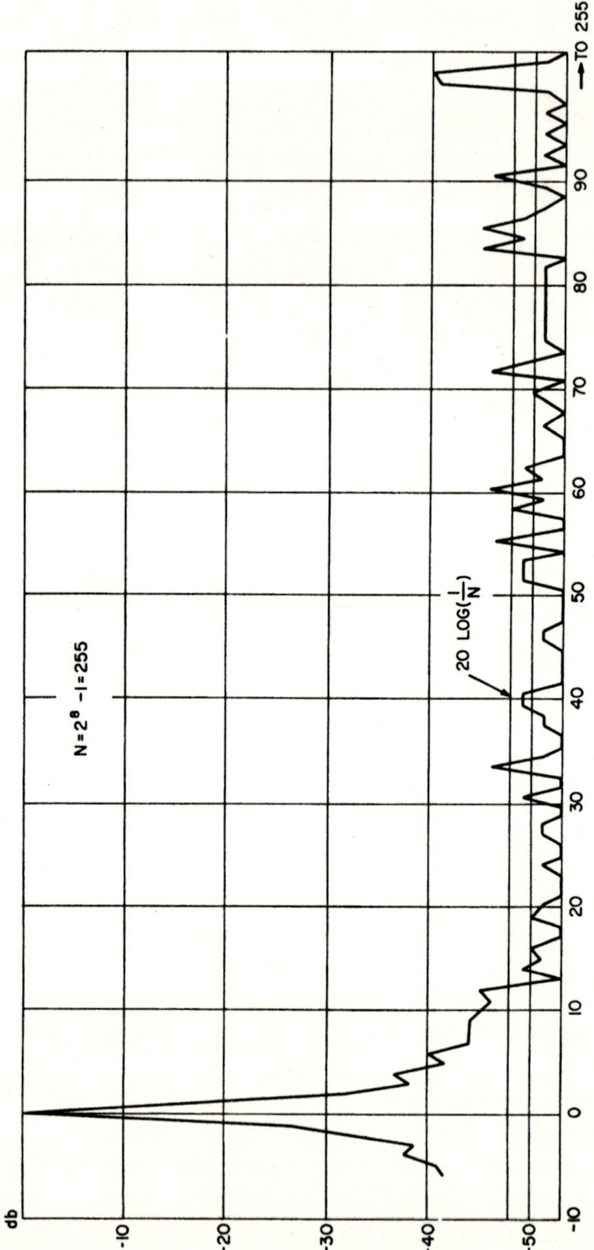

FIG. 8.20   Experimental autocorrelation waveform of periodic pseudorandom binary phase code (from Bernfeld *et al.*, Ref. 4, Chapter 1).

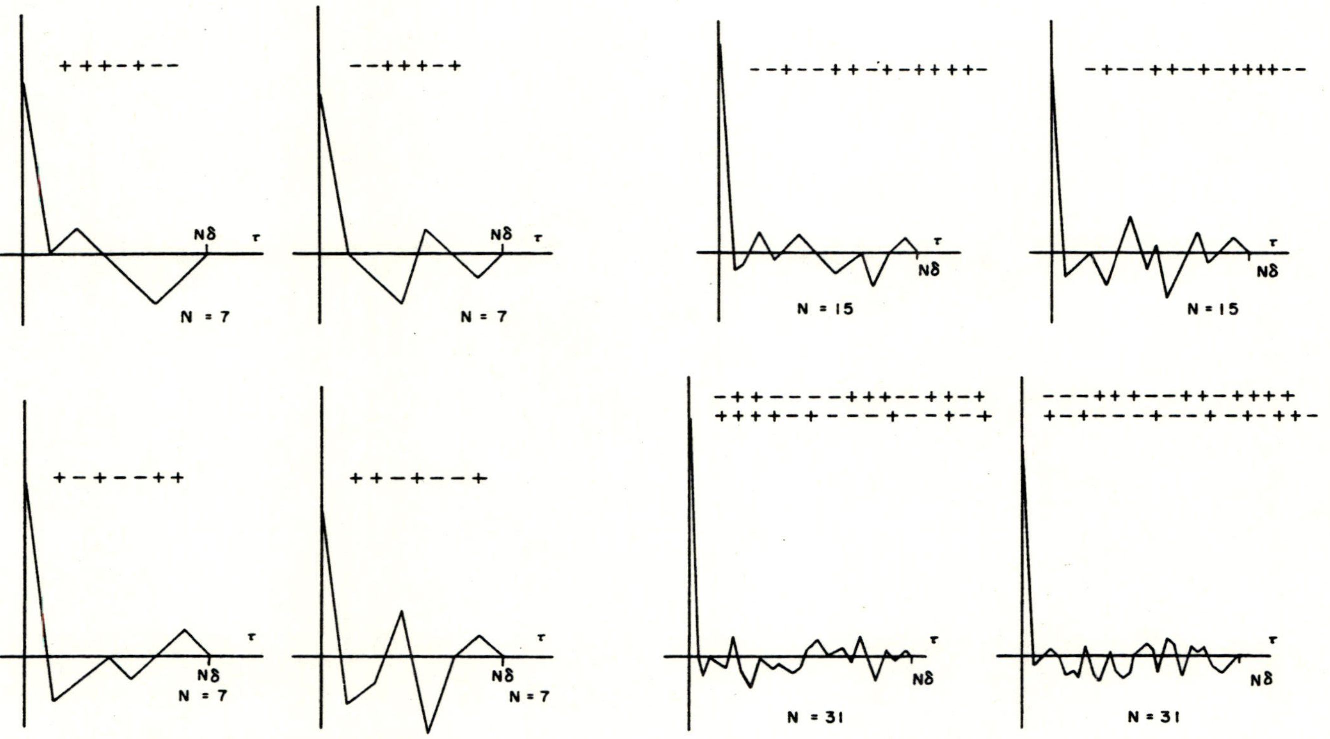

Fig. 8.21 Autocorrelation functions of truncated maximum length sequence binary phase code waveforms; note effect of sequence order on sidelobes. (Courtesy of T. Sakamoto et al. [14].)

Fig. 8.22 Autocorrelation functions of truncated maximum length sequence binary phase code waveforms; note effect of sequence order on sidelobes. (Courtesy of T. Sakamoto et al. [14].)

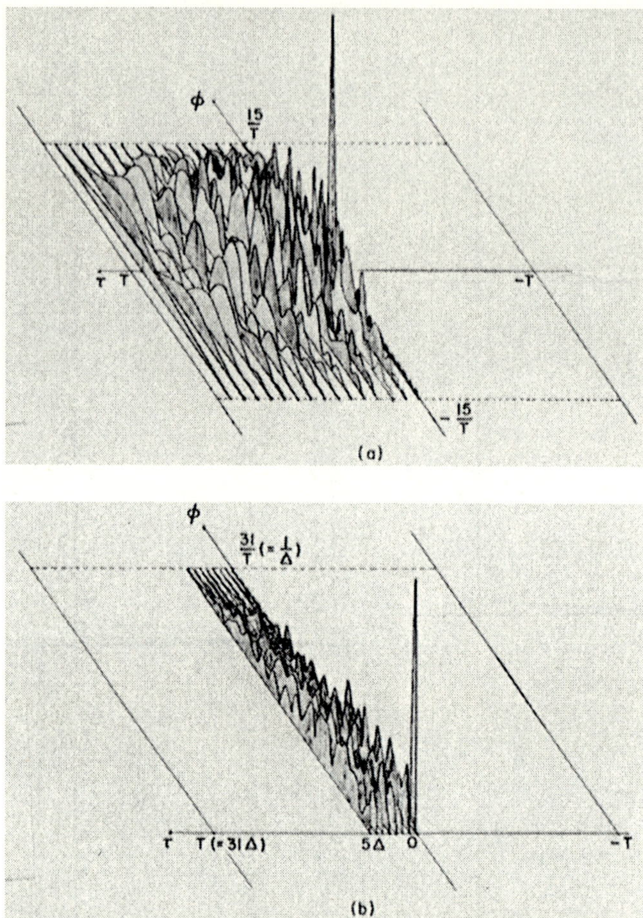

FIG. 8.23 Response functions for truncated binary phase code maximum length sequences. (a) Response function for binary code *M*-sequence, $N = 15$. (b) Response function for binary code *M*-sequence, $N = 31$. (Courtesy of T. Sakamoto *et al.* [14].)

Polyphase Barker Sequences

Sequences that are constructed from alphabets containing two or more phases and possessing the property $|\chi(k, 0)| \leqslant 1$ $(k \neq 0)$, form a general class of Barker sequences investigated by Golomb and Scholtz [19]. Constructions for lengths of from 1 to 9 and different alphabet sizes can be found in the discussion of their investigation. This reference also furnishes an exhaustive tabulation of sequences, derived from a six element alphabet, together with the corresponding $\chi(k, 0)$. The six element alphabet appears to be unique since it has yielded a Barker sequence for every length investigated through $N = 12$; there appear to be no theoretical constraints which would preclude the existence of longer sequences.

In general, the alphabets employed to derive the "generalized Barker sequences" consist of p roots of unity (i.e., ρ^0, $\rho^1 = \exp[j2\pi/p]$, $\rho^2, \ldots, \rho^{p-1}$). An example for length $N = 12$ is:

$$\{c_n\} = 1, 1, \rho, \rho^2, \rho^3, \rho, \rho, \rho^5, 1, \rho^3, 1, \rho^4$$

where for this case $\rho = \exp[j2\pi/6]$. This sequence yields

k	-11	-10	-9	-8	-7	-6	-5	-4	-3	-2	-1	0	1	2	3	4	5	6	7	8	9	10	11
$\chi(k,0)$	ρ^4	ρ^5	ρ^3	0	1	0	ρ^5	ρ^3	ρ^2	ρ^5	ρ	12	ρ^5	ρ	ρ^4	ρ^3	ρ	0	1	0	ρ^3	ρ	ρ^2

$$(8\text{-}48)$$

where the 0 notation in the $\chi(k, 0)$ row represents the absence of any output signal at that point.

Quantized Phase Codes

Heimiller [20], Frank and Zadoff [21], and Frank [22] have described a method that can be used to construct polyphase codes which utilizes a matrix [22] that has the following general structure. (The waveforms that are formed from these sequences are called quantized phase codes. Those based on the matrix below are also called Frank polyphase codes.)

$$
\begin{matrix}
0 & 0 & 0 & \cdot & \cdot & \cdot & 0 \\
0 & 1 & 2 & \cdot & \cdot & \cdot & (N-1) \\
0 & 2 & 4 & \cdot & \cdot & \cdot & 2(N-1) \\
\cdot & & & & & & \\
\cdot & & & & & & \cdot \\
\cdot & & & & & & \cdot \\
\cdot & & & & & & \cdot \\
0 & (N-1) & 2(N-1) & \cdot & \cdot & \cdot & (N-1)^2
\end{matrix}
$$

The matrix reads the same across and downwards. The elements represent multiplying coefficients of a basic phase angle $2\pi p/N$, where p and N are integers and p is relatively prime to N. For this discussion, it will be assumed that $p = 1$. The actual coded sequence is formed by placing the rows (or columns) side by side. This will yield a sequence containing N^2 elements.

For example, suppose $N = 3$, then this matrix yields the sequence[1]

$$\left\{\frac{N\theta_n}{2\pi}\right\} = 0, 0, 0; \quad 0, 1, 2; \quad 0, 2, 1 \tag{8-49}$$

which consists of nine elements. Note that the elements of this sequence are presented mod N and that each of the N groups starts with a zero element. The first group of three elements in (8-49) corresponds to no phase shift; in the next group of three elements the coefficients progress in steps of one and correspond to $0°$, $120°$, and $240°$; the last group of three elements progress in steps of two and correspond to $0°$, $240°$, $120°$ mod $360°$. Because the phases progress to the right in an orderly quadratic manner, the Frank polyphase coded waveforms formed from such sequences have also been described as quantized phase linear FM. The structure of these signals can be compared with linear step-FM signals, another quantized approximation to the analog linear FM signal (discussed in Section 8.5). Other examples of some sequences are

N	N^2	
2	4	(a) $0, 0/0, 1$ (b) $0, 0/1, 0$
4	16	$0, 0, 0, 0/0, 1, 2, 3/0, 2, 0, 2/0, 3, 2, 1$
5	25	$0, 0, 0, 0, 0/0, 1, 2, 3, 4/0, 2, 4, 1, 3/0, 3, 1, 4, 2/0, 4, 3, 2, 1$

The sequences for $N = 2$ are also seen to be Barker binary sequences.

As periodic sequences, these codes provide $\chi(k, 0)$ with zero sidelobes for $k \neq 0$ mod N^2 [20, 21]. This optimum sidelobe performance is not obtained when these sequences are truncated beyond a single period. However, autocorrelation characteristics have been obtained which possess lower sidelobes than the linear FM signals. The general description of the waveforms on which the truncated sequences are impressed is conveniently expressed in terms of two indices n, m, as a double summation given by

$$\psi(t) = \sum_{m=1}^{N} \sum_{n=1}^{N} P[t - \{(n-1)\delta + (m-1)N\delta\}]$$

$$\times \exp\left[j\left\{\omega_0 t + \frac{2\pi}{N}(m-1)(n-1)\right\}\right] \tag{8-50}$$

where the index $n = 1, 2, \ldots, N$ progresses from element to element up to N elements and the index $m = 1, 2, \ldots, N$ progresses from group to group

----

[1] This sequence and other sequences consisting of the cube roots of unity were investigated by DeLong [23].

up to N groups. The total time duration of these waveforms is $N^2 \delta$ and they have an effective bandwidth of $N(1/N\,\delta)$. The time-bandwidth product is therefore N^2.

The spectral description for the waveform given by (8-50) is given, in normalized units, by

$$U(x) = \delta \frac{\sin \pi x/N}{\pi x/N} \sum_{n=1}^{N} \frac{\sin \pi[x - (n-1)]}{\sin \pi[x - (n-1)]/N}$$

$$\times \exp\left[-j2\pi(n-1)\left(x - \frac{N-1}{2N}\right)\right] \tag{8-51}$$

where $x = N\,\delta f$ and $U(f) = \Psi(f + f_0)$. It is evident from (8-51) that $U(x)$ consists of a weighting function of the form $\sin y/y$ and the vector sum of

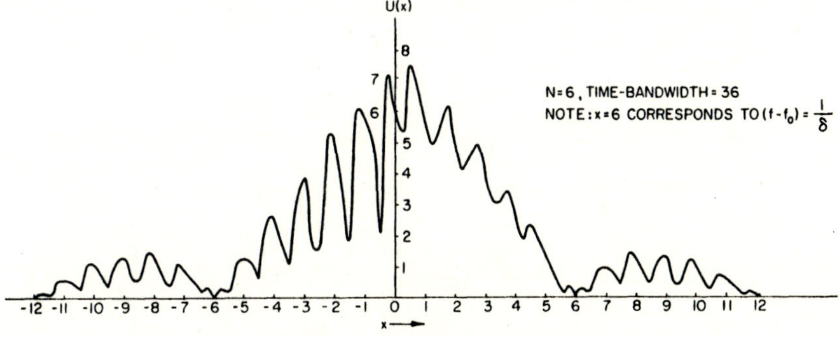

FIG. 8.24(a)

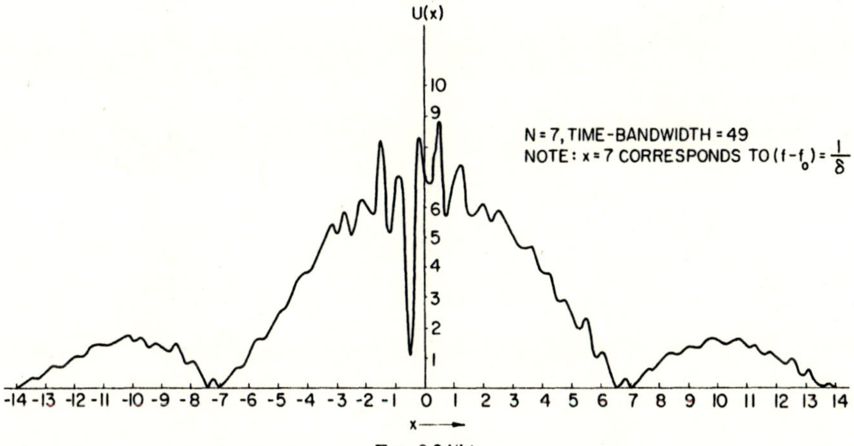

FIG. 8.24(b)

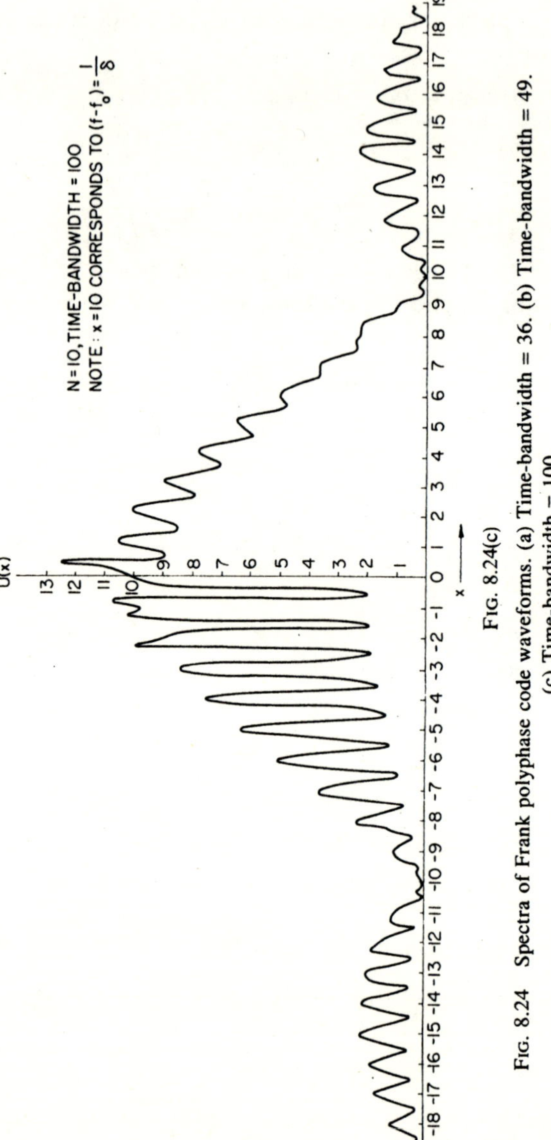

FIG. 8.24(c)

FIG. 8.24   Spectra of Frank polyphase code waveforms. (a) Time-bandwidth = 36. (b) Time-bandwidth = 49. (c) Time-bandwidth = 100.

Note: A more recent and detailed analysis of Frank and Frank-derived codes is given in B. L. Lewis, F. F. Kretschmer, Jr., and W. Shelton, *Aspects of Radar Signal Processing*, Artech House, Norwood, MA, 1986.

functions of the form $\sin Ny/\sin y$ which appear at unit intervals on the normalized scale x or at intervals of $1/N\delta$ on the unnormalized frequency scale. Moreover, it is also evident that the Fourier transform is equal to N times the weighting function at the integer values of x, since at these points along x, all but one of the vector functionals disappear. The modulus $|U(x)|$, evaluated for several cases including $N = 6$, 7, and 10, is shown in Fig. 8.24. An experimental measurement of the spectrum and autocorrelation function for $N = 6$ is shown in Fig. 8.25. It can be seen from these figures that $|U(x)|$ has a ripplelike structure which varies as a function of time-bandwidth product. It has been conjectured that the ambiguity function behavior of the Frank code signals can be explained by a paired echo analysis of the ripples on $|U(x)|$.

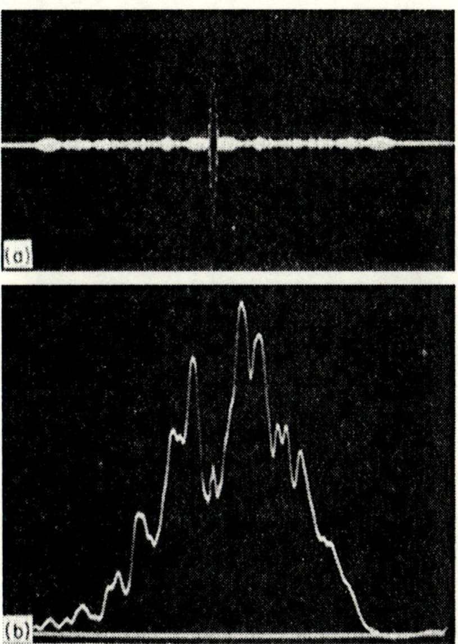

FIG. 8.25 Experimental matched-filter output and signal spectrum for Frank polyphase code of length 36. (a) Matched-filter autocorrelation response. (b) Signal spectrum (detected). (Courtesy of G. F. Miller, Canadian General Electric Co. Ltd., Toronto, Canada.)

The general response function expression for these coded signals is conveniently described by letting $\tau = \tau' + p\delta + kN\delta$, where $0 \leqslant \tau' \leqslant \delta$, and τ' represents continuous time shifts in the intervals $p\delta \leqslant \tau \leqslant (p + 1)\delta$; and $0 \leqslant p \leqslant N - 1$, and p represents discrete time shifts in the intervals $kN\delta \leqslant \tau \leqslant (k + 1)N\delta$, where $0 \leqslant k \leqslant N - 1$. The resulting response function is given by

$$\chi(\tau, \phi) = \chi(\tau' + p\,\delta + kN\,\delta, \phi)$$

$$= \chi(\tau', \phi)\Big[P(p, k, N\phi)Q(p, k, \phi)$$

$$+ P(p - N, k + 1, N\phi)Q(N - p, k + 1, \phi)\Big]\exp[-j\Omega(p, k, \phi)]$$

$$+ \chi(\delta - \tau', \phi)\Big[P(p + 1, k, N\phi)Q(p + 1, k, \phi)$$

$$+ P(p - N + 1, k + 1, N\phi)Q(N - p - 1, k + 1, \phi)\Big]$$

$$\times \exp[-j\Omega(p - 1, k, \phi)] \tag{8-52}$$

where

$$\chi(\tau', \phi) = \delta\left[1 - \frac{\tau'}{\delta}\right]\frac{\sin \pi\phi\,\delta\left[1 - \dfrac{\tau'}{\delta}\right]}{\pi\phi\,\delta\left[1 - \dfrac{\tau'}{\delta}\right]}\exp\left[-j\pi\phi\left(1 - \frac{\tau'}{\delta}\right)\right] \tag{8-53}$$

$$P(p, k, N\phi) = \frac{\sin \pi[N - k]\left[\dfrac{p}{N} + \phi N\,\delta\right]}{\sin \pi\left[\dfrac{p}{N} + \phi N\,\delta\right]}\exp\left[-j\pi(N - 1)\frac{p}{N}\right] \tag{8-54}$$

$$Q(p, k, \phi) = \frac{\sin \pi[N - p]\left[\dfrac{k}{N} + \phi\,\delta\right]}{\sin \pi\left[\dfrac{k}{N} + \phi\,\delta\right]}\exp\left[-j2\pi(N - 1)\frac{k}{N}\right] \tag{8-55}$$

$$\Omega(p, k, \phi) = [N^2 - Nk - (p + 1)]\phi\,\delta \tag{8-56}$$

The following observations can be made from (8-52):

(1)   When $\tau = 0$

$$\chi(0, \phi) = \delta N^2\frac{\sin \pi N^2\phi\,\delta}{\pi N^2\phi\,\delta}$$

(2)   When $\tau' = p = 0$ and $\phi = 0$, then $\chi(Nk\,\delta, 0) = 0$.

(3) $\chi(\tau, 0)$ has the same absolute value for displacements of j and $(N^2 - j)$ where $1 < j < N^2$ (see Fig. 8.26).

(4) The absolute value of the correlation function is unity for a displacement of one more or one less than multiples of N [22].

(5) Frank [22] has conjectured that the maximum sidelobe is the vector sum of $N/2$ (for N even) or $(N + 1)/2$ (for N odd) unit vectors, where the

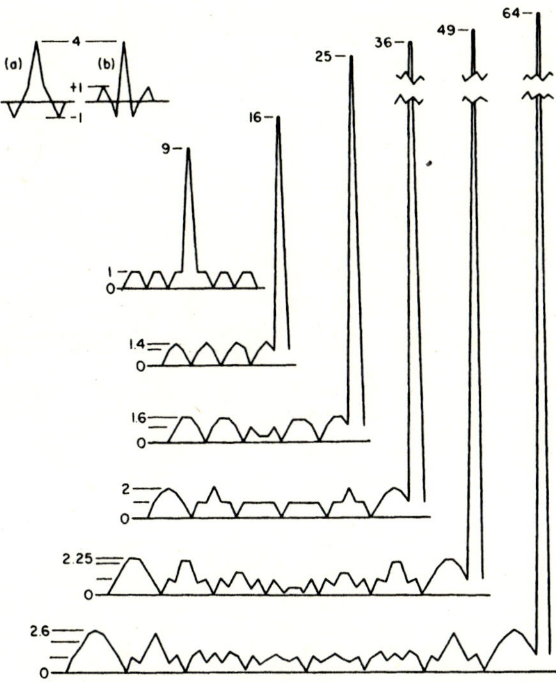

FIG. 8.26 Frank polyphase code autocorrelation functions (courtesy of R. L. Frank [22]).

vectors are separated by $2\pi/N$. This implies that for large N the peak-to-sidelobe ratio asymptotically approaches πN.

The ambiguity characteristics corresponding to a signal for which $N = 10$ are illustrated in Fig. 8.27. The magnitudes of additional characteristics (i.e., secondary maxima, image signal) as a function of Doppler shift are shown in Fig. 8.28. It is evident from these figures that the ambiguity characteristic of this signal grossly resembles that of linear FM signals, since it has the ridgelike characteristic often associated with the linear FM signal. However, it is seen that there are also large (up to 30% of the peak signal output) cyclic ambiguity pop-ups as a function of Doppler shift. These tend to make this particular quantized phase coded waveform nonsuitable when there are large Doppler shifts in an ensemble of received signals. For very large Doppler shifts, the ambiguity function characteristic breaks up into two similar signals separated in time by the unprocessed waveform duration $N^2 \delta$.

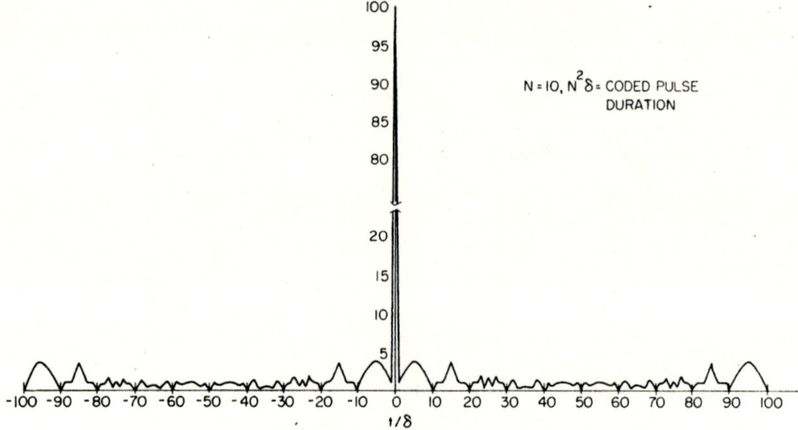

FIG. 8.27(a), $\phi T = 0$.

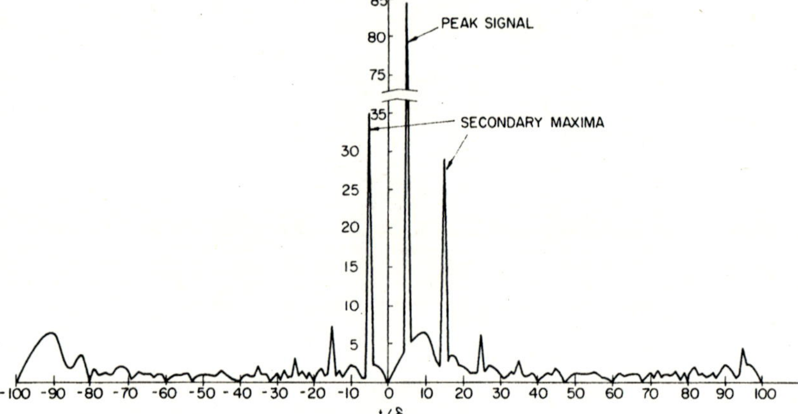

FIG. 8.27(b), $\phi T = 5$.

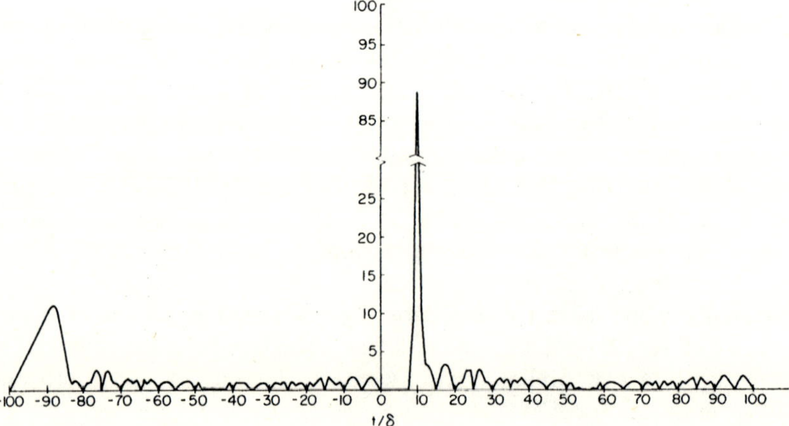

FIG. 8.27(c), $\phi T = 10$.

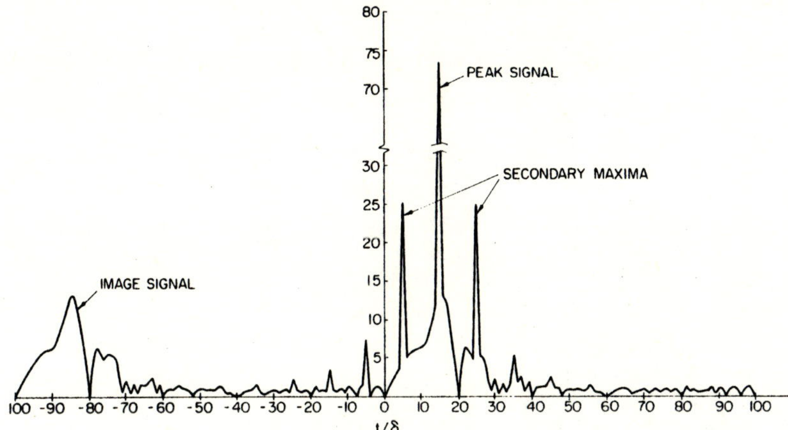

FIG. 8.27(d), $\phi T = 15$.

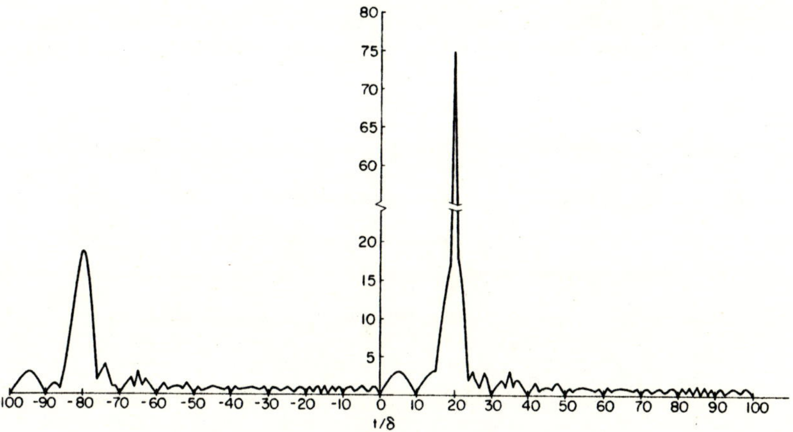

FIG. 8.27(e), $\phi T = 20$.

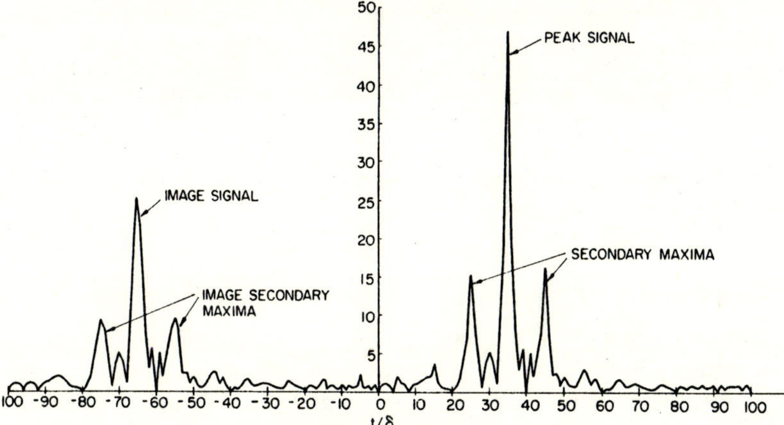

FIG. 8.27(f), $\phi T = 35$.

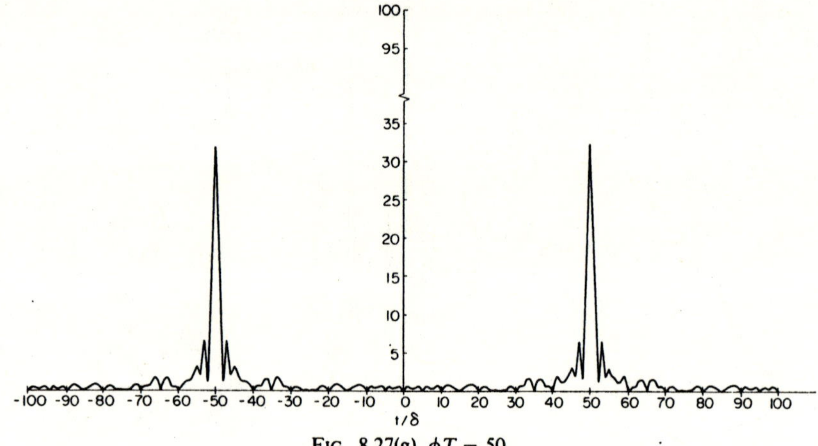

Fig. 8.27(g), $\phi T = 50$.

Fig. 8.27 Frank polyphase code Doppler shifted matched-filter outputs, time-bandwidth = 100; ϕT is normalized Doppler shift. (Courtesy of J. Paolillo, Sperry Gyroscope Co., Great Neck, New York.)

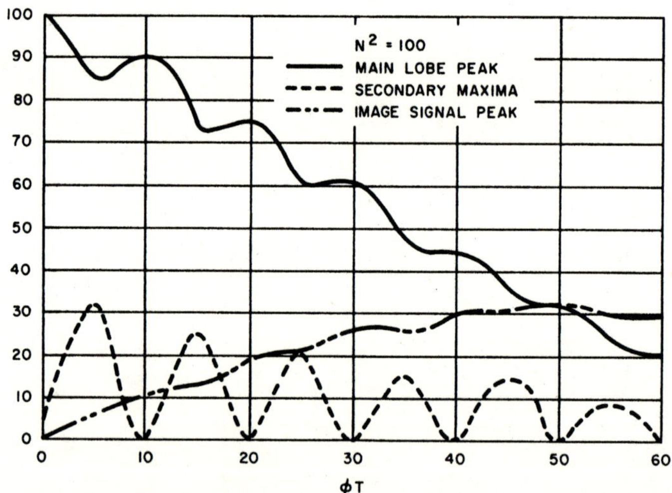

Fig. 8.28 Magnitudes of significant characteristics of Frank polyphase coded waveform vs Doppler shift, time-bandwidth = 100.

Huffman Codes

The preceding examples discussed in this chapter under Group II codes involved only discrete phase sequences. The best peak-to-sidelobe ratio of $\chi(k, 0)$ was N. In each case where this was achieved, the sidelobes were distributed uniformly between $-N\delta < \tau < N\delta$. By removing the restriction $\{a_n = 1\}$ and sacrificing some efficiency in the use of available transmission time, it is possible to derive sequences that provide the property

$$\chi(k,0) = \begin{cases} 2E, & k = 0 \\ 0, & |k| \neq 0, N-1 \\ \pm 1, & |k| = N-1 \end{cases} \tag{8-57}$$

where E denotes the signal energy. The method employed to derive these sequences was introduced by Huffman [24]. Because of the resultant structure of $\chi(k,0)$ Huffman designated them "impulse equivalent" sequences. In general, these sequences consist of amplitude and phase pairs $\{a_n, \theta_n\}$ which may have a continuum of values. For convenience, these sequence pairs will be denoted by $\{b_n\}$ where

$$b_n = a_n \exp[j\theta_n] \tag{8-58}$$

The method of Huffman may be described by starting with the polynomial

$$P(s) = b_1 + b_2 s + b_3 s^2 + \cdots + b_N s^{N-1} \tag{8-59}$$

$$= b_N \prod_{i=1}^{N-1} (s - r_i)$$

where s is an indeterminant. The polynomial corresponding to the matched-filter implementation for the sequence represented by (8-59) is

$$P^*(s) = \bar{b}_1 + \bar{b}_2 s^{-1} + \bar{b}_3 s^{-2} + \cdots + \bar{b}_N s^{-(N-1)} \tag{8-60}$$

The matched-filter response $\chi(k,0)$ is obtained from the product PP^* which is given by

$$P(s)P^*(s) = b_1 \bar{b}_N s^{-(N-1)} + \cdots + (b_1 \bar{b}_1 + b_2 \bar{b}_2 + \cdots + b_N \bar{b}_N)s^0$$

$$+ \cdots + \bar{b}_1 b_N s^{N-1}$$

$$= b_N \bar{b}_N (-s)^{-(N-1)} \prod_{i=1}^{N-1} \bar{r}_i \prod_{i=1}^{N-1} [s^2 - (r_i + \bar{r}_i^{-1})s + r_i \bar{r}_i^{-1}] \tag{8-61}$$

where

$$\chi(k,0) = \text{coefficient of } s^k \tag{8-62}$$

For the Huffman ("impulse equivalent") sequences, this product reduces to

$$PP^* = b_1 \bar{b}_N s^{-(N-1)} + 2E + \bar{b}_1 b_N s^{N-1} \tag{8-63}$$

where $2E = b_1 \bar{b}_1 + b_2 \bar{b}_2 + \cdots + b_N \bar{b}_N$ represents the energy of the coded waveform derived from these sequences. The $2N - 2$ roots of Eq. (8-63) lie on two origin centered concentric circles whose radii are given by

$$R^{N-1} = \frac{-E \pm \sqrt{E^2 - |b_1 \bar{b}_N|^2}}{\bar{b}_1 b_N} \qquad (8\text{-}64)$$

It is readily shown that these radii are mutual reciprocals when

$$|\bar{b}_1 b_N| = 1 \qquad (8\text{-}65)$$

For that case, it is seen by referring to Eq. (8-63) that the peak-to-sidelobe ratio depends only on the energy E. In addition, it is also seen that

$$2E \geqslant b_1 \bar{b}_1 + b_N \bar{b}_N = \frac{(1 - b_N \bar{b}_N)^2}{b_N \bar{b}_N} + 2 \geqslant 2 \qquad (8\text{-}66)$$

Of the $2N - 2$ roots of PP^*, half lie on the inner circle, and half, along the same radii, on the outer circle. At the same time, half of these roots taken from either circle can be used to form $P(s)$ and the remaining half to form the conjugate polynomial. Therefore, there are 2^{N-1} possible impulse equivalent polynomials that can be derived for a given E and sequence length N. The coefficients of these polynomials are usually complex. However, a polynomial with real coefficients can be derived by placing the conjugate pairs of complex roots assigned to $P(s)$ on the same circle.

Although some inefficiency in the use of available transmission time may be acceptable in order to obtain the property given by (8-57), it would certainly be wasteful not to seek the most efficient sequence for an application. Huffman pointed out that the greatest efficiency is derived when the ratio $2E/\text{g.o.}|b_n|^2$ is maximized. This so called "effective energy" ratio has two degrees of freedom; namely, the total energy of the coded waveform, $2E$, and the greatest magnitude of the coefficients of $P(s)$, denoted by g.o.$|b_n|$. By referring to Eq. (8-64), it is seen that E varies as a function of the radii of the concentric circles from which the roots of $P(s)$ are taken. The greatest of $|b_n|$, however, depends on the particular choice of $N - 1$ roots for $P(s)$ as well as these radii. Unfortunately, an analytical method has not been found which leads, without trial and error, to the most efficient Huffman sequence. Huffman has conjectured that the energy ratio for an impulse equivalent sequence is maximized when approximately half of the roots of P are taken from each circle.

The general description of Huffman coded waveforms, obtained from Eq. (8.1) is given by

$$\psi(t) = \sum_{n=1}^{N} b_n P_n(t) \exp[j\omega_0 t], \qquad 0 < t \leqslant N\delta$$

$$= 0, \qquad \text{elsewhere} \qquad (8\text{-}67)$$

The general ambiguity function description for these waveforms is given by (8-28) except for the fact that the c_n's are replaced by b_n's and the shifted coefficient is conjugated.

One example of a Huffman sequence of length 14 which has only real coefficients is [25]

$$\{b_n\} = -0.57, 0.27, -0.56, 0.55, -0.14, -0.28, -0.31, -1, -0.43, 0.5,$$
$$0.037, -0.34, 0.22, 0.43$$

The envelope structure and ambiguity characteristics of this sequence are illustrated in Fig. 8.29. Another sequence of length 9 with almost constant

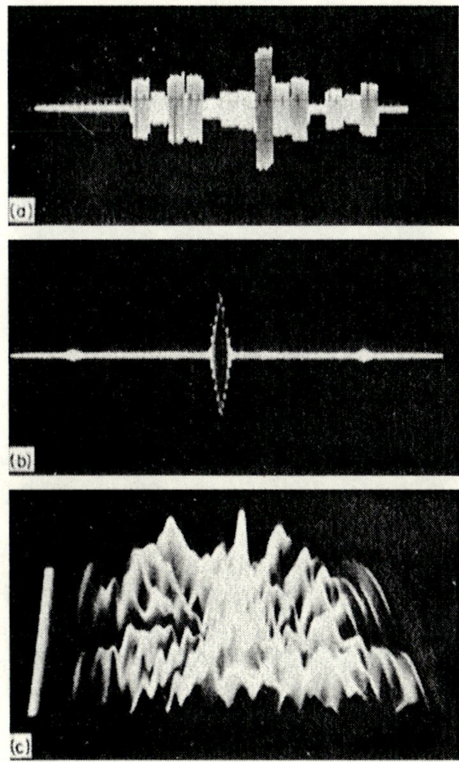

FIG. 8.29   Huffman code waveforms and response surface; code length = 14, peak-to-side-lobe ratio = 13.1. (a) Matched-filter input signal. (b) Matched-filter autocorrelation response. (c) Composite response surface. (Courtesy of C. E. Jagger and R. H. McLaughlin, Canadian General Electric Co., Toronto, Canada.)

amplitude is [25, 26]

$$\{b_n\} = 1\underline{/180°}, \quad 0.97\underline{/157.5°}, \quad 0.90\underline{/76.5°}, \quad 0.95\underline{/60.9°}, \quad 0.89\underline{/165.1°},$$

$$0.95\underline{/195.9°}, \quad 0.90\underline{/346.5°}, \quad 0.97\underline{/202.5°}$$

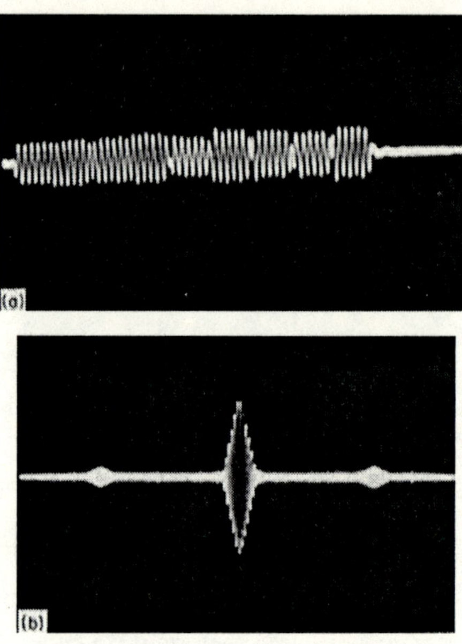

FIG. 8.30   Waveforms for Injeyan code of length 9. The input signal has uniform amplitude; the autocorrelation response is similar to that of a Huffman polyphase code. (a) Matched-filter input signal. (b) Matched-filter autocorrelation response. (Courtesy of C. E. Jagger and R. H. McLaughlin, Canadian General Electric Co., Toronto, Canada.)

The energy ratio of this sequence is 8.125. Injeyan [26] investigated this sequence, but with all the magnitudes set equal to unity and found the deviation from the property given by (8-57) negligible. The matched-filter response for the Injeyan sequence is illustrated in Fig. 8.30.

The effective energy ratio is not the only important consideration in connection with the selection of Huffman sequences. It may also be necessary to consider the peak-to-sidelobe ratio. Since these requirements are usually not compatible, the final selection of a particular Huffman sequence will depend on what trade-off the system application will allow.

An actual implementation to gather ionospheric sounding data was made by Coll and Storey [27], using both Huffman and binary sequences of

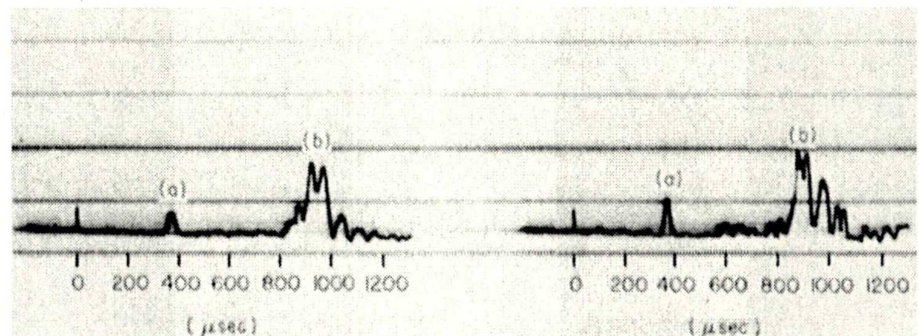

FIG. 8.31 Comparison of sidelobe levels with Huffman and binary sequences. Left: Huffman (a) extraordinary F2 echo, (b) ordinary F2 echo; right: binary (courtesy of D. C. Coll and J. R. Storey [27]).

length 17. Figure 8.31 clearly shows the clearing out of false signals, created by sidelobe responses of the binary signal, when the Huffman code was employed. The sidelobe ratio of the binary sequence employed for this investigation was 17/2.

8.5 Discrete Frequency Sequences (Group III)

Discrete coded waveforms that are derived from frequency sequences $\{\omega_n\}$ form the third group of waveform examples discussed in this chapter. The general description of these waveforms is given by

$$\psi(t) = \sum_{n=1}^{N} u_n(t)\exp[j\omega_0 t] \tag{8-68}$$

where

$$u_n(t) = a_n P_n(t)\exp[j\omega_n t]$$

$$P_n(t) = P[t - (n-1)\delta], \quad (n-1)\delta < t < n\delta$$

$$a_n = \begin{cases} 0 \\ 1 \end{cases}$$

A particular sequence structure which yields what are commonly referred to as stepped FM waveforms has the frequencies spaced $1/\delta$ apart. For a sequence of increasing frequencies, the elements of $\{\omega_n\}$ are given by

$$\omega_m = 2\pi\left[\frac{m-1}{\delta} - \frac{M-1}{2\delta}\right] \tag{8-69}$$

where the index m is a count of the ones in $\{a_n\}$ and M equals the total number of ones. A sequence of decreasing frequencies is obtained by changing the sign of (8-69).

The Fourier transform obtained for $\psi(t)$ by using (8-69) and setting $a_n = 1$ (i.e., $M = N$, the contiguous linear step FM signal) is given by

$$\Psi(f) = \delta \sum_{n=1}^{N} \frac{\sin \pi \delta \left[f - \left(f_0 + \dfrac{n-1}{\delta} \right) \right]}{\pi \delta \left[f - \left(f_0 + \dfrac{n-1}{\delta} \right) \right]}$$

$$\times \exp \left[-j \left\{ \pi \delta \left(f - \left[f_0 + \frac{n-1}{\delta} \right] \right) (2n-1) \right\} \right] \qquad (8\text{-}70)$$

It is instructive to recall the spectral description given by (8-51) for the quantized phase codes. Except for the weighting function outside the summation of (8-51), the structural similarities of (8-51) and (8-70) are clearly evident. One important characteristic is that the time-bandwidth product of both waveforms is N^2, obtained from the products $N^2 \delta(N/N\,\delta)$ for the Frank code and $N\,\delta(N/\delta)$ for the step FM signal, where N is the number of groups in the Frank code and the number of frequencies in the step FM, respectively.

In general, the matched-filter response for the waveforms of Group III can be obtained with a filter configuration shown in Fig. 8.32 that consists of a parallel filter bank, each filter being matched to one subpulse of $\psi(t)$, followed by a coherent summation network. Figure 8.33 roughly illustrates the individual responses of these filters, and the overall matched-filter response for a train of stepped FM pulses ($N = 5$) separated by one pulse length. A description of the general ambiguity response obtained from (8-3) for stepped FM waveforms having no separation between subpulses

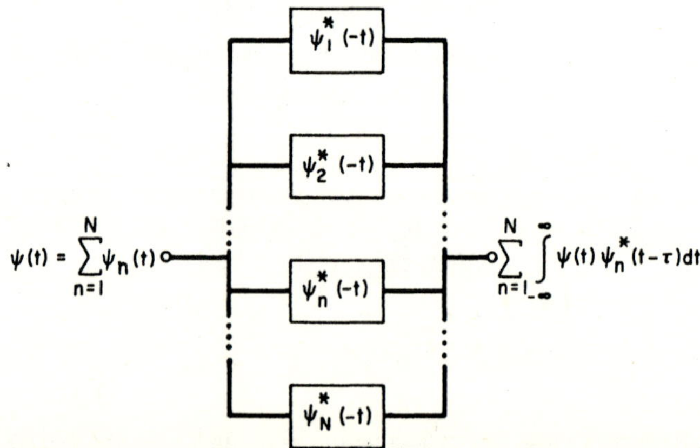

FIG. 8.32 Parallel matched-filter bank for Group III waveforms.

is given by

$$\chi(\tau, \phi) = F(\tau', k, \phi)H(\tau', k, \phi)$$
$$+ F(\delta - \tau', k + 1, \phi)H(\tau' - \delta, k + 1, \phi), \qquad (8\text{-}71)$$
$$0 < \tau < N\delta, \quad -\infty < \phi < \infty$$

where τ' and k are defined by $\tau = \tau' + k\delta$, $0 < \tau' < \delta$, $k = 0, 1, 2, \ldots, N - 1$, and

$$F(\tau', k, \phi) = (\delta - \tau') \frac{\sin \pi \left[\phi + \dfrac{k}{\delta} \right] [\delta - \tau']}{\pi \left[\phi + \dfrac{k}{\delta} \right] [\delta - \tau']} \exp \left[-j\pi \left(\phi + \frac{k}{\delta} \right) N\delta \right] \quad (8\text{-}72)$$

$$H(\tau', k, \phi) = \frac{\sin \pi [N - k] \delta \left[\phi + \dfrac{\tau' + k\delta}{\delta^2} + \dfrac{k}{\delta} \right]}{\sin \pi \delta \left[\phi + \dfrac{\tau' + k\delta}{\delta^2} + \dfrac{k}{\delta} \right]} \exp \left[j\pi \left(\phi + \frac{k}{2\delta} \right) (\tau' + k\delta) \right]$$
$$(8\text{-}73)$$

It should be noted that (8-71) describes $\chi(\tau, \phi)$ for only half the ambiguity plane. The description for the remaining half can be obtained by applying the symmetry property

$$\chi^*(\tau, \phi) = \chi(-\tau, -\phi) \exp[-j2\pi\phi\tau] \qquad (8\text{-}74)$$

An expression similar to (8-71) can also be obtained from (8-3) for the case of separated pulses and is left as an exercise. The gross ambiguity structures for both the contiguous and separated pulse train stepped FM waveforms are illustrated by the level cross sections of $\chi(\tau, \phi)$ shown in Fig. 8.34. The configurations shown would occur at approximately 3 db and 4 db below the peak response at $\tau' = \phi = 0$. Figure 8.35 shows the calculated matched-filter responses obtained for a contiguous train of 10 pulses, stepped in frequency according to (8-69). Figure 8.36 illustrates experimental data for an 8-step contiguous signal. Although not indicated in Fig. 8.34(a), it is seen from Fig. 8.35 that these responses also exhibit cyclic subsidiary peaks. These attain peak values for ϕ odd multiples of $N/2T$ (or $1/2\delta$). However, the height of these pop-ups are much lower than the contour levels of $\chi(\tau, \phi)$ illustrated in Fig. 8.34(a). By comparison, the subsidiary responses for the separated pulse trains (see Figs. 8.37 through 8.40) are much higher and are included in the $\chi(\tau, \phi)$ level contours shown in Fig. 8.34(b). Figures 8.41 and 8.42 demonstrate the effect on these responses when the frequencies are separated by $3/\delta$. This is an example

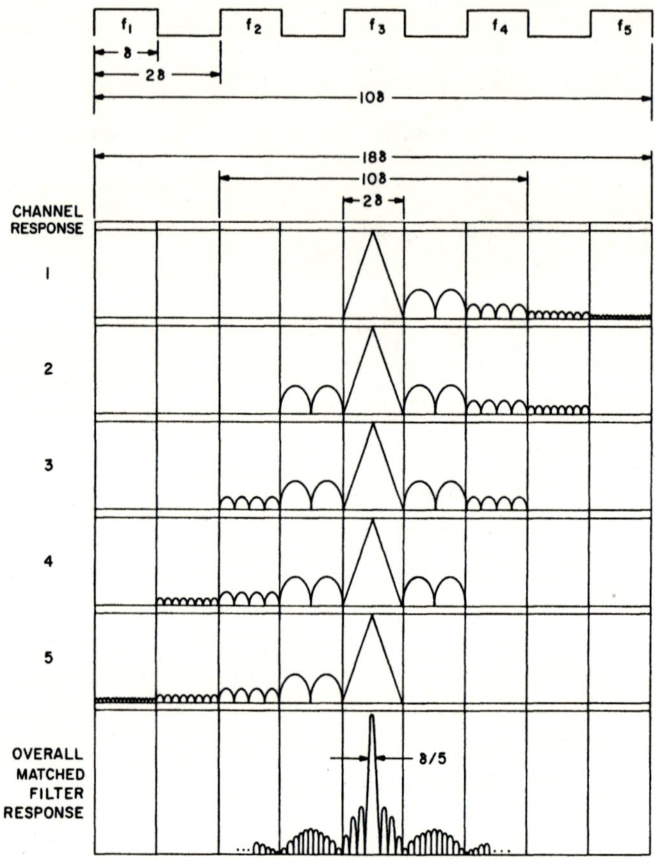

FIG. 8.33   Parallel channel and overall matched-filter response for five pulse stepped FM pulse train.

of the frequency steps not being matched to the component pulse duration (i.e., $\omega_{n+1} - \omega_n > 2\pi/\delta$) [5]. The increased separation of the frequency steps reduces the levels of the noncentral ambiguity regions, but at the expense of increased ambiguity within the central response region. The sliding of the "grating" lobes in the central region as a function of Doppler shift is shown in Fig. 8.42. Figures 8.43 and 8.44 illustrate the case when the frequencies, separated by integer values of $1/\delta$, are randomly placed in a pulse train. For this example, it is shown that the order of the steps does not grossly affect the central autocorrelation response, but does govern the behavior of the noncentral ambiguities and off-Doppler responses.

The characteristic ridgelike formation of $\chi(\tau, \phi)$ seen from inspection of (8-71) to (8-73) and also apparent in the illustrations of this section, is

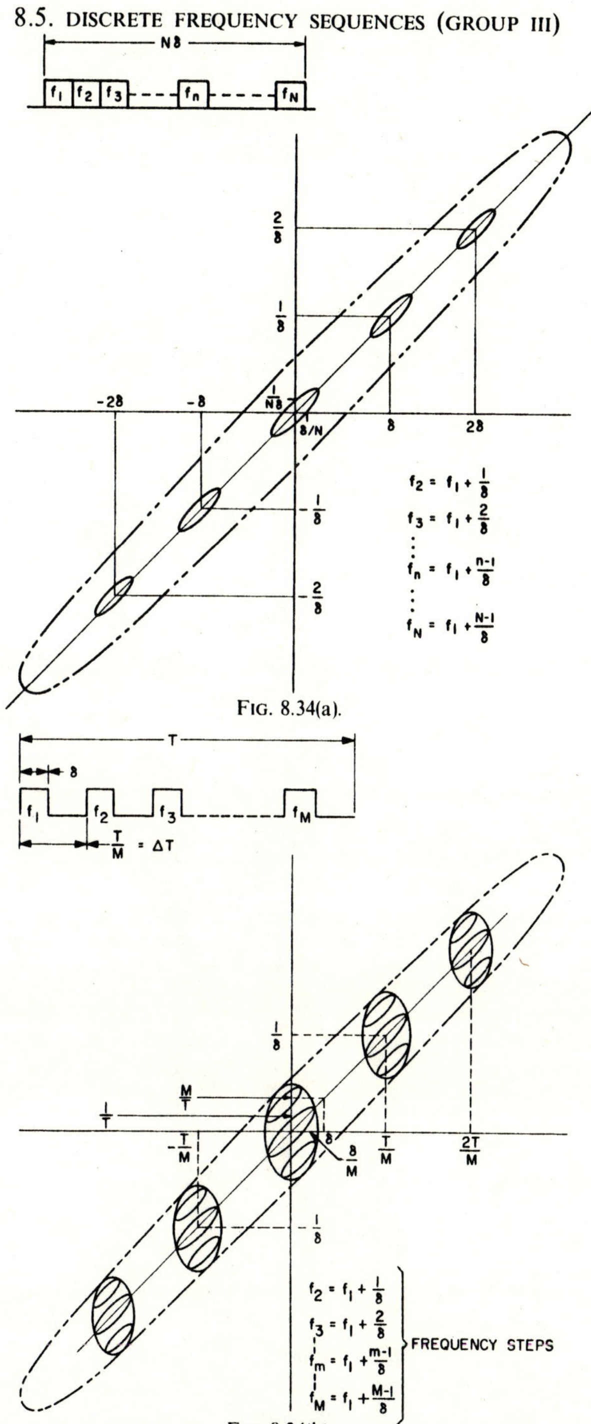

$$f_2 = f_1 + \frac{1}{\delta}$$

$$f_3 = f_1 + \frac{2}{\delta}$$

$$f_n = f_1 + \frac{n-1}{\delta}$$

$$f_N = f_1 + \frac{N-1}{\delta}$$

FIG. 8.34(a).

$$f_2 = f_1 + \frac{1}{\delta}$$

$$f_3 = f_1 + \frac{2}{\delta}$$

$$f_m = f_1 + \frac{m-1}{\delta}$$

$$f_M = f_1 + \frac{M-1}{\delta}$$

FREQUENCY STEPS

FIG. 8.34(b).

FIG. 8.34   Response function level contours of stepped FM signals. (a) Contiguous train of pulses. (b) Separated train of pulses.

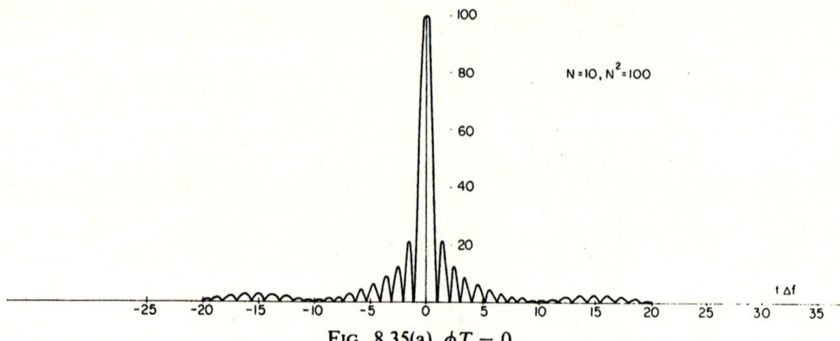

FIG. 8.35(a), $\phi T = 0$.

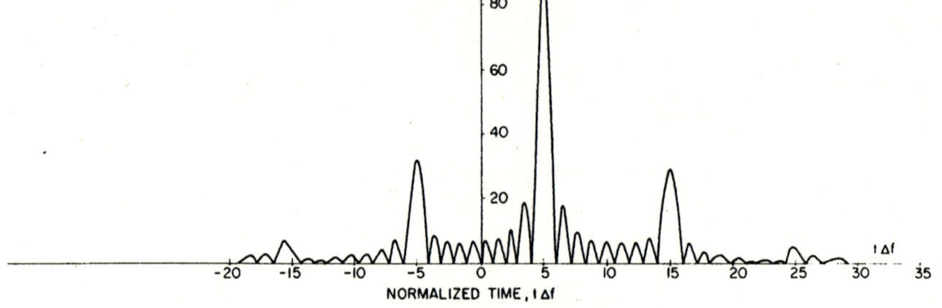

FIG. 8.35(b), $\phi T = 5$.

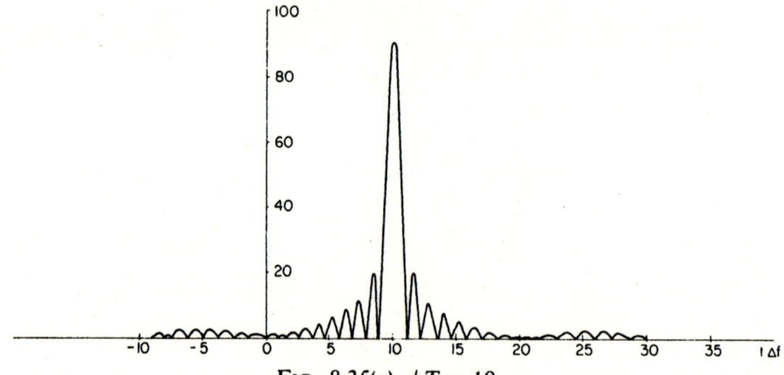

FIG. 8.35(c), $\phi T = 10$.

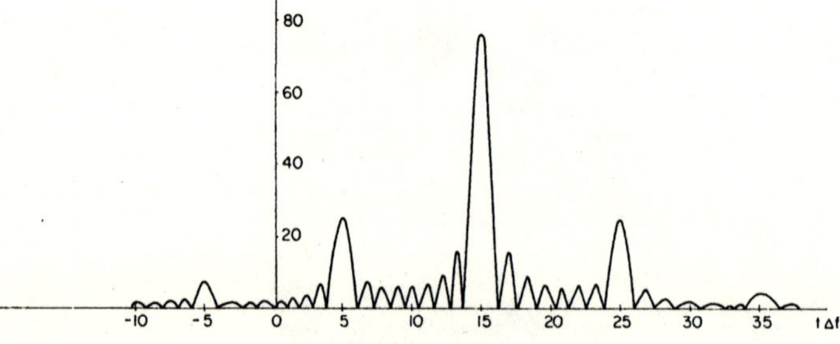

FIG. 8.35(d), $\phi T = 15$.

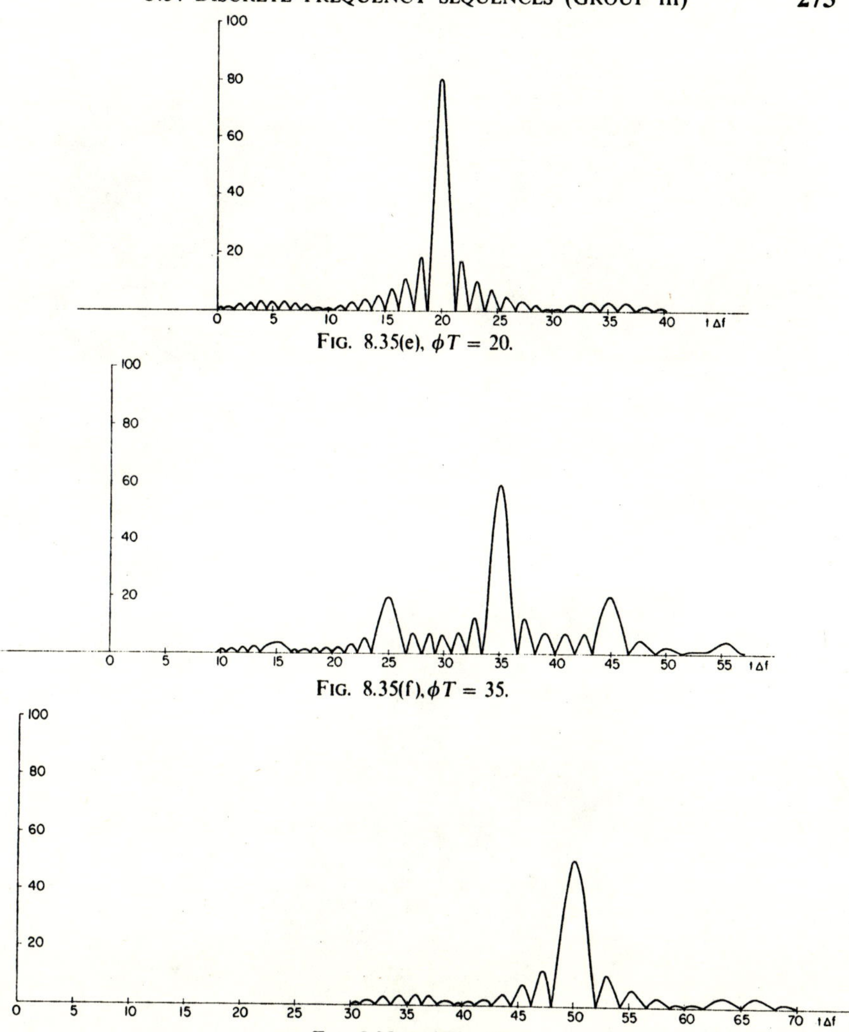

FIG. 8.35(e), $\phi T = 20$.

FIG. 8.35(f), $\phi T = 35$.

FIG. 8.35(g), $\phi T = 50$.

FIG. 8.35   Matched-filter Doppler shifted outputs for the contiguous step FM waveform, $N = 10$.

further indication, in addition to $\Psi(f)$ given by (8-70), that the stepped FM signal belongs to the linear FM family along with the quantized phase codes. This association may be more clearly discernible by considering the continuous phase functions attached to the analog linear FM waveform and the two discrete counterparts described in this chapter. The phase function for the stepped FM waveform may be viewed as a first order quantization of the parabolic (linear FM) phase function. The quantized

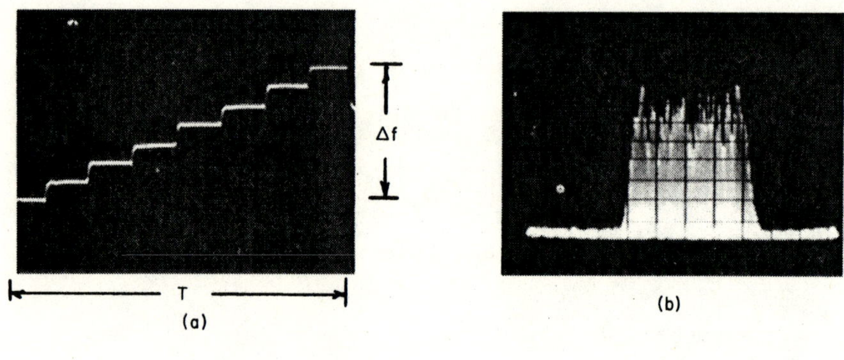

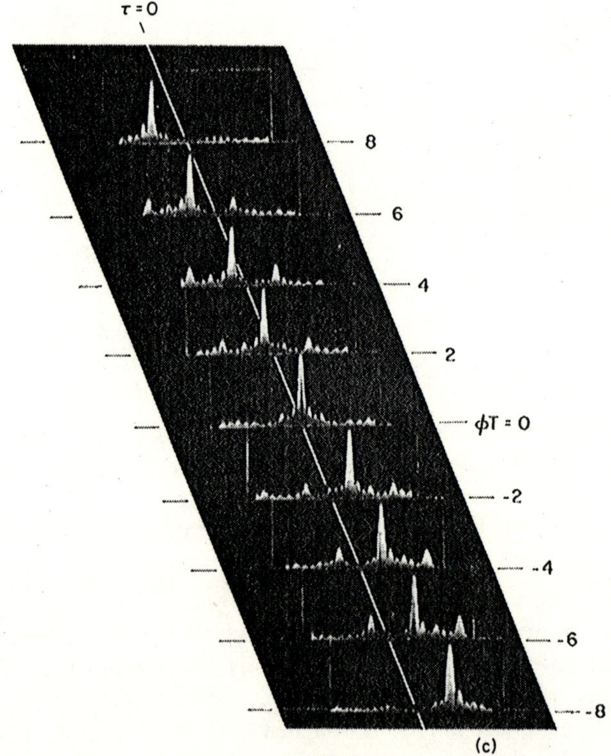

FIG. 8.36   Experimental characteristics of contiguous step FM waveform, $N = 8$. (a) Frequency discriminator response to signal modulation. (b) Spectrum. (c) Response function profiles.

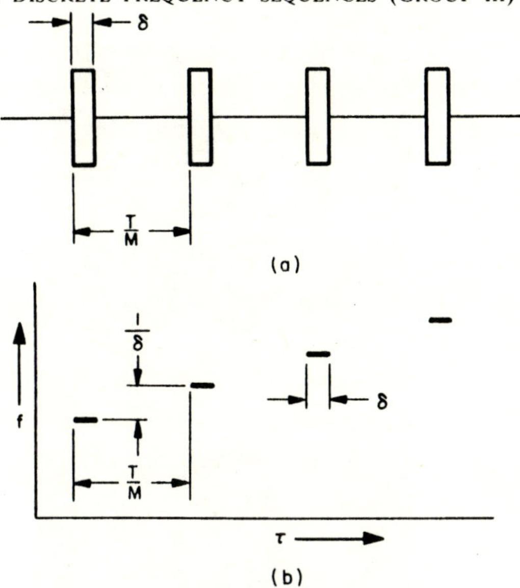

FIG. 8.37 Four pulse linear step FM pulse train: $\Delta f_n = 1/\delta$, $T/M = 7$, (a) Pulse train signal. (b) Frequency function.

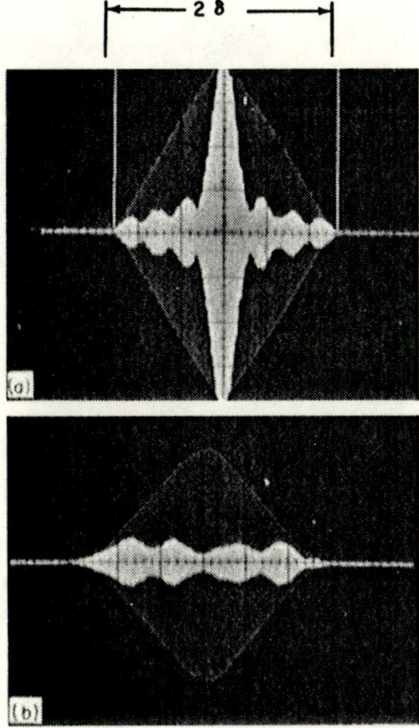

FIG. 8.38 Comparison of details of matched-filter autocorrelation responses of four pulse uniform and linear step FM ($\Delta f_n = 1/\delta$) pulse trains. (a) Central response. (b) First ambiguity sidelobe.

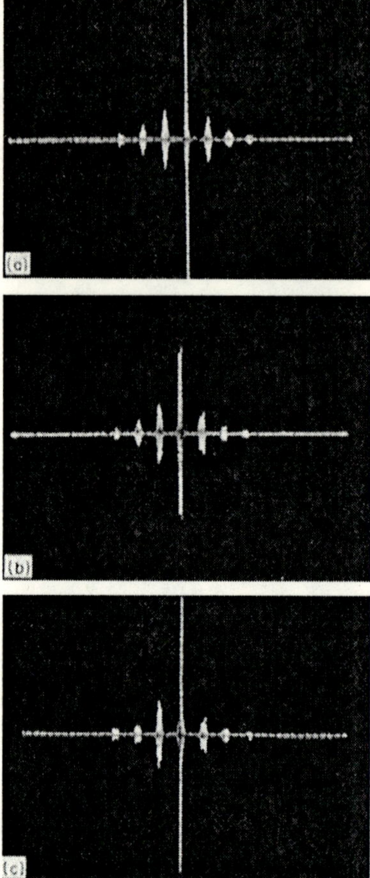

FIG. 8.39   Matched-filter responses for four pulse linear step FM train. (a) Autocorrelation response. (b) Matched-filter response for $\phi = M/2T$. (c) Matched-filter response for $\phi = M/T$.

phase codes, then, represent an additional degree of quantization which is compounded by more periodicity in the basic waveform structure. Thus, as seen from the illustrations for this coding technique, the ambiguity characteristics are more erratic, in contrast with the analog and stepped linear FM waveforms, and also have higher isolated pop-ups.

8.6   Matched Filters for Discrete Coded Signals

The discrete nature of the matched-filter signals discussed in this chapter makes possible matched-filter implementation and signal generation techniques that are based on tapped delay line or digital methods. In contrast

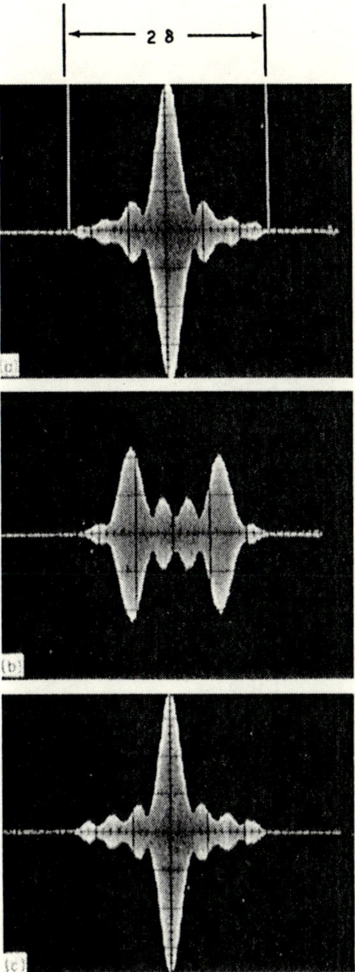

FIG. 8.40 Details of central response of four pulse linear step FM train as a function of Doppler shift, $\Delta f_n = 1/\delta$. (a) $\phi = 0$. (b) $\phi = M/2T$. (c) $\phi = M/T$.

to the dispersive delay line techniques, discussed in detail in Chapters 12–14, an extensive body of literature is available on tapped delay line and digital design techniques and these will not be covered here. This section briefly summarizes their general application to the design of matched filters for discrete coded signals.

In order to simplify the discussion, it will be assumed that the discrete coded signals contain either a phase code or a frequency code; that is, in the general signal formulation of (8-1) either ω_n or θ_n is not a variable

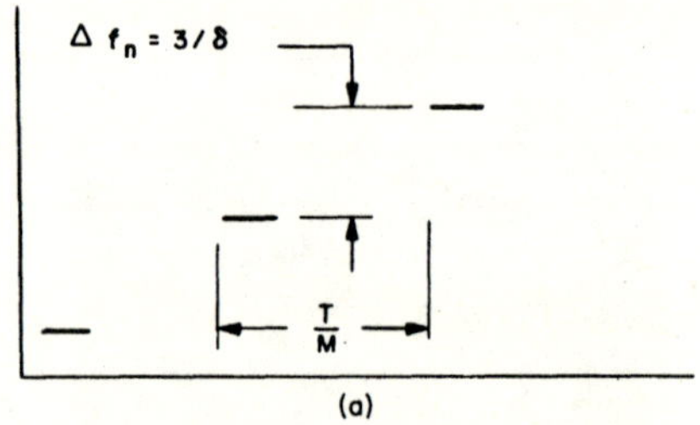

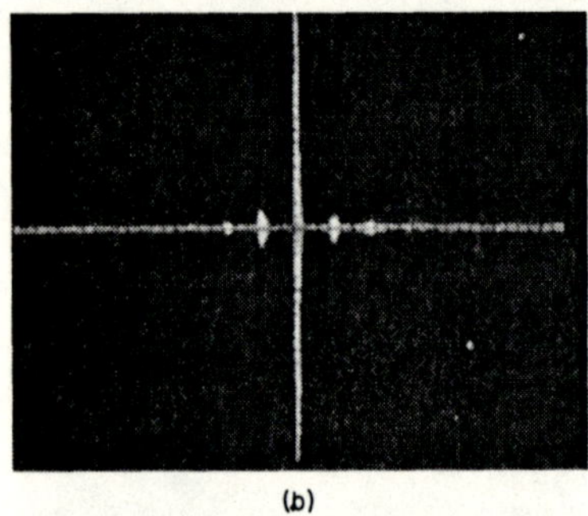

FIG. 8.41   Matched-filter response for four pulse linear step FM train, $\Delta f_n = 3/\delta$. (a) Frequency function. (b) Autocorrelation response.

from one basic subinterval of the sequence to the next. It will also be assumed that the subinterval durations are equal. None of the foregoing assumptions are a fundamental limitation on signal design, and the methods outlined below can be extended to cases in which the frequency, phase and duration of each component pulse subinterval are simultaneous variables in the discrete code sequence.

Taking the case of the phase coded sequences, a general matched-filter configuration is shown in Fig. 8.45. The correct phase conjugation and matching amplitude weighting are placed in each tapped output lead of the delay line. Since the subpulse intervals are equal, only a single matched filter for an uncoded subpulse is required. This is shown placed at the summation output, but could just as well be at the input prior to the tapped delay line. In most cases of practical interest for radar applications the subpulse amplitude weightings a_n will also be equal. When this holds, and the phase code is a binary sequence, then the filter shown can be used as both the matched filter and coded signal generator since each phase component (0 or 180 degrees) is its own conjugate (i.e., $180° = -180°$). To form the matching binary coded signal for this case an impulse is introduced

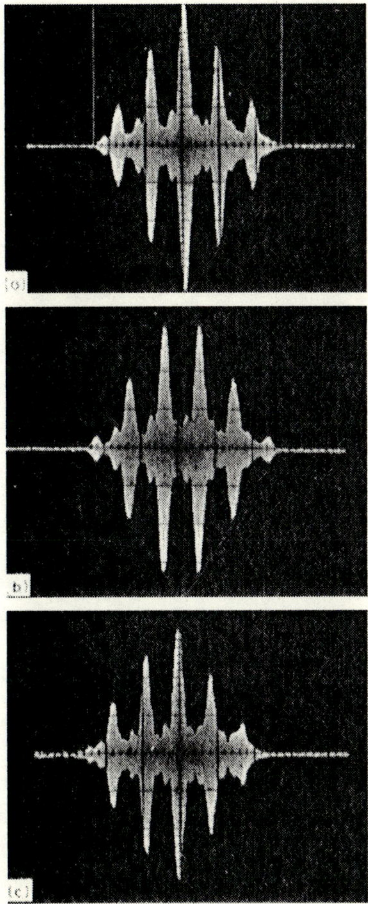

FIG. 8.42   Details of central response for four pulse linear step FM train as a function of Doppler shift, $\Delta f_n = 3/\delta$. (a) $\phi = 0$. (b) $\phi = M/2T$. (c) $\phi = M/T$. (Triangle is central response of uniform pulse train with no step FM).

at the right hand end of the tapped delay line. In practice this can be an i.f. pulse that has a wide bandwidth compared to that of the subpulse duration δ. As the impulse propagates from right to left down the tapped delay line and feeds through the tapped, phase coded outputs a signal of duration $N\delta = T$ is formed. This signal can be designated as $u(-t)$, the impulse response of the matched filter when the delay line is excited from the reverse (or transmitter) end. The signal $u(-t)$ for the binary phase code is the time inverse of the impulse response of the matched filter when

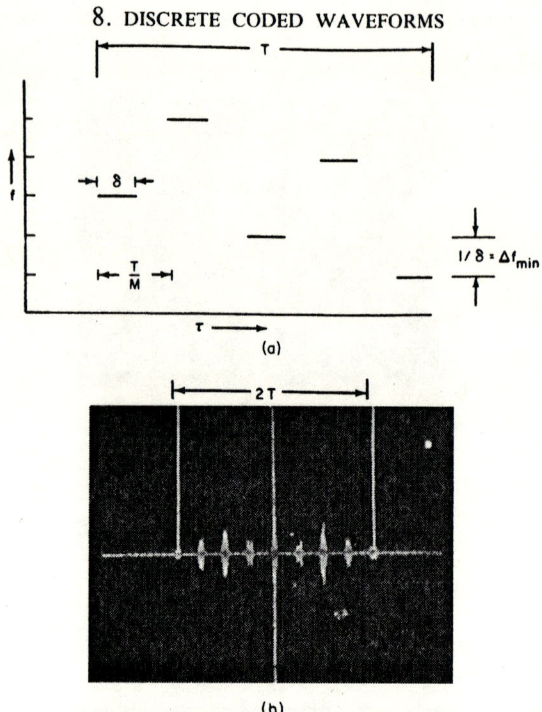

FIG. 8.43   Five pulse random step FM signal, $\Delta f_{min} = 1/\delta$. (a) Frequency function. (b) Auto-correlation response.

viewed from the receiver input, or left-hand end. Thus, when the phase coded signal that is generated in this manner is introduced at the receiver input a "matched" condition exists, and the types of autocorrelation and cross-correlation responses discussed in Section 8.3 are obtained.[1] The same type of filter would be used for the repetitive or periodic binary coded sequences. In this case the lower sidelobe levels are observed at the matched-filter output only after one complete period of the sequence has initially entered the delay line, or when the periodic steady state condition has been established. For the more general nonbinary case this method of signal generation can be applied only if there is a separate set of conjugate phase shifters at the delay line taps that can be switched in, after the coded pulse is formed, to replace the phase shifters used during the signal generation

---

[1] In considering this type of operation the reader can see from Fig. 8.45 that the subpulse described by $a_1 \exp[j\theta_1]$ is the first component of the coded signal to enter the delay line, and thus must travel to the last tap shown in the matched filter to see its matched condition. At this point, the entire coded waveform fills the tapped delay line, and all the component pulses are matched so that they are added in phase by the summation network.

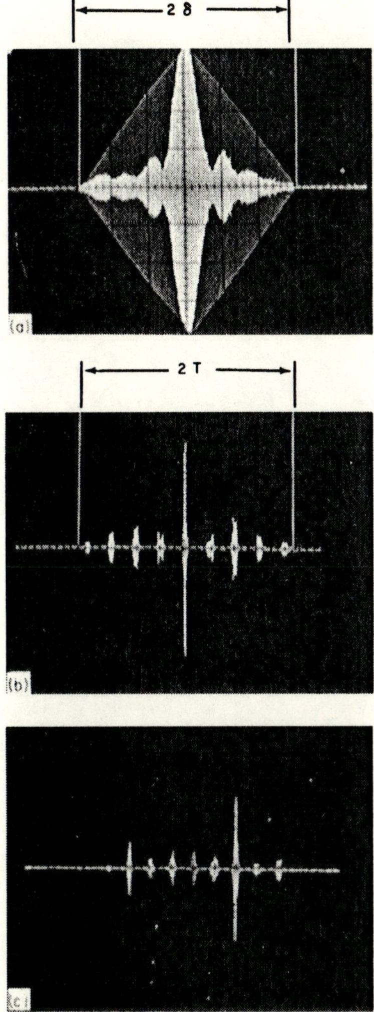

FIG. 8.44   Matched-filter responses for five pulse random step FM train. (a) Detail of central autocorrelation response. (b) Matched-filter response, $\phi = M/2T$. (c) Matched-filter response, $\phi = 1/\delta$.

cycle. This is illustrated in Fig. 8.46. The alternative is either to build a set of conjugate filters or to use an active signal source that generates a signal matched to the receiver filter.

In constructing a tapped delay line matched filter some care must be taken to minimize reflections back into the tapped delay line from the tap

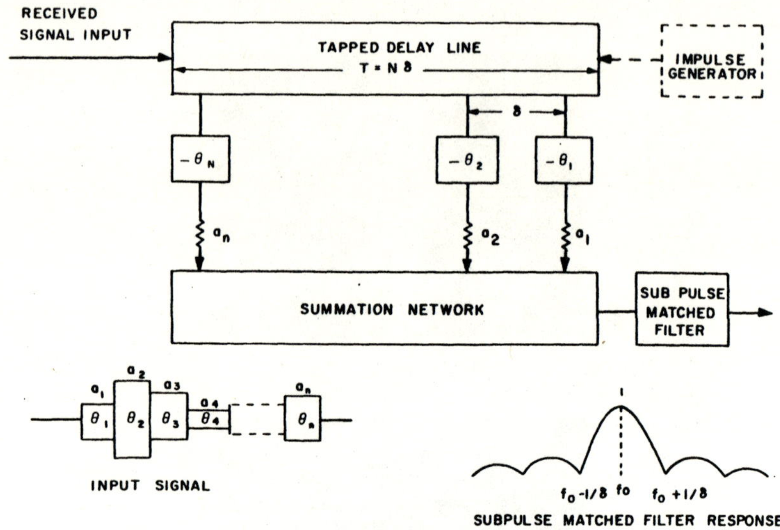

FIG. 8.45   General matched-filter configuration for phase coded waveforms.

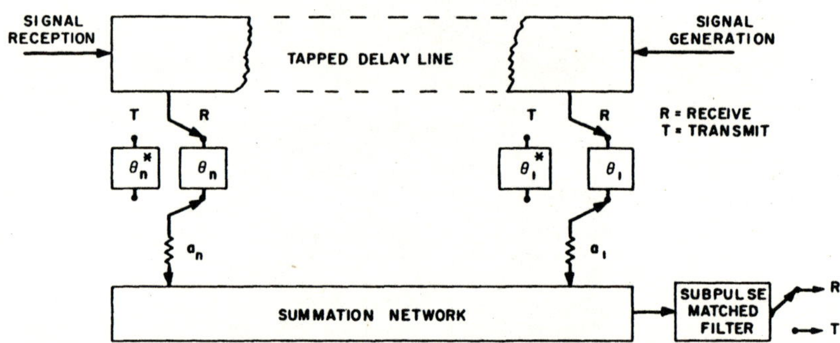

FIG. 8.46   Matched-filter configuration with conjugate phase shifters for signal generation
and signal processing for nonbinary signals.

arms, and isolation amplifiers may be required. It is also important to
control the exact phase of the signal introduced into each tap arm.
Lerner [28] describes the use of a tapped magnetostrictive delay line. In
this type of implementation the taps can be coils wound around the
magnetostrictive rod and placed at uniform intervals apart. The positions
of each coil can be slightly adjusted to provide the required phase control
at the tap outputs. A similar technique, using coiled delay cable as the
delay medium, is described by Benjamin [29]. If a lumped constant tapped
delay line is used then discrete phase control of each tap output can be

obtained if there are several tap points available in each delay subinterval δ. In some cases matched-filter signal processing is done at a coherent video frequency rather than at bandpass frequencies as assumed in the discussion above. In this case a quadrature channel arrangement is required in order to avoid loss of detectability as the phase of the received signal varies. A matched-filter implementation of this type, discussed by Coll and Storey [27], is shown in Fig. 8.47. Allen and Westerfield [30] describe several techniques applicable to digital matched filters.

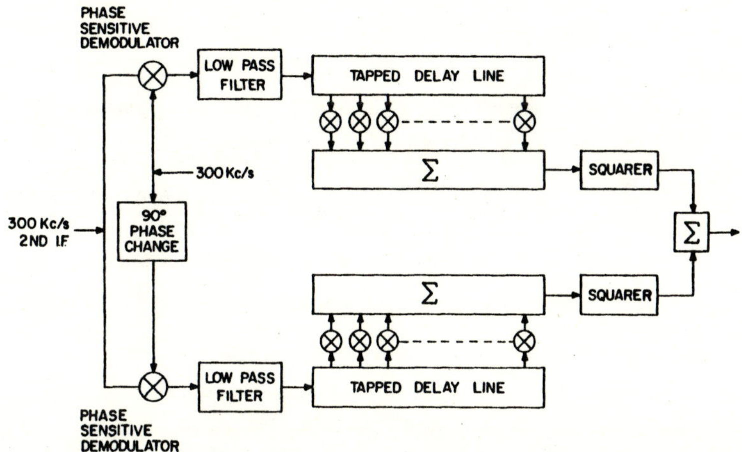

FIG. 8.47   Quadrature matched-filter detector (courtesy of D. C. Coll and J. R. Storey [27]).

A tapped delay line matched filter for a discrete frequency coded signal, such as a stepped FM waveform, can be constructed in an analogous manner to that of the phase coded tapped delay line shown in Fig. 8.45. In this case matched-bandpass filters with different center frequencies would replace the conjugate phase shifters. The disadvantage of this approach is that the tapped delay line must handle the full bandwidth of the signal. This may offer difficulties in the delay line synthesis if the discrete frequency components cover a very wide band (this could be such as to obtain either a uniformly filled spectrum or a "comb" type spectrum). A means of avoiding this particular problem is shown in Fig. 8.48. At the matched-filter input the bandpass filters separate the component frequency bands, each of which is delayed the appropriate amount in its own narrow band delay line. The outputs from each delay line are recombined in the summation network. Depending on the spacing of the component frequencies of the coded waveform and the characteristics of the input bandpass filters, it

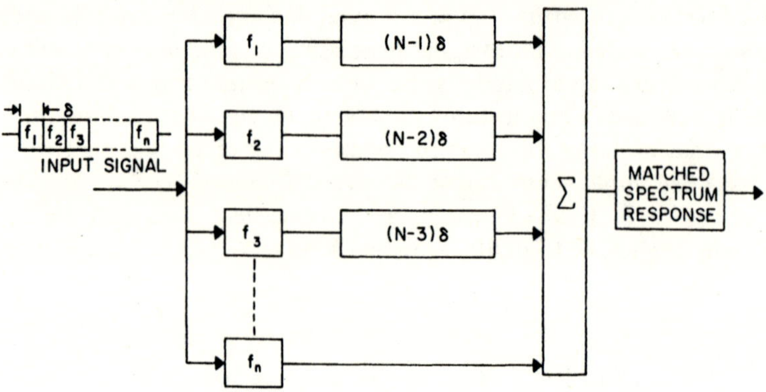

FIG. 8.48   Parallel channel matched-filter for discrete frequency coded signal.

may be desirable to add an overall matched spectrum amplitude response at the output of the summation network. For the linear step FM waveform, where $|f_n - f_{n-1}| = 1/\delta$, the response of the nth bandpass filter should be

$$U_n(f) = \frac{\sin \pi \, \delta(f - f_n)}{\pi \, \delta(f - f_n)} \qquad (8\text{-}75)$$

From the discussion in Section 8.5 it can be seen that for this case N parallel channels provides a time-bandwidth product of N^2. This is a special case of the parallel channel linear FM technique discussed in Section 6.7, where N frequency displaced and contiguous parallel channels, each with a compression ratio of $T_c \Delta f_c$, gave an overall time-bandwidth product of $N^2 T_c \Delta f_c$. The arrangement shown in Fig. 8.48 will also serve as a discrete frequency coded signal generator. To provide a signal matched to the filter shown in Fig. 8.48 the coded signal generation filter should have the fixed delays in reverse order for that shown for the receiver filter. The same filter can be used for signal generation and reception if a switching network is added to connect the fixed delay lines in the correct reverse order after signal transmission.

The implementation for a coded or uncoded pulse train signal follows directly from the implementations discussed above, noting that certain specific values of the amplitude factor are identically zero. Thus, at the corresponding tap points of the delay line no tap arm is included in the design. If each pulse in the train has an associated frequency or phase subcode, then the appropriate tap arm contains a matched filter for that subcode.

In some applications the range sidelobe level of the discrete phase coded signals may be too high. It is possible to reduce the sidelobe level by a transversal filtering method similar to that briefly discussed in Section 7.8. However, this approach has not been exploited for phase coded signals to the same extent as for the linear FM class of signals. Key *et al.* [31] have shown that a Barker code of length 13 can have its range sidelobes reduced from -22 db (13:1 peak-to-sidelobe ratio) to -32.4 db (42:1 peak-to-sidelobe ratio) using a transversal filter. The overall delay of the transversal filter was $26\,\delta$, with 13 taps spaced $2\,\delta$ apart. The tap weightings chosen by Key *et al.* were such as to yield zero sidelobe level in the time interval where the matched-filter sidelobes originally existed, with residual sidelobes occurring outside of this region. The mismatch loss associated with this method was 0.25 db.

8.7   Doppler Correction of Discrete Coded Signals

Chapters 2 and 5 discussed the simultaneous measurement of the range and velocity of a radar target (parameter estimation) as one of the fundamental applications of large time-bandwidth signals. This is accomplished with the least amount of measurement error when the waveform range-Doppler coupling factor (see Eqs. (5-46) and (5-47)) is zero. The range-Doppler coupling factor for various classes of matched-filter signals is discussed in detail in Chapter 9. It is also desirable that the matched-filter response function $\chi(\tau, \phi)$ be uniformly low except for the main peak at $\tau = 0, \phi = 0$. These requirements are met if the response function has the thumbtack type of characteristic, as shown in Fig. 4.8 as an example. The aperiodic truncated maximum length binary phase code signal (pseudo-random sequence) has a response or ambiguity function that approximates this form of behavior, and thus can be considered as an example of a waveform for parameter estimation applications.

Generally, the required implementation for parameter estimation is the bank of matched filters illustrated in Fig. 2.3 or Fig. 9.1 of the next chapter. The center frequencies of the matched filters in the bank are spaced to cover the expected Doppler frequency spread. For a continuous band of Doppler frequencies a nominal center frequency spacing is $1/T$, the reciprocal of the pulsed envelope duration (assuming a rectangular pulse function). However, it may be desirable to space the center frequencies of the matched filters more closely in order to fill in the responses between the $1/T$ spacing and provide a more uniform normalized peak signal output from the matched-filter bank as the Doppler shifts of the received signals vary. Since each matched filter for a large time-bandwidth signal can be a complex and expensive item, the construction of a bank of matched filters can be a costly part of the radar receiver. A method of avoiding this cost and complexity for a bandpass tapped delay line matched filter is to combine a resistor weighting matrix with a single matched filter to simulate an entire bank of matched filters. Such a weighting matrix is often called a Doppler matrix. The characteristics of the resistor weighting matrix can be obtained from the complex narrow band representation of the Doppler shifted received signal given by (see Eq. (4-23))

$$\psi_d(t) = u(t)\exp[j2\pi(f_0 + \phi)t] \qquad (8\text{-}76)$$

Invoke the real part operator on (8-76); this yields

$$s_d(t) = u(t)\cos 2\pi(f_0 + \phi)t$$

$$= u(t)\cos 2\pi f_0 t \cos 2\pi\phi t - u(t)\sin 2\pi f_0 t \sin 2\pi\phi t \qquad (8\text{-}77)$$

The term $u(t)\cos 2\pi f_0 t$ represents the zero-Doppler received signal. The binary phase coded class of signals is described by

$$u(t) = \sum_{n=1}^{N} u_n(t), \qquad 0 < t < N\delta$$

$$= 0, \qquad \text{elsewhere}$$

where $u_n(t)$ is either $a_n P_n(t)\exp[j0]$ or $a_n P_n(t)\exp[j\pi]$ and $P_n(t)$ is defined by Eq. (8-2).

Noting that $\sin 2\pi f_0 t = -\cos(2\pi f_0 t + \pi/2)$, (8-77) can finally be written as

$$s_d(t) = \sum_{n=1}^{N} u_n(t)\cos 2\pi f_0 t \cos 2\pi\phi t$$

$$+ \sum_{n=1}^{N} u_n(t)\cos(2\pi f_0 t + \pi/2)\sin 2\pi\phi t \qquad (8\text{-}78)$$

Lerner [28] points out that for the normal Doppler ranges of interest $\cos 2\pi\phi t$ and $\sin 2\pi\phi t$ are slowly varying compared to the subpulse interval, δ, since ϕ is usually measured in terms of the overall sequence duration, i.e., $\phi = m/T, m \geqslant 0$. Thus, (8-78) can be approximated by

$$s_d(t) = \sum_{n=1}^{N} u_n(t)\cos 2\pi f_0 t \cos 2\pi mn/N$$

$$+ \sum_{n=1}^{N} u_n(t)\cos(2\pi f_0 t + \pi/2)\sin 2\pi mn/N \qquad (8\text{-}79)$$

The equivalent matched-filter operations for this signal are clearly indicated by (8-79). For any specified Doppler shift $\phi = m/T$, the output taps of a zero-Doppler matched filter are weighted, after appropriate phase conjugation, by the factors $\cos 2\pi mn/N$, to yield the first summation of (8-79).

The weightings as a function of tap position are designated as the cosine bus by Lerner. The same tapped outputs are weighted by the factors $\sin 2\pi mn/N$, summed and shifted by 90° to yield the second summation of (8-79). These two terms are then added coherently to reconstruct the equivalent matched-filter response for the Doppler shifted signal.

The implementation of this weighting technique is shown in Fig. 8.49. The weighting factors can be realized by resistor combinations. For those tap points where $\cos 2\pi mn/N$ or $\sin 2\pi mn/N$ is negative, a tap output

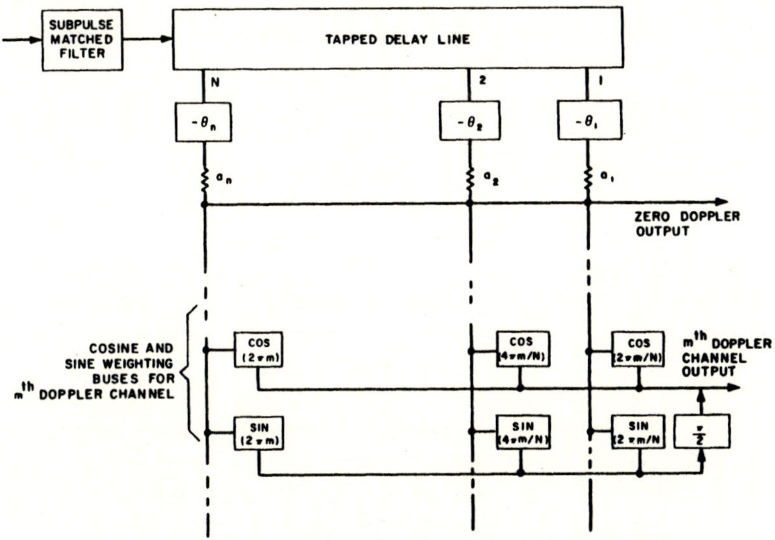

FIG. 8.49 Doppler matrix simulation of a bank of matched filters.

shifted by 180° is required. This can be accomplished by designing each delay line output as a center tapped coil, from which are obtained the appropriate bandpass realizations of the required positive or negative signals. To cover an expected band of Doppler shifts (i.e., $1 \leqslant m \leqslant M$), an M by N resistor weighting matrix is required that has the indicated cosine and sine weighting bus in each Doppler row. Lerner indicates that little degradation from the ideal is obtained if the outputs from the cosine and sine bus are directly detected and added as video signals. This would eliminate the need for the 90° phase shifters and coherent addition of the component signals in each Doppler row of the matrix. The Doppler matrix described by Lerner simulated a bank of 18 matched filters, including the zero-Doppler channel, for a binary phase coded signal (a_n's equal) for which $N = 100$. The results obtained with this matched-filter bank simulator were very close to that theoretically achievable with a bank of true matched filters. The Doppler correction matrix technique described above can be applied to any type of narrow band signal for which the matched filter can be realized in a tapped delay line form.

The Doppler correction and matched-filter simulation technique discussed above can be applied when the Doppler dispersion factor $(2v/c)T\Delta f$ is less than unity. The effect of large values of $(2v/c)T\Delta f$ on the compressed-pulse output of the linear FM matched filter was discussed in Section 6.8. It will be recalled that when $(2v/c)T\Delta f > 1$, the effect of the target velocity on the time scale expansion or contraction of the received signal cannot be ignored. For this case the received signal is more correctly written as

$$s_d(t) = \sqrt{\alpha}u(\alpha t) \tag{8-80}$$

where

$$\alpha = 1 + \frac{2v}{c}\bigg/\left(1 + \frac{v}{c}\right) \doteq 1 + \frac{2v}{c}$$

and the signal response function has the form given by Eq. (6-69). The effect of the factor α on the outputs of the various types of discrete coded waveforms has not been calculated with the same amount of detail as it has for the linear FM signal. However, Remley[32] has calculated the effect of this Doppler dispersion on a signal that is a large, but finite, sample of a stationary random process. Remley draws the conclusion that the formulation of the problem in this manner leads to results that are generally applicable to the pseudorandom phase reversal class of signals. The central result of this investigation is the description of the average (expected) value of the matched-filter response function as

$$|E\{\chi_d(\tau, \Delta\phi)\}| = \left|\frac{1}{T}\int_{-T/2}^{T/2} \chi(t + \delta\tau, 0)\exp[-j2\pi\,\Delta\phi\,\tau]\,d\tau\right| \tag{8-81}$$

where $\delta = 2v/c/(1 + v/c)$. The notation $\Delta\phi$ indicates that the matched filter has been designed to compensate for the mean Doppler translation ϕ, and (8-81) describes the matched-filter response as the Doppler shift varies about the mean Doppler by the amount $\Delta\phi$. As before, $\chi(\tau, 0)$ is the auto-correlation function of the zero-Doppler matched-filter response and T is the signal duration. Assuming a rectangular signal spectrum (i.e., $\chi(\tau, 0)$ a $(\sin x)/x$ function), Remley calculates the effects of the Doppler dispersion on various matched-filter output parameters as shown in Figs. 8.50–8.52. Figure 8.50 plots the reduction in peak output signal voltage as a function of the dispersion product $TW|\delta|$, where δ is defined in (8-81) and $W = \Delta f$, the general bandwidth measure used in this volume. This

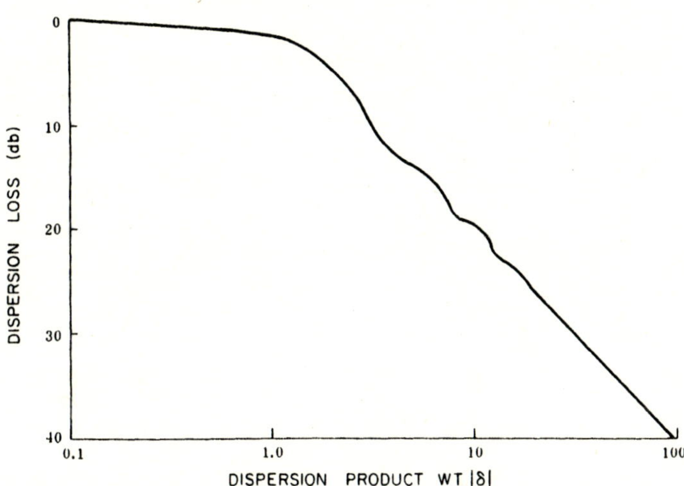

FIG. 8.50   Doppler dispersion loss in peak matched-filter output. Curve based on rectangular spectra (courtesy of W. R. Remley [32].

figure shows that for $TW|\delta| > 1$ the peak output voltage is proportional to $1/TW|\delta|$. Figure 8.51 demonstrates the effect of the factor $|\delta|$ on the matched-filter processing gain relative to receiver noise as a function of the time-bandwidth product. This result shows that the maximum obtainable processing gain is approximately $1/|\delta|$, and occurs when $WT|\delta| \doteq 1$. Figure 8.52 plots the degradation of the 6 db widths of $\chi(\tau, 0)$ and $\chi(0, \Delta\phi)$, which are a measure of the time and frequency resolution capability of the Doppler distorted output signal. A result associated with this figure is that when $WT|\delta|$ is large the distorted pulse width is given by $T_p = |\delta|T$, and the frequency resolution is given by $W_p = |\delta|W$. From this one can deduce that the ratio of the uncertainty ellipse area for the cases of large Doppler dispersion product and no (or very small) Doppler dispersion is approximately $(WT|\delta|)^2$. The results obtained by Remley are quite similar, although not identical, to the results obtained for the linear FM signal. While exact only for the specific case calculated, they support the general conclusion that Doppler dispersion does not have a significant effect on the matched-filter output when $(2v/c)T\Delta f$, or $WT|\delta|$, is less than unity. When this factor is greater than unity the resulting effects on the matched-filter output signal-to-noise ratio and resolution capability can be serious. These effects are independent of the transmission frequency of the signal. When this latter condition prevails the radar designer may wish to consider the use of a receiver matched-filter function that meets the general description given by Eq. (4-29). This type of filter design would compensate not only for the Doppler frequency translation, but also for the signal time scale

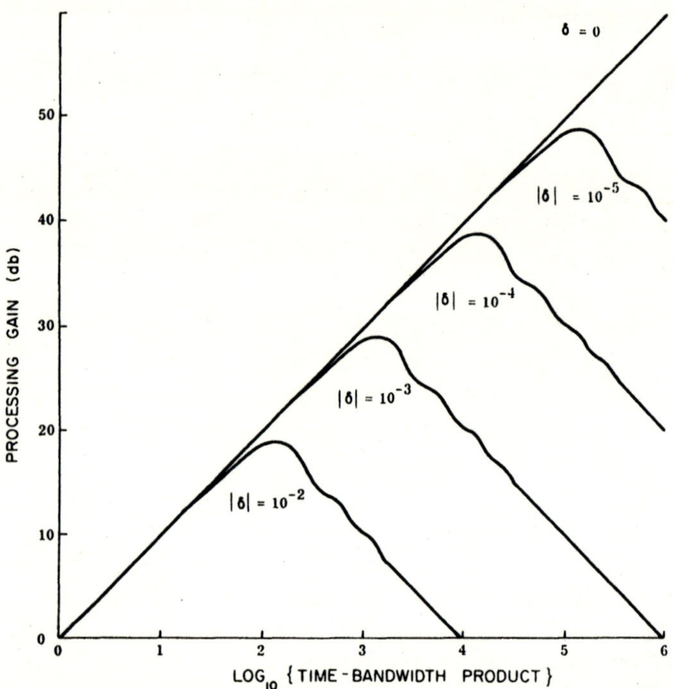

Fig. 8.51   Matched-filter processing gain with Doppler dispersion and rectangular spectra (courtesy of W. R. Remley [32]).

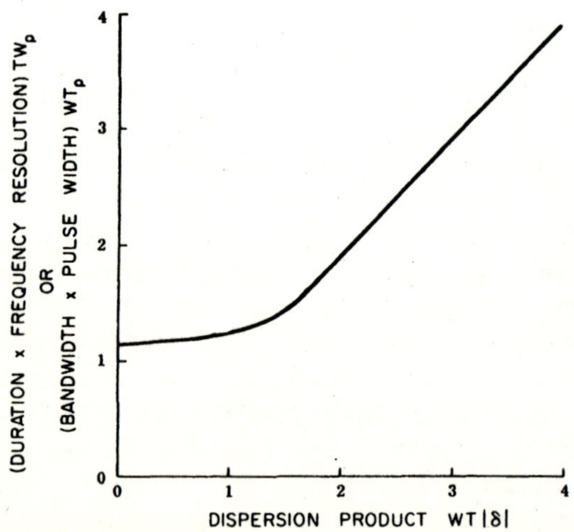

Fig. 8.52   Matched-filter resolution in time and frequency for various dispersion products (courtesy of W. R. Remley [32]).

contraction (approaching target) or expansion (receding target). Matched-filter implementations of this type are discussed by Benjamin [29] and by Kelly and Wishner [33].

8.8 Summary

This chapter has shown that discrete coded waveforms afford another way to shape the responses observed at the matched-filter output. The different distributions which the discrete coded waveforms provide are clearly evident from the illustrations furnished in this chapter. The prevailing characteristic is seen to be the formation of discrete areas of high pop-ups, a feature not necessarily undesirable. Figure 8.2 provides a good example of this type of characteristic. The degree of isolation attained by the pop-ups is determined by how much separation there is between the code elements. The discussion of the maximum volume free area in Section 4.7 furnishes an upper bound for the degree of isolation of these pop-ups.

The utilization of discrete coding provides $3N$ degrees of coding freedom, in contrast with the analog codes which depend on one, or perhaps two, parameters. The $3N$ degrees of freedom are derived from the ability to independently code the amplitude, phase and frequency parameters of the signal. How these $3N$ degrees of freedom are applied must be determined by the needs of a specific application. An attempt has been made in this chapter to show the results of various design objectives. Thus, the Barker codes and the staggered pulse trains provide examples of the use of N of these degrees of freedom to obtain maximum peak-to-sidelobe ratios on $\chi(\tau, 0)$. A comparison of the peak-to-sidelobe ratios for different phase coded waveforms is shown in Fig. 8.53. The Huffman codes utilize another N degrees of freedom, combining amplitude and phase sequences, to achieve an "impulse equivalent" $\chi(\tau, 0)$ response. Each of the examples mentioned above is a step toward the "thumbtack" response function, an objective often cited for applications involving simultaneous measurement of range and velocity.[1] Another example of the use of the $3N$ degrees of freedom is described by Venier [35], who has investigated frequency coded, nonuniform pulse trains. His objective was to maintain the sidelobes of $\chi(\tau, 0)$ as low as possible, and at the same time allow overlapping input signals of widely different amplitudes to be seen when a limiter precedes the matched filter. For some applications, especially when the

---

[1] Kaiteris and Rubin [34] describe a method of achieving a "video thumbtack ambiguity function" by noncoherently combining the outputs of a bank of filters that are each matched to one set of an ensemble of interleaved staggered pulse trains, in which each component train is disjoint in its frequency spectrum from all the others.

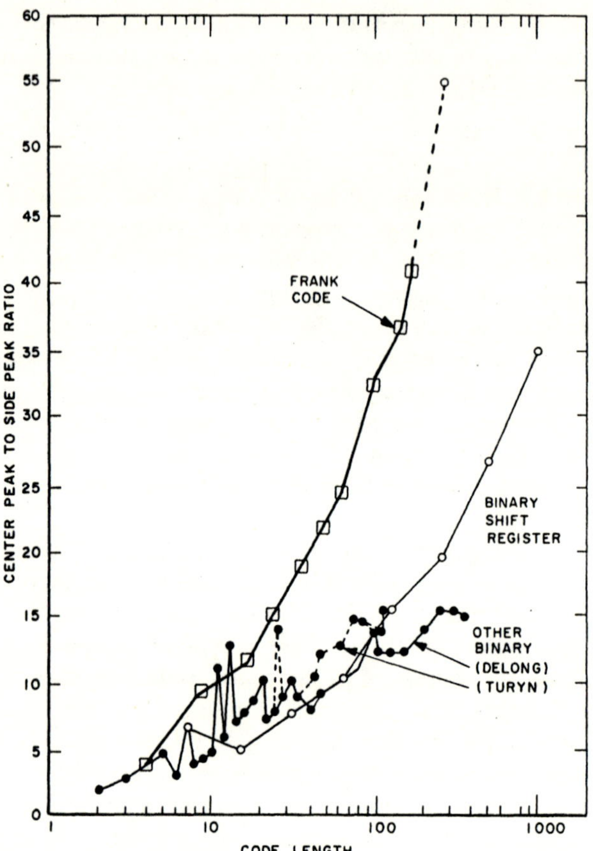

FIG. 8.53   Comparison of sidelobe levels for various phase code waveforms (courtesy of R. L. Frank [22]).

benefits derived from digital implementations of matched filters are important, the design objective may be to approximate the response characteristic of a given analog coded waveform that happens to be best suited for a particular application. Howard [36] describes the combination of frequency step FM and staggered pulse positions to achieve pulse train ambiguity functions that have central responses in the region about $\tau = 0$, $\phi = 0$ similar to those of specified analog FM waveforms. Fitting the response characteristics of the often desirable linear FM waveform is another example of a discrete code design objective. The characteristic ridgelike $\chi(\tau, \phi)$ response of the linear FM waveform is closely approximated by the stepped FM and Frank quantized phase codes that have been discussed in this chapter. Whatever the application may be, the waveform design approach using discrete coding appears to offer a degree of flexibility which has many advantages in terms of shaping matched-filter responses and implementing the matched-filter processing.

This chapter has presented a unified treatment of discrete coding techniques. The principle objective here has been to convey a logical mathematical discipline regarding the description of discrete coded waveforms and the derivation of their matched-filter response characteristics. It will be noted that attention has not been focused on comparing the waveform examples presented. Rather, as in other discussions of the different waveform examples in this volume, the theme of this chapter basically is that all waveforms are equally good (or bad) until the given waveform application is specified. In this respect, Chapters 9 and 10 discuss some important applications in which the waveform characteristics are critical factors. The reader motivated by the need to solve a particular problem should find sufficient information in this chapter which will enable him to form value judgements about the merits of the individual waveforms that have been discussed. At a minimum, the reader will have added to his store of knowledge some important discrete coding techniques that have found use in radar and other applications. Furthermore, it is hoped by virtue of the many illustrations of different discrete coded waveforms that the reader will have gained some measure of insight that will enable him to construct other waveforms based on discrete coding techniques to suit his specialized needs.

REFERENCES

1. P. M. Woodward, "Probability and Information Theory with Applications to Radar." Pergamon Press, Oxford, 1953.
2. W. M. Siebert, A radar detection philosophy, *IRE Trans.* **IT-2**, 204–221, (1956).
3. E. N. Fowle, E. J. Kelly and J. A. Sheehan, Radar system performance in a dense target environment, *IRE Intern. Conv. Record, Pt.* 4, 136–145 (1961).
4. J. B. Resnick, High resolution waveforms suitable for a multiple target environment, M.S. Thesis, Mass. Inst. Tech., Cambridge, Massachusetts (June, 1962).
5. A. W. Rihaczek, Radar resolution properties of pulse trains, *Proc. IEEE* **52**, 153–164 (1964).
6. C. Kaiteris and W. L. Rubin, Pulse trains with low residue ambiguity surfaces that minimize overlapping target echo suppression in limiting receivers, *Proc. IEEE* **54**, 438–439 (1966).
7. G. J. Simmons, A factorization technique for binary autocorrelation functions, *Proc. IEEE (Letters)* **54**, 794–795 (1966).
8. M. Bernfeld, A property of binary sequences, *Proc. IEEE (Correspondence)* **52**, 744 (1964).
9. R. H. Barker, Group synchronization of binary digital systems, *in:* "Communication Theory" (W. Jackson, ed.), pp. 273–287. Academic Press, New York and London, 1953.
10. J. E. Storer and R. Turyn, Optimum finite code groups, *Proc. IRE (Correspondence)* **46**, 1649 (1958).
11. R. Turyn, On Barker codes of even length, *Proc. IRE (Correspondence)* **51**, 1256 (1963).

12. B. Elspas, A radar based on statistical estimation and resolution considerations, Stanford Electronics Lab., Stanford Univ., Stanford, California, Tech. Rept. 361–1 (August, 1955).

13. S. E. Craig, W. Fishbein, and O. E. Rittenbach, Continuous-wave radar with high range resolution and unambiguous velocity determination, *IRE Trans.* **MIL-6**, 153–161 (1962).

14. T. Sakamoto, Y. Taki, H. Miyakawa, H. Kobayashi, and T. Kanda, Coded pulse radar system, *J. Fac. Eng. Univ. Tokyo* **27**, 119–181 (1964).

15. S. W. Golomb, Sequences with randomness properties, Glenn L. Martin Co., Baltimore, Maryland, Final Rept. on Contract SC–54–33611 (June, 1955).

16. N. Zierler, Linear recurring sequences, *J. Soc. Ind. Appl. Math.* **7**, 31–48 (1959).

17. W. W. Peterson, "Error Correcting Codes," M.I.T. Press, Cambridge, Massachusetts, 1961.

18. D. A. Huffman, The synthesis of linear sequential coding networks, *in*: "Information Theory," (C. Cherry, ed.), Academic Press, New York, 1956.

19. S. W. Golomb and R. A. Scholtz, Generalized Barker sequences, *IEEE Trans.* **IT-11**, 533–537 (1965).

20. R. C. Heimiller, Phase shift codes with good periodic correlation properties, *IRE Trans.* **IT-7**, 254–257 (1961).

21. R. Frank and S. Zadoff, Phase shift codes with good periodic correlation properties, *IRE Trans. (Correspondence)* **IT-8**, 381–382 (1962).

22. R. L. Frank, Polyphase codes with good nonperiodic correlation properties, *IEEE Trans.* **IT-9**, 43–45 (1963).

23. D. F. DeLong, Jr., Three phase codes, Mass. Inst. Technol., Lincoln Lab., Lexington, Massachusetts, Group Rept. 47–28 (July, 1959).

24. D. A. Huffman, The generation of impulse-equivalent pulse trains, *IRE Trans.* **IT-8**, S10–S16 (1962).

25. C. E. Jagger and R. H. McLaughlin, A compilation of ambiguity surfaces of some short codes, Canadian General Electric Co., Ltd., Toronto, Canada, Tech. Memo. RQ 65EE32 (August, 1965).

26. J. Injeyan, Generation of Huffman codes, Canadian General Electric Co., Ltd., Toronto, Canada, Tech. Memo. RQ 65EE2 (October, 1965).

27. D. C. Coll and J. R. Storey, Ionospheric sounding using coded pulse signals, *J. Res. Natl. Bur. Std., Radio Sci.* **68D**, 1155–1159 (1964).

28. R. M. Lerner, A matched-filter detection system for complicated Doppler shifted signals, *IRE Trans.* **IT-6**, 373–385 (1960).

29. R. Benjamin, Recent developments in radar modulation and processing techniques, *Proc. IEE (London)* **111**, 2002–2015 (1964).

30. W. B. Allen and E. C. Westerfield, Digital compressed-time correlators and matched filters for active sonar, *J. Am. Acoust. Soc.* **36**, 121–139 (1964).

31. E. L. Key, E. N. Fowle, and R. D. Haggarty, A method of sidelobe suppression in phase coded pulse compression systems, Mass Inst. Technol., Lincoln Lab., Lexington, Massachusetts, Tech. Rept. TR-209 (August, 1959).

32. W. R. Remley, Doppler dispersion effects in matched-filter detection and resolution, *Proc. IEEE* **54**, 33–39 (1966).

33. E. J. Kelly and R. P. Wishner, Matched-filter theory for high velocity accelerating targets, *IEEE Trans.* **MIL-9**, 56–69 (1965).

34. C. Kaiteris and W. L. Rubin, A noncoherent signal design technique for achieving a low residue ambiguity function, *IEEE Trans.* **AES-2**, 468–471 (1966).

35. G. O. Venier, Pulse compression using frequency-coded nonuniform pulse trains, Defence Research Board, Ottawa, Canada, DRTE Rept. 1149 (September, 1965).

36. T. B. Howard, The application of some linear FM results to frequency-diversity waveforms, *RCA Rev.* **26**, 75–105 (1965).

CHAPTER 9

The Measurement Accuracies of Matched-Filter Radar Signals— Waveform Design Criteria

9.1 Introduction

The discussion of the radar ambiguity function in Chapter 4 indicated that three characteristics of the radar signal—resolution, ambiguity, and accuracy—are of interest to the radar designer. Resolution and ambiguity are most often linked together for, as Woodward states, the requirement that the signal autocorrelation function $|\chi(\tau, 0)|$ be as small as possible except near $\tau = 0$ is a necessary condition for signals at different ranges to be distinguishable at the receiver [1]. Thus, emphasis is placed on the design of matched-filter waveforms that are sharply peaked and narrow about $\tau = 0$ and have extremely low level range sidelobes (often contradictory requirements that must be resolved by compromise). Several criteria for resolution and ambiguity were presented in Chapter 4. Measurement accuracy, on the other hand, is usually developed on the assumption of a single signal present at the receiver, and as shown in Chapter 5 is related to the signal-to-noise ratio and the behavior of the two-dimensional correlation function $|\chi(\tau, \phi)|$ in the small region about $\tau = 0, \phi = 0$. The measurement accuracy criteria were derived in Chapter 5 using the Cramér–Rao inequality, a familiar tool in the field of statistical estimation. It was shown that the system implementation that leads to the realization of these error variances is the maximum likelihood receiver, composed of a bank of matched filters illustrated in Fig. 9.1.

In many applications only one parameter (range) is of interest to the radar engineer in the context of the individual waveform characteristics, since velocity measurement would be derived from ancillary techniques that are not inherently dependent on the signal matched-filter behavior. In other areas there is considerable interest in being able to perform accurate, simultaneous measurements of range and velocity. In this case,

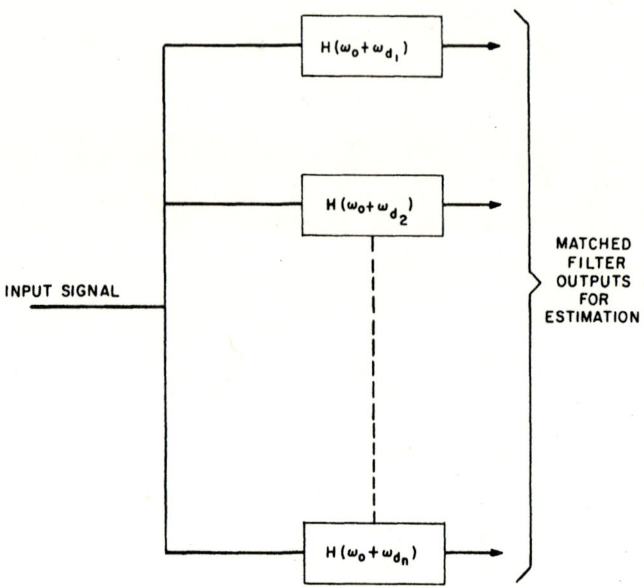

FIG. 9.1   Matched filter bank for range-Doppler measurement.

as shown in Chapter 5, the waveform time-frequency (or range-Doppler) coupling factor becomes an important consideration. In general, the waveform design criterion for minimum measurement errors, particularly for joint measurements, is not always compatible with achieving the degree of resolution and ambiguity necessary in a multiple radar target environment.

The measurement accuracies of a number of the large time-bandwidth waveforms discussed in earlier chapters are compared in this chapter, and the effect of the time-frequency coupling factor is evaluated. In the broader sense, these waveforms fall into two categories: (1) those that maintain their time resolution properties at the matched-filter output in the face of a Doppler shift (i.e., knife-edge ambiguity functions) and (2) those that break up or decompose as a result of a Doppler shift (i.e., thumbtack ambiguity functions, or rough approximations thereto). The latter class of signals tend to have minimum time-frequency coupling, and a general condition for rectangular envelope FM signals is presented for zero coupling. As various examples are presented it will become clear that the general theory of parameter estimation, as developed in Chapter 5, is useful primarily in making qualitative comparisons of signals, rather than being a means for obtaining quantitive data for explicit radar signals. This results from the fact that the mathematical formulation of the majority of radar waveforms does not meet the condition of regularity cited in Chapter 5.

9.2 Minimum Time and Frequency Measurement Error Variances for Some Large Time-Bandwidth Radar Signals

The minimum time and Doppler frequency measurement error variances are (from Chapter 5)

$$(\sigma_\tau^2)_{min} = \frac{N_0}{2E\beta^2} \tag{9-1}$$

$$(\sigma_\phi^2)_{min} = \frac{N_0}{2E\alpha^2} \tag{9-2}$$

where E is the received signal energy, N_0 the single sided noise power density, and 2β and 2α are the effective bandwidth and time duration of the signal, defined by the expressions

$$\beta^2 = \frac{4\pi^2 \int_{-\infty}^{\infty} f^2 |U(f)|^2 \, df}{\int_{-\infty}^{\infty} |U(f)|^2 \, df} = \frac{4\pi^2}{2E} \int_{-\infty}^{\infty} f^2 |U(f)|^2 \, df \tag{9-3}$$

$$\alpha^2 = \frac{4\pi^2 \int_{-\infty}^{\infty} t^2 |u(t)|^2 \, dt}{\int_{-\infty}^{\infty} |u(t)|^2 \, dt} = \frac{4\pi^2}{2E} \int_{-\infty}^{\infty} t^2 |u(t)|^2 \, dt \tag{9-4}$$

where $u(t)$ is the complex representation of the signal $= a(t) \exp[j\theta(t)]$, and $|U(f)|$ is the magnitude of the complex spectrum $= 2|S_+(f + f_0)|$, where $S_+(f)$ is the positive frequency component of the real signal (assumed to be narrow band as defined in Chapter 4), and f_0 is the mean frequency.

From (9-1) and (9-2) are obtained the rms range error $\delta_r = \frac{1}{2}c\sigma_\tau$ and the rms velocity error $\delta_v = \frac{1}{2}\lambda\sigma_\phi$, where c is the propagation velocity and λ the transmitted wavelength. The factors β and α are in radian dimensions, so that the defined bandwidth and duration are expressed as $2\beta/2\pi$ and $2\alpha/2\pi$, respectively. Equations (9-3) and (9-4) have the form of second moments, and are not equivalent to the standard definitions of 3 db bandwidth and time duration.

If only the positive frequencies of the spectrum of the real signal are considered then (9-3) becomes

$$\beta^2 = \frac{8\pi^2}{E} \int_{0}^{\infty} (f - f_0)^2 |S(f)|^2 \, df \tag{9-5}$$

The minimum error variances given by (9-1) and (9-2) hold if one of the parameters (ϕ or τ) is assumed known and the other is to be estimated, or if the waveform has zero time-frequency coupling. The values obtained for β and α provide a comparative indication of the range and velocity measurement accuracies of the signals. However, care must be taken in interpreting the values obtained in some instances.

For several of the waveforms to be considered $|u(t)|$ will be assumed to have unit amplitude, and a time duration of T seconds. The energy E associated with this signal is

$$E = \tfrac{1}{2}\int_{-T/2}^{T/2} |u(t)|^2\, dt = T/2 \tag{9-6}$$

and thus

$$\alpha^2 = \frac{4\pi^2}{T}\int_{-T/2}^{T/2} t^2|u(t)|^2\, dt = \frac{\pi^2 T^2}{3} \tag{9-7}$$

This applies to all rectangular envelope signals, regardless of the phase modulation function $\theta(t)$ contained under the envelope. The analogous result for a rectangular spectrum bandlimited over the interval Δf is $\beta^2 = \pi^2\,\Delta f^2/3$. An alternate expression for β^2 was shown to be

$$\beta^2 = \frac{1}{2E}\int_{-\infty}^{\infty} |u'(t)|^2\, dt + \frac{1}{2E}\int_{-\infty}^{\infty} |u(t)|^2[\theta'(t)]^2\, dt \tag{9-8}$$

The first term of (9-8) implies that $\beta = \infty$ for any waveform with envelope discontinuities, such as the rectangular envelope. For this envelope function the spectrum skirts eventually decay as $1/f$, which if inserted into (9-3) would lead to the same result. This result is not meaningful in any practical sense, and in some of the examples of this and the next section a modified spectrum skirt function is discussed as one approach to obtain finite values of β. An alternate method based on truncating the spectrum at arbitrary limits is described by Skolnik [2]. In essence, these are means of adapting parameter estimation theory to accommodate signals that do not meet the regularity condition, such as the rectangular pulse envelope.

Using (9-3) or (9-5) and (9-4) one can calculate the values of β and α for some of the pulse-compression signals described in earlier chapters. From these the minimum rms measurement errors δ_r and δ_v can be computed. The Gaussian envelope, linear FM signal [3] will be developed as a first example since both its time and frequency distributions are regular functions that do not encounter the problem inherent in (9-8).

Gaussian Envelope, Linear FM Signal

This signal is assumed to be generated by impulsing one of a pair of conjugate filters that has a normalized amplitude response given by $|U(f)|$, where

$$|U(f)| = 2\exp[-4f^2/\Delta f^2], \qquad -\infty < f < \infty \tag{9-9}$$

and a linear dispersive delay of T seconds over the spectrum interval Δf (the e^{-1} spectrum width), or $-d\Phi/d\omega = Tf/\Delta f$, where $\Phi(\omega)$ is the spectrum phase characteristic. The normalized envelope function of the linear FM output of this filter is given by [3]

$$|u(t)| = \exp\left[4t^2 \Big/ \left(T^2 + \frac{16}{(\pi T \Delta f)^2}\right)\right] = \exp\left[-\frac{4t^2}{T_0^2}\right] \tag{9-10}$$

From (9-9)

$$E = 2\int_{-\infty}^{\infty} \exp\left[-\frac{8f^2}{\Delta f^2}\right] df = \sqrt{\frac{\pi}{2}}\,\Delta f \tag{9-11}$$

and

$$\beta^2 = \frac{8\pi^2}{E}\int_{-\infty}^{\infty} f^2 \exp\left[-\frac{8f^2}{\Delta f^2}\right] df = \frac{\pi^2 \Delta f^2}{4} \tag{9-12}$$

Similarly, from (9-10)

$$\alpha^2 = \frac{\pi^2 T_0^2}{4} = \frac{\pi^2 T^2}{4}\left[1 + \frac{16}{(\pi T \Delta f)^2}\right] \tag{9-13}$$

The $\beta\alpha$ product of a waveform is a measure of the limit of the accuracy of simultaneous range and velocity estimates, with a large $\beta\alpha$ product being one of the conditions for accurate joint estimates of range and velocity. It was shown in Chapter 4 that $\beta\alpha \geq \pi$, with equality holding for the case of the constant frequency Gaussian envelope pulse. From (9-12) and (9-13), which are exact solutions for β and α, the $\beta\alpha$ product for the linear FM Gaussian envelope signal is

$$\beta\alpha = \frac{\pi}{4}[16 + (\pi T \Delta f)^2]^{1/2} \tag{9-14}$$

Considering (9-14), $\beta\alpha = \pi$ for $T\Delta f = 0$. This interesting result indicates that for frequency modulated pulse-compression signals $T\Delta f = 0$ implies the limiting case of the constant frequency pulse, or $\Delta f = 0$ in the context of the frequency modulation parameters. When $\Delta f = 0$ in this sense then of course the value of β is related to the Fourier transform of $|u(t)|$, and would not be given by (9-12), taking the Gaussian envelope pulse as an example. The matched-filter output rms bandwidth is $2\beta/2\pi = \Delta f/2$ for the Gaussian envelope linear FM pulse. This is the interval between the $e^{-1/2}$ points of the spectrum (0.607 amplitude), as compared to the nominal 3 db definition of bandwidth.

The most common class of large time-bandwidth radar signals has a rectangular pulse envelope that, from (9-8) would require $\beta = \infty$. The most well-known of these signals is the linear FM rectangular envelope pulse-compression signal.

Linear FM—Rectangular Envelope

Figure 6.6 showed that the linear FM spectrum approached a rectangular distribution for large $T\Delta f$. Thus, from (9-5) the limiting value of $\beta^2 = \pi^2\,\Delta f^2/3$, neglecting the fact that the rectangular pulse envelope theoretically leads to an infinite β. Skolnik [2] demonstrates the limiting value of β^2 by truncating the linear FM spectrum over the interval $-\Delta f/2$ to $\Delta f/2$, as indicated by the Fresnel region in Fig. 9.2, obtaining for integer values of $T\Delta f$:

$$\beta_1^2 = \left[\frac{\pi^2\,\Delta f^2}{3}\right]\left[1 + \frac{3S(\pi T\,\Delta f)}{(2T\,\Delta f)^{3/2}}\right] \Big/ \left[1 + \frac{C(\pi T\,\Delta f)}{(2T\,\Delta f)^{1/2}}\right] \qquad (9\text{-}15)$$

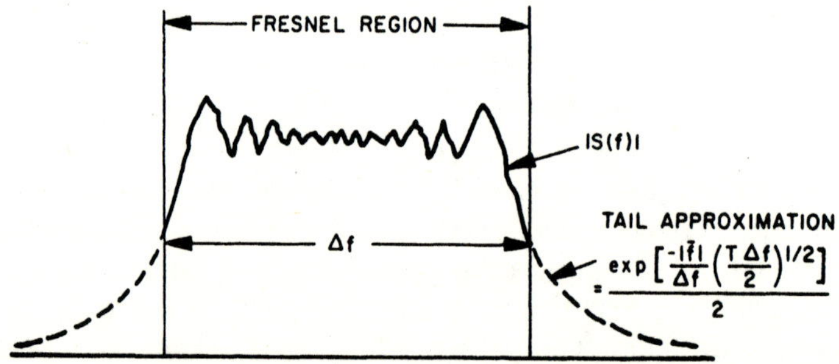

FIG. 9.2 Linear FM spectrum tail approximation ($\tilde{f} = |f| - \Delta f/2$).

where $S(\pi T\,\Delta f)$ and $C(\pi T\,\Delta f)$ are the usual Fresnel functions. Equation (9-15) yields $\beta^2 = 0.88\pi^2\,\Delta f^2/3$ for $T\Delta f = 1$; $0.95\pi^2\,\Delta f^2/3$ for $T\Delta f = 10$; and $0.965\pi^2\,\Delta f^2/3$ for $T\Delta f = 1000$. It can be seen that the desired limiting value of β^2 is being approached from below for this case. This will yield negative values of measurement error variances when the range-Doppler coupling effect is considered. Assuming the limiting value of β^2 results in infinitely large rms measurement errors under the same condition of range-Doppler coupling (see Section 9.3). One method of avoiding this situation is to assume a spectrum skirt function that can be handled analytically and that does not result in $\beta^2 = \infty$. One such function, shown by the dashed lines in Fig. 9.2, is the normalized spectrum skirt given by

$$|S(f)|_{\text{skirt}} = \frac{1}{2}\exp\left[-\left\{\left(|f| - \frac{\Delta f}{2}\right)\Big/ \Delta f\right\}\sqrt{T\,\Delta f/2}\right], \qquad (9\text{-}16)$$

$$\Delta f/2 < |f| < \infty$$

Figure 9.3 compares (9-16) to the average level of the theoretical rectangular envelope linear FM spectrum skirts for $T\,\Delta f = 30$ and $T\,\Delta f = 90$, indicat-

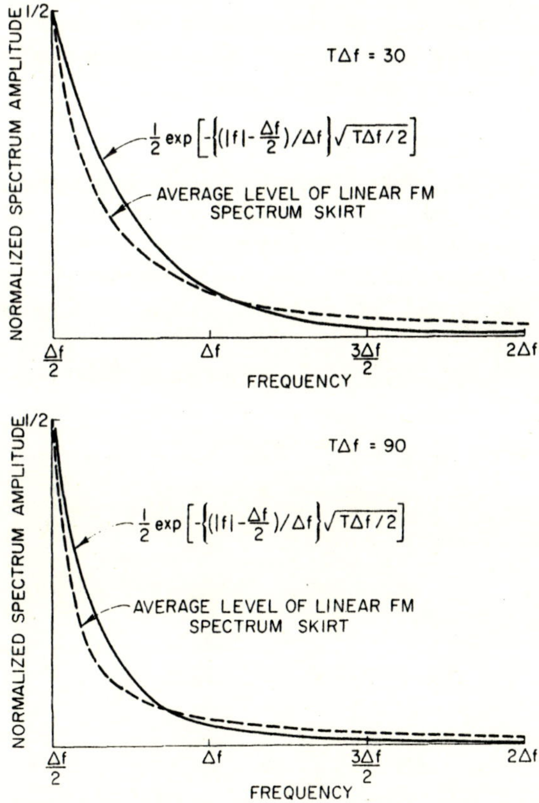

FIG. 9.3   Comparison of Eq.(9-16) with the average level of the linear FM spectrum skirt.

ing that a fair approximation is obtained. The combination of (9-16) with Skolnik's result, using (9-5), yields an expression for the linear FM β^2

given by

$$\beta^2 = \beta_1^2 + \frac{\pi^2 \Delta f^2}{3}\left[\frac{3\sqrt{2}}{(T\Delta f)^{3/2}} + \frac{3\sqrt{2}}{4(T\Delta f)^{1/2}}\right] \qquad (9\text{-}17)$$

(9-17) results in $\beta^2 = 6.2\,\pi^2\,\Delta f^2/3$ for $T\Delta f = 1$, and $\beta^2 \to \pi^2\,\Delta f^2/3$ from above as $T\Delta f \to \infty$. This better fits the intuitive idea of the behavior of β^2 for the linear FM spectrum as $T\Delta f$ increases, assuming that β^2 should become large for the limiting case of the constant frequency pulse.

Other spectrum skirt functions could be chosen to improve the approximation. However, as will be indicated in the next section, this approach also does not yield satisfactory results when one desires to account for range-Doppler coupling. It can be concluded that either of the methods outlined above (i.e., spectrum truncation, or spectrum skirt construction) can be used primarily to obtain approximate values of β^2 from which to calculate the minimum rms measurement errors.

Parabolic FM—Rectangular Envelope [4, 5]

This signal was originally proposed as producing a better approximation to the thumbtack ambiguity function than the linear V-FM signal. Its characteristics as a unidirectional FM waveform are of interest in comparing the effect of nonlinear FM curvature on the comparative theoretical measurement accuracies of several nonlinear FM pulse-compression waveforms.

The unidirectional parabolic FM waveform, illustrated in Fig. 9.4 is defined by

$$s(t) = \cos[\omega_l t + \lambda(t + T/2)^3], \qquad -T/2 < t < T/2 \qquad (9\text{-}18)$$

and the FM function is

$$\omega(t) = \omega_l + 3\lambda\left(t + \frac{T}{2}\right)^2 = \omega_l + \frac{2\pi\,\Delta f}{T^2}\left(t + \frac{T}{2}\right)^2 \qquad (9\text{-}19)$$

The spectrum of the parabolic FM signal, assuming $f_1 = \omega_l/2\pi = 0$, can be obtained by the principle of stationary phase, yielding [5]

$$|S(f)| = \frac{1}{2\sqrt{2}}\sqrt{\frac{T}{(\Delta f)^{1/2}}}\,f^{-1/4} \qquad (9\text{-}20)$$

where $0 < f < \Delta f$. From the first term of (9-8) it can be seen that waveforms with similar envelope functions will have similar contributions to β^2 in the spectrum skirt region. Thus, the same spectrum skirt expression used for the rectangular envelope linear FM signal could be employed for this signal

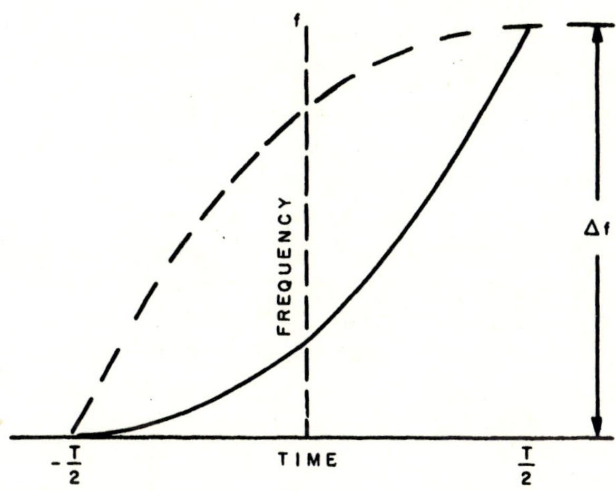

FIG. 9.4   Unidirectional parabolic FM functions.

if desired. Assuming such a symmetrical spectrum skirt distribution about f_0, the mean frequency of the parabolic FM spectrum can be calculated from (9-20) alone:

$$f_0 = \frac{2}{E} \int_0^{\Delta f} f |S(f)|^2 \, df = \frac{1}{2(\Delta f)^{1/2}} \int_0^{\Delta f} f^{1/2} df = \frac{\Delta f}{3} \qquad (9\text{-}21)$$

The contribution to β^2 from the spectrum (9-20) is

$$\beta^2 = \frac{8\pi^2}{E} \int_0^{\Delta f} (f - f_0)^2 |S(f)|^2 \, df$$

$$= \frac{2\pi^2}{(\Delta f)^{1/2}} \left\{ \int_0^{\Delta f} f^{3/2} \, df - \frac{2\,\Delta f}{3} \int_0^{\Delta f} f^{1/2} \, df + \frac{\Delta f^2}{9} \int_0^{\Delta f} f^{-1/2} \, df \right\}$$

$$= \frac{16\pi^2 \, \Delta f^2}{45} \qquad (9\text{-}22)$$

Use the exponential spectrum skirt equation (9-16); then the value of β^2 for the unidirectional parabolic FM signal can be expressed for large values of time-bandwidth product as

$$\beta^2 = \frac{\pi^2 \, \Delta f^2}{3} \left[1.067 + \frac{3\sqrt{2}}{4(T \, \Delta f)^{1/2}} \right] \qquad (9\text{-}23)$$

Other Pulse-Compression Waveforms

In Chapters 3 and 7 several techniques were described for shaping the spectrum at the matched-filter output in order to reduce the ambiguity associated with the compressed-pulse range sidelobes. These took the form of amplitude weighting of the transmitted pulse envelope, weighting the receiver frequency response (a mismatch technique), or using a non-linear FM function under a rectangular envelope [6]. Reducing the range sidelobe levels by these methods results in a broadening of the compressed pulse width. This will be reflected in the calculated value of β. For the case of receiver mismatch frequency response weighting, the factor $|S(f)|W(f)$ replaces $|S(f)|^2$ in (9-5), $|S(f)|$ being the spectrum distribution at the mismatched-filter input and $W(f)$ the weighting function. An example of the nonlinear FM signals is the one that yields a cosine-squared spectrum at the matched-filter output (i.e., -32 db range sidelobes [7]). The spectrum of the signal is described by (for $f_0 = 0$)

$$|S(f)|^2 = \cos^2 \frac{\pi f}{\Delta f} = \frac{1}{2}\left[1 + \cos \frac{2\pi f}{\Delta f}\right], \qquad \frac{-\Delta f}{2} < f < \frac{\Delta f}{2} \qquad (9\text{-}24)$$

As before, $\alpha^2 = \pi^2 T^2/3$. For the normalized spectrum shown, $E = \Delta f$, and β^2 is calculated from (9-5) as

$$\beta^2 = \frac{4\pi^2}{\Delta f}\int_{-\Delta f/2}^{\Delta f/2} f^2 \left(1 + \cos \frac{2\pi f}{\Delta f}\right) df$$

$$= \frac{\pi^2 \Delta f^2}{3} + \frac{\Delta f^2}{2\pi}\int_{-\pi}^{\pi} x^2 \cos x \, dx = \frac{\pi^2 \Delta f^2}{3}\left[1 - \frac{6}{\pi^2}\right] \qquad (9\text{-}25)$$

Values of β^2 and α^2 for $|S(f)|^2$ or $|u(t)|^2$ a Hamming function or cosine function are entered in Table 9-I. For the case of $|S(f)|^2$ being the result of a nonlinear FM function under a rectangular envelope, the entries shown represent the limiting value of β^2 for large time-bandwidth products.

Uniform Pulse Train Envelopes or Frequency Train Spectra

The value of α^2 for the uniform train of N pulses shown in Fig. 9.5 has been calculated by Kelly [8], as

$$\alpha_{\text{p.t.}}^2 = \pi^2 k_t T^2 \qquad (9\text{-}26)$$

where

$$k_t = \frac{1}{3}\frac{N+1}{N-1}\left[1 - \frac{2\tau}{T} + \frac{2N}{N+1}\cdot\frac{\tau^2}{T^2}\right]$$

TABLE 9-I

ACCURACY PARAMETERS OF UNIDIRECTIONAL FM PULSE-COMPRESSION SIGNALS[a]

Function	β^2	α^2	A_{12}	Limiting value of σ^2 for large $T\,\Delta F$ product[c]
A. *Linear FM*: rect (t/T) envelope	$\dfrac{\pi^2\,\Delta f^2}{3}$	$\dfrac{\pi^2 T^2}{3}$	$\dfrac{\pi^2 T\,\Delta f}{3}$	$\dfrac{(\pi T\,\Delta f)^2}{18}\sigma^2_{\min}$
Gaussian envelope: $T_0 = e^{-1}$ pulse width[b] $\Delta f = e^{-1}$ bandwidth	$\dfrac{\pi^2\,\Delta f^2}{4}$	$\dfrac{\pi^2 T^2}{4}\left[1 + \left(\dfrac{4}{\pi T\,\Delta f}\right)^2\right]$	$\dfrac{\pi^2 T\,\Delta f}{4}$	$\left(\dfrac{\pi T\,\Delta f}{4}\right)^2\sigma^2_{\min}$
$\sqrt{\text{Hamming}}$ envelope: $\Delta f = $ base bandwidth	$\dfrac{\pi^2\,\Delta f^2}{3}\left[1 - \dfrac{5.5}{\pi^2}\right]$	$\dfrac{\pi^2 T^2}{3}\left[1 - \dfrac{5.5}{\pi^2}\right]$	$\dfrac{\pi^2 T\,\Delta f}{3}\left[1 - \dfrac{5.5}{\pi^2}\right]$	Not evaluated
$\cos(\pi f/T)$ envelope $\Delta f = $ base bandwidth	$\dfrac{\pi^2\,\Delta f^2}{3}\left[1 - \dfrac{6}{\pi^2}\right]$	$\dfrac{\pi^2 T^2}{3}\left[1 - \dfrac{6}{\pi^2}\right]$	$\dfrac{\pi^2 T\,\Delta f}{3}\left[1 - \dfrac{6}{\pi^2}\right]$	Not evaluated
B. *Parabolic FM*: rect (t/T) envelope	$\dfrac{16\pi^2\,\Delta f^2}{45}$	$\dfrac{\pi^2 T^2}{3}$	$\dfrac{\pi^2 T\,\Delta f}{3}$	$16\sigma^2_{\min}$
C. *Nonlinear FM*: Cosine power spectrum, rect (t/T) envelope	$\dfrac{\pi^2\,\Delta f^2}{3}\left[\dfrac{3\pi^2 - 24}{\pi^2}\right]$	$\dfrac{\pi^2 T^2}{3}$	$\dfrac{\pi^2 T\,\Delta f}{4}$	$100\sigma^2_{\min}$
$(\cos)^2$ power spectrum, rect (t/T) envelope	$\dfrac{\pi^2\,\Delta f^2}{3}\left[1 - \dfrac{6}{\pi^2}\right]$	$\dfrac{\pi^2 T^2}{3}$	$\dfrac{\pi^2 T\,\Delta f}{3}\left[1 - \dfrac{15}{4\pi^2}\right]$	$40\sigma^2_{\min}$
Hamming power spectrum, rect (t/T) envelope	$\dfrac{\pi^2\,\Delta f^2}{3}\left[1 - \dfrac{5.5}{\pi^2}\right]$	$\dfrac{\pi^2 T^2}{3}$	$\dfrac{\pi^2 T\,\Delta f}{3}\left[1 - \dfrac{3.49}{\pi^2}\right]$	$17.7\sigma^2_{\min}$

TABLE 9-I—ACCURACY PARAMETERS OF UNIDIRECTIONAL FM PULSE-COMPRESSION SIGNALS[a]—*continued*

Function	β^2	α^2	A_{12}	Limiting value of σ^2 for large $T\,\Delta F$ product[c]
Time limited uniform pulse train	Depends on modulation	$\pi^2 k_t T^2$	—	—
Band limited uniform frequency "train"	$\pi^2 k_f T^2$	Depends on time function	—	—
cw-FM-cw frequency jump signal	$\dfrac{\pi^2\,\Delta f^2}{3}[3-2k]$	$\dfrac{\pi^2 T^2}{3}$	$\dfrac{\pi^2 T\,\Delta f}{3}\left[\dfrac{3-k^2}{2}\right]$	$\dfrac{4(3-2k)\sigma^2_{\min}}{4(3-2k)-(3-k^2)^2}$
Linear frequency step FM—rect (t/T) envelope	Same as linear FM for large $T\,\Delta f$	$\dfrac{\pi^2 T^2}{3}$	$\dfrac{\pi^2 T\,\Delta f}{3}\left[\dfrac{T\,\Delta f-1}{T\,\Delta f}\right]$	Similar to linear FM

[a] Values of β^2 for symmetrical bidirectional FM functions are the same as obtained with the unidirectional FM that is functionally similar and covers the same band Δf.

[b] $T_0 = T\{1 + (4/\pi T\Delta f)^2\}^{1/2}$. T is the dispersive delay over band Δf.

[c] Note: $(\sigma_\tau^2)_{\min} = N_0/2E\beta^2$, $(\sigma_\phi^2)_{\min} = N_0/2E\alpha^2$.

$$k_t = \frac{1}{3}\frac{N+1}{N-1}\left[1 - \frac{2\tau}{T} + \frac{2N}{N+1}\cdot\frac{\tau^2}{T^2}\right]$$

$$k_f = \frac{1}{3}\frac{N+1}{N-1}\left[1 - \frac{2\Delta f_i}{\Delta f} + \frac{2N}{N+1}\cdot\frac{\Delta f_i^2}{\Delta f^2}\right]$$

$(k_t)_{\max} = 1 = (k_f)_{\max}$ for $N = 2$ and τ or $\Delta f_i \to 0$

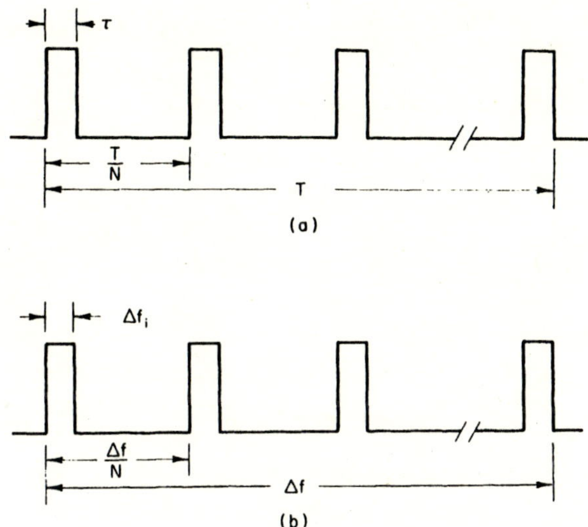

FIG. 9.5   (a) Uniform train of N pulses. (b) Uniform train of N frequency bands.

τ is the component pulse width and T the total pulse train duration. When $\tau = T/N$, the long rectangular pulse envelope is obtained and $k_t = \frac{1}{3}$. This same value of k_t is also obtained for $N \to \infty$ and each component pulse an impulse (i.e., $\tau = 0$).

For a spectrum composed of N uniformly spaced bands having the same form factor as the pulse train in Fig. 9.5, the value of β^2 is

$$\beta^2_{\text{f.t.}} = \pi^2 k_f \, \Delta f^2 \qquad (9\text{-}27)$$

where

$$k_f = \frac{1}{3} \frac{N+1}{N-1} \left[1 - \frac{2\Delta f_i}{\Delta f} + \frac{2N}{N+1} \cdot \frac{\Delta f_i^2}{\Delta f^2} \right]$$

Δf_i is the subinterval bandwidth. For $N = 2$, and $(\Delta f_i/\Delta f) = 0$, $k_f = 1$ and $\beta^2 = \pi^2 \, \Delta f^2$. This is the maximum value of β^2 that can be obtained for a spectrum bandlimited to the interval Δf. Similarly, two narrow pulses separated by T yield $\alpha^2 = \pi^2 T^2$. Again, this is the maximum value of α^2 for a waveform limited to the interval T. These two last cases result in minimum values of theoretical rms range error δ_r and rms velocity error δ_v for equal energy band limited and time limited signals, respectively.

9.3 The Effect of Range-Doppler Coupling on Theoretical Measurement Errors

The range-Doppler coupling factor ρ derived in Chapter 5 can be expressed as $\rho = A_{12}/\beta\alpha$, where

$$A_{12} = \frac{2\pi \int_{-\infty}^{\infty} t\theta'(t)|u(t)|^2\, dt}{\int_{-\infty}^{\infty} |u(t)|^2\, dt} = \frac{2\pi}{2E} \int_{-\infty}^{\infty} t\theta'(t)|u(t)|^2\, dt \qquad (9\text{-}28)$$

or

$$A_{12} = -\frac{2\pi}{2E} \int_{-\infty}^{\infty} f\, \frac{dB(f)}{df}\, |U(f)|^2\, df \qquad (9\text{-}29)$$

and $\theta(t)$ is the phase modulation function, $dB(f)/df = -2\pi\, d\Phi(\omega + \omega_0)/d\omega$, and $-d\Phi(\omega)/d\omega$ is the group time delay of matched filter.

The range-Doppler coupling factor becomes important when range and velocity must be estimated simultaneously. In this case the measurement error variances are given by

$$\sigma_\tau^2 = \frac{N_0}{2E\beta^2(1 - \rho^2)} \qquad (9\text{-}30)$$

$$\sigma_\phi^2 = \frac{N_0}{2E\alpha^2(1 - \rho^2)} \qquad (9\text{-}31)$$

The values of A_{12}, and its effect on the theoretical measurement errors, for unidirectional FM signals considered in the previous section are examined below. The resulting measurement errors, in most cases, will represent limiting conditions for large values of time-bandwidth product. As will be shown for the linear FM, rectangular envelope signal, the assumption concerning a spectrum skirt behavior can lead to acceptable approximations for values of β as a function of $T\Delta f$, but is not adequate for quantitative evaluation of the effect of time-frequency coupling as the compression ratio increases. The following constraint, given by

$$1 - \frac{A_{12}^2}{\beta^2\alpha^2} \geq 0 \qquad (9\text{-}32)$$

must apply to all waveforms in order that (9-30) and (9-31) always be positive. Expression (9-32) is useful as a check for calculated values of β, α, and A_{12}.[1]

---

[1] In addition, as was shown in Chapter 4, $\beta^2\alpha^2 - A_{12}^2 \geq \pi^2$.

Linear FM—Gaussian Envelope

The coupling term for this signal is most easily evaluated from the alternate definition (9-29), where $dB(f)/df = 2\pi Tf/\Delta f$. Thus,

$$A_{12} = -\frac{4\pi}{E} \int_{-\infty}^{\infty} f^2 \frac{2\pi T}{\Delta f} \exp\left[-\frac{8f^2}{\Delta f^2}\right] df = \frac{\pi^2 T \Delta f}{4} \tag{9-33}$$

From (9-12) and (9-13) one obtains

$$1 - \rho^2 = 1 - \frac{A_{12}^2}{\beta^2 \alpha^2} = \frac{16}{(\pi T \Delta f)^2 + 16} \tag{9-34}$$

and the error variances

$$\sigma_\tau^2 = (\sigma_\tau^2)_{\min} \frac{(\pi T \Delta f)^2 + 16}{16} \tag{9-35a}$$

$$\sigma_\phi^2 = (\sigma_\phi^2)_{\min} \frac{(\pi T \Delta f)^2 + 16}{16} \tag{9-35b}$$

where (9-12) and (9-13) are used to calculate the $(\sigma^2)_{\min}$.

For time-bandwidth products greater than 10 (9-35) can be approximated by

$$\sigma^2 = \left(\frac{\pi T \Delta f}{4}\right)^2 \sigma_{\min}^2 \tag{9-36}$$

For the case of the time estimate error variance

$$(\sigma_\tau^2)_{\min} = \frac{4N_0}{2E\pi^2 \Delta f^2} \tag{9-37}$$

and as a result (9-36) yields

$$\sigma_\tau^2 = \frac{T^2}{4} \frac{N_0}{2E} \tag{9-38a}$$

Similarly, one can obtain

$$\sigma_\phi^2 = \frac{\Delta f^2}{4} \frac{N_0}{2E} \tag{9-38b}$$

Equations (9-38a) and (9-38b) show that the actual theoretical rms range and velocity errors for the joint estimation case are independent of the time-bandwidth product. Thus, once a signal duration T is fixed, the theoretical rms range error does not depend on the bandwidth of the frequency modulation, as would be the case for a signal for which $A_{12} = 0$.

It is interesting to note that $T^2/4$ and $\Delta f^2/4$ are the second moments of the projection of the locus of the peaks of the linear FM, Gaussian envelope response function on vertical planes passing through the time and frequency axes, respectively, of the ambiguity plane, as shown in Fig. 9.6. This is a useful relationship that can be applied to evaluate the error variances of other linear FM matched-filter signals.

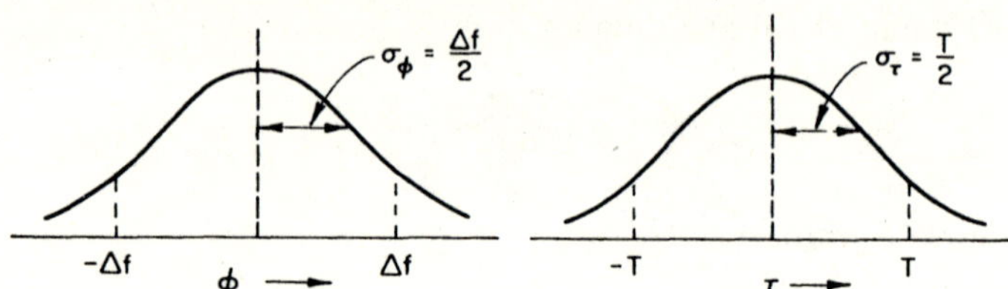

FIG. 9.6   Projection of $|\chi(\tau, \phi)|_{\text{max}}$ for large $T\Delta f$ of Gaussian envelope linear FM on vertical planes through time and frequency axes.

Linear FM—Rectangular Envelope

The range-Doppler coupling factor for this signal can be calculated by noting that $|u(t)| = 1$ for $|t| \leq T/2$, $2E = T$ and $d\theta(t)/dt = 2\pi\,\Delta f\, t/T$. Then from (9-28)

$$A_{12} = \frac{4\pi^2\,\Delta f}{T} \int_{-T/2}^{T/2} t^2\, dt = \frac{\pi^2 T\,\Delta f}{3} \tag{9-39}$$

If the value of β^2 based on the exponential spectrum skirt approximation given by (9-17) is used, the following expressions for the measurement error variances are obtained for large $T\Delta f$:

$$\sigma_\tau^2 = \frac{N_0}{2E\beta^2} \frac{4(T\Delta f)^{1/2}}{3\sqrt{2}} \tag{9-40a}$$

$$\sigma_\phi^2 = \frac{N_0}{2E\alpha^2} \frac{4(T\Delta f)^{1/2}}{3\sqrt{2}} \tag{9-40b}$$

Equations (9-40a) and (9-40b) are of interest, since they do not follow the expected behavior of the σ's increasing with $T\Delta f$ (referenced to $\sigma_{\min}$). These illustrate the problem of attempting to obtain numerical solutions by developing an approximate expression for β when the transmitted pulse envelope has discontinuities. If the result associated with (9-8) (i.e., $\beta = \infty$) is interpreted literally, then there would be zero range measurement error regardless of $T\Delta f$ and $2E/N_0$. This, of course, reflects the shortcomings of adopting the second moment of $|S(f)|^2$ as a measure of bandwidth, and is the reason for the various approximations for β

developed by Manasse [9], Skolnik [2], and in the previous section. As the example of the exponential skirt shows, such an approximation can be chosen to meet the requirements of a good fit to the theoretical spectrum and the constraint given by (9-32), and still not be useful for quantitative evaluation of joint estimation properties. Other measures of bandwidth and time duration were described in Chapter 4, but these have been primarily applied to the resolution of objects in either time or frequency, and have not been related to the estimation problem where the effects of range-Doppler coupling are significant. In the practical sense, β can never be infinite since realizable envelope functions do not exhibit step discontinuities. For most large time-bandwidth signals there is a limiting distribution for $|S(f)|^2$ that can be used to obtain finite values of β and for making comparative evaluations among the measurement accuracies achievable with various waveforms.

In the case of the rectangular envelope linear FM signal a more realistic dependence of the error variances on $T\Delta f$ can be obtained by analogy with the results obtained for the Gaussian envelope case. The projection of the locus of the response function peak along the time and frequency dimensions for the rectangular envelope signal are the triangular functions shown in Fig. 9.7. The second moments of these functions are $T^2/6$ and

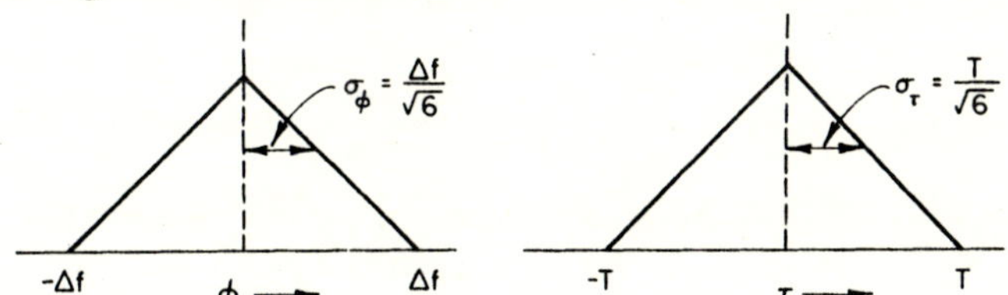

FIG. 9.7 Projection of $|\chi(\tau, \phi)|_{\max}$ for rectangular envelope linear FM signal.

$\Delta f^2/6$, respectively. Following (9-38a) and (9-38b), then one obtains for large $T\Delta f$ (i.e., $T\Delta f > 10$)

$$\sigma_\tau^2 = \frac{T^2}{6}\frac{N_0}{2E} \tag{9-41a}$$

$$\sigma_\phi^2 = \frac{\Delta f^2}{6}\frac{N_0}{2E} \tag{9-41b}$$

Note that $(\sigma_\tau^2)_{\min} = 3N_0/2E\pi^2\,\Delta f^2$ and $(\sigma_\phi^2)_{\min} = 3N_0/2E\pi^2 T^2$; (9-41a) and (9-41b) then become

$$\sigma_\tau^2 = \frac{(\pi T\Delta f)^2}{18}(\sigma_\tau^2)_{\min} \tag{9-42a}$$

$$\sigma_\phi^2 = \frac{(\pi T\Delta f)^2}{18}(\sigma_\phi^2)_{\min} \tag{9-42b}$$

These yield the expected dependence on $T\Delta f$, showing that the actual theoretical rms errors, from (9-41a) and (9-41b), are not improved by the linear FM waveform. The reasoning above can be extended to other linear FM matched-filter waveforms by calculating the locus of the response function peaks, obtaining the second moments of its projection along the time and frequency axes, and then obtaining equivalent expressions to (9-38) or (9-41).

The error variance expressions for the linear FM waveforms rest on the implied assumption that the interval of Doppler shifts is of the same order of magnitude as the signal bandwidth. This is rarely the case, and a Doppler frequency limit $\ll \Delta f$ will lead to greatly reduced range measurement errors for the rectangular envelope linear FM signal. This is considered in Section 9.5.

Parabolic FM (Unidirectional)

The frequency modulation function for the unidirectional parabolic FM waveform is

$$\omega(t) = \omega_0 + \left[\frac{2\pi \Delta f}{T^2}\left(t + \frac{T}{2}\right)^2 - \frac{2\pi \Delta f}{3} \right] \tag{9-43}$$

where $\omega_0 = \omega_l + (2\pi \Delta f / 3)$. The expression in brackets above is $\theta'(t)$. Using this in (9-28) one obtains

$$A_{12} = \frac{\pi^2 T \Delta f}{3} \tag{9-44}$$

This is the same value of A_{12} as that of the linear FM signal.[1] The alternate parabolic FM function shown by the dashed line in Fig. 9-4 results in (9-44) also. Using (9-7) and (9-22) as the large $T\Delta f$ limit for β^2 leads to the following relation for the error variances when the signal time-bandwidth product is large:

$$\sigma^2 \doteq 16\sigma_{\min}^2 \tag{9-45}$$

From (9-45), the error variances of this signal are bounded (relative to the $\sigma_{\min}^2$), and the actual measurement errors will decrease as T or Δf increases.

Nonlinear FM—Cosine-Squared Power Spectrum

The matched filter group time delay for this nonlinear FM signal is

$$-\frac{d\Phi(\omega + \omega_0)}{d\omega} = T\left[\frac{\omega}{\Delta \omega} + \frac{1}{2\pi}\sin\frac{2\pi\omega}{\Delta\omega} \right], \qquad -\frac{\Delta\omega}{2} < \omega < \frac{\Delta\omega}{2} \tag{9-46}$$

---

[1] A large value of $A_{12}^2/(\alpha\beta)^2$ is often cited as being a measure of the linear FM content of a signal. As the parabolic FM example shows (i.e., $A_{12}^2/(\alpha\beta)^2 = 0.933$), a large value of this parameter is a necessary but not sufficient condition for the frequency modulation of a signal to have a predominant linear FM component.

and thus

$$\frac{dB(f)}{df} = \frac{2\pi Tf}{\Delta f} + T \sin \frac{2\pi f}{\Delta f}, \qquad -\frac{\Delta f}{2} < f < \frac{\Delta f}{2} \qquad (9\text{-}47)$$

Using (9-46) and (9-47) in (9-29), one has

$$A_{12} = -\frac{4\pi^2 T}{\Delta f^2} \int_{-\Delta f/2}^{\Delta f/2} f^2 \, df - \frac{T\Delta f}{2\pi} \int_{-\pi}^{\pi} (x^2 \cos x + x \sin x) \, dx$$

$$+ \frac{T\Delta f}{4\pi} \int_{-\pi}^{\pi} \sin^2 x \, dx = -\frac{\pi^2 T\Delta f}{3} \left[1 - \frac{15}{4\pi^2} \right] \qquad (9\text{-}48)$$

From (9-7), (9-25) and (9-48), $1 - \rho^2 = 0.025$, and one obtains for large $T\Delta f$

$$\sigma^2 \doteq 40\sigma_{\min}^2 \qquad (9\text{-}49)$$

Again, the nonlinear FM function results in bounded expressions for the error variances. The larger bound for this waveform, compared to that of the unidirectional parabolic FM, can be attributed to the strong linear FM component in the nonlinear frequency modulation that gives a $(\text{cosine})^2$ power spectrum.

The range-Doppler coupling factors for rectangular envelope nonlinear FM signals that have cosine and Hamming function power spectra are listed in Table 9-I. The bounds on the error variances for these waveforms are also indicated. The expression for dB/df for the Hamming power spectrum signal can be derived from Eq. (3-74), using $k = 0.08$.

9.4 Nonlinear FM and Frequency Jump Signals with Small rms Measurement Errors

Small rms measurement errors can be obtained with certain nonlinear FM and frequency jump waveforms. One such type of FM function is shown in Fig. 9.8. For the case where the swept frequency jump Δf is large the spectrum contribution of the constant frequency sections can be approximated by two narrow band regions separated by Δf. Following the development of Kelly [8] (see (9-27)), the β^2 contribution of this portion of the signal is proportional to $\pi^2 \Delta f^2$. The β^2 contribution of the linear FM segment is proportional to $\pi^2 \Delta f^2/3$. The constants of proportionality are determined by the energy ratios of the different parts of the signal, and the approximate expression for β^2 is

$$\beta^2 = \frac{\pi^2 \Delta f^2}{3}[3 - 2k], \qquad k \leqslant 1 \qquad (9\text{-}50)$$

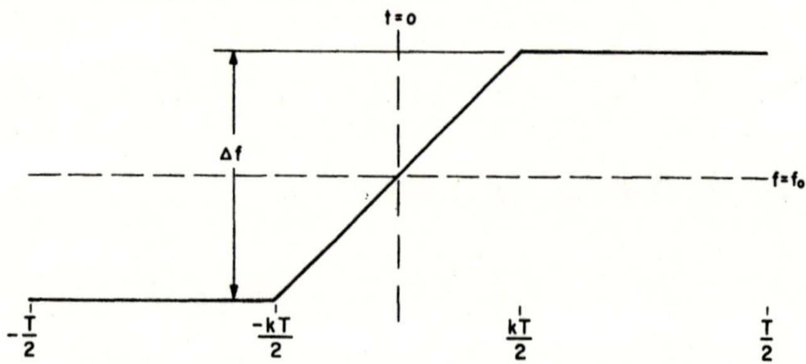

FIG. 9.8   An example of a swept frequency jump function.

Noting that for the first interval $\theta'(t) = -\pi \Delta f$, and for the last interval $\theta'(t) = \pi \Delta f$, the following is obtained for the coupling factor

$$A_{12} = \frac{\pi^2 T \Delta f}{3} \left[\frac{3 - k^2}{2} \right] \tag{9-51}$$

For a rectangular envelope signal (9-50) and (9-51) yield

$$\sigma^2 = \frac{4(3 - 2k)}{4(3 - 2k) - (3 - k^2)^2} \sigma^2_{min}, \quad k < 1 \tag{9-52}$$

For the limiting case of $k = 0$, $\sigma^2 = 4\sigma^2_{min}$. Figure 9.9 illustrates the autocorrelation and crosscorrelation waveforms, as well as the spectrum, for $k = \frac{1}{3}$. For this case $\sigma^2 = 9.1\sigma^2_{min}$. It can be seen that reduction of the measurement errors with this type of nonlinear FM signal results in increased waveform ambiguity. The waveform that programs a frequency jump between $f_0 - \Delta f/2$ and $f_0 + \Delta f/2$ can be made to have zero range-Doppler coupling by timing the frequency jumps as shown in Fig. 9.10(a). This and related power law modulation function signals are discussed in detail by Schweppe [10], who shows that this type of signal also makes the measurements independent of acceleration. Figure 9.10(b) illustrates the autocorrelation function of this waveform. When the time-bandwidth product is large (i.e., Δf much greater that the inverse of the signal time duration) this function yields a time limited approximation to the optimum range accuracy signal, discussed by Kelly, that is bandlimited over the interval Δf for which $\beta^2 = \pi^2 \Delta f^2$.

In general, the nonlinear FM pulse-compression signals yield measurement errors that are bounded (as a function of $T \Delta f$) and much less than the theoretical measurement errors of the linear FM signals. However, the nonlinear FM matched-filter signals that have the lowest theoretical measurement errors tend to have the disadvantages illustrated by the waveforms of Figs. 9.9 and 9.10.

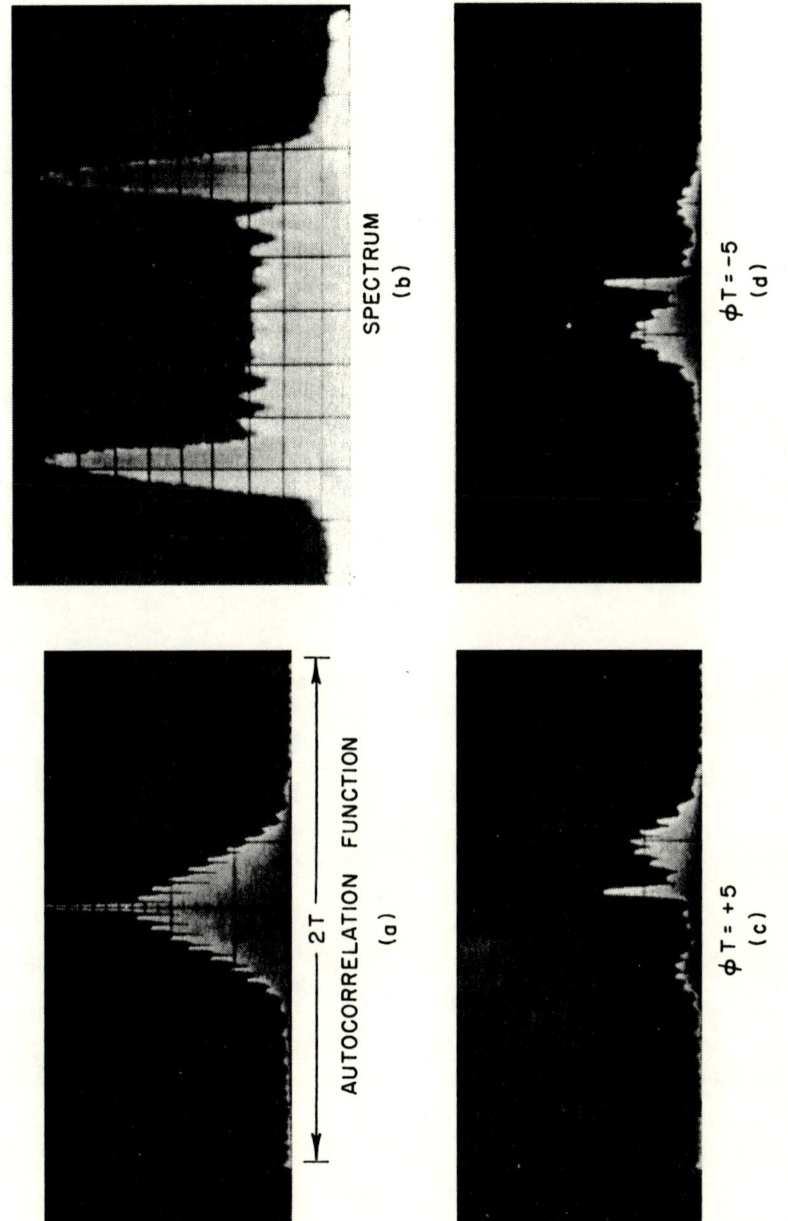

Fig. 9.9 Waveforms and spectrum for FM function of Fig. 9.8, $k = \frac{1}{3}$.

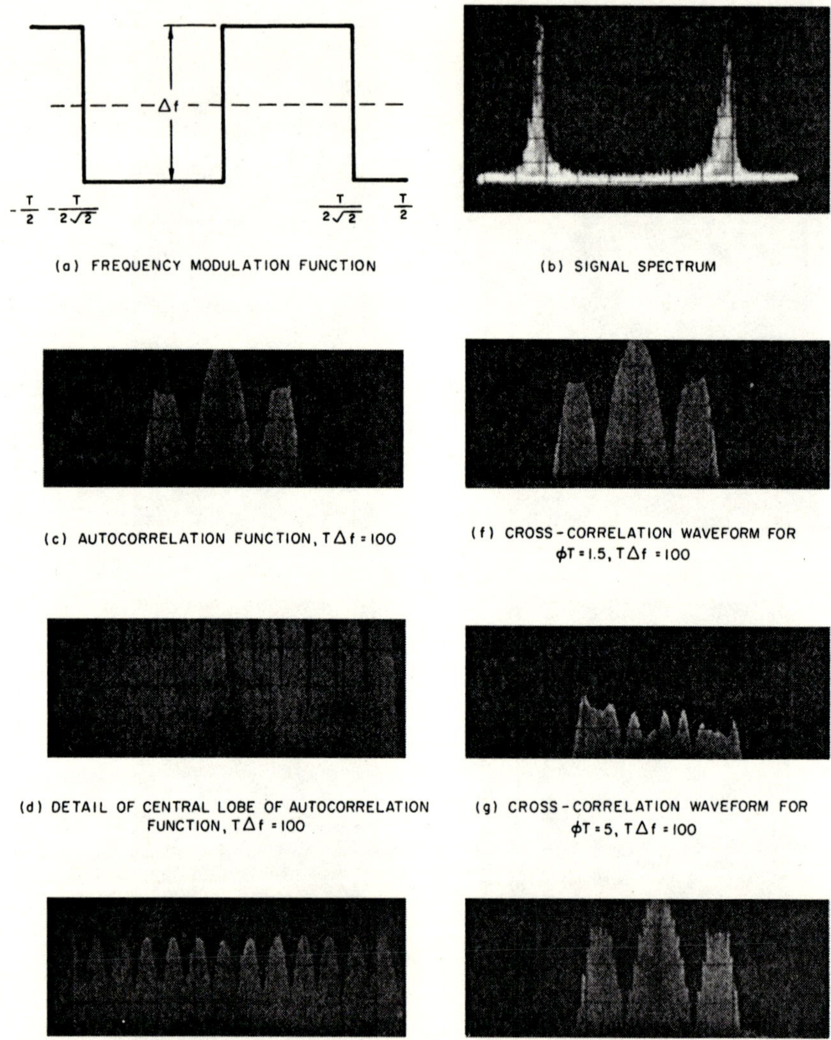

FIG. 9.10   Waveforms for three frequency jumps over the band Δf.

As noted previously, the estimation error variances are related to the behavior of the ambiguity function about the origin. Those signals having the smaller error variances exhibit ambiguity functions that fall off more rapidly in this region. This is illustrated by Fig. 9.11, in which are plotted

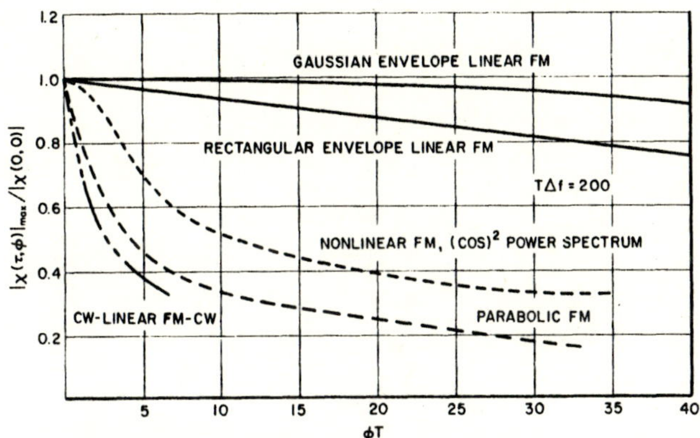

Fig. 9.11   Behavior of response function peak as a function of Doppler shift for unidirectional FM pulse-compression signals.

the maximum values of the response functions of several signals as a function of normalized Doppler shift ϕT. This maximum value coincides with the peak of the shifted matched-filter output in most cases. This shows that the comparative data tabulated in Table 9-I are in agreement with the respective ambiguity function properties.

9.5 Considerations for Improved Measurement Accuracy with Unidirectional FM Waveforms

It was shown in the previous section that, in the framework of the classical estimation system, the measurement errors of the linear FM signals referenced to the minimum error increased without bound as the time-bandwidth product of the signal increased. Signals that contain nonlinear FM functions approached a finite limit for rms measurement errors as $T \Delta f$ increased. These results applied to the extraction of information based on a single received signal. In the context of the classical estimation system (i.e., bank of matched filters as in Fig. 9.1) these measurement errors can be improved by performing the estimate based on the reception of a larger number of signals. If N pulses reflected from an object are used for the measurement then the error variances become [11]

$$(\sigma^2)_N = \sigma^2/N \tag{9-53}$$

This assumes that the parameter being measured, and the signal-to-noise ratio $2E/N_0$ per pulse, do not change over the interval of N pulses; also that the observations of each of the N pulses are statistically independent with normally distributed error curves. The resulting rms errors for several of the waveforms considered in the previous sections are plotted in Fig. 9.12 as a function of N. These represent pessimistic upper bounds, since more practical measures of measurement accuracy can be postulated for some undirectional FM signals, most notably the linear FM waveform.[1] These measures are based on a system instrumentation designed on principles other than that of the classical parameter estimation system depicted in Fig. 9.1. Two examples of this are discussed below.

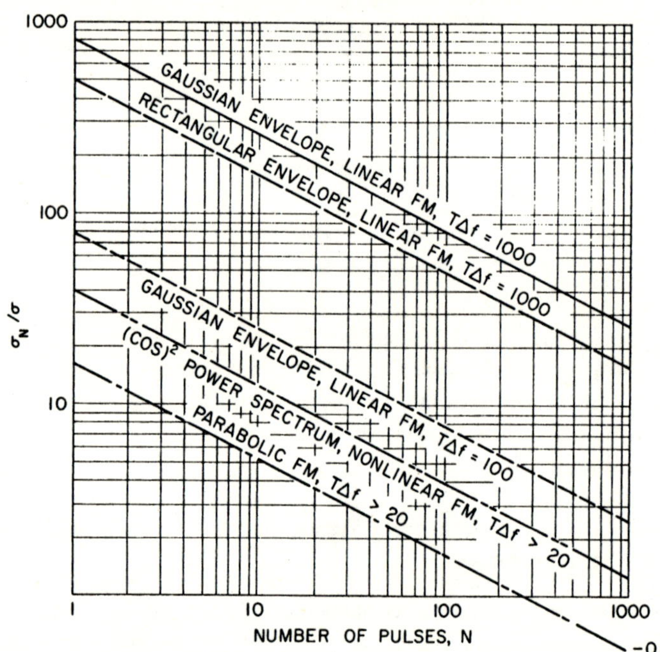

FIG. 9.12   Effect of N observations on rms measurement errors of unidirectional FM pulse-compression signals.

Example 1: Linear FM Range Prediction Property

If the linear FM signal has a positive FM slope, a Doppler shift ϕ at the input to the matched filter (negative delay slope) causes a relative time shift at the matched filter given by $t_s = -T\phi/\Delta f$. Since $\phi = 2vf_0/c$ ($v =$ velocity

---

[1] For the case of velocity measurement (9-53) does not reflect the additional advantage of the longer coherent observation time (i.e., signal duration) in which the N pulses are received. (9-53) is essentially derived from considering the increased energy factor $2NE$ in N received signals.

of object, f_0 = transmitter carrier frequency and c = velocity of light), the time shift at the matched-filter output is described by

$$t_s = -2Tvf_0/c\,\Delta f \tag{9-54}$$

and the associated range error ΔR is

$$\Delta R = ct_s/2 = -vTf_0/\Delta f \tag{9-55}$$

Taking ΔR as the incremental distance covered by the object in a time increment Δt, or $\Delta R = -v\,\Delta t$, Klauder et al.[12] show that for the case of the positive slope linear FM signal an ensemble of outputs from the matched filter correctly locates the positions of the moving objects Δt seconds *after* the time of true reflection, where

$$\Delta t = Tf_0/\Delta f \tag{9-56}$$

The relationship given by (9-56) is a constant in $\Delta R = -v\,\Delta t$, independent of velocity.

Rihaczek [13] points out that this interpretation leads to a minimum range measurement error variance given by (9-1) referenced to the object's position Δt seconds after actual time of reception of the signals, eliminating the velocity uncertainty from the range estimate. This is based on the assumption that the velocities of the objects of interest are constant over the interval Δt, and that the Doppler shift ϕ is a small enough fraction of the signal bandwidth so that negligible pulse spreading distortion results. If these hold true, then (9-56) provides the basis for accurate ranging with the linear FM signal based on the predicted range position, since the prediction time Δt can be accounted for in the range tracking computer. In this application velocity would be determined as a by-product of the range tracking program, deriving range-rate by observing the changes in the object's position over a space of N pulses.

If the transmitted signal has a negative FM slope then (9-55) and (9-56) are interpreted as defining the true position of an object Δt seconds *before* the time of true reflection. In a number of applications it is more useful to determine where an object is going rather than where it has been.

Example 2: Range Location Error with Only a priori *Knowledge of Doppler Interval*

If the Doppler shift is known (or assumed) to be uniformly distributed in the interval $-\Delta\phi/2$ to $\Delta\phi/2$, then from the discussion above the location of a compressed-pulse signal can be uniformly distributed over an interval about its true position defined by

$$\Delta t_s = T\,\Delta\phi/\Delta f \tag{9-57}$$

Assuming that the Doppler interval $\Delta\phi$ is small compared to Δf, so that there is negligible amplitude variation of the compressed-pulse signal for any given Doppler shift, then for larger signal-to-noise ratios the uniform distribution of time positions given by (9-57) yields an *a priori* variance about the true target time delay given by

$$\sigma_{t_d}^2 = \frac{\Delta t_s^2}{12} = \frac{T^2(\Delta\phi)^2}{12\,\Delta f^2} \tag{9-58}$$

and an rms range location uncertainty of

$$\delta_r = \tfrac{1}{2} c\sigma_{t_d} = \frac{cT\,\Delta\phi}{4\sqrt{3}\,\Delta f} \tag{9-59}$$

It can be seen from (9-59) that the range location measurement error decreases as Δf increases, and thus can be within acceptable limits if Δf is large enough. Range-rate would be measured as outlined in Example 1, with the possibility of using this information to reduce the *a priori* range error. Both of these examples point out instances in which the linear FM signal can have acceptable range measurement errors. In many situations the use of this waveform may be preferable to the use of one that exhibits a theoretical minimum measurement error, but produces interfering self-clutter and ambiguous responses.

9.6 Conditions for Minimum-Error Joint Estimates of Range and Velocity

The unidirectional FM pulse-compression waveforms considered in Sections 9.3 and 9.4 illustrated the effect of the range-Doppler coupling factor, (9-28) or (9-29), on increasing the theoretical rms measurement errors when several parameters are being jointly estimated, or when a single parameter is being estimated without *a priori* knowledge of the values of other parameters that influence the measurement. Minimum rms measurement errors occur when this coupling term is zero. The most obvious example of this is the signal in which $\theta'(t) = 0$, or the constant frequency pulse. However, the range and velocity measurements cannot be simultaneously accurate with this type of signal, although the measurement accuracy can be improved by making the measurement over a number of signal returns. In order to obtain good measurement accuracy in both range and velocity one must consider those classes of large time-bandwidth product signals for which

$$\frac{\pi}{E}\int_{-T/2}^{T/2} t\theta'(t)|u(t)|^2\,dt = 0 \tag{9-60}$$

Assuming that $|u(t)|$ is an even function (usually rectangular), this condition is met when

$$\int_{-T/2}^{0} t\theta'_1(t)\,dt = -\int_{0}^{T/2} t\theta'_2(t)\,dt \tag{9-61}$$

or

$$\theta'_1(-t) = \theta'_2(t) \tag{9-62}$$

(9-62) describes a frequency modulation characteristic that is an even function of time. The property of this type of signal to yield minimum-error estimates has been noted by several authors [8,9,14], and the V-FM and quadratic FM waveforms are specific examples that meet this criterion. The condition imposed by (9-60) for which (9-62) represents a special case, can be extended to the more general types of nonsymmetrical bidirectional FM waveforms. Several examples of these are shown in Fig. 9.13. The

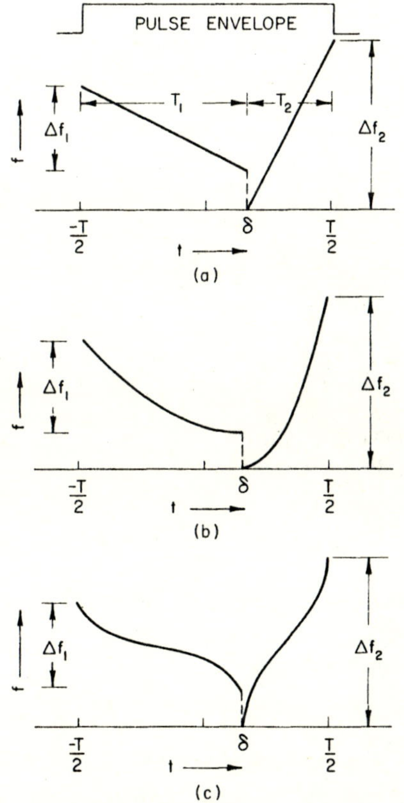

FIG. 9.13   Nonsymmetrical bidirectional FM functions.

envelope function $|u(t)|$ of these waveforms is rectangular and, for the moment, the two segments of each example are functionally identical (i.e., linear, parabolic, etc.) and are assumed to have the same mean frequency. The zero coupling condition is now

$$\frac{\pi}{E}\left[\int_{-T/2}^{\delta} t\theta_1'(t)\,dt + \int_{\delta}^{T/2} t\theta_2'(t)\,dt\right] = 0 \tag{9-63}$$

where $\theta_1'(t)$ and $\theta_2'(t)$ are referenced to the spectrum mean frequency. The relationship among $\Delta f_1, \Delta f_2, T$, and δ, as illustrated in Fig. 9.13, that satisfies (9-63) is derived below for the two linear FM segment modulation function shown in Fig. 9.13(a). The linear FM function of each segment, referenced to their respective mean frequencies are

$$\theta_1'(t) = \frac{4\pi\,\Delta f_1(t - \delta)}{(T + 2\,\delta)} - \pi\,\Delta f_1 \tag{9-64a}$$

$$\theta_2'(t) = \frac{4\pi\,\Delta f_2(t - \delta)}{(T - 2\,\delta)} - \pi\,\Delta f_2 \tag{9-64b}$$

The contribution to A_{12} by $\theta_1'(t)$ is

$$(A_{12})_1 = \frac{8\pi^2\,\Delta f_1}{T(T + 2\,\delta)}\int_{-T/2}^{\delta} (t - t\,\delta)\,dt - \frac{2\pi^2\,\Delta f_1}{T}\int_{-T/2}^{\delta} t\,dt \tag{9-65a}$$

$$= -\frac{\pi^2\,\Delta f_1}{12T(T + 2\,\delta)}\left[T^3 + 6T^2\,\delta + 12T\delta^2 + 8\,\delta^2\right]$$

$$= -\frac{\pi^2\,\Delta f_1}{12T}(T + 2\,\delta)^2 \tag{9-65b}$$

Similarly, using (9-64b) in equation (9-28), the coupling term contribution for the second linear FM segment is

$$(A_{12})_2 = \frac{\pi^2\,\Delta f_2}{12T}(T - 2\,\delta)^2 \tag{9-66}$$

Let the duration of the first segment $(T/2 + \delta) = T_1$, and that of the second segment $(T/2 - \delta) = T_2$, then one obtains from (9-65b) and (9-66) the following condition for a zero coupling factor

$$\Delta f_2/\Delta f_1 = (T_1/T_2)^2 \tag{9-67}$$

From (9-67) it can be seen that the time-bandwidth products of the segments are inversely proportional to the ratio of their durations. Considering as a second example the parabolic FM segments waveform shown in Fig. 9.13(b), the respective FM functions are

$$\theta_1'(t) = 8\pi\,\Delta f_1\left[\frac{t-\delta}{T+2\delta}\right]^2 - \frac{2\pi\,\Delta f_1}{3}, \qquad -T/2 < t < \delta \qquad (9\text{-}68\text{a})$$

$$\theta_2'(t) = 8\pi\,\Delta f_2\left[\frac{t-\delta}{T-2\delta}\right]^2 - \frac{2\pi\,\Delta f_2}{3}, \qquad \delta < t < T/2 \qquad (9\text{-}68\text{b})$$

or

$$\theta_3'(t) = \frac{2\pi\,\Delta f_2}{3} - 8\pi\,\Delta f_2\left[\frac{t-T/2}{T-2\delta}\right]^2, \qquad \delta < t < T/2 \qquad (9\text{-}68\text{c})$$

where $\theta_3'(t)$ is for the FM function similar to that shown by the dotted line in Fig. 9.4. Evaluating the coupling terms from the above yields the same values as for the linear FM segments (9-65b) and (9-66), meeting the same condition given by (9-67). In general, the coupling term of a segment shown in Fig. 9.13 can be obtained by the relation

$$(A_{12})_{\text{segment}} = \frac{A_{12}}{4T^2}(T \pm 2\,\delta^2) \qquad (9\text{-}69)$$

where A_{12} is the coupling factor for a functionally similar unidirectional FM signal in which $|u(t)|$ is rectangular, and the same bandwidth is swept out over the total duration T. Using (9-69), the zero range-Doppler coupling condition for various combinations of linear and nonlinear FM functions can be developed using Table 9-I. Figure 9.14 illustrates a nonsymmetrical bidirectional FM function that meets the requirement for uncoupled range and velocity estimates. Figure 9.15 shows the autocorrelation and off-Doppler cross-correlation waveforms for the FM function shown in Fig. 9.14. It can be deduced from these waveforms that the ambiguity

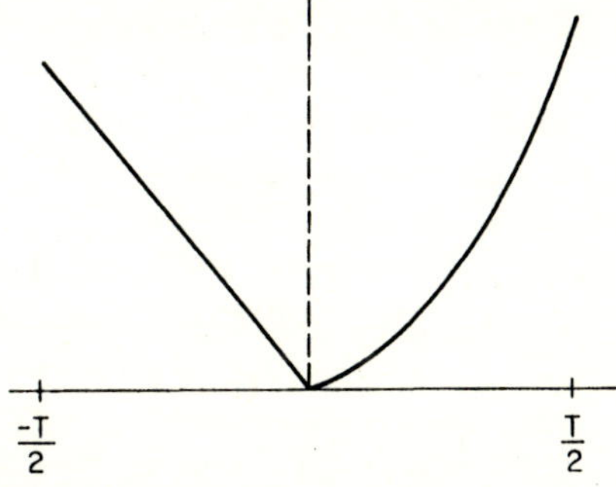

FIG. 9.14   Nonsymmetrical V-FM function, combination linear and parabolic FM.

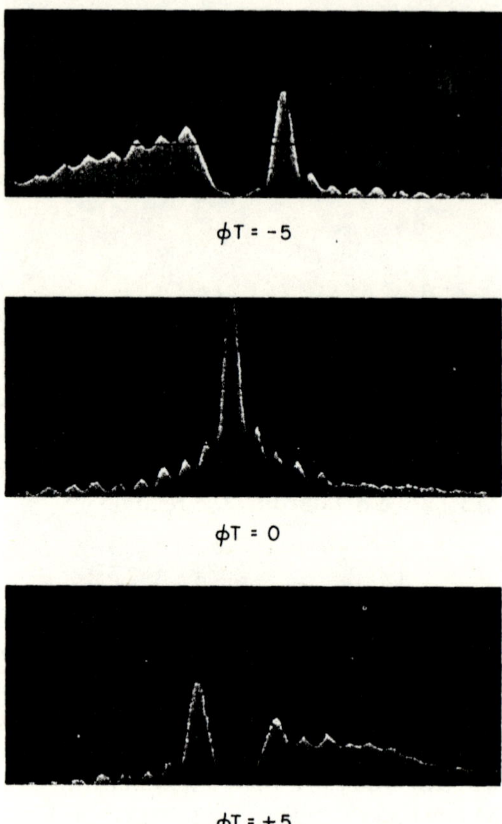

FIG. 9.15   Waveforms for FM function of Fig. 9.14.

functions of the nonsymmetrical bidirectional FM signals have a form of odd symmetry. For sufficiently large signal-to-noise ratios this interesting property can be made use of to distinguish between on-Doppler and off-Doppler signals in a particular channel (assuming a nondense reflector environment). If a single matched filter is being used to obtain radial velocity information on an isolated moving object, then the separation of the signal lobes is a measure of the magnitude of the velocity. The configuration of the lobes, as indicated in Fig. 9.15, indicates the vector sense of the radial velocity. The linear V-FM signal has been suggested as a velocity sensing signal for isolated objects. The use of a nonsymmetrical bidirectional FM signal offers an opportunity to extract additional information under certain conditions.

The effect on the zero-range-Doppler coupling condition when the FM segments have different mean frequencies can be derived using the general relationships employed in the examples of this chapter. For this case the zero coupling condition becomes

$$\frac{K_1 T \Delta f_1 T_1^2}{4T^2} - \frac{K_2 T \Delta f_2 T_2^2}{4T^2} + \frac{2\pi^2 f_s T_1 T_2 T}{T^2} = 0 \qquad (9\text{-}70)$$

where f_s is the spread between the mean frequencies and K_1 and K_2 are spectrum form factors that can be determined from Table 9-I for the waveforms discussed in this chapter. Equation (9-70) reduces to (9-67) for $f_s = 0$ and the FM segments functionally similar (i.e., $K_1 = K_2$).

If f_s is small its effect is negligible. For $f_s \gg \Delta f_1$ and Δf_2 then the situation approaches that of the two narrow frequency regions separated by a large interval discussed in Section 9.4. Thus, one can determine that the effect of f_s on theoretical measurement errors is not large, although it will have a critical effect on the calculated value of β^2 and the creation of waveform ambiguities. Finally, it should be noted that if the two segments of the waveform are transmitted simultaneously rather than sequentially then f_s has no effect on the coupling factor. For both segments of long duration and of constant frequency this latter condition approaches the optimum range accuracy signal confined to a bandlimited region, as discussed by Kelly [8]. For this case the FM functions for each segment will have small effect on the range measurement accuracy, this being a function of f_s for $f_s \gg \Delta f_1$ and Δf_2. However, the FM functions can contribute to reduced ambiguity for Doppler shifted signals that are processed through the matched filter, although this would not be a dramatic reduction for most cases.

9.7 Range-Doppler Coupling for Discrete Coded Waveforms

Various members of the class of discrete coded waveforms were described in Chapter 8. It is of interest to determine the effect of the form of the discrete code sequence on the theoretical measurement accuracies, as reflected in the value of the range-Doppler coupling factor. It was shown in Chapter 8 that the binary phase coded signals have ambiguity functions that approximate the so-called thumbtack shape for the case of large time-bandwidth signals. It would appear from the considerations of earlier sections that the factor A_{12} for these signals would automatically equal zero as a result of $\theta'(t) = 0$ in each sub-interval of the coded waveform. This is not necessarily a safe assumption in all cases, since this could also be applied to the Frank polyphase coded signal, which grossly resembles the linear FM signal, and would be expected to have a relatively large coupling factor.

An approach to the treatment of discrete phase coded signals is shown in Fig. 9.16(a) for a binary coded waveform. The assumption of a finite switching time from one phase state to the other gives rise to the impulse-like FM function shown in Fig. 9.16(b). Based on this model the coupling for a maximum length sequence code can be written:

$$A_{12} = \frac{4\pi^2}{T\delta} \sum_n (-1)^{|n|} \int_{t_n-(\delta/2)}^{t_n+(\delta/2)} t\,dt \tag{9-71}$$

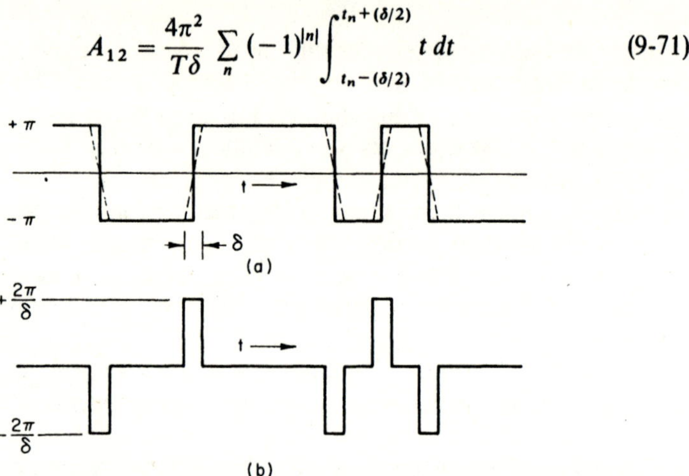

(a)

(b)

FIG. 9.16 (a) Binary phase code with finite switching time (dotted lines). (b) Equivalent impulse FM function.

where $-(2^{k-1} - 1) < n < 2^{k-1} - 1$, and $2^k - 1$ is the number of code elements and $2^k - 2$ the number of phase reversals. As k increases the value of A_{12} given by (9-71) becomes small, and this form of discrete coded waveform meets the requirements for applications where range and velocity measurements must be uncoupled, or independent. From a rigorous viewpoint, this same conclusion is obtained from the intrinsic property of

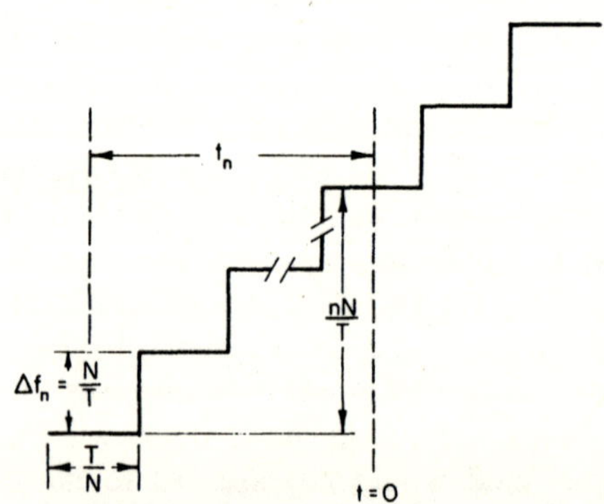

FIG. 9.17 Linear step FM function.

symmetry regarding $|\chi(\tau, \phi)|$ for all binary codes given by Eqs. (8-15) and (8-17) of the previous chapter.

Figure 9.17 illustrates the parameters for analysis of the range-Doppler coupling factor of the linear step FM waveform. The general expression for the frequency modulation is

$$\theta'_n = 2\pi(f_0 + n\,\Delta f_n) \tag{9-72}$$

and the location of the t_n is given by $t_n = nT/N$, where $T = $ total duration, N is the number of frequency steps and

$$-(N - 1)/2 < n < (N - 1)/2, \qquad N \text{ odd or even}$$

$$\Delta f_n = N/T \qquad \text{and} \qquad \Delta f = N\,\Delta f_n = N^2/T$$

The time-bandwidth product of this signal is N^2. The range-Doppler coupling factor is given by

$$A_{12} = \frac{2\pi}{T}\sum_n \int_{t_n-(T/2N)}^{t_n+(T/2N)} \frac{2\pi n N t_n\,dt}{T} = \frac{4\pi^2 N}{T}\sum_n \frac{n t_n T}{N} \tag{9-73}$$

Note that $t_n = nT/N$ and $\Delta f = N^2 T$; then (9-73) becomes

$$A_{12} = \frac{\pi^2 T \Delta f}{3}\frac{12}{N^3}\sum_n n^2 \tag{9-74}$$

The summation in (9-74) can be shown to be equal to $(N^3 - N)/12$ so that the coupling factor can finally be written as

$$A_{12} = \frac{\pi^2 T\,\Delta f}{3}\left[\frac{N^2 - 1}{N^2}\right] = \frac{\pi^2 T\,\Delta f}{3}\left[\frac{T\,\Delta f - 1}{T\,\Delta f}\right] \tag{9-75}$$

Equation (9-75) approaches the value of the coupling factor for the linear FM signal as N^2, or $T\,\Delta f$, increases. Assuming that the value of β^2 is the same for both signals, then it would be expected from (9-75) that the theoretical measurement accuracies of the step FM signal would be slightly better than that of the linear FM signal. This can be related in physical terms to the behavior of the ambiguity function of each signal about its peak point. From Fig. 9.18 one can see that the step FM response function starts to fall off more rapidly as a function of Doppler shift. As shown in Chapter 8, this is caused by transfer of energy from the main signal at the matched-filter output into the cyclic sidelobes located at $\pm N/\Delta f$ units on either side of the main signal.

The Frank polyphase code offers a third example of a discrete coded waveform. The phase sequence of this signal is shown in Fig. 9.19(a). Using the same type of switching time model shown in Fig. 9.16(a), an equivalent impulse FM function for this polyphase code is shown in Fig.

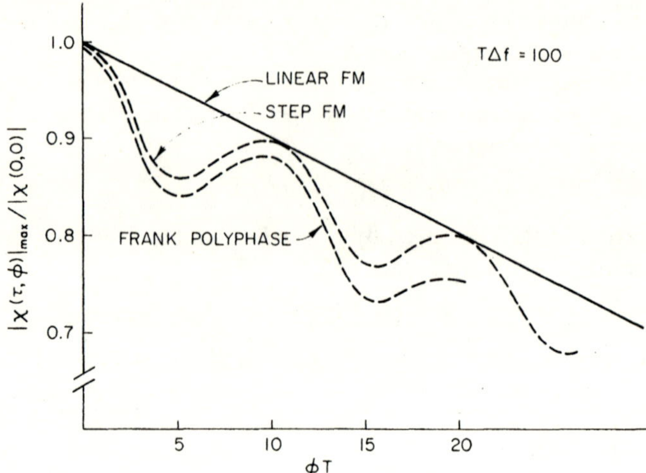

FIG. 9.18 Behavior of response function peaks for linear FM, step FM, and Frank polyphase signals.

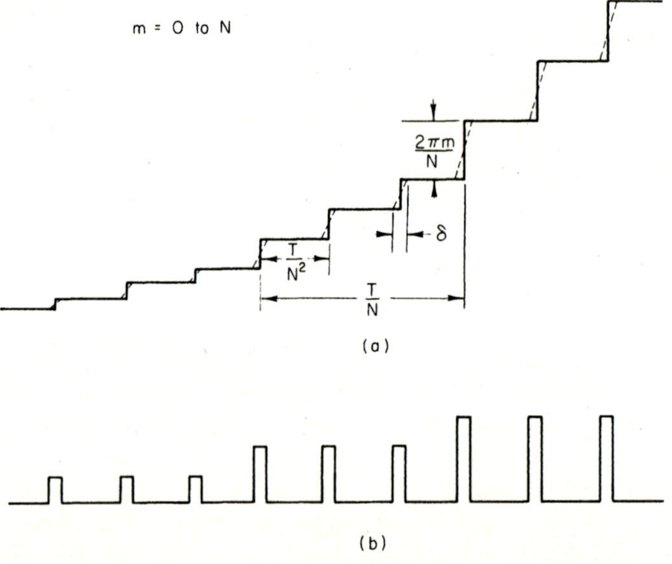

FIG. 9.19 (a) Frank code phase progression. (b) Equivalent step impulse FM function.

9.19(b). Assuming the interval δ and the total signal duration remain fixed as the number of code elements increases, the function of Fig. 9.19(b) approaches the step FM signal. On this basis, the same general behavior would be expected. The discussion in Chapter 8 showed that the Doppler shifted matched-filter outputs for the polyphase code have a greater number of Doppler induced cyclic sidelobes. This results in a greater rate of change of the magnitude of the ambiguity function about the origin of the ambiguity plane as a function of Doppler shift. This is indicated in Fig. 9.18. As a practical matter, one would expect little difference in actual measurement accuracies among the linear FM, step FM and Frank coded waveforms, with the discussion of Section 9.5 also being applicable to the two latter signals.

A great number of variations of discrete frequency and phase coded waveforms are possible. Using the methods outlined in this section the range-Doppler coupling properties can be obtained by either exact or approximate analysis. However, as many of the examples in this chapter have demonstrated, the waveforms that yield minimum measurement errors often have large levels of zero-Doppler or off-Doppler ambiguity associated with them. Thus, designing a radar signal for the criterion of minimum error in the simultaneous measurement of range and velocity does not in all cases result in the most useful signal when all the projected applications of the radar system are taken into consideration.

A final word is appropriate concerning the use of the criterion of minimum measurement errors (as discussed here) as a waveform design consideration. This is only one of several evaluation criteria that can be applied to various types of radar signals. This particular criterion can be related to the behavior of the ambiguity function about its peak value at the origin of the time-frequency ambiguity plane (see Chapters 4 and 5). Thus, it conveys little information about the ambiguity function response at other locations in the plane. Several of the examples discussed in this chapter have shown that waveforms that exhibit a minimum measurement error quality can also have large zero-Doppler or off-Doppler ambiguity responses, and would not be acceptable for use in any but a single target environment. In a multiple scatterer environment the radar designer will wish to balance the need for accurate single pulse measurement against the need to minimize the interference among several signals of interest, as represented by their respective responses at the matched-filter output. The multiple and dense scatterer problem is discussed in the next chapter.

In any event, no absolute judgement can be attached to the fact that a radar signal does or does not possess a "good" measurement accuracy capability according to the Cramér-Rao, or related, conditions. As a matter of record, some of the least favorable signals for general radar use are those that have theoretically optimum measurement accuracies. On the other hand, signals that may be judged poorly in the light of the Cramér–Rao condition are, when placed in the context of practical radar considerations, much more suitable; the discussion of the linear FM signal range measurement accuracy in Section 9.5 is one such example. The choice of a waveform for a particular application may be subject to many compromises. The measurement accuracy criterion can provide useful data for relative evaluations among prospective signals. The other pertinent factors must be added to this before an intelligent choice can be made.

REFERENCES

1. P. M. Woodward, "Probability and Information Theory, with Applications to Radar," Pergamon Press, Oxford, 1953.
2. M. I. Skolnik, Theoretical accuracy of radar measurements, *IRE Trans.* **ANE-7**, 123–129 (1960).
3. E. N. Fowle, D. R. Carey, R. E. Vander Schuur, and R. C. Yost, A pulse-compression system employing a linear FM Gaussian signal, *Proc. IEEE*, 304–312 (1963).
4. F. P. Callahan, Exposition of the theory of enhanced accuracy radar, General Atronics Co., Philadelphia, Rept. 897–98–1391 (October, 1959).
5. D. Richman, Superresolution, *Proc. Pulse-Compression Symposium*, Rome Air Development Center, Griffiss A.F. Base, New York, RADC TR–59–161, (September, 1959).
6. E. L. Key, E. N. Fowle, and R. D. Haggarty, A method of pulse compression employing nonlinear frequency modulation, Mass. Inst. Technol., Lincoln Lab., Lexington, Massachusetts, Tech. Rept. 207 (August, 1959).
7. C. E. Cook, A class of nonlinear FM pulse-compression signals, *Proc. IEEE (Correspondence)* **52**, 1369–1371 (1964).
8. E. J. Kelly, The radar measurement of range, velocity, and acceleration, *IRE Trans.* **MIL-5**, 51–58 (1961).
9. R. Manasse, Range and velocity accuracy from radar measurements, Mass. Inst. Technol., Lincoln Lab., Lexington, Massachusetts, Group Rept. 312–26 (February, 1955).
10. F. C. Schweppe, Radar frequency modulations for accelerating targets under a bandwidth constraint, *IRE Trans.* **MIL-9**, 25–32 (1965).
11. H. Crámer, "Mathematical Methods of Statistics." Princeton Univ. Press, Princeton, New Jersey, 1946.
12. J. R. Klauder, A. C. Price, S. Darlington, and W. J. Albersheim, The theory and design of chirp radars, *Bell Syst. Tech. J.* **39**, 745–807 (1960).
13. A. W. Rihaczek, Range accuracy of chirp signals, *Proc. IEEE (Correspondence)* **53**, 412–413 (1965).
14. P. Bello, Joint estimation of delay, Doppler, and Doppler rate, *IRE Trans.* **IT-6**, 330–341 (1960).

CHAPTER 10

Waveform Design Criteria for Multiple and Dense Target Environments

10.1   Introduction

The preceding chapter presented data concerning the effect of the design factors (i.e., bandwidth, duration, modulation function) of a number of pulse-compression matched-filter signals on the range and velocity measurement accuracies that can be theoretically obtained with these waveforms. However, the theory of parameter estimation, as applied to radar system analysis, is generally based on the assumption that the extraction of the information about a single object is not influenced by the presence of other reflecting objects in the same vicinity. The examples discussed in the previous chapter showed that many different types of matched-filter waveforms can meet the conditions for uncoupled, or independent, measurement of the range and velocity parameters. These signals can be classified as Doppler sensitive waveforms, since their ambiguity functions usually have a well-defined peak about the origin of the ambiguity plane. In some cases the ambiguity functions of this broad class of signals approximate the thumbtack shape (see Fig. 4.8), which is in itself an approximation to the ideal, but unobtainable, spike ambiguity function. The use of these Doppler sensitive large time-bandwidth signals can result in the spreading of signal generated self-clutter energy into range cells far removed from the location of the actual signal, thus having a net harmful effect on detection, resolution, and parameter measurement of other signals not located at the same range as the interfering signal. The question may then be asked, what should the waveform design criteria be when the radar must operate in a dense or semidense environment of scatterers? This is an area of continuing interest in the field of waveform design. Many of the suggested criteria are based on studies addressed to specific applications and depend strongly on the assumed statistical properties of the dense scattering environment (terrain, ocean, chaff, etc.), as well as on the extent and distributions in range and velocity that must be examined by the radar. The material of this chapter discusses in basic terms some of the situations that may confront the radar designer in his choice of a waveform or waveforms for a particular system.

One of the most severe problems that is often encountered, and in which the waveform design factors can markedly affect the operational results of the system, is that of a desired signal obscured by the overlapping returns from many scatterers, or clutter. For the general case the solution to this problem can be expressed in terms of the ambiguity function properties of the radar signal. The desired operational results and the constraints imposed by the ambiguity function, as discussed in Chapter 4, may often be in conflict. Consequently, the radar designer may be forced to a compromise solution in the choice of a waveform.

10.2 Waveform Considerations for Extended Distributions of Semi-Isolated Moving Scatterers

A basic example is presented in this section to indicate the possible interplay between the radar environment and the various factors that can enter into the design of the radar waveform. An ensemble of semi-isolated reflecting objects is assumed in which the mean spacing between the objects is less than the range interval corresponding to the transmitted pulse duration T (i.e., uncompressed-pulse width), but the overall range extent of all the objects of interest is many times that which corresponds to the duration T. The objects are assumed to be in motion with respect to each other, and it is desired that the system be able to perform detection and unambiguous measurement of range and velocity on the several objects. From the latter constraint it can be readily seen that the ambiguities of the uniform pulse train signal are not tolerable, thus eliminating one type of signal from consideration. The required signal should therefore have a continuous spectrum envelope (neglecting the fine structure caused by the repetition rate), as well as a continuous time envelope over the interval T (i.e., rectangular, Hamming weighted, etc.). If relative velocity differences are to be measured the system designer has the choice of using one of the Doppler sensitive waveforms, such as V-FM, parabolic FM, binary phase code sequence, etc., or a waveform that is not markedly sensitive to Doppler shift, except for a time displacement (linear FM, tangent FM).[1] In this latter case other methods are relied on to extract the velocity information than that of observing the outputs from a bank of matched filters. If the

---

[1] Doppler sensitivity, used in this context, refers to the effect of a Doppler shift on the time distribution of the energy of a large time-bandwidth signal at the matched-filter output.

further constraint is added that the range of Doppler frequency shifts of the received signals is relatively small compared to the signal bandwidth (a not unrealistic assumption for wideband signals), then all signals at the matched-filter output will be at essentially maximum energy, although generally of different form depending on the type of signal and the magnitudes of the Doppler shifts involved. For purposes of illustration it is assumed that the possible Doppler sensitive waveform has an ideal thumbtack ambiguity function for which the thumbtack pedestal is uniformly spread over the correlation interval $2T$. Figure 10.1 shows the two extremes

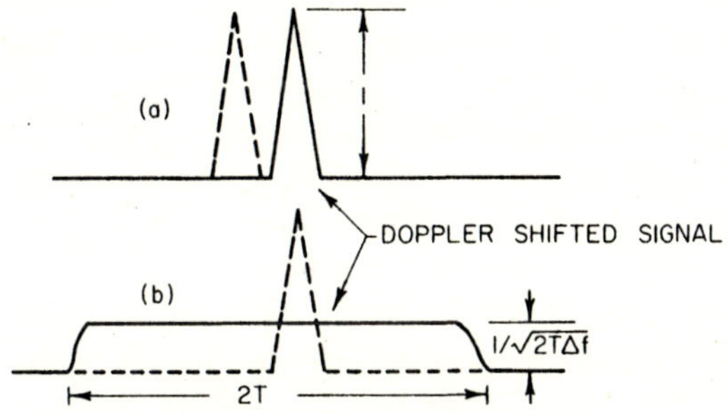

FIG. 10.1 Two extremes of Doppler sensitivity.

of Doppler sensitivity for this form of thumbtack ambiguity signal and a linear FM signal, as an example. The assumption of a uniform spread of signal energy over the interval $2T$ for the former case leads to an off-Doppler pedestal voltage level of $(2kT\Delta f)^{-1/2}$ relative to the peak output of the zero-Doppler signal, where k is a function of Doppler shift and the spectrum power density shape.

The effect of both types of Doppler shift behavior on the relative interference between a small and large signal is shown in Fig. 10.2. If the signals

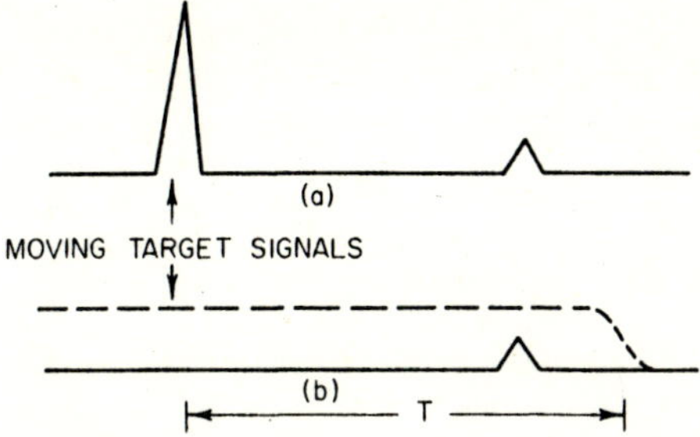

FIG. 10.2 Effect of Doppler shift behavior on relative interference between signals at different range locations.

of interest can vary over a 40 to 50 db range of amplitudes, then a Doppler sensitive waveform having the ideal thumbtack ambiguity function would require a time-bandwidth product in the order of 10^4 to 10^5 to yield an average pedestal response that is no larger than the peaks of the smallest zero-Doppler output. The practical problems involved in constructing a matched filter for time-bandwidth products of this magnitude can be severe, and could be an important factor in the decision. For a more moderate range of signal amplitudes, say 30 db or less, the radar designer may profitably use a waveform that exhibits an ambiguity function with an effective thumbtack quality in the type of semidense or nondense radar target environment outlined above.[1] Waveforms with intermediate types of ambiguity functions such as those of the V-FM and parabolic FM signals generally would not be suited for the multiple target application where energy spreading (spillover) or large ambiguity in adjacent range cells from Doppler shifted matched-filter inputs can be a problem. When the dynamic range of the received signal levels is large in the nondense target environment it is preferable to avoid the Doppler spillover into adjacent range cells that can occur with the Doppler sensitive signals, accepting any loss of smaller signals in the range cell where the larger signal actually occurs to maintain good detectability for smaller signals at all other range locations. Minimum interference between the two signals is obtained when they are designed to minimize the buildup of sidelobe energy of one signal at all possible locations of the mainlobe of the other. (Criteria for self-generated waveform interference are discussed in Sections 10.4 and 10.5.) For the criterion of maintaining detectability in range cells not at the strong signal location the system operation must then determine velocity as a by-product of the range tracking operation, or by other pulse-to-pulse velocity measuring techniques. As noted in the last chapter, the linear FM signal is well suited to this application.

A modification of the environment described above is the case of semi-isolated objects scattered over a relatively large range interval ($\gg T$), in which are also contained regions of densely packed scatterers (rain, for example) that have a mean velocity that is different from that of the targets of interest. It will be assumed that the system matched filter or filters are

---

[1] The thumbtack ambiguity function illustrated in Fig. 10.2 does not meet the self-transform property required of ambiguity functions and is therefore not realizable. Binary phase code M-sequence waveforms have an approximate thumbtack ambiguity function for very large time-bandwidth products. As a general rule, waveforms that have a smooth and fairly uniform ambiguity distribution away from the ambiguity plane origin have not been found. It should be noted that the thumbtack ambiguity function illustrated in Fig. 10.1 differs somewhat from that shown in Fig. 4.8. The difference results from the respective criteria used to define the thumbtack ambiguity functions.

tuned to the Doppler shift of the desired signals, so that in relative terms "zero Doppler shift" refers to the expected Doppler shift of the signals of interest. For this example it can also be assumed that the minimum separation between the semi-isolated objects is either smaller or greater than T; the latter case is chosen, since the point to be illustrated is related to the regions containing the densely packed scatterers. Subject to conditions that are discussed more fully in the next section, the matched-filter output in time corresponding to the actual densely packed regions will be noiselike in character. Depending on whether a waveform with Doppler sensitivity is used or not, this noiselike interference can spillover into adjacent regions, as shown in Fig. 10.3, again resulting in unwanted interference at range

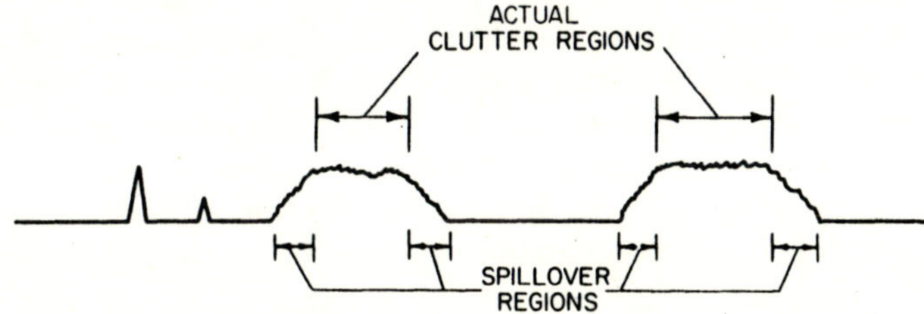

FIG. 10.3   Clutter spillover caused by Doppler sensitive waveform.

locations far removed from the densely packed regions. If one of the functions of the radar is to track the objects of interest, making use of a memory mode to track through regions of interference, then any signal generated spillover clutter can be detrimental in reacquiring the target track in the shortest possible time. Of course, this becomes relatively less of a factor as the range extent of the densely populated regions becomes much larger than the equivalent range duration of the transmitted signal.

The examples illustrated above are concerned with waveforms that are designated as compact, or solid, waveforms as opposed to pulse train waveforms that may be used when the range interval of interest is much smaller than the total waveform duration. The discussion above is not meant to imply that Doppler sensitive pulse-compression waveforms are of no utility, but rather to point out that judgement must be exercised by the radar designer in assessing the overall potential of a waveform with regard to the environment in which it is expected to function. There are a number of tactical requirements that can profitably use simultaneous range and velocity information from truly isolated (range separation greater than T) objects. In this case the multiple matched-filter receiver of Fig. 9.1 would be a possible instrumentation technique.

10.3 Waveform Optimization for Stationary or Slowly Moving Clutter

The example of a radar target or targets of interest located in an extended region of densely packed scatterers was briefly considered in the previous section. This is now examined in greater detail for the case where the relative Doppler shift between the desired signals and that of the dense scattering region, or clutter return, is small; that is $|\phi|_{max} \leqslant 1/T$, where ϕ is the clutter Doppler shift relative to the mean Doppler shift of the desired signals and T is the transmitted waveform duration. These conditions apply to many radar situations for which the transmitted pulse length is not too long and/or relative velocity differences are not extreme, as might be the case if the radar targets were missiles. The radar targets are taken to be "in" the clutter region in the sense that the respective returns from targets and clutter occur simultaneously at the receiver. Thus, physically the target may be in the clear (as an aircraft flying above terrain), but because of the radar geometry and the finite antenna beam size and spatial sidelobe configuration may still have its reflected signal obscured by clutter returns. In other cases the target may be physically located in the clutter region, as when an aircraft is flying through rain or a vehicle is located on the ground.

It will be assumed that the clutter region is composed of small independent broadband point scatterers, for which the size and location of any particular scattering element is a random variable. It is further assumed that the individual scattering elements are spaced much closer than the effective waveform resolution capability at the matched-filter output. For these assumptions the composite clutter signal in any one range cell has the characteristics of random noise as the result of the overlap of many individual signals that are present. If the further restriction is imposed that the inverse fourth power of range can be ignored (either because of the limited extent of the clutter region, or its distant location from the radar site), then the overall clutter return will essentially resemble a stationary Gaussian random process. This clutter noise signal will be independent of the type of waveforms used (for example, the ones discussed in the preceding section), since if the signal entering the matched filter is sufficiently noiselike then the matched-filter transfer function will not materially affect the noiselike condition of the signal. This is subject to the condition that, at the matched-filter output, the individual scatterers are not resolvable. Fowle *et al.* [1] note that as the signal resolution is increased the matched-filter clutter output tends to become less random in nature as a result of the smaller number of contributing signals in any one range cell.

Manasse [2], who was the first to consider the optimization of coded waveform parameters in the context of the clutter model described above, visualized the combination of the radar transmitter and clutter space as a series of filters excited by a driving impulse as shown in Fig. 10.4(a), where $s(t)$ is the impulse response of the transmitter filter (i.e., the coded waveform) and $w_c(t)$ is the impulse response of the filter representing the space through which the transmitted waveform is propagated and reflected. Since $s(t)$ and $w_c(t)$ are outputs of linear filters they may be transposed as shown in Fig. 10.4(b). Manasse employs an equivalent shot noise model, based on letting the clutter scatterers become infinitely dense and uniformly distributed and infinitesimal in size, to show that the input to the receiver in Fig. 10.4(b) has the form

$$v(t) = As(t - t_d) + n_c(t) \tag{10-1}$$

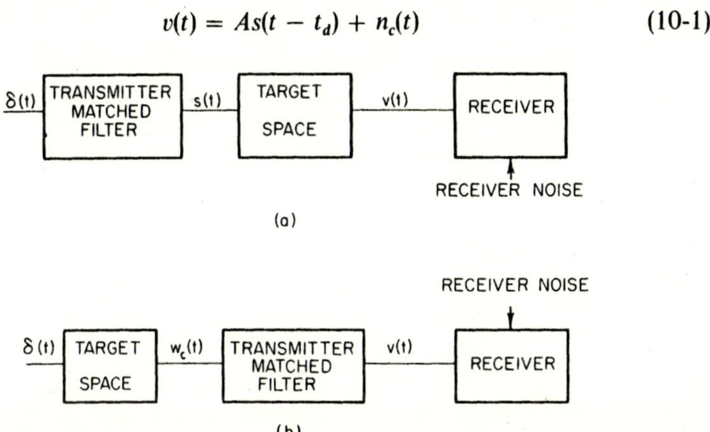

FIG. 10.4 Equivalent filter models of transmitter matched filter and target space (after Manasse [2]).

The first term of (10-1) represents a desired signal immersed in the clutter region. The second term is the clutter noise, a colored Gaussian noise that has a spectral power density given by

$$N_c(f) = k_c|S(f)|^2 \tag{10-2}$$

where $S(f)$ is the Fourier transform of $s(t)$. The proportionality factor k_c depends on the clutter density and the average size of the clutter scatterers. When white Gaussian receiver noise of spectral power density $N_0/2$ is added to the clutter noise the total noise density seen at the matched-filter input is

$$N(f) = N_0/2 + k_c|S(f)|^2 \tag{10-3}$$

The optimum receiver discussed in Chapter 2 has the transfer function

$$H(f) = S^*(f)/N(f) \tag{10-4}$$

so that the signal-to-noise ratio at the optimum filter output for $N(f)$ given by (10-3) is

$$\frac{S}{N} = A^2 \int_{-\infty}^{\infty} \frac{|S(f)|^2 \, df}{(N_0/2) + k_c|S(f)|^2} \tag{10-5}$$

It can be noted that when $k_c = 0$ (i.e., no clutter) then (10-4) and (10-5) reduce to the usual matched-filter relationships. When $N_0 = 0$ then $H(f) = 1/k_c S(f)$, the inverse filter specified by Urkowitz [3] for optimum signal-to-clutter ratio for the noiseless case. Manasse considers the optimization of (10-5) for the constraint of a fixed system bandwidth Δf. Using the calculus of variations,[1] Manasse determined that the spectrum function that optimizes (10-5) for the given constraint is flat over the interval Δf such that

$$|S(f)|^2 = E/2 \, \Delta f, \qquad f_0 - \Delta f/2 < |f| < f_0 + \Delta f/2$$

$$= 0, \quad \text{elsewhere} \tag{10-6}$$

Inserting this flat spectrum into (10-5) yields the optimum signal-to-noise ratio

$$\left(\frac{S}{N}\right)_{\text{opt}} = \frac{2A^2E}{N_0 + k_c E/\Delta f} = \frac{2A^2E}{(N_0)_{\text{eff}}} \tag{10-7}$$

Equation (10-7) illustrates the importance in this case of increased bandwidth in minimizing the effect of the clutter. Physically, the reason for this can be seen by recognizing that the clutter return is a power limited signal, depending on the transmitted energy per pulse and the average clutter particle size. Thus, the effect of increasing the signal bandwidth while keeping the transmitted power constant is to lower the effective "noise" power density associated with the clutter return; that is, the same "noise" power is spread over a greater bandwidth. This is equivalent to stating that as the signal resolution increases the total clutter density per range cell decreases. Of course, for this improvement to be useful the signals of interest should be larger than the average clutter return in a range cell. The limiting value of useful resolution is reached when the majority of the desired signals are detected, or when the resolution achieved results in the signals of interest breaking up and appearing at the matched-filter output as the independent returns from the subreflectors that comprise the radar targets being sought. The optimum bandlimited signal derived by Manasse can be approximately realized with the linear FM pulse-compression waveform. As brought out in the opening chapter, one of the major advantages

---

[1] An alternate, but related, approach to this derivation is outlined in Section 4.11.

of a pulse-compression system is that it permits the radar designer to choose the system resolution capabilities independently of the transmitted signal duration, which can then be chosen to maximize the energy per pulse. Thus for the type of clutter environment described above the signal bandwidth may be increased to any (practical) desired value without incurring any detection degradation for the case of receiver noise only as the interference.

10.4 Comparison of the Effectiveness of Different Waveforms for the Stationary or Slowly Moving Clutter Case

In developing the waveform optimization criterion based on the signal spectrum characteristics Manasse remarks that this is equivalent to minimizing the integral of the squared autocorrelation function. On a normalized basis, this is a minimization of the Woodward time resolution constant T_r, for the constraint of a bandlimited signal. From Chapter 4, the time resolution constant is defined as

$$T_r = \frac{\int_{-\infty}^{\infty} |\chi(\tau, 0)|^2 \, d\tau}{|\chi(0, 0)|^2} \tag{10-8a}$$

$$= \frac{\int_{-\infty}^{\infty} |S(f)|^4 \, df}{\left[\int_{-\infty}^{\infty} |S(f)|^2 \, df\right]^2} \tag{10-8b}$$

Physically interpreted, in this sense T_r is a measure of the spread of energy of the matched-filter output outside of the interval near $\tau = 0$. If T_r is large this implies a greater spread of energy, or signal generated interference (self-clutter), over the correlation interval $2T$. Since the $(\sin x)/x$ compressed-pulse waveform obtained with the bandlimited flat spectrum is not desired for general application[1] it is of interest to compare the relative effectiveness of some of the weighted pulses with reduced sidelobes discussed in Chapter 7 (see Table 7-I), applying the criterion of (10-8a) or (10-8b). The pertinent data are tabulated in Table 10-I for members of the (cosine)^n power spectra and the general $k + (1 - k)\cos^2$ weighting function. For the general (cosine)^2-on-pedestal weighted spectrum at the matched-filter output described by

$$|S(f)|^2 = k + (1 - k)\cos^2(\pi f/\Delta f)$$

---

[1] For the few target case peak sidelobe levels are of more concern than the total energy associated with all the sidelobes.

TABLE 10-I

TIME RESOLUTION (SELF CLUTTER) COMPARISON OF BANDLIMITED SPECTRA

Spectrum power density function	Range sidelobes (peak) [db]	Normalized time resolution function $(\Delta f\,T_r)$	Pulse width $(-3\,\mathrm{db})$	Time resolution function for all signals normalized to equal 3 db pulse widths
rect $(f/\Delta f)$	-13	1.00	1.00	1.00
$\cos(\pi f/\Delta f)$	-23	1.23	1.34	0.918
$\cos^2(\pi f/\Delta f)$	-32	1.50	1.62	0.938
$\cos^3(\pi f/\Delta f)$	-39	1.73	1.87	0.925
$\cos^4(\pi f/\Delta f)$	-47	1.93	2.2	0.877
$0.08 + 0.92 \times \cos^2(\pi f/\Delta f)$	-42.8	1.36	1.47	0.92
$0.16 + 0.84 \times \cos^2(\pi f/\Delta f)$	-34	1.26	1.41	0.89

the normalized time resolution function is given by

$$\Delta f\,T_r = \frac{3k^2 + 2k + 3}{2(k^2 + 2k + 1)} \tag{10-9}$$

Equation (10-9) is plotted in Fig. 10.5.

The last column in Table 10-I gives the equivalent value of T_r when the bandwidth of each weighted spectrum is increased so that all the compressed pulses have equal 3 db pulse widths. It is seen that there is not a great degree of difference in the clutter performance characteristics of the various spectrum weighting functions. If the weighting functions are associated with a mismatching filter following a rectangular envelope

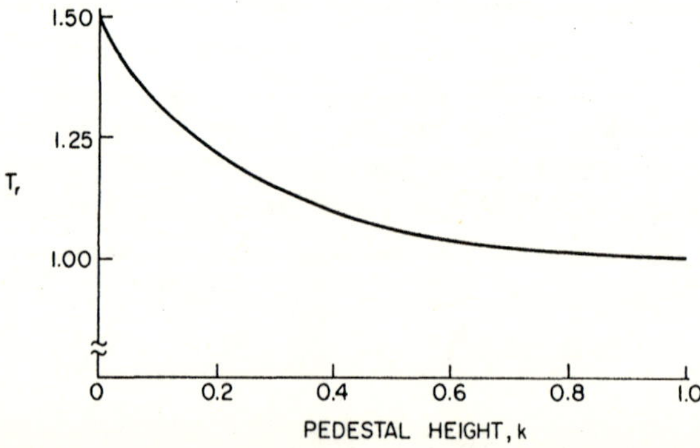

FIG. 10.5 Effect of pedestal height on T_r for $\cos^2$-on-pedestal power density spectrum.

linear FM matched filter then the signal-to-clutter degradation is equal to the signal-to-noise mismatch loss derived in Chapter 7 (see Eq. (7-53) and Table 7-I). This can be seen from the fact that (referring to Fig. 10.4(b)) the clutter noise entering the receiver is essentially white Gaussian noise that becomes colored Gaussian noise due to the bandlimiting response of the mismatched receiver. Since the flat spectrum signal is the optimum bandlimited waveform for this model, the mismatch loss for the non-optimum filter can be treated as equivalent to that obtained for the case of white Gaussian receiver noise only.

The clutter effectiveness of other types of waveforms can be evaluated by using (10-8b). Thus, for the triangular pulse shown in Fig. 10.6 that is the matched-filter output for a constant frequency, rectangular envelope

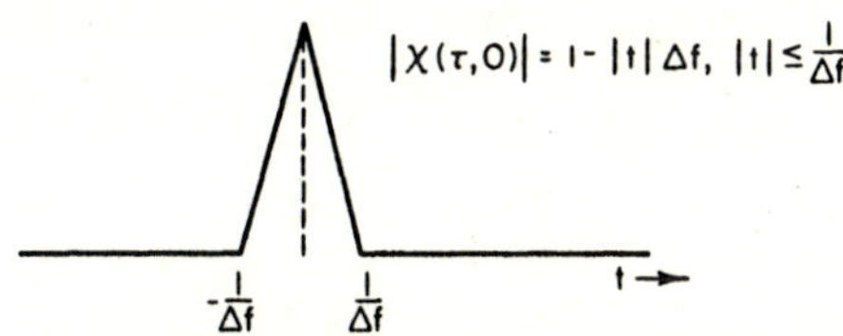

FIG. 10.6   Autocorrelation function of rectangular pulse waveform.

transmitted signal one obtains $\Delta f\, T_r = \frac{2}{3}$. This is less than the normalized value of unity shown in Table 10-I for the optimum signal discussed by Manasse. This is not contradictory, since the spectrum functions considered in Table 10-I are bandlimited, whereas the spectrum for the triangular pulse is not bandlimited.[1] Figure 10.7 illustrates two types of coded waveform autocorrelation functions that have extended baseline pedestal formations. The waveform shown in Fig. 10.7(a) approximates that of the binary phase coded maximum length sequence, and yields $\Delta f\, T_r = 8/3$. The waveform of Fig. 10.7(b) has a pedestal level of $(2T\,\Delta f)^{-1/2}$ rather than $(T\,\Delta f)^{-1/2}$, and is associated with the ideal thumbtack ambiguity function discussed in Section 10.2. The normalized clutter factor for this signal is $\Delta f\, T_r = \frac{5}{3}$. In the uniformly distributed clutter case these numbers imply that signals of these types will have poorer signal-to-clutter noise ratios by at least 3 to 4 db compared to the equivalent optimum bandlimited signal ($\Delta f\, T_r = 1$), plus any additional loss caused by small Doppler shifts ($\phi \le 1/T$) that yield lower peak matched-filter outputs (i.e.,

---

[1] The value of $\Delta f\, T_r = \frac{2}{3}$ is based on the assumption that the first null points of the $(\sin x)/x$ and triangular waveforms occur at $\pm 1/\Delta f$, with Δf being equal for both cases. If these signals are designed to have equal 3 db pulse widths then their time resolution constants will be approximately the same.

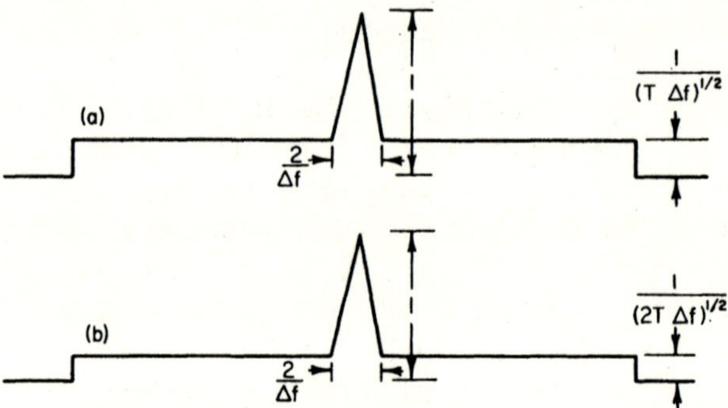

FIG. 10.7 Autocorrelation functions (approximate) for two types of thumbtack ambiguity functions.

thumbtack type ambiguity function). The periodic maximum length sequence autocorrelation function is shown in Fig. 10.8. If the length of the sequence is greater than the range interval of interest, so that attention can be confined to an unambiguous interval, as illustrated, then application of (10-8a) over one cycle of the autocorrelation function yields $\Delta f\, T_r = (\frac{2}{3} + 1/N)$. For $N > 3$ this value is also less than unity, and this particular signal also can be considered useful when the extent of the clutter region is less than the sequence length. In general, T_r can be taken as an inverse measure of the radar system target handling capability for the case of stationary or slowly moving interference.

The autocorrelation function for the uniform pulse train is shown in Fig. 10.9. If the range interval is confined to an unambiguous interval, as discussed above, then $\Delta f\, T_r = \frac{2}{3}$. However, if the entire correlation interval is considered then

$$\Delta f\, T_r = \frac{2}{3}\left[1 + \frac{2}{N^2}\{1^2 + 2^2 + 3^2 + \cdots + (N-1)^2\}\right]$$

$$= \frac{2}{3}\frac{4N^2 + 2}{6N} \doteq \frac{4N}{9}, \qquad N \text{ large} \tag{10-10}$$

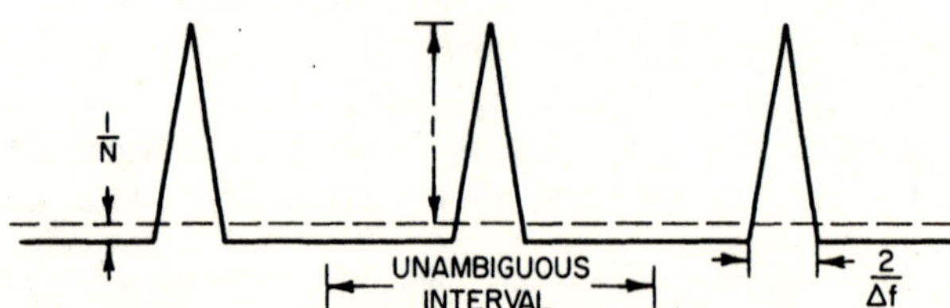

FIG. 10.8 Autocorrelation function for periodic M-sequence binary phase coded waveform.

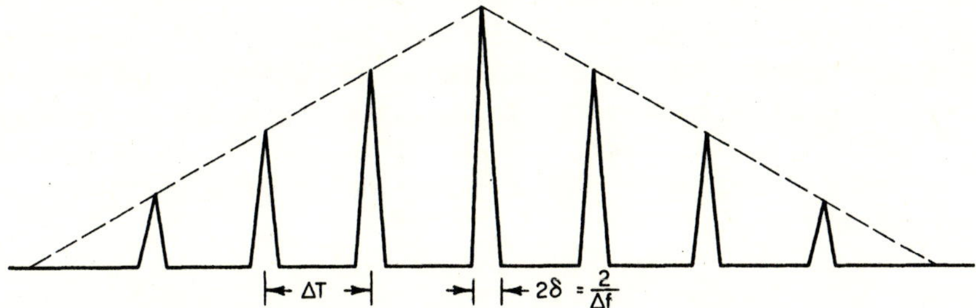

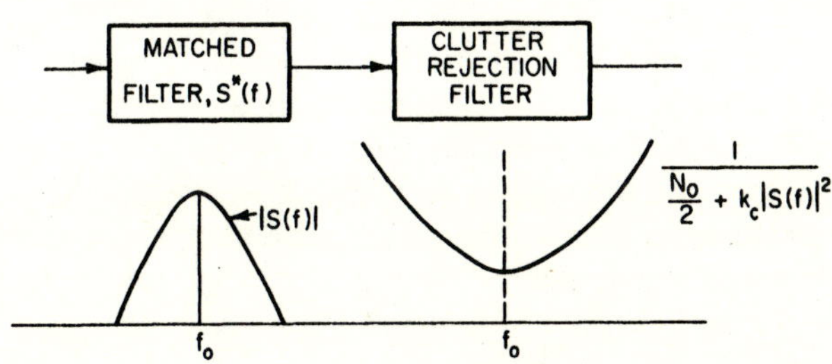

FIG. 10.9   Autocorrelation function of uniform pulse train.

10.5   The "Optimum" Clutter Rejection Filter [4, 5]

The time resolution constants, or self-clutter figures of merit of Table 10-I apply for the matched-filter implementation except as noted for the mismatched linear FM case. The filter that optimizes the signal-to-clutter noise ratio for the general nonflat spectrum case is given by, from (10-4)

$$H(f)_{\text{opt}} = \frac{S^*(f)}{(N_0/2) + k_c|S(f)|^2} \qquad (10\text{-}11)$$

Rubin and Kaiteris [5] point out that the optimum filter of (10-11) is equivalent to a matched filter followed by a clutter rejection filter, as shown in Fig. 10.10. The phase response for this combination is the same

FIG. 10.10   Composition of optimum filter for clutter rejection.

as for the matched filter alone. Alternately, Rihaczek [4] approximates the optimum filter with a parallel combination of a matched filter and an Urkowitz filter; the former cascaded with a bandstop filter of width $2f_c$ and the latter cascaded with a bandpass filter of equal width, where $k_c|S(f_c)|^2 = N_0/2$ defines f_c.

For the rectangular spectrum the matched filter and optimum filter are the same. From (10-11) it can be seen that the design of the general optimum filter requires a knowledge of the ratio of the clutter noise to receiver noise. For most cases this ratio is not predictable, and the design of the optimum filter (other than for the rectangular spectrum case) would be difficult to implement. Rihaczek considers the loss resulting from using a matched filter rather than an optimum filter for the specific case of a signal that has a spectrum of the form

$$|S(f)| = \frac{|S(f_0)|}{1 + (\bar{f}/\Delta f)^2}, \qquad \bar{f} = f_0 - f$$

The loss resulting from the use of a matched filter for this spectrum is plotted in Fig. 10.11 as a function of clutter noise-to-receiver noise ratio. Other examples are considered by Rubin and Kaiteris [5], with the losses

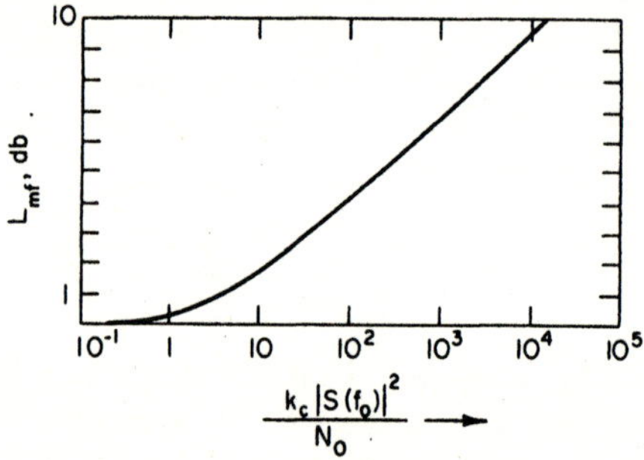

FIG. 10.11   Performance loss of the matched filter as compared to the optimum filter (courtesy of A.W. Rihaczek [4]).

calculated being roughly similar to that shown, except for a scale factor that depends on the particular waveform. Rubin and Kaiteris also consider the case where the clutter rejection filter is designed for some expected mean level of clutter. This type of filter performs poorly compared to the matched filter when clutter is absent. In general, the comparison of the optimum filter and the matched filter is somewhat academic for several reasons. The clutter factor k_c is a function of the radar resolution capability and the system designer can, in most cases for which the individual clutter scatterers are small, insure operation on the low loss portion of the characteristic plotted in Fig. 10.11 (or its equivalent for other waveforms) by having sufficient bandwidth available. As noted above, if a weighted spectrum is desired in order to have a matched-filter output with low range sidelobes then the mismatched linear FM signal, with clutter losses relative to the optimum filter as indicated in Table 10-I for all ratios of clutter noise-to-receiver noise, is superior to the general nonflat spectrum matched filter discussed by Rihaczek. For the mismatched linear FM case both clutter noise and receiver noise are equally weighted by the spectrum shaping function. For the general matched-filter case the clutter noise is weighted by $|S(f)|^2$ while the receiver noise is only weighted by $|S(f)|$, where $|S(f)|$ is the matched-filter amplitude response.

As the clutter factor k_c in (10-11) becomes large the optimum filter approaches the Urkowitz inverse filter. This has been noted as another approach for increasing the bandwidth to obtain higher resolution [3–5]. Since there are practical obstacles to achieving an optimum filter for different levels of clutter,[1] the design effort to increase the bandwidth to combat clutter interference normally should be confined to the matched-filter implementation. This will be optimum for the case of receiver noise only, and nearly optimum for a broad range of clutter interference.

It was shown, for the case of densely packed scatterers that are not moving rapidly with respect to the targets of interest (i.e., $\phi \leqslant 1/T$), that the waveform design calls for a wideband signal. Within the class of band-limited signals the spectrum function should be rectangular, although the degradation resulting from other weighted spectra is not great. If the weighted spectra represent the outputs of a rectangular envelope linear FM filter mismatched to achieve low level range sidelobes then this degradation factor from the performance obtained with the optimum filter is independent of the clutter noise-to-receiver noise ratio. It is of interest to note that a nonbandlimited signal, such as the constant frequency rectangular pulse (triangular pulse autocorrelation function), is superior to the rectangular spectrum signal for equivalent measures of bandwidth as defined by the null points of the matched-filter output. Physically, this derives from the fact that the former signal does not have any range sidelobes to contribute to self-clutter buildup in distant range cells. The problem encountered with this signal, of course, is that the narrow pulse may not have sufficient energy content for long range detection when only receiver noise is present. Matched filtering, rather than optimum filters, for the uniform pulse train and the periodic maximum length sequence can also perform well in a dense clutter environment if all of the clutter is confined to the unambiguous range interval associated with these signals.

10.6 Waveform Design Considerations for Dense Distributions of Moving Scatterers

When the dense multiple target or clutter environment has a significant Doppler shift relative to the targets of interest then there exists an added dimension for maximizing the signal-to-clutter noise ratio at the receiver

---

[1] When the clutter strength dependence on range cannot be ignored the optimum filter becomes a time varying filter, as it also does if the clutter density changes as a function of time.

output. For this case, as an extension of Eq. (10-5), the signal-to-noise ratio obtained at the output of the optimum filter, as noted by Rihaczek [4], is given by

$$\frac{S}{N} = A^2 \int_{-\infty}^{\infty} \frac{|S(f)|^2 \, df}{(N_0/2) + k_c |S_d(f)|^2} \tag{10-12}$$

where $S_d(f)$ is the Doppler shifted clutter noise spectrum. For most cases of interest the spread of Doppler frequencies is usually narrow compared to the mean Doppler, so that $S_d(f)$ can be replaced by $S(f + \bar{\phi})$, where $\bar{\phi}$ is the mean Doppler shift. When this situation prevails (10-12) is maximized by having $|S(f)|$ and $|S(f + \bar{\phi})|$ disjoint; that is the peaks of the spectral responses of the Doppler shifted spectrum occur at or close to the minima of the spectra of the desired signals. An example of a waveform that meets this requirement is the uniform pulse train if the interpulse spacing ΔT can be adjusted so that $\phi = 1/2\Delta T$ (the uniform pulse train matched-filter response for $\phi = 1/2\Delta T$ is shown in Fig. 8.2). As before, the optimum filter is a matched filter followed by a clutter rejection filter as shown in Fig. 10.10, with the clutter rejection filter now being inversely "tuned" to $|S(f + \bar{\phi})|$ rather than $|S(f)|$ as in the stationary clutter case. Unfortunately, the interfering clutter may occur at several discrete bands, or be spread over a sufficiently wide band that it may not be possible to design a waveform that has a spectrum that is properly disjoint to that of the Doppler shifted clutter noise while still retaining the other waveform characteristics that are desirable for the tactical situation at hand. Additionally, the other design problems cited in the previous section that are associated with the optimum filter do not become any less severe for the general moving clutter case. It can be assumed that in the majority of cases it will be the matched filter rather than the optimum filter that will be implemented. For most cases in which these two types of filters have been studied for specific waveforms and Doppler shifted interference signals it has been shown that the advantage of the optimum filter over the matched filter becomes much less as the transmitted signal bandwidth increases. Again, this is a direct result of the effect of bandwidth on the clutter factor k_c.

The formulation of a matched-filter signal-to-interference ratio for the general case of dense target environments has been considered by Stewart and Westerfield [6], Westerfield *et al.* [7], and Fowle *et al.* [1]. These workers have shown that for the general situation the signal-to-interference (or clutter noise) ratio is closely related to the matched-filter waveform characteristics. As might be expected, when the clutter or dense target scatterers are composed of elements that are not individually resolvable then the composite interference output from the matched filter depends in general on the total energy contribution from each scatterer rather than on the details of each individual Doppler shifted time response. Since the normalized Doppler shifted matched-filter output is described by the ambiguity slice $\chi(\tau, \phi)$, the energy in this matched-filter output is given by

$$\int_{-\infty}^{\infty} |\chi(\tau, \phi)|^2 \, d\tau \equiv T_r(\phi) \tag{10-13}$$

$T_r(\phi)$ is denoted here as the generalized Woodward time resolution constant, being an extension of (10-8a) for the Doppler shift case. However, a more restricted interpretation must be placed on $T_r(\phi)$ as compared to $T_r(0)$. $T_r(0)$ has the significance of being a measure of the spread of energy in the autocorrelation waveform $\chi(\tau, 0)$ away from $\tau = 0$ since $\chi(\tau, 0)$ has its peak value at $\tau = 0$. No such shape dependency can, in general, be associated with $T_r(\phi)$. This function is a measure of the energy caused to appear at the matched-filter output by a signal that is mismatched in Doppler shift to the desired matched-filter channel. Westerfield et al. [7] were the first to specifically point out the importance of this function in determining signal-to-clutter noise ratios for the general case.[1]

$T_r(\phi)$ is of central importance in evaluating the moving clutter rejection effectiveness of different waveforms that are processed through a matched filter. Various forms of $T_r(\phi)$ were derived in Chapter 4, and are listed below:

$$T_r(\phi) = \frac{\int_{-\infty}^{\infty} |\chi(\tau, \phi)|^2 \, d\tau}{|\chi(0, 0)|^2} \tag{10-14a}$$

$$= \frac{\int_{-\infty}^{\infty} |\chi(\tau, 0)|^2 \exp[j2\pi\tau\phi] \, d\tau}{|\chi(0, 0)|^2} \tag{10-14b}$$

$$= \frac{\int_{-\infty}^{\infty} |S(f)|^2 |S(f + \phi)|^2 \, df}{\left[\int_{-\infty}^{\infty} |S(f)|^2 \, df\right]^2} \tag{10-14c}$$

For the normalized case $|\chi(0, 0)| = \int_{-\infty}^{\infty} |S(f)|^2 \, df = 1$.

---

[1] Westerfield et al. did not give what is called here $T_r(\phi)$ a specific designation. The authors have adopted this notation to indicate its historical dependence on the earlier work of Woodward in defining $T_r(0)$. Richman [8] has, in his own work, designated the same function as $Q(\phi)$. This can be confused with the well-known Q-function associated with radar detection studies. Richman was one of the first to systematically apply the generalized time resolution function to the study of the effectiveness of various waveforms in an environment of dense moving targets.

Equation (10-14b) indicates the dependence of $T_r(\phi)$ on the signal auto-correlation function, and hence on the magnitude of the signal spectrum. Thus, the Doppler shifted energy content of a signal at the matched-filter output cannot be specified independently of the waveform autocorrelation function. This relationship, first noted by Westerfield *et al.* [7], can lead to some conflict in the specifications that may be considered for the desired matched-filter output. A related implication is that $T_r(\phi)$ is not influenced by the individual waveform parameters (i.e., envelope shape and the form of the phase or frequency modulation) that combine to yield a specific autocorrelation function.

How $T_r(\phi)$ enters into the determination of the clutter rejection effectiveness of a waveform can be seen by considering the formulation of the signal-to-interference ratio derived by Fowle *et al.* [1] for the case of a large number of independent, broadband scatterers distributed in both range and velocity. For this case it is shown that the mean power associated with the interference at the matched-filter output is given by

$$\sigma_c^2(t_0, f_0) = n\bar{b}\int\int_{-\infty}^{\infty} p(\tau, \phi)|\chi(t_0 + \tau, f_0 + \phi)|^2 \, d\tau \, d\phi \tag{10-15}$$

where n is the number of scatterers, $\bar{b}$ the average scatterer cross section, and $p(\tau, \phi)$ the joint probability distribution in range and velocity of the scatterers; t_0 and f_0 represent the reference delay and frequency at the matched-filter output. For the case of particular interest we may take $t_0 = f_0 = 0$; that is, the signal of interest is at f_0, to which the matched filter is tuned, and (10-15) represents the interference from other signals that are distributed in time about the true target location t_0, excluding in the calculation that response that represents the signal of interest. The cross section of this signal is denoted as b_T and its squared matched-filter output at t_0 is $b_T|\chi(0, 0)|^2$. Since the foregoing discussion is based on a normalized ambiguity function representation (i.e., $|\chi(0, 0)|^2 = 1$), the peak signal power-to-mean interference ratio becomes

$$\frac{S}{I} = \frac{b_T}{n\bar{b}\iint\limits_{R=\Delta_\phi\Delta_\tau} p(\tau, \phi)|\chi(\tau, \phi)|^2 \, d\tau \, d\phi} \tag{10-16}$$

In (10-16) R represents the region of range-Doppler space over which the interfering scatterers are distributed, as shown in Fig. 10.12. Fowle *et al.*, as do Westerfield *et al.*, assume that the distribution in range and velocity are independent so that $p(\tau, \phi) = p(\tau)p(\phi)$. Furthermore, it is often assumed

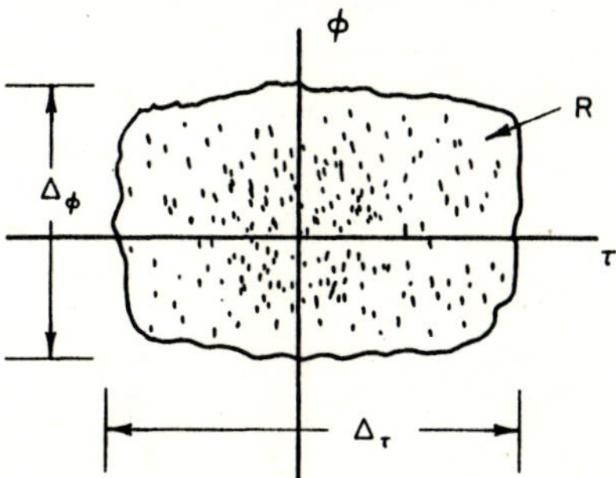

FIG. 10.12   Distribution of interfering scatterers over region R of range-Doppler space.

that in a restricted range interval Δ_τ ($>2T$) the distribution of scatterers is uniform (this includes the tacit assumption that the range dependence of the strength of the individual returns can be ignored). Under these conditions (10-16) becomes

$$\frac{S}{I} = \frac{b_T \, \Delta_\tau}{n\bar{b} \int_{\Delta_\phi} p(\phi) \int_{\Delta_\tau} |\chi(\tau, \phi)|^2 \, d\tau \, d\phi} \tag{10-17}$$

The use of (10-13) yields

$$\frac{S}{I} = \frac{b_T \, \Delta_\tau}{n\bar{b} \int_{\Delta_\phi} p(\phi) T_r(\phi) \, d\phi} \tag{10-18}$$

If $p(\phi)$ is uniform over the interval Δ_ϕ then (10-18) reduces to

$$\frac{S}{I} = \frac{b_T \, \Delta_\tau \, \Delta_\phi}{n\bar{b} \int_{\Delta_\phi} T_r(\phi) \, d\phi} \tag{10-19}$$

For the case where the Doppler band is narrowly spread about a mean Doppler shift $\bar{\phi}$, (10-18) yields [5]

$$\frac{S}{I} = \frac{b_T \, \Delta_\tau}{n\bar{b} T_r(\bar{\phi})} \tag{10-20}$$

Equation (10-20) is subject to the condition that $T_r(\phi)$ does not change radically in the interval about $\bar{\phi}$. All of the expressions given above indicate the importance of the generalized time resolution constant $T_r(\phi)$. The behavior of $T_r(\phi)$ as a function of Doppler shift is thus of primary importance to the radar designer. Figure 10.13 plots this function for three different types of regular (or solid) spectra that have been discussed in terms of matched-filter output signals. Taking the $T_r(\phi)$ of the rectangular

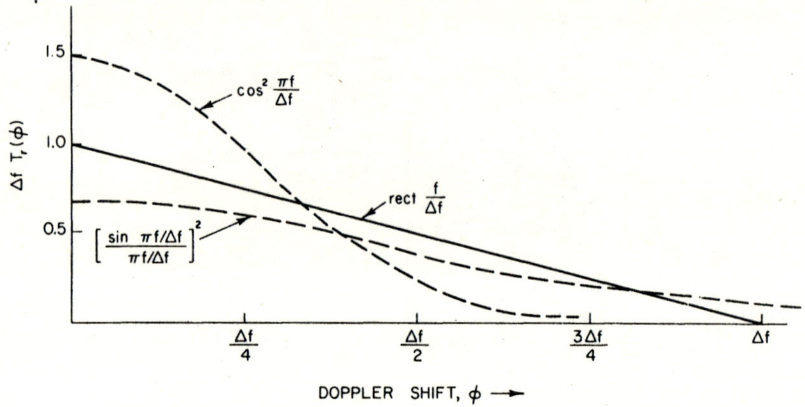

FIG. 10.13 $T_r(\phi)$ for three continuous spectrum functions.

spectrum as a reference, it is seen that this function for other signals cannot be either uniformly above or below this reference for all values of ϕ. This is a consequence of the ambiguity principle, since

$$\int\!\!\!\int_{-\infty}^{\infty} |\chi(\tau, \phi)|^2 \, d\tau \, d\phi = \int_{-\infty}^{\infty} T_r(\phi) \, d\phi = 4E^2 \qquad (10\text{-}21)$$

$$(= 1, \text{ for the normalized case})$$

A more striking example of this is given by the $T_r(\phi)$ function for the uniform pulse train, shown in Fig. 10.14(a). The interference appearing at the matched-filter output for this signal has alternate maxima and minima as a function of Doppler shift. This waveform will provide good (S/I) ratios if the Doppler shifts of the interfering signals are narrowly located about odd multiples of $\phi = M/2T$ (M = number of pulses in the train).[1] At even multiples of this value large amounts of interfering signal energy appear at the matched-filter output, reflecting the large number of regions of peak ambiguity inherent in this type of waveform. Figure 10.14(b) illustrates the $T_r(\phi)$ functions for two types of approximate thumbtack ambiguity waveforms that are representative of signals associated with the requirements for simultaneous range and velocity measurements. It can be seen that they are uniformly poor for moderate Doppler shifts with regard to performance in dense clutter compared to other types of waveforms, as observed in the discussion of Section 10.2.

---

[1] M as used here and in Figs. 10.14 and 10.15 is equivalent to $M - 1$ appearing in the analytic development of the uniform pulse train presented in Chapter 8.

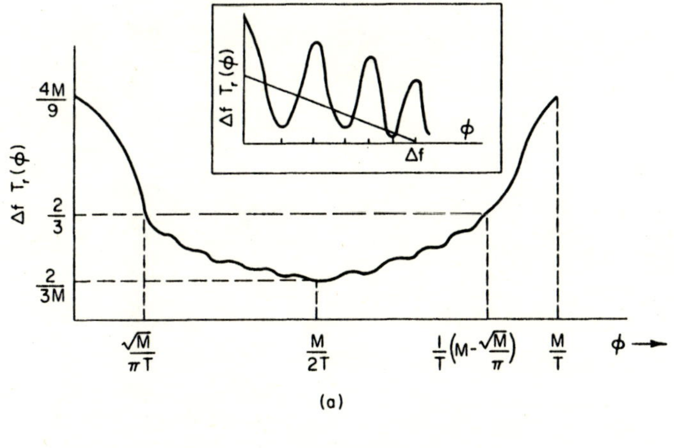

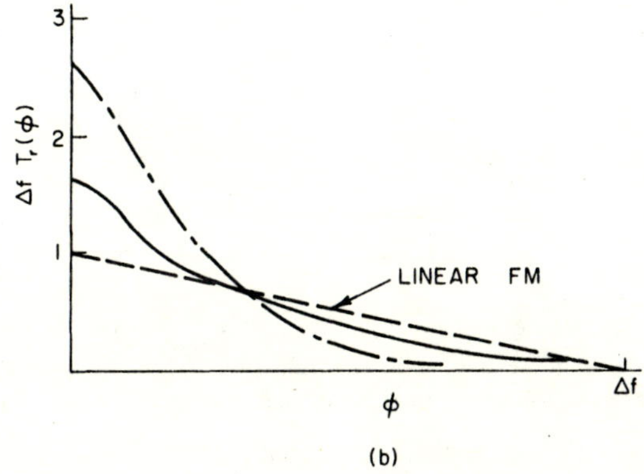

FIG. 10.14   Behavior of $T_r(\phi)$ for pulse train and other waveforms. (a) $T_r(\phi)$ of uniform pulse train (courtesy of W. L. Rubin and C. Kaiteris [5]). (b) $T_r(\phi)$ for thumbtack type ambiguity functions (average levels).

If the targets of interest are such that they are contained in the un-ambiguous central region of the uniform pulse train ambiguity plane, as shown in Fig. 10.15(a), then this waveform can be used for detection and resolution in the (restricted) dense target environment. For this application the equivalent $T_r(\phi)$ in the central region is shown in Fig. 10.15(b). This particular application has been studied by Fowle *et al.* [1]. Their approach is to scale the pulse train parameters so that the interpulse spacing ΔT matches the time extent of the dense target environment. They then recommend setting the carrier frequency so that the spread of Doppler shifts fills the Doppler dimension $1/\Delta T$ of the unambiguous clear area of Fig.

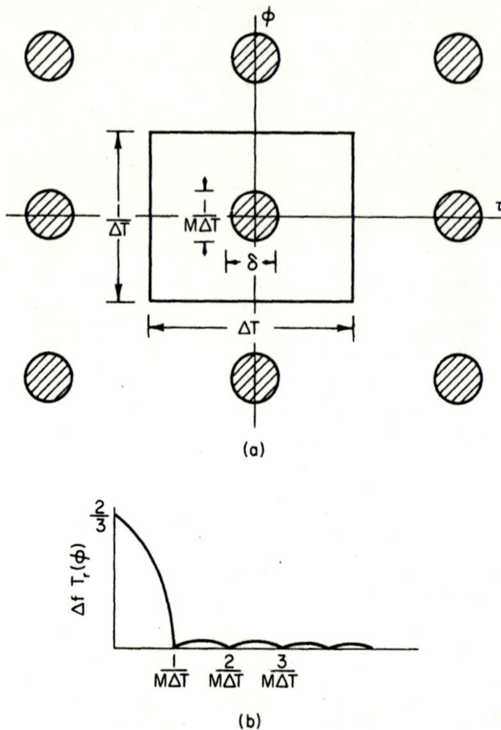

FIG. 10.15 Properties of uniform pulse train. (a) Dimensions of unambiguous clear area of uniform pulse train ambiguity function. (b) $T_r(\phi)$ of central region of uniform pulse train ambiguity function.

10.15(a). Thus

$$f_0 = c/2v_r\Delta T \tag{10-22}$$

where v_r is the radial velocity spread of the interfering targets.

The condition given by (10-22) may not lead to practical values for the carrier frequency f_0. As an example, if $\Delta T = 20$ microseconds and $v_r = 300$ meters/sec then $f_0 = 25$ gigahertz (GHz), which is fairly high for most types of systems. Equation (10-22) can be used to construct a family of curves for various combinations of f_0, v_r, and ΔT if desired. As noted, this may not represent a practical solution in many cases, particularly if ΔT must be made very large.

Of interest to the radar designer is the number of objects that can be separately detected or resolved in the face of the interference caused by a dense distribution of scatterers. The expression derived from (10-16) by

Fowle *et al.* for the case of uniform distributions in range and velocity is

$$n_{\max} = \frac{\Delta_\tau \, \Delta_\phi}{(\bar{b}/b_{\min})(S/I)_{\min} \displaystyle\iint_{R=\Delta_\tau \Delta_\phi} |\chi(\tau, \phi)|^2 \, d\tau \, d\phi} \qquad (10\text{-}23)$$

where $b_{\min}$ is the minimum target size and $(S/I)_{\min}$ the minimum signal-to-interference ratio required for detection.

Equation (10-23) is for the case of the interference much larger than the receiver thermal noise. For the uniform pulse train designed to confine the targets to the unambiguous central clear area $n_{\max}$ can be considered to be the number of resolution elements in the clear area. Fowle *et al.* [1] show that for this case

$$n_{\max} = \frac{\text{clear area}}{\text{area of central spike}} = \frac{1}{\delta(1/M \, \Delta T)} = \frac{M \, \Delta T}{\delta} \qquad (10\text{-}24)$$

Since $M \, \Delta T$ is the total signal duration and $\delta \doteq 1/\Delta f$, where Δf is the nominal spectrum width, then $n_{\max}$ is related to the time-bandwidth product of the pulse train. The $n_{\max}$ given in (10-24) is based on counting multiple targets that fall in the same range-Doppler cell as a single target. If this is not satisfactory in an expected situation (i.e., too many objects not resolved) then the single-pulse duration δ can be decreased or the total number of pulses in the train can be increased, or both, to provide adequate range-Doppler separation of all the objects of interest.

The discussion above has been based on the assumption that the range location and extent of the dense target environment is such that the range dependence of returned signal strength is negligible. Van Trees [9] considers the case in which uniform distribution in range cannot be assumed, so that the clutter noise is treated as a nonstationary random process. Taking the linear FM Gaussian envelope signal (see Chapter 9) as an example, Van Trees calculates the difference between the matched-filter and the optimum-filter detection performance. This difference becomes smaller as the signal bandwidth increases, as would be expected from the discussion in previous sections. Rubin and Kaiteris [5] consider a larger variety of waveforms, as well as the specific effects of range dependence and extent on clutter signal strength for different types of clutter. They also conclude that in general the performance of a matched filter approaches that of the optimum filter for a well designed waveform. In each case the behavior of the generalized time resolution constant $T_r(\phi)$ is important in determining the signal-to-interference ratio for a specific waveform.

Under certain conditions the return from clutter areas or from extended targets will give the appearance of a combination of noiselike and impulsive signals. The impulsive signals, which tend to behave like signal returns of interest, may be caused by large individual subreflectors in the environment or by coherent phase reinforcement in the signal returns of smaller sub-reflectors or clutter particles.[1] The degree to which these correlated clutter signals stand out from the general clutter background will be a function of signal bandwidth, transmitter frequency and clutter cross section properties. These can be accounted for in the general formulation of the signal-to-clutter interference ratio given by Eq. (10-16). However, expressions incorporating these relationships for meaningful target or clutter models pose considerable difficulty and have, in general, not been developed except for certain applications of restricted utility.

10.7   Some Summary Remarks Concerning the General Signal Design Problem

It was stated in Chapter 1 that much of the impetus to progress in the development of radar waveform design technology resulted from Woodward's monograph [10]. Woodward closed his work with the remark that he felt some disappointment in not being able to unequivocally answer the question "What should the radar waveform be?" This question has motivated over a decade of research in the attempt to provide a definitive answer. It is clear that, in the broadest sense, there is no clear cut answer unless the question is hedged about with restrictions on the environment in which the radar waveform is expected to perform. Many of the wave-form examples described in this volume have been discussed in the context of a particular function that the waveform appears to be suited for. How-ever, the examples provided tend to indicate that no one waveform is uniquely optimum in a particular situation. More likely than not, a study of a situation will reveal what ought *not* to be done rather than what is the best thing to do. A good illustration of this is given in the example discussed by Westerfield *et al.* [7]. They treat the case of a Gaussian shaped power spectrum, for which $|\chi(\tau, 0)|^2$ is given by

$$|\chi(\tau, 0)|^2 = \exp[-\pi W^2 \tau^2] \tag{10-25}$$

---

[1] When there are very large undesired signal returns present in the receiver the system designer may consider the use of a limiter to suppress the large signals. When the limiter precedes the matched filter, as is usually the desired form of operation, then there exists the possibility of the suppression of desired signals over extended range intervals caused by the limiter output being controlled by strong large time-bandwidth signals. Reference 5 of the Bibliography following Chapter 14 presents data on detection loss resulting from overlapping large time-bandwidth signals present at a limiter input. In this situation the use of the staggered pulse train signals offers a method of minimizing limiter signal suppression of wanted signals by choosing the pulse train positions so that there is minimum individual pulse coincidence as a function of arbitrary relative time displacements between two such signals. This application of staggered pulse trains is discussed in References 34 and 35 of Chapter 8.

W is the nominal frequency span of the signal in the notation of Westerfield *et al.* They also assume a uniform distribution in range for the dense clutter region, and a Gaussian distribution of clutter signal energy as a function of velocity such that

$$p(\phi_c) = 1/W_q \exp[-\pi\phi_c^2/W_q^2] \tag{10-26}$$

where $\phi_c = (\bar{\phi} + \phi)$ is the Doppler shift relative to the mean Doppler and W_q the Doppler frequency spread.

Westerfield *et al.* analyzed the effect of the various parameters for their waveform and clutter space model; that is, the ratio of the mean Doppler to Doppler spread, $\bar{\phi}/W_q$ and the ratio of signal bandwidth to Doppler spread, W/W_q. Their results are plotted in Fig. 10.16 as processing gain (in decibels) of the desired signal relative to clutter noise at the matched-filter output for various combinations of the ratios given above. When the mean Doppler shift is large compared to the Doppler spread the curves of Fig. 10.16 show that a long duration waveform is best, that is $W/W_q < 1$. When the Doppler spread is large compared to the mean Doppler, or there is little or no relative Doppler shift between the desired signal and the clutter noise, then a very short duration waveform performs best, that is $W/W_q \gg 1$. The curves clearly show that there is a range of values of W/W_q that should be avoided. These results are another way of stating that the waveform design goal in the dense target case is to have as little intersection as possible between the waveform ambiguity contours and the distribution in the range-Doppler plane of the interfering signals. This is illustrated in Fig. 10.17 for the two extremes discussed above. When the mean Doppler is small a signal such as the linear FM waveform performs nearly as well as the narrow pulse.

The criterion of distributing waveform ambiguity in regions of the range-Doppler plane where *a priori* targets are not expected is a good general design guide. However, if the target distributions in the range-Doppler plane result in disconnected or complex shaped regions then this design guide may become of limited usefulness because of the realizability constraints on the ambiguity function. One of the common objectives in waveform design is to achieve clear areas in the distribution of ambiguity. The uniform pulse train has been proposed for this application. However, one of the objectionable properties of the uniform pulse train signal is the large amount of peak ambiguities associated with it. This is the penalty that must be paid for obtaining a clear area around the central spike location. Various methods for reducing the peak magnitudes of the pulse train ambiguities were discussed in Chapter 8. It was seen that techniques such as staggering the positions of the pulses in the train or of imposing an FM or phase code on each pulse would disperse the ambiguity volume

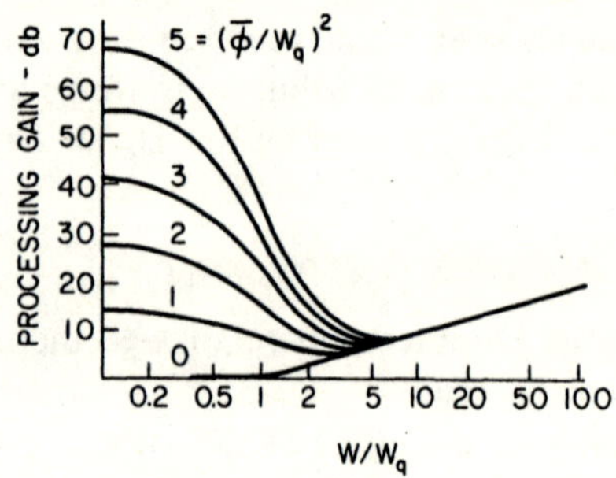

FIG. 10.16 Processing gain of Gaussian pulse against dense clutter (courtesy of E. C. Westerfield *et al.* [7]).

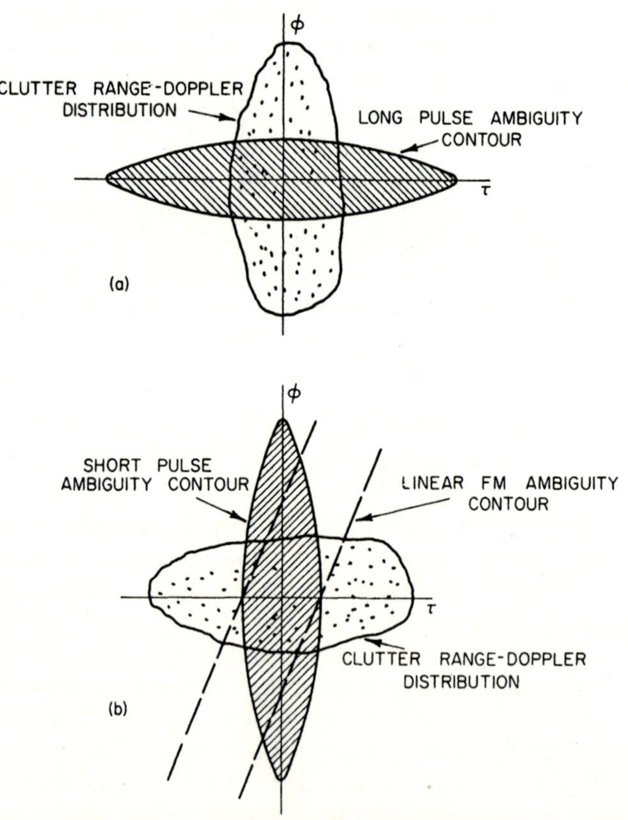

FIG. 10.17 Examples of minimizing intersection of waveform ambiguity contours and clutter range-Doppler distribution (after Stewart and Westerfield [6]).

of the undesired spikes into the otherwise ambiguity free regions. This tends to create the thumbtack type of ambiguity function which provides a large amount of waveform generated clutter at all values of range and Doppler shift of practical interest. Thus, as Fowle *et al.* [1] note, this procedure is not recommended for dense target distributions in a compact region. One possible method of clearing away large ambiguities is to change the frequency of each pulse in the train in a linear or random manner. This will tend to decorrelate the ambiguity regions in the range and Doppler intervals of interest, depending on the magnitude of the frequency change from pulse-to-pulse. However, this technique will not yield a large clear area around the central response. The effect of a linear change in frequency with $\Delta f_n = 3/\delta$ was illustrated in Chapter 8, showing that the central ambiguity response is broken up into a fine grained ambiguous structure. In general, this structure will be such that resolution among different signals will correspond to the envelope of the response rather than its fine grained structure. Thus, the cost of removing ambiguity by this method is an increase of transmitter and receiver bandwidths with no equivalent increase, for all practical purposes, of the range resolution capability of the signal. As Rihaczek [11] points out, depending on individual circumstances this may or may not be a reasonable price to pay.

It is of interest to examine signal design criteria for the general dense target case, in the light of the statement made in Chapter 4 that with absolutely no prior knowledge concerning the particular radar application or environment, all signals must be considered as potentially equally good (or bad). This point can be illustrated by taking the examples of the thumbtack ambiguity function discussed earlier in Section 10.2 and the linear FM ambiguity function for an assumed target distribution that is uniformly dense and extending in time and frequency shift over the limits of the ambiguity plane, which are taken as $-T < \tau < T$ in time and $-\Delta f < \phi < \Delta f$ in frequency. T is the transmitted signal duration, and the bandwidth Δf is assumed to be a well defined modulation bandwidth rather than the second moment (rms) bandwidth β which, as noted in Chapter 9, often does not conform to the measures of bandwidth employed by equipment engineers. The fact that signal duration and bandwidth cannot both be truncated is of secondary importance in this discussion, and it will be assumed for these large time-bandwidth signals that well defined limits can be placed on the ambiguity plane in both directions.

The thumbtack response described in Section 10.2 has a pedestal level relative to $|\chi(0,0)| = 1$ of $(2kT\,\Delta f)^{-1/2}$. For the bounds on the ambiguity plane given above the average value of $k = 2$ (required to give unit volume to the pedestal function of $|\chi|^2$, since the central spike has negligible volume). The average peak power level at the matched-filter output from a single target located in the plane (excluding the central spike) is $1/4T\,\Delta f$. Since there are $2T\,\Delta f$ resolution elements along both the τ and ϕ axes, there are $(2T\,\Delta f)^2$ resolution cells in the ambiguity plane. If the further assumption is made that essentially equal amplitude returns are associated with all of the resolution elements, then the signal-to-interference ratio observed at the matched-filter output for a desired signal at $\tau = 0$, $\phi = 0$ is

$$\frac{S}{I} = \frac{1}{T\,\Delta f} \tag{10-27}$$

An essentially identical result is shown by Rihaczek [12] for the type of ambiguity function discussed in Section 4.12, where the time and frequency extent of the pedestal (assumed to be uniform) are defined by α and β, the second movement measures of signal duration and bandwidth, rather than $2T$ and $2\,\Delta f$ as above.[1]

Considering next the linear FM signal, for which the Doppler shifted responses are weighted in amplitude by $1 - |\phi/\Delta f|$, the average voltage level of a single matched-filter output for the same target distribution assumed above is one-half, and the average peak power is one-quarter. As shown in Fig. 10.18, the linear FM ambiguity ridge can be considered to enclose approximately $2T\,\Delta f$ full resolution cells along the ridge line and $4T\,\Delta f$ half resolution cells. Signals occurring in any of these locations in time and frequency will contribute to the interference in the desired resolution cell at the origin of the ambiguity plane. Assuming that the effective number of resolution cells enclosed by the ambiguity ridge is $4T\,\Delta f$ gives a signal-to-interference ratio identical to (10-27) when this waveform is applied to the same target distribution. This heuristic discussion illustrates the fact that for this case, which is equivalent to having no prior target distribution information, no particular signal is more advantageous to use than another, a result directly derivable analytically from the general ambiguity function properties. A significant point brought out by these examples is that an increase in the signal time-bandwidth product will only serve to degrade the signal-to-interference ratio, since it has the result of encompassing more resolution cells that can reflect ambiguity into the desired cell under observation. The only means for combating this situation is to increase the carrier frequency of the transmitted signal, a solution derived by Fowle *et al.* [1]. Taking the linear FM ambiguity function as an example, this has the effect of increasing the Doppler shift of the received signals, and thus moving them out along the ambiguity ridge. Since the ambiguity function bounds are related to the signal modulation parameters and not the carrier frequency, a doubling of the carrier

---

[1] If the Doppler shift $\Delta\phi < \Delta f$, then (10-27) becomes $S/I = k_{av}/2T\,\Delta\phi$, where $-\Delta\phi < \phi < \Delta\phi$ and $k_{av} \doteq 1$ for $\Delta\phi \lesssim \Delta f$.

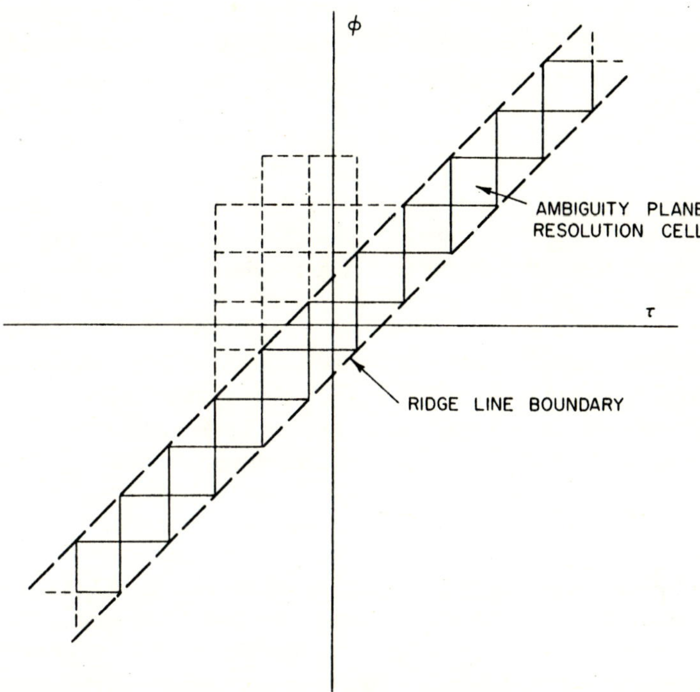

FIG. 10.18   A definition of the number of effective resolution cells enclosed by linear FM ambiguity ridge.

frequency for the same physical distribution of objects in range-velocity space will halve the number of signals that appear in the time-frequency space defined by the ambiguity function. This results in a halving of the interference level observed at $\tau = 0$, $\phi = 0$. In any event, the target handling capacity of the radar system will still be quite limited. For this type of uniformly dense target environment the results of Fowle *et al.* [1] and Rihaczek [12] point out that the "effective" resolution cell size, independent of the waveform used, is of the order of unity rather than $1/T\,\Delta f$ if acceptable signal-to-interference ratios are to be obtained.

Fortunately, Doppler spreads of $2\,\Delta f$ in magnitude are extremely unlikely, and more reasonable Doppler shift distributions will normally prevail. Here, the radar designer may attempt to match the ambiguity distribution to the unoccupied signal space. Alternatively, he may try to make the ambiguity distribution orthogonal to the target distribution in time delay and frequency shift, as shown by the example taken from Westerfield *et al.* [7]. For compact target distributions a waveform design can be attempted that results in a clear ambiguity region, as with the uniform pulse train. It is also theoretically possible to achieve an essentially clear area of unity surrounding the origin in an otherwise thumbtacklike ambiguity function (see Chapter 4 and Rihaczek [12] for a discussion of this). The optimum and suboptimum stagger pulse trains discussed in Chapter 8 are one means of obtaining this kind of behavior.

For the case where desired signals and interfering clutter are in separately resolvable narrow Doppler bands the system engineer may consider the use of some of the more complex pulse train coding schemes to minimize the ambiguity volume in a specific strip parallel to the zero-Doppler axis. If the desired signals are relatively few for this situation, then the smearing of ambiguity volume into the clear areas obtained with the uniform pulse train may be tolerable from the interference viewpoint, and the complex coded pulse train waveform can be used over a wider range interval than is practical with the uniform train. One can readily appreciate that pulse train waveforms provide less than the maximum amount of transmitted energy for a specified signal duration T and a given peak power limit. Thus, the application of such signals would be chiefly found where reception is interference, rather than receiver noise, limited.

A word of caution must also be sounded in the use of the generalized time resolution constant $T_r(\phi)$ as a design criterion. This function was seen to be important in determining radar performance in a dense clutter environment. The results obtained by several investigators lead to the conclusion that for this case a waveform with little effective Doppler resolution, such as the linear FM signal, is preferable to a waveform that provides Doppler resolution and waveform generated self clutter as a by-product. Waveforms that have $T_r(\phi)$ functions comparable to the linear FM signal include the linear step FM and Frank polyphase signals. However, when the relative target density becomes thinner (this could be a result of a larger signal bandwidth) the individual range sidelobes of the Doppler shifted matched-filter output start to become more important than the total energy in all of the sidelobes. Thus, the radar designer should carefully consider the expected Doppler shift environment before making a choice among the linear FM, step FM, and Frank code signals, as an example.

As noted in Chapter 4, and implicit in the discussion of this chapter, $T_r(0)$ is a measure of the range resolution capability of a matched-filter waveform as well as of its relative level of sidelobe generated interference. From a practical viewpoint these two are closely tied together since, in an environment in which the received signal strengths have large variations, the minimum resolvable range separation of two or more signals will depend on the ability of the weaker signals to compete against the range sidelobe background of the stronger signal. The measure of resolution provided by $T_r(0)$

is also called the "equivalent rectangle of resolution" since, from the formulation given by Woodward, it defines the duration of a rectangular pulse that contains the same amount of energy as $|\chi(\tau, 0)|$.

Other analytic measures of resolution can be applied to $\chi(\tau, 0)$. One such important measure is the time spread of the matched-filter response as defined by the standard deviation σ_T of $|\chi(\tau, 0)|$, where

$$\sigma_T^2 = \frac{\int_{-\infty}^{\infty} \tau^2 |\chi(\tau, 0)| \, d\tau}{\int_{-\infty}^{\infty} |\chi(\tau, 0)| \, d\tau} \tag{10-28}$$

Eq. (10-28) defines an effective half-width of $|\chi(\tau, 0)|$, so that $2\sigma_T$ is the measure of total time duration spread. A comparison of $T_r(0)$, $2\sigma_T$ and half-amplitude widths for some elementary pulse shapes is given in Table 10-II.

TABLE 10-II

COMPARISON OF MEASURES OF TIME RESOLUTION

Pulse shape	$T_r(0)$	$2\sigma_T$	-6 db width
rect (t/T)	T	$0.64\,T$	T
$(1 - |t|/T)$, $\quad |t| \leqslant T$	$2T/3$	$0.815\,T$	T
$\cos(\pi t/2T)$, $\quad |t| \leqslant T$	T	$1.67\,T$	$1.33\,T$
$(\sin \pi t/T)/\pi t/T$	T	∞	$1.2\,T$

The major disadvantage of Eq. (10-28) is that it gives excessive weight to low skirt and/or sidelobe levels of $\chi(\tau, 0)$. Thus, it does not appear to be an appropriate measure of resolution for large time-bandwidth signals that have extended range sidelobe regions, unless the sidelobe regions are arbitrarily excluded from the calculation. The interested reader may, using (10-28), verify for himself the strong dependence of σ_T on time-bandwith product for the "thumbtack" versions of $\chi(\tau, 0)$ discussed in Section 10.4. By contrast, the $T_r(0)$ for these functions are independent of time-bandwith product. If the observed signal is $|\chi(\tau, 0)|^2$ rather than $|\chi(\tau, 0)|$, as it would be if the matched filter is followed by a large dynamic range square-law device, then (10-28) can be used equivalently with the substitution of $|\chi(\tau, 0)|^2$ for $|\chi(\tau, 0)|$. This measure of spread is discussed further in Section 11.8 in the context of resolution degradation caused by distortion. In both cases the effect of extended range sidelobes does not recommend either of these forms for many of the large time-bandwidth signals that are of interest to the radar designer. In any event, the system designer will not rely solely on any of the analytic measures of resolution just presented, but will add to these his own subjective criteria derived from the problem at hand. As a final thought in this context, one of the favorable features of the class of linear FM signals is that they retain their range resolution capability (for practical resolution criteria) for most Doppler ranges of interest.

It would appear that the linear FM, binary phase code, and uniform pulse train signals are among the most important of the class of large time-bandwidth signals for radar application. However, it should be abundantly clear from the examples discussed in this and previous chapters that no one waveform is the ideal signal for detection, resolution, and estimation under all circumstances. In this regard the system designer should determine for himself which kinds of target distributions and which portions of the ambiguity plane are his chief concern. These may briefly be described as falling into the following broad categories:

(1)   The entire ambiguity plane (for all practical purposes).
(2)   Restricted regions in range and Doppler.
(3)   Restricted Doppler region or regions.
(4)   Restricted range region or regions.
(5)   Single, multiple, or dense targets for each of the above.

For each of these applications one or more waveforms may be suitable. The choice might depend on whether there is the requirement for simultaneous extraction of range and velocity information. Other practical factors that may enter into the determination are ease of implementation for a particular signal, susceptibility to distortion, target densities and characteristics, the desirability of using more than one type of waveform, transmitter limitations, cost, etc. Since one cannot, in the abstract, define an "optimum" waveform, it is hoped that the basic examples provided in the preceding chapters will give the system designer a frame of reference from which to examine his own needs and devise his own solutions. In the last analysis, this is *his* responsibility. The authors hope that they have fulfilled their intent of providing for the system engineer a knowledge of the various types of waveforms at his command, as well as an appreciation of some of the theoretical and practical problems associated with their use.

REFERENCES

1. E. N. Fowle, E. J. Kelly, and J. A. Sheehan, Radar system performance in a dense target environment, *IRE Intern. Conv. Record*, Pt. 4, 136–145 (1961).
2. R. Manasse, The use of pulse coding to discriminate against clutter, Mass. Inst. Technol., Lincoln Lab., Lexington, Massachusetts, Rept. 312–12 (August, 1957).
3. H. Urkowitz, Filters for detection of small signals in clutter, *J. Appl. Phys.* **24**, 1024–1036 (1953).
4. A. W. Rihaczek, Optimum filters for signal detection in clutter, *IEEE Trans.* **AES-1**, 297–299 (1965).

5. W. L. Rubin and C. Kaiteris, Waveform design for detection of signals in clutter, Sperry Gyroscope Co., Great Neck, New York, Rept. RD-5262-0611 (June, 1965). To be published in *Proc. IEE* (*London*), (1967).

6. J. L. Stewart and E. C. Westerfield, A theory of active sonar detection, *Proc. IRE.* **47**, 872–881 (1959).

7. E. C. Westerfield, R. H. Prager, and J. L. Stewart, Processing gains against reverberation (clutter) using matched filters, *IRE Trans.* **IT-6**, 342 (1960).

8. D. Richman, Resolution of multiple targets in range, and range-rate, Rome Air Develop. Center, Griffiss AF Base, New York, Rept. RADC-TDR-62-580 (1963) (not generally available).

9. H. L. Van Trees, Optimum signal design and processing for reverberation limited environments, *IEEE Trans.* **MIL-9**, 212–229 (1965).

10. P. M. Woodward, "Probability and Information Theory with Applications to Radar." Pergamon Press, Oxford, 1953.

11. A. W. Rihaczek, Radar resolution properties of pulse trains, *Proc. IEEE* **52**, 153–164 (1964).

12. A. W. Rihaczek, Radar signal design for target resolution, *Proc. IEEE* **53**, 116–128 (1965).

CHAPTER 11

Effects of Distortion on Matched-Filter Signals

11.1 Introduction

The radar systems engineer can count, along with the finite human span and taxes [1], distortion and its undesirable effects as being among the certainties. The problem that must be solved, or made amenable to some suitable compromise, by the system designer is that of identifying the sources of distortion and then making sure that the contribution of each to the overall distortion is reduced to a minimum. Because of the nature of matched-filter waveforms, many of which exhibit the phenomenon of range sidelobes, it has sometimes been erroneously assumed that in radar applications the presence of distortion effects are peculiar to matched-filter techniques. This is, of course, not true, and it can be assumed that early, and even more recent, radar systems had distortion effects present that undoubtedly limited the dynamic range and/or weak signal detection and resolution capabilities more severely than implied by the system specifications. What *can* be stated as a peculiarity of matched-filter systems is that the results of time modulation distortion (amplitude and phase) can often be separately observed apart from the desired signal, providing in some instances rather unique and interesting results.

It was shown in Chapter 7, in connection with range sidelobe reduction and weighting function design, that a rippling variation in a spectrum function can contribute the equivalent of distortion sidelobes to the matched-filter waveform. It is the presence of the range sidelobes of a matched-filter signal that has caused a renewed interest in distortion as a criterion of radar system design and the application of a distortion analysis technique that has long been used in telephone and television applications. The following sections will review the basic paired-echo distortion analysis technique and its relation to radar system and waveform design.

11.2 Paired-Echo Distortion Analysis

In many applications it is desirable to reduce the largest range sidelobes of a pulse-compression matched-filter signal to the -30 db to -40 db level relative to the peak amplitude of the compressed pulse. The discussion that follows will show that the presence of amplitude and phase distortion in the receiver can contribute distortion lobes, or the so-called paired-echo signals, that are similar in appearance to the range sidelobes, but which cannot be easily removed by straightforward frequency weighting techniques. In the context of a practical matched-filter radar application, it is meaningless to talk about low range sidelobe levels unless permissible levels of distortion in all parts of the system are specified, or unless practical distortion compensation techniques are incorporated into the system.

Figure 11.1(a) illustrates the ideal, distortionless filter through which a signal is passed. The output spectrum is $S(\omega)$, so that the undistorted output time function is

$$s_1(t) = \frac{1}{2\pi} \int_{-\infty}^{\infty} S(\omega) \exp[j\omega t]\, d\omega \tag{11-1}$$

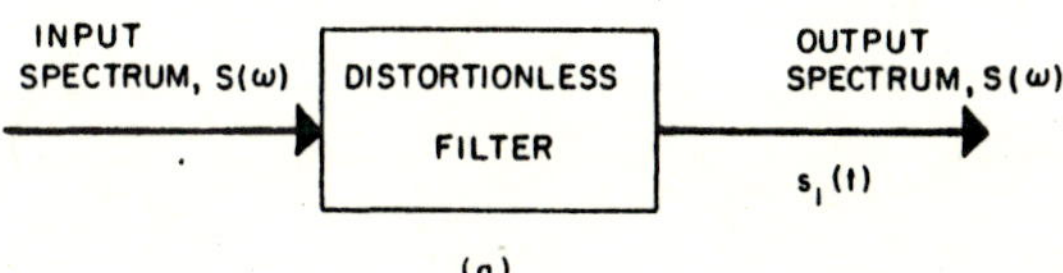

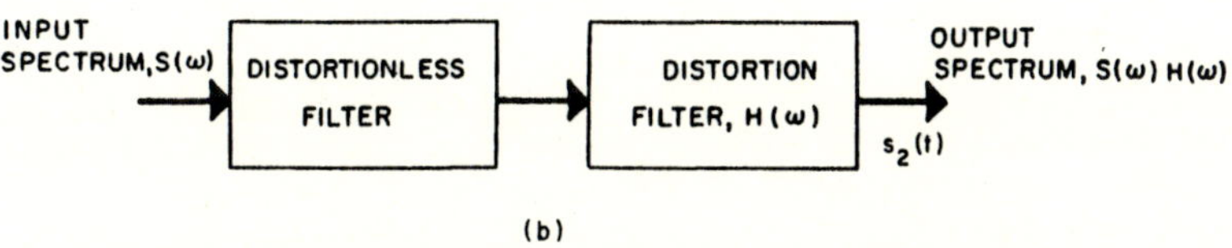

FIG. 11.1   Ideal and distorted filters for distortion analysis.

Equation (11-1) represents the ideal, undistorted time waveform. If, as shown in Fig. 11.1(b), a distortion filter is connected in tandem with the ideal filter, the output signal will be affected by the characteristic of the distortion filter, so that

$$s_2(t) = \frac{1}{2\pi} \int_{-\infty}^{\infty} S(\omega)H(\omega) \exp[j\omega t]\, d\omega \tag{11-2}$$

MacColl [2] and Wheeler [3] both approached the analysis of distortion by assuming sinusoidal functions for the characteristics of the distortion filter. The respective amplitude and phase components of a single frequency component of the Fourier expansion·of the distorted transfer function were chosen as

$$A(\omega) = a_0 + a_1 \cos C_1 \omega \tag{11-3a}$$

$$B(\omega) = b_0 \omega - b_1 \sin C_1 \omega \tag{11-3b}$$

These basic distortion components are illustrated in Fig. 11.2. The composite distortion function is

$$H(\omega) = A(\omega) \exp[-jB(\omega)] \tag{11-4}$$

Using (11-4) in the expression for the distorted output signal, one has

$$s_2(t) = \frac{1}{2\pi} \int_{-\infty}^{\infty} S(\omega)(a_0 + a_1 \cos C_1 \omega) \exp[j\omega t' + jb_1 \sin C_1 \omega] \, d\omega \tag{11-5}$$

where $t' = t - b_0$.

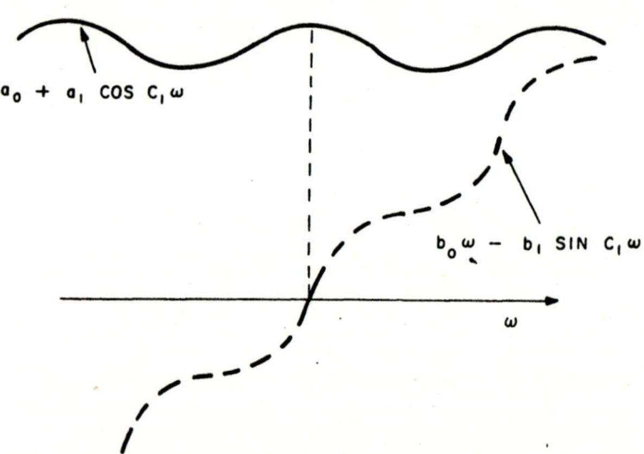

FIG. 11.2   Sinusoidal distortion components of $H(\omega)$.

For the simple case of $b_1 = 0$, and writing the amplitude distortion term as

$$a_1 \cos C_1 \omega = \frac{a_1}{2}(\exp[jC_1 \omega] + \exp[-jC_1 \omega]) \tag{11-6}$$

the expression for amplitude distortion only becomes

$$s_2(t) = \frac{a_0}{2\pi}\int_{-\infty}^{\infty} S(\omega)\exp[j\omega t']\,d\omega + \frac{a_1}{2}\left[\frac{1}{2\pi}\int_{-\infty}^{\infty} S(\omega)\exp[j\omega(t' + C_1)]\,d\omega\right.$$

$$\left. + \frac{1}{2\pi}\int_{-\infty}^{\infty} S(\omega)\exp[j\omega(t' - C_1)]\,d\omega\right] \tag{11-7}$$

Noting that the constant C_1 in the imaginary exponent implies a shift in the time origin for that part of the output signal, (11-7) can be expressed as

$$s_2(t) = a_0\,s_1(t') + \frac{a_1}{2}[s_1(t' + C_1) + s_1(t' - C_1)] \tag{11-8}$$

where $s_1(t' + C_1)$ and $s_1(t' - C_1)$ represent advanced and delayed replicas of the undistorted signal, weighted by the factor $a_1/2$. As a result of the symmetry evidenced by the distortion terms in Eq. (11-8), the designation "paired echo" was coined for these distortion signals.

When there is only phase distortion present (i.e., $a_1 = 0$), the output of the distortion filter is given by

$$s_2(t) = \frac{a_0}{2\pi}\int_{-\infty}^{\infty} S(\omega)\exp[j\omega t']\exp[jb_1 \sin C_1\omega]\,d\omega \tag{11-9}$$

The phase distortion function can be replaced by a Bessel function expansion, yielding

$$\exp[jb_1 \sin x] = J_0(b_1) + \sum_{n=1}^{\infty} J_n(b_1)\{\exp[jnx] + (-1)^n \exp[-jnx]\} \tag{11-10}$$

Substitute this into Eq. (11-9) and perform the indicated integration; this results in

$$s_2(t) = a_0\left\{J_0(b_1)s_1(t')\right.$$

$$\left. + \sum_{n=1}^{\infty} J_n(b_1)[s_1(t' + nC_1) + (-1)^n s_1(t' - nC_1)]\right\} \tag{11-11}$$

For the case of pure phase distortion there exists an infinite set of paired echoes. However, if b_1 is small, say less than 0.5 radians, the following approximations can be made:

$$J_0(b_1) = 1, \qquad J_1(b_1) = b_1/2, \qquad J_n(b_1) = 0, \qquad n \geq 2$$

and Eq. (11-11) reduces to

$$s_2(t) = a_0\left\{s_1(t') + \frac{b_1}{2}[s_1(t' + C_1) - s_1(t' - C_1)]\right\} \tag{11-12}$$

Thus, under the condition of relatively small phase errors, the results of phase distortion are similar to that of amplitude distortion, except that the echo terms have opposite polarity. Equations (11-8) and (11-12) can be combined, so that the effect of the composite amplitude and phase errors as given in (11-3a) and (11-3b) is described by

$$s_2(t) = a_0\left[s_1(t') + \left(\frac{a_1}{2a_0} + \frac{b_1}{2}\right) s_1(t' + C_1) + \left(\frac{a_1}{2a_0} - \frac{b_1}{2}\right) s_1(t' - C_1)\right] \quad (11\text{-}13)$$

If the phase distortion is not small, then the exact result of performing the integration of Eq. (11-5) is

$$\begin{aligned}
s_2(t) = a_0\Bigg\{ &\frac{a_1}{2a_0} J_0(b_1)[s_1(t' + C_1) + s_1(t' - C_1)] + J_0(b_1)s_1(t') \\
&+ \sum_1^\infty J_n(b_1)[s_1(t' + nC_1) + (-1)^n s_1(t' - nC_1)] \\
&+ \frac{a_1}{2a_0} \sum_1^\infty J_n(b_1)[s_1(t' + nC_1 + C_1) + (-1)^n s_1(t' - nC_1 + C_1)] \\
&+ \frac{a_1}{2a_0} \sum_1^\infty J_n(b_1)[s_1(t' + nC_1 - C_1) + (-1)^n s_1(t' - nC_1 - C_1)]\Bigg\}
\end{aligned}$$

$$(11\text{-}14)$$

where $t' = t - b_0$ as before.

The expansion of the terms in the summations indicated above results in combinations of the following type:

$$\frac{a_1}{2}[J_{n-1}(b_1) + J_{n+1}(b_1)]s_1(t' \pm nC_1) \quad (11\text{-}15)$$

so that by making use of the Bessel function recursion formula

$$\frac{2n}{x}J_n(x) = J_{n-1}(x) + J_{n+1}(x) \quad (11\text{-}16)$$

one can obtain the following expression for the distorted output function:

$$\begin{aligned}
s_2(t) = a_0\Bigg\{ &J_0(b_1)s_1(t') + \sum_1^\infty J_n(b_1)\left[\left(1 + \frac{na_1}{a_0 b_1}\right) s_1(t' + nC_1)\right. \\
&\left. + (-1)^n\left(1 - \frac{na_1}{a_0 b_1}\right)s_1(t' - nC_1)\right]\Bigg\}
\end{aligned}$$

$$(11\text{-}17)$$

This represents the complete formulation of the distorted output signal for network phase and amplitude errors of the same frequency. For the case of small phase distortion Eq. (11-17) reduces to (11-13), using the Bessel function approximations given above.

When the network distortion function is not expressible as a simple harmonic function it can be treated as consisting of the components of the Fourier expansion of the complex distortion function. The complex distortion function can then be expressed as

$$A(\omega) = a_0 + \sum_{n=1}^{N} a_n \cos C_n \omega \qquad (11\text{-}3c)$$

$$B(\omega) = b_0 \omega - \sum_{n=1}^{N} b_n \sin C_n \omega \qquad (11\text{-}3d)$$

To obtain the Fourier components, a_n and b_n, of the network distortion function it is necessary to assume a truncated bandpass function, and then treat the distortion functions as being periodic in the frequency domain. The truncation of the bandpass, and thus the distortion, function can be imposed at any arbitrary frequency limits. A logical set of limits could be the frequencies at which the signal spectrum has fallen to half its mid-band amplitude. However, the error involved would be related to the rate at which the spectrum is falling off in amplitude, or to the amount of spectrum energy excluded by the truncation process (for the analogous case of complex time distortion functions, the application to pulsed radar signals results in a truncated time function so that this error is not an important consideration in time domain distortion analysis). By proper choice of the frequency reference points for the Fourier expansion, and by the assumption of the type of periodic behavior of the distortion function outside the truncation limits (i.e., odd or even symmetric behavior about the truncation limits) one can force the Fourier decomposition of the amplitude and phase distortion functions to conform to Eqs. (11-3c) and (11-3d), as well as minimize the number of significant terms in the expansion.

For the amplitude case, the individual components of the distortion function will contribute independent sets of paired echoes. For the phase distortion terms, however, the combined effect is that of a multiplication of imaginary exponential sine functions. From Eq. (11-10) this results in a multiplication of Bessel function expansions. For small phase errors (less than 0.5 radian) the effects of the various phase error terms can be treated as being essentially independent. For larger phase error terms there is the possibility of significant cross-product echo signals. The section on modulation distortion presents the basic methods of treating this latter situation.

In the particular example of the matched-filter application the analytical method outlined above can be made more general to cover two basic cases of interest. The first of these results from the fact that the matched-filter delay characteristic is usually designed independently of any amplitude characteristic (see Chapters 12 and 13), so that the frequencies of major amplitude and phase errors may be completely different. This case has been treated by Liebman [4], and follows directly from the procedures outlined above, with the distortion network functions now being given by

$$A(\omega) = a_0 + a_1 \cos C_1 \omega \tag{11-18a}$$

$$(C_1 \neq C_2)$$

$$B(\omega) = b_0 \omega - b_1 \sin C_2 \omega \tag{11-18b}$$

The exact expression is similar to Eq. (11-14), except that the phase distortion echo terms are displaced by the factor $\pm nC_2$ about the locations of the desired term located at $t' = 0$, and about the amplitude paired-echo terms located at $t' = \pm C_1$. Assuming again that the phase distortion is small, the composite distorted output function is

$$
\begin{aligned}
s_2(t) = a_0 \Big\{ &s_1(t') + \frac{b_1}{2}[s_1(t' + C_2) - s_1(t' - C_2)] \\[2mm]
&+ \frac{a_1}{2a_0}[s_1(t' + C_1) + s_1(t' - C_1)] \\[2mm]
&+ \frac{a_1 b_1}{4a_0}[s_1(t' + C_1 + C_2) - s_1(t' + C_1 - C_2) \\[2mm]
&+ s_1(t' - C_1 + C_2) - s_1(t' - C_1 - C_2)] \Big\}
\end{aligned}
\tag{11-19}
$$

In many cases the last four terms will be negligible, and only the first five terms (the main signal plus four echoes) need be considered. It is interesting to note that at no point in the preceding development was it necessary to specify the functional form of the signal $s_1(t)$, illustrating that the locations of the paired-echo terms, and their waveform shapes, are independent of the type of signal used. This is not the case when time modulation errors are considered for general matched-filter signals, although an approximate one-to-one relationship between network paired echoes and modulation paired echoes can be made for the linear FM matched-filter signal.

The second case of general interest is also related to the fact that the matched-filter delay characteristic is designed separately from the system amplitude response function. Because of this there is no reason for the phase errors to assume the particular sinusoid form given in Eq. (11-3b). A more general expression for the sinusoidal form of phase error would include an arbitrary phase constant [5], so that

$$B(\omega) = b_0\omega - b_1 \sin(C_1\omega + \Phi_0) \tag{11-20}$$

Expressing the input to the distortion filter as an i.f. signal, or

$$s_1(t) = \overline{s_1(t)} \cos \omega_0 t \tag{11-21}$$

where $\overline{s_1(t)}$ denotes the matched-filter output envelope, then the effect of the phase constant Φ_0 can be determined. The phase distorted output signal for this case can be expressed, in complex notation, as

$$s_2(t) = \frac{1}{2\pi} \exp[j\omega_0 t] \int_{-\infty}^{\infty} S(\omega) \exp[j\{\omega t' + b_1 \sin(C_1\omega + \Phi_0)\}] \, d\omega \tag{11-22}$$

The Bessel function expansion of Eq. (11-10) will now have the imaginary exponential terms

$$\exp[jn(C_1\omega + \Phi_0)] \quad \text{and} \quad \exp[-jn(C_1\omega + \Phi_0)]$$

Using these, the equivalent expression to Eq. (11-11) for the phase distorted output is

$$\begin{aligned}
s_2(t) = a_0\Big\{ &J_0(b_1)\overline{s_1(t')} \cos \omega_0 t \\
&+ \sum_1^{\infty} J_n(b_1)[\overline{s_1(t' + nC_1)} \cos(\omega_0 t + n\Phi_0) \\
&+ (-1)^n \overline{s_1(t' - nC_1)} \cos(\omega_0 t - n\Phi_0)]\Big\}
\end{aligned} \tag{11-23}$$

Here, one can see that the effect of the phase constant Φ_0 is carried over into the carrier frequency terms of the paired-echo signals as phase constants $\pm n\Phi_0$. Depending on the value of C_1 and Φ_0, the combination of the overlapping portions of the desired signal and the distortion signals can cause the composite waveform to exhibit nonsymmetrical (about $t' = 0$) cancellation and reinforcement effects. Because of the correspondence of network and modulation distortion for the linear FM case, the modulation distortion waveforms shown in Figs. 11.12 to 11.17 can be considered as typical of the effects of network distortion on any matched-filter output, since it was shown that the form of the results of network distortions were independent of the waveform used.

When the desired level of range sidelobes and the frequency weighting network have been determined for the matched-filter system, then the paired-echo signals should be no larger, and preferably at least 3 db less, than the range sidelobes. This establishes the permissible distortion level in the system. Figure 11.3 plots the paired-echo amplitude level in db relative to the peak undistorted signal as a function of amplitude and phase distortion magnitudes. It can be seen that only small distortions can be tolerated when paired echoes are to be kept 30 db or more below the desired signal. That the paired-echo problem can be a serious limitation on system dynamic range capability in the vicinity of a large received signal

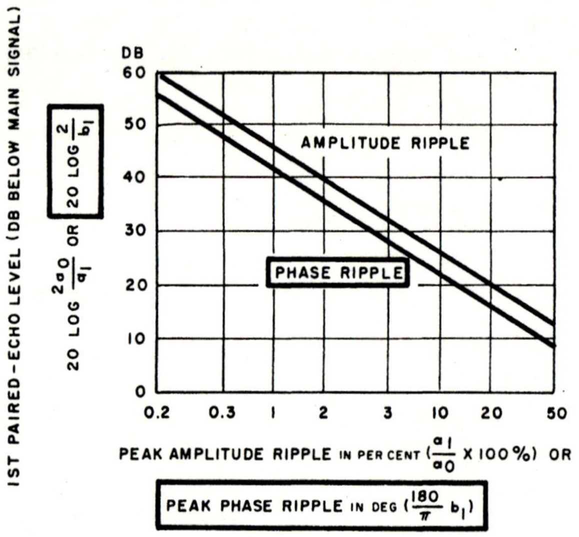

FIG. 11.3   Amplitude of first paired echo as a function of amplitude and phase distortion magnitude.

is shown in Fig. 11.4, which is reproduced from the basic distortion study reported by DiFranco and Rubin [6]. Here are shown, for idealized rectangular pulses, the paired-echo effects of amplitude and phase distortions of network functions that are commonly used in receiving systems. The 3 db signal spectrum width is over the interval -1 to $+1$ shown, with the transfer function 3 db points, $\omega_{3\,db}$, being as indicated in the transfer function column. A receiver bandwidth that is considerably wider than the signal spectrum is not unusual, especially in matched-filter systems in which a component of the matched filter, such as the sidelobe reduction bandshaping network, establishes the effective signal and noise bandwidths of the system. In order to eliminate such distortion signals, amplitude and phase correction, or cancellation based on transversal filter techniques, must be used. In addition, a different approach to the design of the receiver transfer function may be necessary.

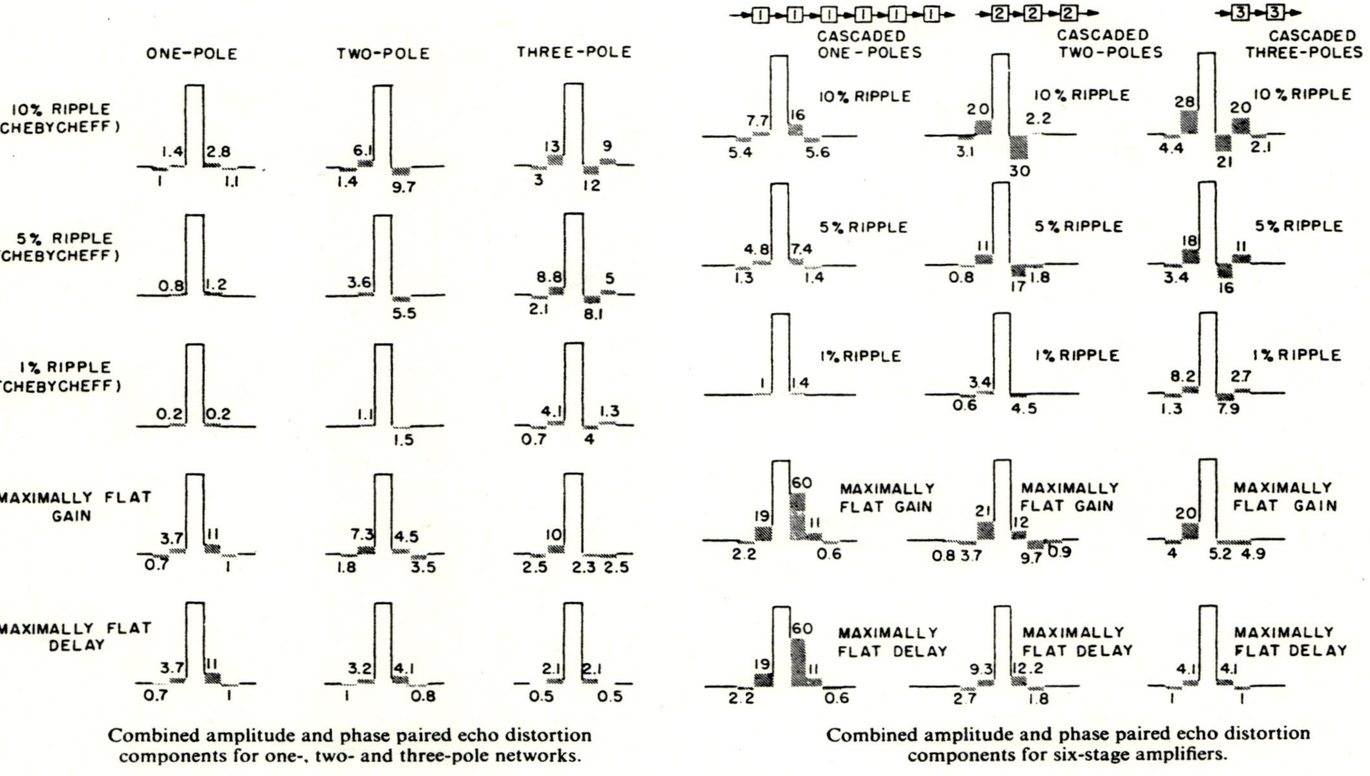

Combined amplitude and phase paired echo distortion components for one-, two- and three-pole networks.

Combined amplitude and phase paired echo distortion components for six-stage amplifiers.

Fig. 11.4(a) Bar graph representation of effect of receiver response amplitude and phase distortion of pulsed signals (courtesy of J. V. DiFranco and W. L. Rubin [6]).

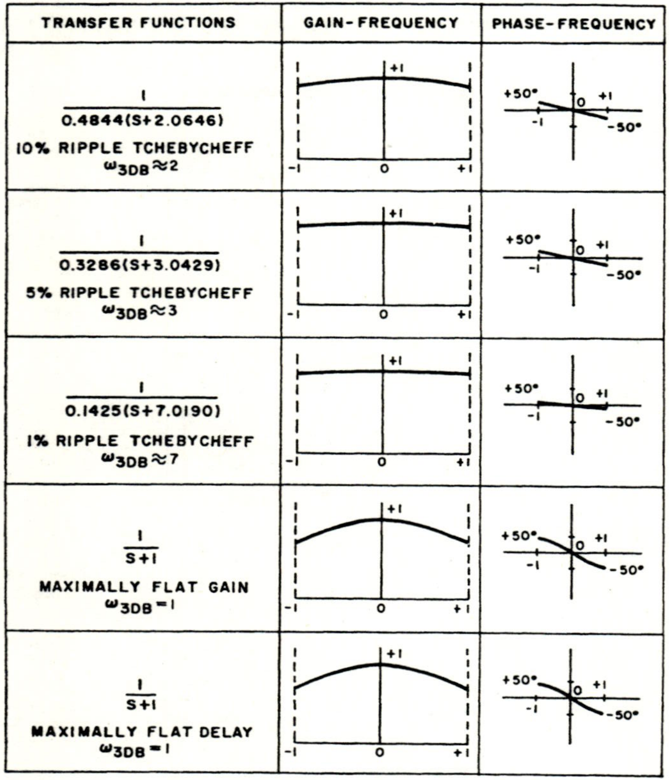
TRANSFER FUNCTIONS
GAIN-FREQUENCY
PHASE-FREQUENCY
1 / 0.4844(S+2.0646)
10% RIPPLE TCHEBYCHEFF
ω3DB≈2
1 / 0.3286(S+3.0429)
5% RIPPLE TCHEBYCHEFF
ω3DB≈3
1 / 0.1425(S+7.0190)
1% RIPPLE TCHEBYCHEFF
ω3DB≈7
1 / S+1
MAXIMALLY FLAT GAIN
ω3DB=1
1 / S+1
MAXIMALLY FLAT DELAY
ω3DB=1
+50°
-50°

One-pole transfer functions.

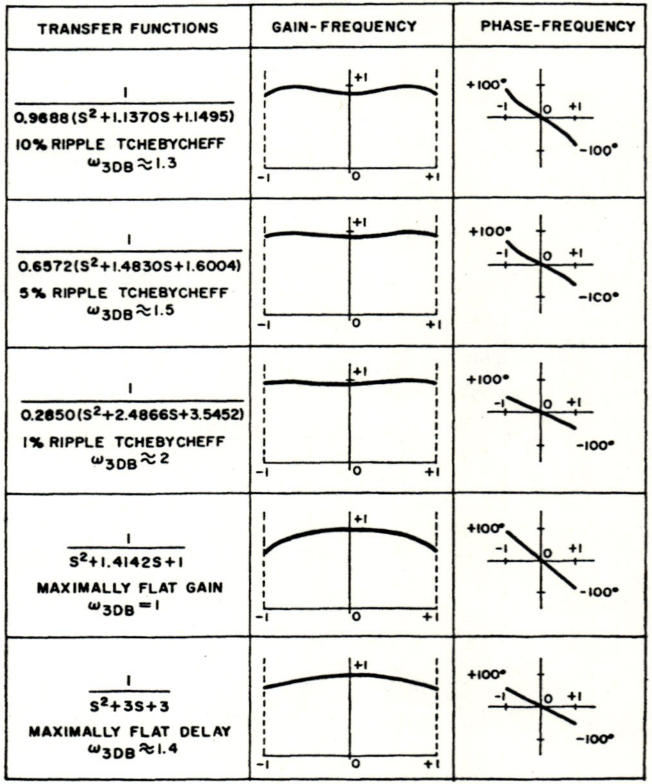
TRANSFER FUNCTIONS
GAIN-FREQUENCY
PHASE-FREQUENCY
1 / 0.9688(S²+1.1370S+1.1495)
10% RIPPLE TCHEBYCHEFF
ω3DB≈1.3
1 / 0.6572(S²+1.4830S+1.6004)
5% RIPPLE TCHEBYCHEFF
ω3DB≈1.5
1 / 0.2850(S²+2.4866S+3.5452)
1% RIPPLE TCHEBYCHEFF
ω3DB≈2
1 / S²+1.4142S+1
MAXIMALLY FLAT GAIN
ω3DB=1
1 / S²+3S+3
MAXIMALLY FLAT DELAY
ω3DB≈1.4
+100°
-100°

Two-pole transfer functions.

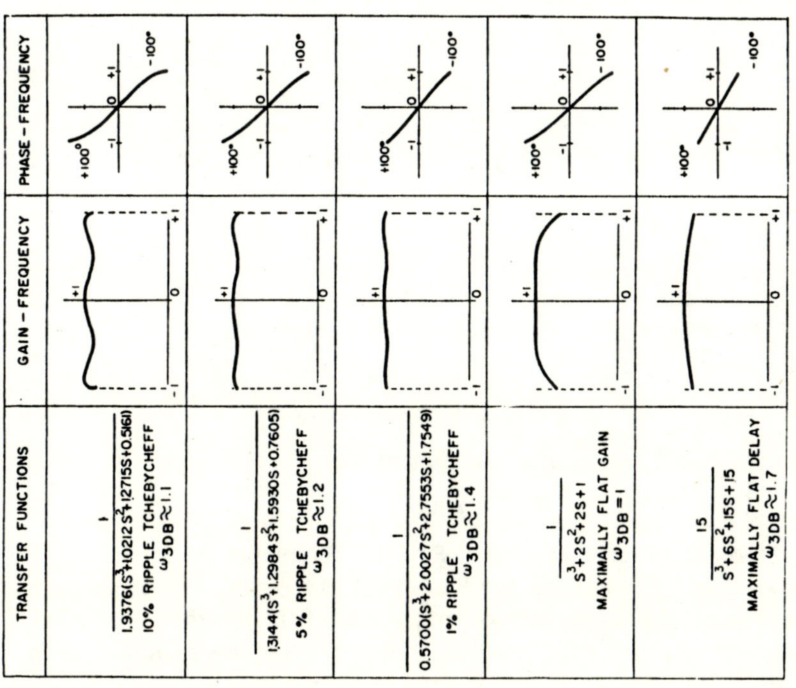

Three-pole transfer functions.

Fig. 11.4(b). Receiver response functions used to calculate waveforms shown in Fig. 11.4(a) (courtesy of J. V. DiFranco and W. L. Rubin [6]).

11.3 Delay Distortion as a Measure of Phase Distortion

The treatment of distortion effects developed in the previous section was based on amplitude and phase error considerations. As will be discussed in Chapters 12 and 13, the measurement most commonly used to evaluate the properties of pulse-compression dispersive delay lines is that of the group time delay characteristic. In the chapters on dispersive delay line design the use of the time delay error as a design criterion is developed. In order to use this error function as a means of estimating the level of the paired-echo distortion signals, the time delay error must be converted to phase error. This is done by making use of the relationship between group delay and phase shift. Thus,

$$t_g = dB(\omega)/d\omega \tag{11-24}$$

where t_g is the group time delay and $B(\omega)$ the phase function, given by (11-3b). Neglecting constants of integration, the phase function can in turn be written

$$B(\omega) = \int t_g \, d\omega \tag{11-25}$$

Assuming that the delay error curve can be expressed as a combination of a finite number of sinusoidal components, the group delay of each component is

$$(t_g)_n = (b_n C_n/\Delta\omega) \cos C_n\omega/\Delta\omega \tag{11-26}$$

and the associated phase error component is

$$B_n = b_n \sin C_n\omega/\Delta\omega \tag{11-27}$$

where C_n is the number of cycles of the error component in the signal band, $\Delta\omega = 2\pi \, \Delta f$. From this the location of the paired-echo signals can be placed at $\pm(C_n/\Delta f)$ seconds on either side of the desired output. The magnitude of the error signal for peak phase errors b_n less than 0.5 radian is $b_n/2$, or in decibel measurement the paired-echo magnitude is $20 \log(b_n/2)$.

From Eq. (11-26) can be seen the basic importance of the group delay error curve as a design criterion. The peak time delay error is

$$(\hat{t}_g)_n = b_n C_n/\Delta\omega \tag{11-28}$$

Since b_n is the critical factor in determining the paired-echo magnitude, time delay error correction to reduce these phase error distortion signals need not require a reduction in the peak time delay error for, as the equation above points out, a time delay error correction technique that results in an increased number of error cycles C_n over the band Δf will reduce the peak phase error even if the peak delay error remains fixed. If delay error correction reduces the peak delay error and increases the number of error cycles, then the peak phase error will be reduced much faster than from either factor alone. An example of this approach to delay error correction is presented in Chapter 12. An illustration of the relationship between time delay and phase errors is given in Figs. 11.5 and 11.6. Figure 11.5 shows the normal and distorted time delay function of a linear delay pulse-compression filter. The distorted time delay curves were obtained by adding and removing appropriate bridged-T sections to yield approximately

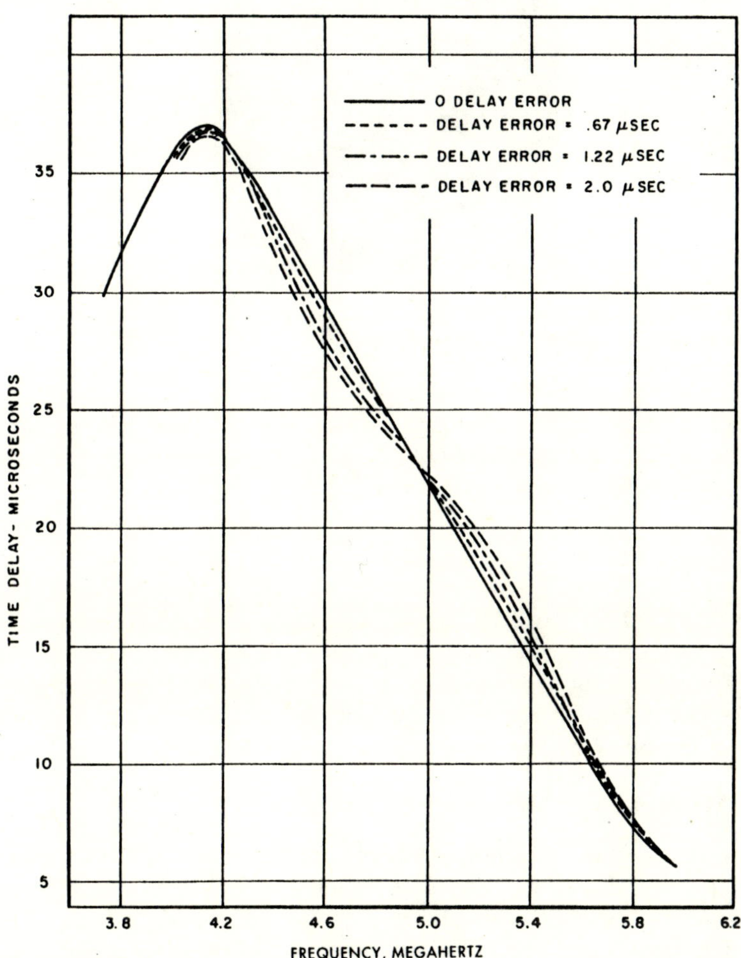

FIG. 11.5   Normal and distorted linear delay functions (from "Modern Radar—Analysis, Evaluation, and System Design" (R. S. Berkowitz, ed.), Wiley, New York, 1965).

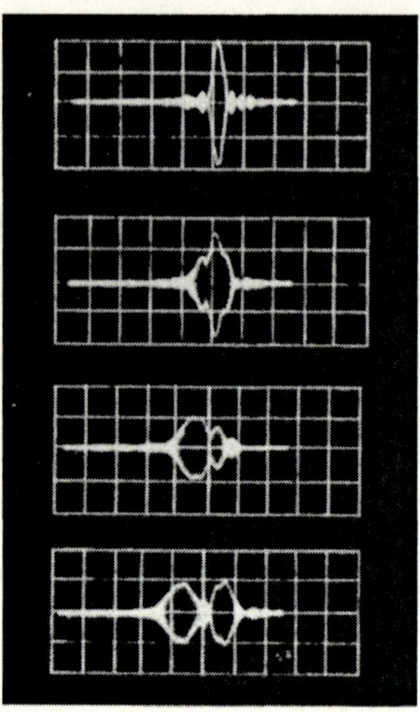

FIG. 11.6   Waveforms showing effect of time delay distortion (from "Modern Radar—Analysis, Evaluation, and System Design" (R. S. Berkowitz, ed.), Wiley, New York, 1965).

one sinusoid cycle of delay error over a band of 1.6 Mc. Peak delay deviations from the linear curve were 0.67, 1.22, and 2.0 microseconds, respectively. From Eq. (11-28) the peak phase error is

$$b_n = \frac{(\hat{t}_g)_n}{C_n}$$

where $C_n = 1/(1.6 \times 10^6)$. Using the peak delay errors given above, peak phase errors of 1.07, 1.95, and 3.2 radians are obtained. The waveform distortions that result from phase errors of these magnitudes are illustrated in Fig. 11.6, in which the effects of delay or phase errors in a matched filter, or any other, system are graphically illustrated.

11.4   Matched-Filter Modulation Distortion

The paired-echo distortion analysis technique can also be used to determine the effects of time modulation distortion on matched-filter signals. As will be shown in a more general discussion in a later section, the assumption of a matched-filter signal permits the use of the matched-filter response function as a means of determining the distortion bounds caused by passive and active sources of distortion. Although the method of distortion analysis is identical to that employed in earlier sections of this chapter, and some of the results are similar in form, the paired signals arising from modulation distortion can be considered primarily to exist in the frequency domain in the form of modulation sidebands. These distortion modulation sidebands are present in any system to a greater or lesser degree. However, for the special application involving the use of large time-bandwidth matched-filter signals, the matched-filter outputs associated with the distortion sidebands often can be observed as equivalent time displaced paired-echo range sidelobes that are similar to the true paired echoes resulting from network distortions. In this respect, the linear FM pulse-compression signal is unique in that there is an approximate one-to-one correspondence between the results of modulation distortion and network distortion observed at the matched-filter output. Amplitude and phase modulation distortion of the linear FM signal, which are relatively easy to implement and control in the laboratory, yield matched-filter outputs that are similar to those resulting from network distortions for any type of signal, and for this reason the linear FM pulse-compression signal can be used to instrument an analog computer technique to analyze general network distortions. The detailed analysis of this section will largely be based on the illustrative examples that are readily obtained with the linear FM signal.

A general form of the basic modulation distorted matched-filter signal seen at the radar transmitter is given by

$$s_m(t) = a(t)\{(a_0 + a_1 \cos \omega_{m_1} t) \exp[j(\omega_0 t + \theta(t) + b_1 \sin \omega_{m_2} t)]\},$$

$$-T/2 \leqslant t \leqslant T/2 \quad (11\text{-}29)$$

where $\theta(t)$ is the coded modulation function and $\omega_{m_{1,2}}$ are the distortion modulation rates.

The expression (11-29) is seen to be similar to the representation of Eq. (11-5). Considering the effects of amplitude modulation alone, the term

$$a_1 \cos \omega_{m_1} t$$

can be put in its exponential form, and the amplitude modulated signal can then be written as

$$s_m(t) = a_0 \Bigg(\exp[j\{\omega_0 t + \theta(t)\}] + \frac{a_1}{2a_0} \exp[j\{(\omega_0 + \omega_{m_1})t + \theta(t)\}]$$

$$+ \frac{a_1}{2a_0} \exp[j\{(\omega_0 - \omega_{m_1})t + \theta(t)\}] \Bigg), \qquad -T/2 \leqslant t \leqslant T/2$$

$$(11\text{-}30)$$

where $a(t) = 1$ for the high power case.

The effect of sinusoidal amplitude modulation is to produce the expected sideband signals, and in general it can be stated that the results of modulation distortion are paired sidebands. The amplitude modulation and, as will be shown, phase modulation does not effect the coded modulation function, $\theta(t)$. When the signal described by Eq. (11-30) is introduced into the matched filter at the receiver the matched-filter output will be established by its response to frequency shifted input signals. The general case can be described in terms of the matched-filter response function. For the special case of the linear FM pulse-compression signal the sideband signals centered at $\omega_0 \pm \omega_{m_1}$ will have their associated output signals shifted in time by $\pm(\omega_{m_1}/\Delta\omega)T$ seconds, where $\Delta\omega$ is the linearly swept bandwidth. Let $\overline{s_1(t)}$ denote the compressed-pulse envelope function. Then, when ω_{m_1} is small compared to $\Delta\omega$ the output waveform can be approximated by

$$s_0(t) = a_0\left[\overline{s_1(t)}\cos\omega_0 t + \frac{a_1}{2a_0}\overline{s_1\left(t + \frac{\omega_{m_1}}{\Delta\omega}T\right)}\cos\left(\omega_0 + \frac{\omega_{m_1}}{2}\right)t\right.$$

$$\left. + \frac{a_1}{2a_0}\overline{s_1\left(t - \frac{\omega_{m_1}}{\Delta\omega}T\right)}\cos\left(\omega_0 - \frac{\omega_{m_1}}{2}\right)t\right] \tag{11-31}$$

Considering the envelope terms of this equation, one can note the similarity to the envelope function for network amplitude distortion given by Eq. (11-8), illustrating the analog relationship between the effects of modulation distortion on the linear FM pulse-compression signal output and the paired-echo effects of network distortion for a general signal. Figure 11.7 illustrates the creation of the linear FM paired echoes for two different rates of amplitude modulation. Figure 11.8 shows the non-symmetry of the paired-echo amplitudes when the amplitude modulation distortion is larger on one end of the linear FM input signal than on the other. The larger paired echo occurs before the compressed pulse if the larger amplitude ripple is at the leading edge of the input signal and after the compressed pulse if the larger ripple is at the trailing edge of the input signal. The effect of complex amplitude modulation functions can be treated by working with the major Fourier components of the modulation function, and the general procedure is the same as that outlined in the development that follows for phase modulation distortion.

For the case of the high power transmitter that limits it can be assumed that the amplitude modulation distortion factors will be converted to phase modulation distortion before the signal is transmitted.[1] Thus, in Eq. (11-29) a_1 can be set equal to zero. Given this condition, a general

---

[1] If there are appreciable runs of waveguide between the final transmitter tube and the radiating antenna, amplitude modulation may still be possible as the result of mismatches in the waveguide, even though limiting has occurred [7].

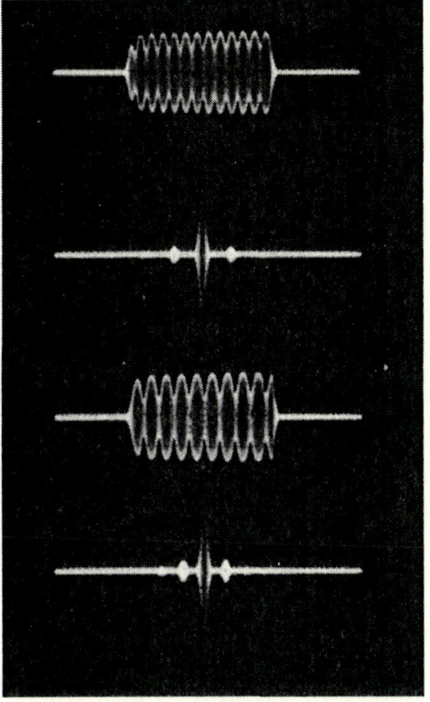

FIG. 11.7   Paired echoes for two different amplitude modulation frequencies.

expression for the phase modulation distorted time function is (using the exponential form to simplify the subsequent analysis)

$$s_m(t) = s(t) \exp[\,jb_1 \sin(\omega_m t + \theta_0)] \qquad (11\text{-}32)$$

where $s(t)$ is the undistorted signal function, b_1 the peak phase error (radians), $\omega_m = 2\pi f_m$ the radian frequency of phase error, and θ_0 the arbitrary phase positioning factor.

The inclusion of the time phase factor θ_0 in expression (11-32) serves the same function in making the paired-echo phase modulation distortion analysis more general as did the equivalent network phase factor in Eq. (11-20).

Let

$$s(t) = a(t) \exp[\,j\{\omega_0 t + \theta(t)\}]$$

where $a(t)$ is the signal envelope function, and use the Bessel function expansion formula given in Eq. (11-10); a general result for the phase

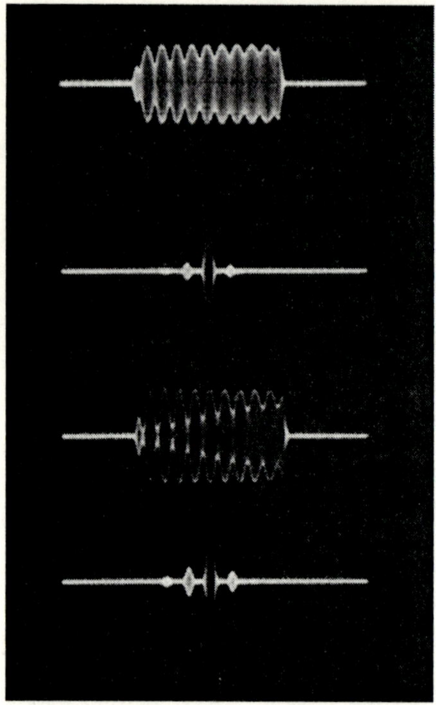

FIG. 11.8   Effect on amplitude modulation paired echoes when a_1 varies as function of time.

modulation error given by (11-32) that is equivalent to the general network phase error result embodied in Eq. (11-23) is

$$s_m(t) = a(t)J_0(b_1)\exp[j\{\omega_0 t + \theta(t)\}] + a(t)\sum_n J_n(b_1)\Big(\exp[j\{(\omega_0 + n\omega_m)t$$

$$+ \theta(t) + n\theta_0\}] + (-1)^n \exp[j\{(\omega_0 - n\omega_m)t + \theta(t) - n\theta\}]\Big) \quad (11\text{-}33)$$

As noted previously, the time function phase modulation error produces "paired sidebands" rather than "paired echoes", as in the network distortion case, with the paired-spectrum frequency displacement factor $n\omega_m$ replacing the waveform time displacement factor. The effect of θ_0 is to establish the relative phases of the carrier frequencies of the signals associated with the paired-spectrum signals.

In the normal, or uncoded, pulse signal (i.e., $\theta(t) = 0$) the additional signals associated with the "paired sidebands" occur during the same time interval as the undistorted time function $s(t)$. The unique feature obtained when $s(t)$ is a pulse compression, or coded, waveform is that the paired-sideband signals are generally dispersed in time. Under certain conditions they are completely separable, and thus individually observed and measured.

Equation (11-33) can be applied to the case where $s(t)$ is the linear FM pulse-compression signal described in Chapter 6. Making use of the linear FM matched-filter output for frequency shifted inputs given by Eq. (11-33), one can obtain the following approximate expression for the matched-filter output of the phase error modulated linear FM input:

$$s_0(t) = J_0(b_1)\overline{s_1(t)}\cos\omega_0 t$$

$$+ \sum k_n J_n(b_1)\left\{\overline{s_1\left(t + \frac{n\omega_m}{\Delta\omega}T\right)}\cos\left[\left(\omega_0 + \frac{n\omega_m}{2}\right)t + n\theta_0\right]\right.$$

$$+ (-1)^n\overline{s_1\left(t - \frac{n\omega_m}{\Delta\omega}T\right)}\cos\left[\left(\omega_0 - \frac{n\omega_m}{2}\right)t - n\theta_0\right]\right\} \qquad (11\text{-}34)$$

where $\overline{s_1(t)}$ is the compressed-pulse envelope function, k_n the weighting factor of the filter frequency response, T the uncompressed-pulse width, and $\Delta\omega/2\pi = \Delta f$, the linear FM swept bandwidth.

A more precise formulation of the compression-filter output would include the waveform modifications based on the frequency-shifted cross correlation waveforms, such as the envelope shape distortion [8] that would be observed for the paired signals at the filter output (see Fig. 6.2). However, Eq. (11-34) is useful for most practical applications, including the case where frequency response weighting is used to reduce the range sidelobes.

Equation (11-34) indicates that there is a centrally positioned signal flanked by symmetrically and evenly spaced pairs of signals having amplitudes governed by the respective Bessel function $J_n(b_1)$. A time separation factor can be defined as

$$\bar{t} = \frac{\omega_m}{\Delta\omega}T = \frac{f_m}{\Delta f}T$$

Since $(1/\Delta f) \cong \tau$, the compressed-pulse width, this spacing factor in terms of the compressed-pulse width is

$$\bar{t} \cong (f_m T)\tau \qquad (11\text{-}35)$$

$f_m T$ represents the number of error modulation cycles over the interval T. Thus, the spacing between the paired signals, expressed in normalized compressed-pulse width units, depends only on the number of cycles of error modulation that occur in the duration T. For example, if there are five error modulation cycles in the time T, then the first set of paired signals observed at the pulse-compression filter output will be approximately five compressed-pulse widths from the center pulse location (J_0 term).

The factor θ_0 becomes important when $f_m T$ is of such a value that the first set of signals overlap the central pulse. These signals will combine in a coherent manner at the compression-filter output. For this case it becomes possible to obtain various patterns of cancellation and reinforcement in the composite waveform depending on the value of b_1, the phase modulation error magnitude, and the error modulation positioning factor θ_0.

Figure 11-9 illustrates the basic phase modulation error parameters described in the discussion above. Controlled laboratory experiments

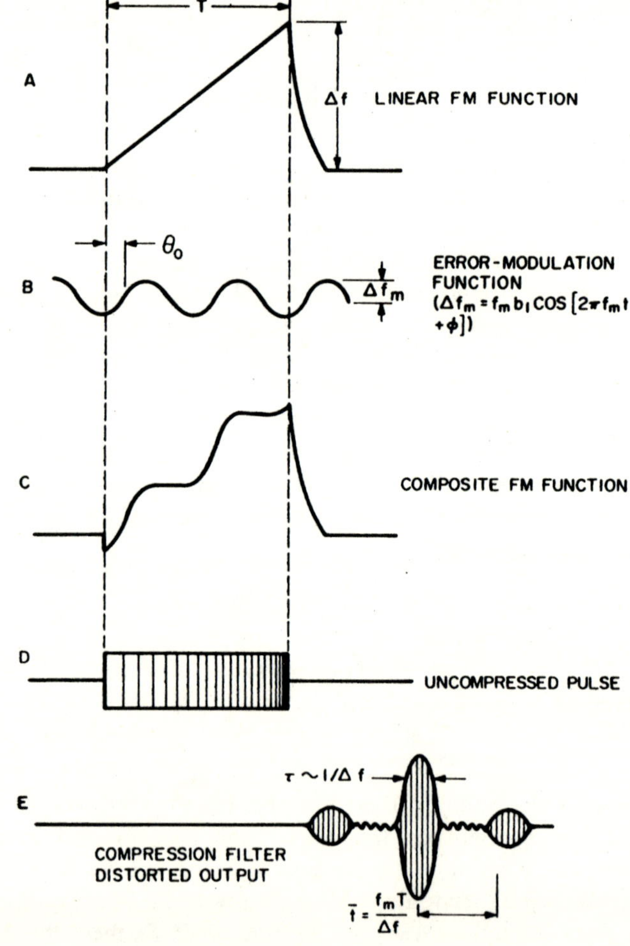

FIG. 11.9   Modulation-induced paired-echo parameters (from Cook [9]).

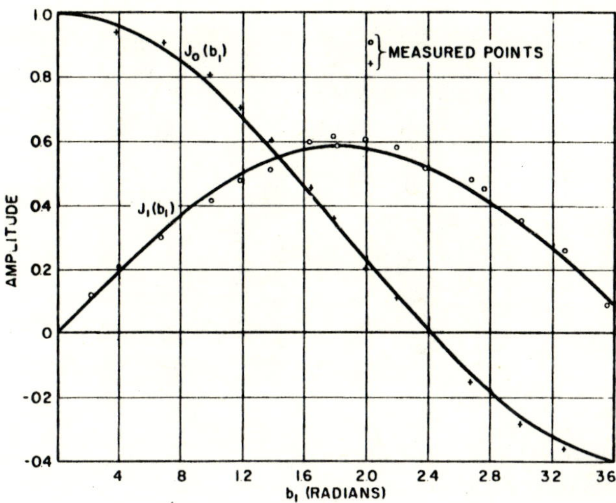

FIG. 11.10 Test-equipment calibration curves (from Cook [9]).

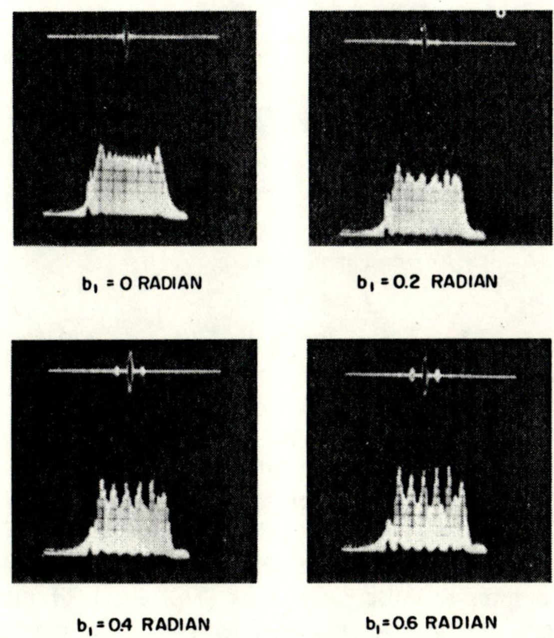

FIG. 11.11 Compression-filter waveforms and spectra for single sinusoidal modulation component $f_m T = 6$ (from Cook [5]).

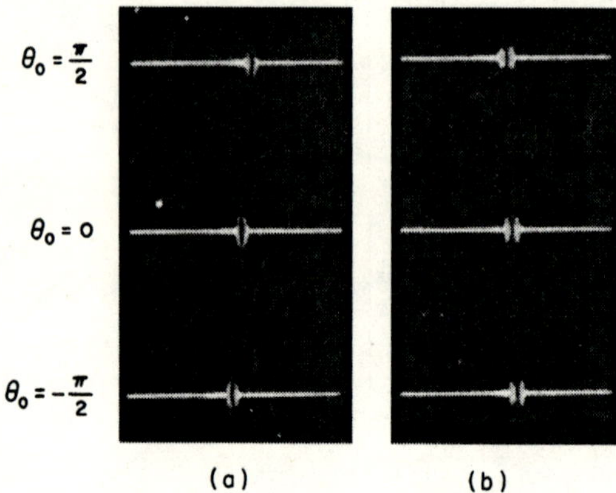

FIG. 11.12   Effect of θ_0 on overlapping and nonoverlapping paired echoes. (a) Overlapping case: $f_m = 2$ cycles/input pulse width. (b) Nonoverlapping case: $f_m = 3$ cycles/input pulse width. (From Cook [9].)

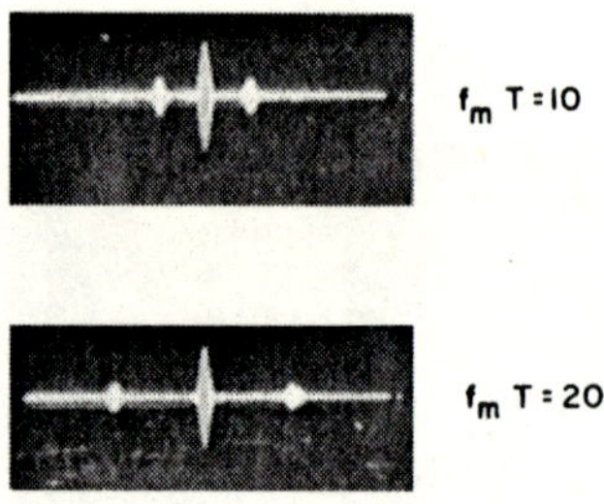

FIG. 11.13   Waveforms illustrating paired-echo spacing as a function of f_m, phase error modulation rate, for $b_1 = 0.6$ radian (from Cook [9]).

based on the use of simple and complex phase modulation distortion of linear FM signals are discussed by Cook [5, 9]. Figure 11.10 compares the measured and theoretical variation in the amplitudes of the central compressed-pulse signal (J_0 term of Eq. (11-34)) and the first set of paired signals (J_1 term). The compressed-pulse signal was weighted to yield -25 db sidelobes. It can be seen that the weighting factor k_1 had negligible effect on the J_1 term. Figure 11.11 illustrates the spectrum of the error modulated linear FM signal for various levels of modulation error, as well as the associated compressed-pulse waveforms. The effect of the modulation timing factor θ_0 on the compression-filter output for the case when $f_m T$ is small is shown in Fig. 11.12, and the displacement of the paired signals as $f_m T$ increases is shown in Fig. 11.13. For this latter figure, it should be noted that the spacing factor is in terms of the unweighted linear FM compressed pulse (i.e., matched-filter signal), and not the weighted compressed-pulse width.

11.5 Complex Modulation Distortion Functions

The analysis of the elementary case of sinusoidal phase modulation can be extended to more complex distortion modulation functions that may arise from more than one source of distortion. The examples presented below are used to indicate the analytical approach and establish the basis for the application of the superposition principle, which can greatly simplify any evaluation of active sources of phase modulation errors.

For the situation where there are present two sources of sinusoidal phase modulation of different frequency, the general phase distorted signal can be written (assuming $\theta(t) = 0$ for convenience)

$$s_m(t) = a(t) \exp[j(\omega_0 t + b_1 \sin \omega_1 t + b_2 \sin \omega_2 t)] \qquad (11\text{-}36)$$

Using the relationships shown earlier, the expression equivalent to Eq. (11-36) becomes

$$a(t) \exp[j\omega_0 t]\left(J_0(b_1) + \sum_1^\infty J_n(b_1)\{\exp[jn\omega_1 t] + (-1)^n \exp[-jn\omega_1 t]\}\right)$$

$$\times \left(J_0(b_2) + \sum_1^\infty J_n(b_2)\{\exp[jn\omega_2 t] + (-1)^n \exp[-jn\omega_2 t]\}\right) \qquad (11\text{-}37)$$

If $b_1, b_2 \leqslant 0.5$ radian, then the Bessel function approximations given earlier may be used. For this condition $s_m(t)$ becomes

$$a(t) \exp[j\omega_0 t]\left(1 + \frac{b_1}{2}(\exp[j\omega_1 t] - \exp[-j\omega_1 t])\right)$$

$$\times \left(1 + \frac{b_2}{2}(\exp[j\omega_2 t] - \exp[-j\omega_2 t])\right) \qquad (11\text{-}38)$$

Expansion of (11-38) yields

$$s_m(t) = a(t)\left[\exp[j\omega_0 t] + \frac{b_1}{2}\left(\exp[j(\omega_0 + \omega_1)t] - \exp[j(\omega_0 - \omega_1)t]\right)\right.$$

$$\left. + \frac{b_2}{2}\left(\exp[j(\omega_0 + \omega_2)t] - \exp[j(\omega_0 - \omega_2)t]\right)\right]$$

$$+ a(t)\left[\frac{b_1 b_2}{4}\left(\exp[j(\omega_0 + \omega_1 + \omega_2)t] - \exp[j(\omega_0 - \omega_1 + \omega_2)t]\right.\right.$$

$$\left.\left. + \exp[j(\omega_0 - \omega_1 - \omega_2)t] - \exp[j(\omega_0 + \omega_1 - \omega_2)t]\right)\right] \qquad (11\text{-}39)$$

If the further restriction is added that the primary paired-echo terms be no larger than -20 db relative to the peak value of $s(t)$, i.e., $(b_1/2)_{max}$ and $(b_2/2)_{max}$ no greater than 0.1, then the cross product term

$$(b_1 b_2/4)_{max} \leqslant 0.01$$

and may be disregarded. Under this assumption the latter four terms of Eq. (11-39) drop out. Thus, for $b \leqslant 0.2$ radian (approximately 11.5° one-way peak) each individual phase error modulation component produces a single set of paired-echo (actually paired-sideband) signals that can be treated independently of other modulation-induced paired echoes.

Applying the above to a linear FM pulse-compression signal, in which n modulation components are present, a composite approximation of the output signal function is obtained.

$$s_0(t) = \overline{s_1(t)} \cos \omega_0 t + \sum_1^n \frac{b_i}{2} \overline{s_1\left(t + \frac{\omega_i}{\Delta\omega}T\right)} \cos\left[\left(\omega_0 + \frac{\omega_i}{2}\right)t + \theta_i\right]$$

$$- \sum_1^n \frac{b_i}{2} \overline{s_1\left(t - \frac{\omega_i}{\Delta\omega}T\right)} \cos\left[\left(\omega_0 - \frac{\omega_i}{2}\right)t - \theta_i\right] \qquad (11\text{-}40)$$

where θ_i is the reference phase of each distortion modulation frequency ω_i. It is seen that each of the paired echoes of (11-40) have the following characteristics:

$$\text{Time shift} \quad \pm\frac{\omega_i}{\Delta\omega}T$$

$$\text{Frequency shift} \quad \pm\frac{\omega_i}{2}$$

$$\text{Phase reference} \quad \pm\theta_i$$

Figure 11.14 shows the set of paired echoes caused by two sinusoidal phase modulation error functions applied to the test equipment. The frequency of the second sinusoid is twice that of the first. The peak phase error of the second signal is constant, while that of the first is varied, illustrating the independence of the paired-echo terms for peak phase errors less than that discussed above. Under such conditions, any complex error modulation function can be expressed in terms of its Fourier components. The composite distorted waveform can then be obtained from the superposition of the distortions caused by each component. For larger peak phase errors the more exact expressions, either Eq. (11-39) for $b_i \leqslant 0.5$ radian or Eq. (11-37) must be used.

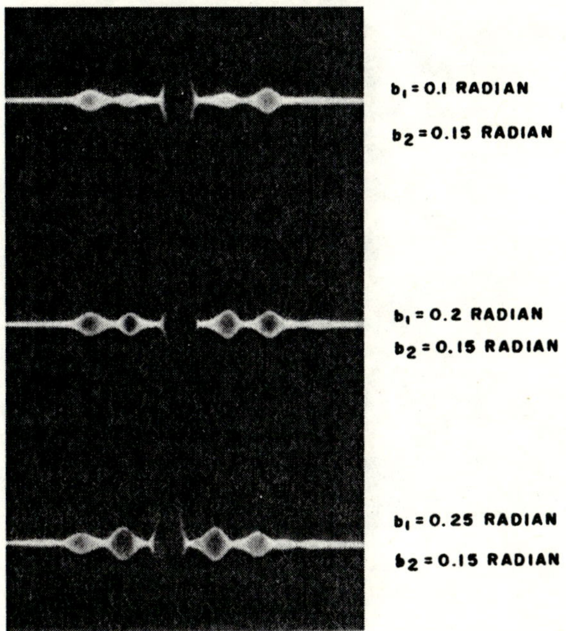

FIG. 11.14 Compression-filter output waveform for modulation-distortion function composed of two sinusoidal components $\omega_2 = 2\omega_1$ (from Cook [9]).

The results of Eq. (11-40) can be extended to the type of distortion function that occurs over a restricted time interval as shown in Fig. 11.15. This type of distortion function can be represented as a set of discrete phase modulation distortion spectrum components, given by

$$A_n \sin(2\pi n f_r + \theta_n) \qquad (11\text{-}41)$$

where f_r is the signal repetition rate, $\theta_n = 2\pi t_0 n f_r$, t_0 is the delay from the start of the linear FM sweep to the center of the distortion function, and A_n the nth Fourier component of the distortion function spectrum.

The phase modulation associated with each component given by (11-41) is

$$\frac{A_n}{2\pi n f_r} \cos(2\pi n f_r t + \theta_n) \qquad (11\text{-}42)$$

For the type of function shown in Fig. 11.15 the peak phase modulation falls off faster than $1/f$ as n increases, since A_n also decreases with an increase in frequency. Thus, in most cases, a finite number of terms of the distortion function spectrum can be employed on the basis of superposition as outlined by Eq. (11-40). It should be noted that inverting the distortion function is equivalent to adding 180 degrees of phase shift to each θ_n. This

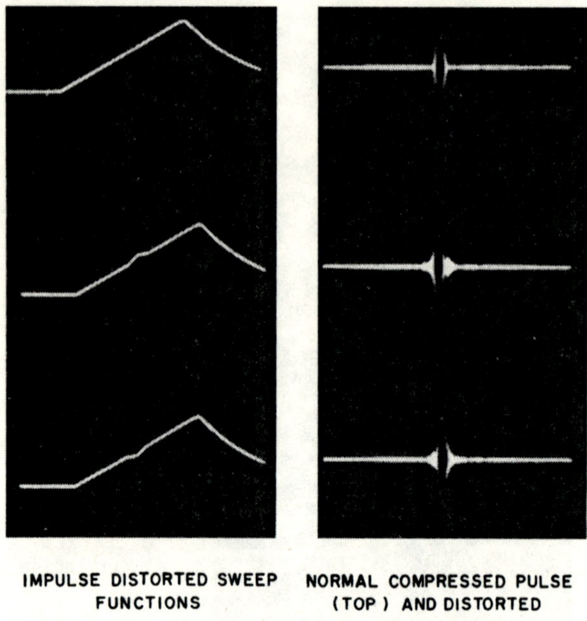

FIG. 11.15   Compressed-pulse waveform distortion caused by a local error function (from Cook [9]).

causes an interchange of certain regions of cancellation and reinforcement in the composite matched-filter output, as illustrated in Fig. 11.15.

Although it has been implied that amplitude modulation should be small in a high power, hard limiting system, there are applications of matched-filter techniques in which this does not apply. Should this be the case, then it is quite possible that simultaneous amplitude and phase modulation will exist. Assuming that the phase modulation magnitude is small so that the truncation of the Bessel function expansion of Eq. (11-10) can be limited to the first pair of terms, the expression for the transmitted coded signal is similar to the first five terms of Eq. (11-19) for the combined network distortion case, being

$$
\begin{aligned}
s_m(t) = a(t)\Bigg[a_0 \Bigg\{ &\exp[\,j\{\omega_0 t + \theta(t)\}] \\
&+ \frac{a_1}{2a_0}\Big(\exp[\,j\{(\omega_0 + \omega_1)t + \theta(t)\}] + \exp[\,j\{(\omega_0 - \omega_1)t + \theta(t)\}]\Big) \\
&+ \frac{b_1}{2}\Big(\exp[\,j\{(\omega_0 + \omega_2)t + \theta(t)\}] - \exp[\,j\{(\omega_0 - \omega_2)t + \theta(t)\}]\Big)\Bigg\}\Bigg]
\end{aligned}
$$

$$(11\text{-}43)$$

For the case of the linear FM matched-filter signal, the matched-filter output becomes (assuming $a_0 = 1$)

$$s_0(t) = \overline{s_1(t)} \cos \omega_0 t + \frac{a_1}{2}\left[\overline{s_1\left(t + \frac{\omega_1}{\Delta\omega}T\right)} \cos\left(\omega_0 + \frac{\omega_1}{2}\right)t \right.$$
$$\left. + \overline{s_1\left(t - \frac{\omega_1}{\Delta\omega}T\right)} \cos\left(\omega_0 - \frac{\omega_1}{2}\right)t \right]$$
$$+ \frac{b_1}{2}\left[\overline{s_1\left(t + \frac{\omega_2}{\Delta\omega}T\right)} \cos\left(\omega_0 + \frac{\omega_2}{2}\right)t \right.$$
$$\left. + \overline{s_1\left(t - \frac{\omega_2}{\Delta\omega}T\right)} \cos\left(\omega_0 - \frac{\omega_2}{2}\right)t \right] \qquad (11\text{-}44)$$

Figure 11.16 presents an example of the linear FM matched-filter output when the amplitude modulation and phase modulation distortions occur

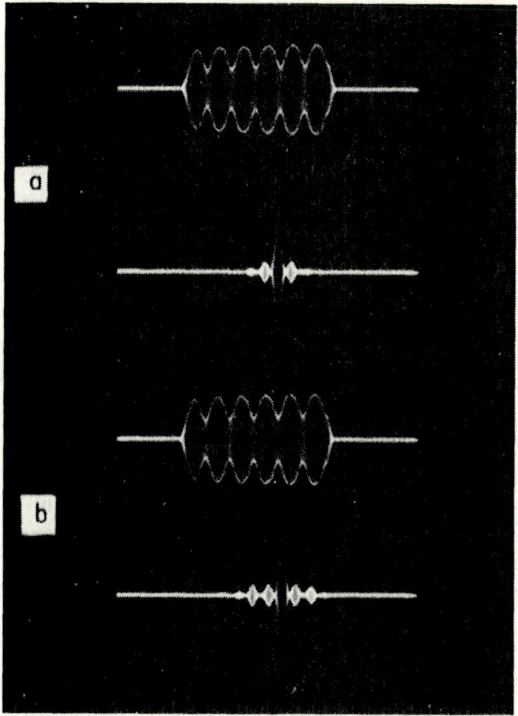

FIG. 11.16 Effects of combined amplitude and phase modulations of different frequencies. (a) Amplitude modulation only. (b) Combined modulation, $\omega_{m_p} > \omega_{m_a}$. (From Cook [9].)

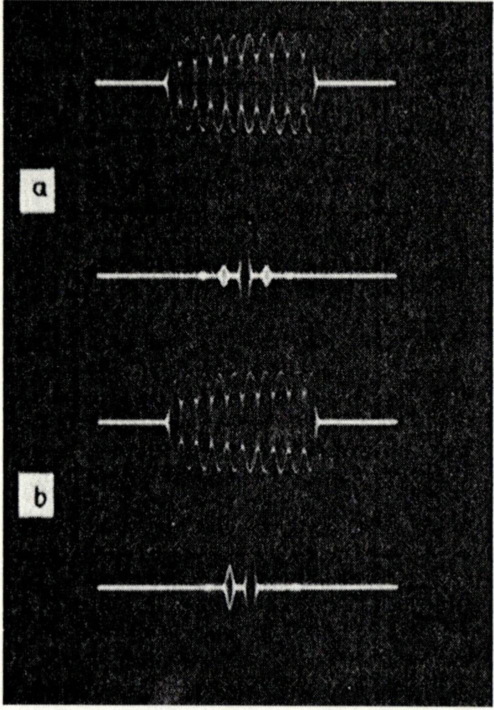

FIG. 11.17   Effects of combined amplitude and phase modulations of the same frequency. (a) Amplitude modulation only. (b) Combined modulation, $\omega_{m_p} = \omega_{m_a}$. (From Cook [9].)

at different frequencies. When $\omega_1 = \omega_2$, and if the magnitudes of a_1 and b_1 are equal there can exist the special case of the paired echoes completely canceling on one side of the desired output and adding in phase on the other side. This is illustrated in Fig. 11.17. A change in the phase reference of either ω_1 or ω_2 by 180 degrees results in the interchange of the locations of cancellation and addition. The presence of an arbitrary phase reference can lead to conditions of partial cancellation and reinforcement of the paired-echo amplitudes, and can be handled by the general techniques outlined previously.

11.6   Sources of Phase Modulation Distortion

The sources of modulation distortion that are critical in a pulse-compression matched-filter application are generally derived from time dependent gains and delays in a high power transmitter chain. As DiFranco and Rubin [6] have pointed out, this is a multiplicative process. However, under the conditions outlined earlier, superposition can be applied if the peak phase error is not too large. Typical sources of modulation errors are the ripple and droop characteristics found in modulators, power supply capacitor discharge, load mismatches from stage to stage and within modulator pulse forming networks, and transformer ripples. Some of these distortion sources will result in amplitude modulation of the linear FM signal. The limiting action of the high power stages will generally convert the amplitude modulation to phase modulation. The total effect of the various sources of phase error will depend on the error sensitivity of the high power electron devices used in the transmitter chain. Liebman [4] has tabulated typical sources of transmitter chain phase error modulation (see Table 11-I).

TABLE 11-I

TRANSMITTER TUBE PHASE RIPPLE

Amplifier type	Approximate magnitude of phase ripple, b_1 [a]
Klystron	$\frac{1}{2}\left(\frac{\Delta E}{E}\right)\theta_l$
Traveling wave tube	$\frac{1}{3}\left(\frac{\Delta E}{E}\right)\theta_l$
Crossed field amplifiers	$\frac{1}{25}\left(\frac{\Delta E}{E}\right)\theta_l$
Triode	$50\left(\frac{\Delta I}{I}\right)$
Tetrode	$(1 \text{ to } 50)\left(\frac{\Delta I}{I}\right)$
Amplitron	$40\left(\frac{\Delta I}{I}\right)$

[a] θ_l is the electrical length of device, $\Delta E/E$ the voltage ripple ratio, and, $\Delta I/I$ the current ripple ratio.

The use of this data for a linear FM signal can be illustrated for a high power klystron tube in which a 100 kV beam voltage has a 1 kV ripple arising from the beam-pulse modulator pulse forming network. If the electrical length of the tube is 70 radians, then the value of the peak phase modulation ripple is

$$b_1 = \tfrac{1}{2}\tfrac{1}{100} \times 70 = 0.35 \text{ radian}$$

or approximately 20 degrees. The amplitude of the paired echo signals in db is

$$20 \log(b_1/2) = -15 \, \text{db}$$

Assuming that there are six ripple cycles over the transmitted pulse interval, and that the swept bandwidth is 1 Mc, then these paired echoes would occur at ± 6 microseconds relative to the center of the desired matched-filter output. Paired echoes of this magnitude are clearly undesirable. One method of reducing the effect of the beam voltage ripple is to use a high voltage clipper. Assuming that such a device could reduce the beam voltage ripple to 0.1 kV, then the phase modulation paired echoes from this source would be

$$20 \log(0.035/2) = -35 \, \text{db}$$

This would be acceptable in many applications.

As noted above, the results of phase modulation distortion described in these sections indicate the need for establishing firm control over the sources of modulation distortion in a high power transmitter that is used in a matched-filter system. The following comments are directly applicable to the linear FM matched-filter signal. They can also be applied, with suitable modification, to many other matched-filter waveforms, particularly to those that have a unidirectional frequency modulation or combinations of linear FM segments.

The number of cycles of a sinusoidal phase error modulation over the uncompressed-pulse interval will determine the location of the distortion signals relative to the nominal compressed-pulse position. Thus, if the error modulation frequency is fixed, an increase in the system compression ratio ($T \Delta f$) by increasing T will result in a greater time displacement away from the main signal for the paired-echo distortion signals.

The system designer who can exercise control over the error-producing subunits may well consider having the error modulation frequencies as high as possible, consistent with other system considerations. For example, a modulator waveform containing a large number of ripples may be preferable to one that has a few ripples, ripple amplitude being constant. The higher error modulation frequency will in general move the distortion signals further away from the main signal, thus not interfering with the system resolution and detection capability within the main signal region. In addition, a higher modulation frequency will move the offending signals further out of the frequency pass-band of the matched filter, yielding an additional measure of distortion signal suppression. If the modulation mechanism is one that produces frequency modulation rather than phase modulation, such as ripples on the linear sweep function, then a higher modulation frequency produces less phase modulation, and thus less distortion, for a fixed amount of error frequency deviation. In addition, if exact control of the modulation distortion parameters b_1, $f_m T$, and θ_0 can be achieved, then partial cancellation of particularly objectionable distortion paired echoes can be obtained. However, if the modulation error is sufficient to produce a marked deterioration in the signal, then an alternate transmitter component design or physical rearrangement should be considered, as should the use of distortion compensation techniques.

11.7 The Representation of Distorted Output Time Functions for General Matched-Filter Signals

In Sections 11.4 to 11.6 the development of the effects of modulation distortion, using the linear FM matched-filter pulse-compression signal as an example, provided waveform illustrations that also served to show the types of waveform distortion that could occur as a result of network distortion. The radar system response function, briefly described in Chapter 3, and in greater detail in Chapter 4, has been used by DiFranco and Rubin [6] as a means for deriving the upper bound of the envelope of the composite distorted output of the matched filter.

It was shown in Section 11.2 that for frequency domain (i.e., filter) distortion, the form of the matched-filter output was independent of the coding function associated with the matched-filter signal. For each amplitude or phase distortion component existing in the matched-filter receiver, the distorted output could be expressed as a combination of time-shifted replicas of the desired signal. Each of these time-shifted signals can be thought of as being a shifted cross section of the matched-filter response function, $|\chi(\tau, \phi)|$, as illustrated in Fig. 11.18. For a complex distortion function it was shown that suitable truncation of this function would permit its decomposition into a number of Fourier components, with each component of the distortion function producing a set of paired echoes, or shifted response function cross sections. Along the time axis the cross sections of the ensemble of response functions have a common frequency parameter. Thus, for a given amplitude distortion Fourier component a_n, the paired response function cross section is described by

$$\frac{a_n}{2}\left|\chi\left(\tau \pm \frac{n_a}{\Delta f}, \phi\right)\right| \tag{11-45}$$

where a_n is the magnitude of nth amplitude ripple component, n_a the number of ripple cycles in the band Δf, and $|\chi(\tau, \phi)|$ the undistorted matched-filter response function.

Similarly, for phase distortion components, the magnitudes of each set of paired response function cross sections, assuming small phase errors

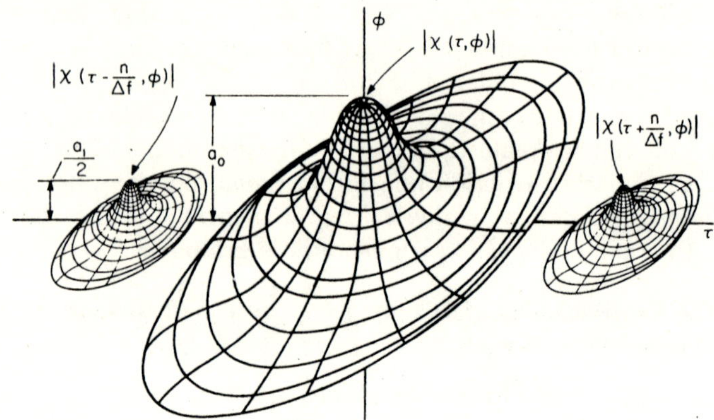

FIG. 11.18   Illustration of paired response functions caused by frequency domain distortion.

(i.e., $b_n \leqslant 0.5$ radian), is given by (letting $a_0 = 1$)

$$\frac{b_n}{2}\left|\chi\left(\tau \pm \frac{n_b}{\Delta f}, \phi\right)\right| \tag{11-46}$$

where b_n is the magnitude of nth phase ripple component and n_b the number of phase ripple cycles in the band Δf.

Similarly, for the case of time domain modulation distortion it was shown in Section 11.4 that paired sidebands are created as the analog of the network distortion paired echoes. Each of these paired sideband signals produces at the matched-filter output a frequency shifted signal that can also be expressed in terms of the signal response function. This is illustrated by Fig. 11.19. For each amplitude modulation component, the paired response function cross sections are given by

$$\frac{a_m}{2}|\chi(\tau, \phi \pm f_{m_a})| \tag{11-47}$$

and each phase-modulation created pair of response function cross sections are given by (for $b_m \leqslant 0.5$ radian)

$$\frac{b_m}{2}|\chi(\tau, \phi \pm f_{m_b})| \tag{11-48}$$

where a_m is the magnitude of mth amplitude modulation component, b_m the magnitude of mth phase modulation component, f_{m_a} the frequency of mth amplitude component, and f_{m_b} the frequency of mth phase component.

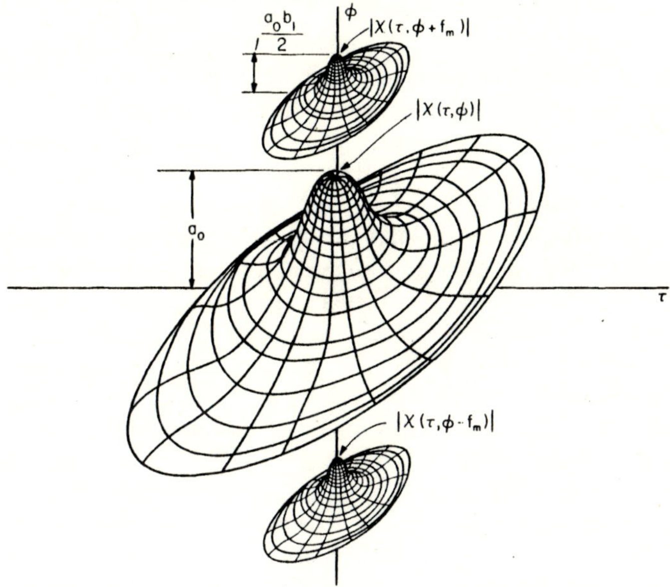

FIG. 11.19   Illustration of paired response functions caused by time domain distortion.

The representations given by Eqs. (11-45)–(11-48) are based on the assumption that the distortions are small, so that cross-product terms between amplitude and phase distortion are negligible as well as the higher order Bessel function terms associated with phase distortions. Equations (11-45)–(11-48) can be combined to obtain a composite upper bound expression for the distorted matched-filter response function cross section. Denoting this upper bound of the distorted matched-filter output by $|\chi_D|$, one obtains

$$|\chi_D(\tau, \phi)| \leqslant \sum_{n=-\infty}^{\infty} C_n |\chi(\tau + n/\Delta f, \phi)|$$

$$+ \sum_{m=-\infty}^{\infty} C_m |\chi(\tau, \phi + f_m)| \qquad (11\text{-}49)$$

where $C_n = \frac{1}{2}(a_n + b_n)$, $C_n = C_{-n}$; $C_m = \frac{1}{2}(a_m + b_m)$, $C_m = C_{-m}$. That this expression does represent an upper bound on the distorted matched-filter output can be recognized by recalling that, by definition, the $|\chi|$ function has only a magnitude, so that the terms in (11-49) represent an addition of magnitudes, whereas, as shown by examples in previous sections, the actual matched-filter outputs can add or subtract depending on the relative phases of the carrier frequencies of the paired-echo signals. The use of Eq. (11-49) permits a calculation of upper bounds for the effects of distortion on waveform accuracy, ambiguity, and resolution parameters.[1] The discussion of DiFranco and Rubin points out that waveform accuracy is degraded by a widening of the desired main lobe of the matched-filter distorted output waveform. This can occur when there are low frequency distortion terms that result in paired-echo signals that overlap the desired signal. This causes a widening of this signal when the phases of the paired-echo signal are such that they add to, rather than subtract from, the desired matched-filter output. The specific case of quadratic phase errors (i.e., linear delay mismatch) was treated in Chapter 6 for linear FM signals, and offers one example of the effect of a low frequency phase error on matched-filter waveform distortion and the associated loss in resolution and accuracy.

Waveform ambiguity results if major distortion paired echoes are large enough to exceed the detection threshold and thus be incorrectly identified as apparent signals at a range location displaced from that of the true signal. This can be a serious problem in a dense target environment. Similarly, resolution among a number of signals of varying amplitude can be severely degraded if the distortion paired echoes are large enough to mask the smaller signal returns. As noted previously, each of these considerations provides an incentive for the reduction of distortion in a radar system.

In a number of situations in which many overlapping paired-echo signals are possible, such as in the example illustrated by Fig. 11.15 the expression given by Eq. (11-49) may be too conservative and result in the placing of unrealistic restraints on the system designer as far as allowable distortion tolerances are concerned. For this case it may be more desirable to consider the average bound on the distorted matched-filter output. This average bound can be obtained from the rms value of the combined distortion created paired-response functions. Denoting this rms distorted matched-filter response function by $|\bar{\chi}_D|$, then

$$|\bar{\chi}_D(\tau, \phi)| = \left[\sum_{n=-\infty}^{\infty} C_n^2 \left| \chi\left(\tau + \frac{n}{\Delta f}, \phi\right) \right|^2 \right.$$

$$\left. + \sum_{m=-\infty}^{\infty} C_m^2 |\chi(\tau, \phi + f_m)|^2 \right]^{1/2} \qquad (11\text{-}50)$$

---

[1] In an actual situation involving phase and amplitude errors of the same frequency, $C_n = (a_n \pm b_n)/2$ and $C_{-n} = (a_n \mp b_n)/2$, so that these terms would not be simultaneously equal. The condition for which one of these coefficients is the larger will depend on the relative phase (± 90 degrees) between the amplitude and phase error components. In the general case this may not be predictable, so that letting $C_{-n} = C_n = (a_n + b_n)/2$ provides the maximum upper bound for all cases.

11.8 Measures of Resolution Loss for Random Phase Errors

It is possible to locate in range and frequency the peak and average bounds of distortion sidelobes at the matched-filter output from the expressions given by (11-49) and (11-50). This assumes that the major Fourier components of the distortion functions are known. When the distortions are arbitrary or random functions one is confined to obtaining a measure of the spread of the response or ambiguity function in the region surrounding the main ambiguity peak. Brown and Palermo have discussed two basic measures of the loss in resolution resulting from random phase errors [10, 11]. One such measure is based on the second moment, or radius of gyration, of the distorted ambiguity function. For the case of a random phase modulation error $\theta_e(t)$, the distorted response function is given by (from Eq. (3-83)):

$$\chi_D(\tau, \phi) = \int u(t)u^*(t + \tau) \exp[-j\{2\pi\phi t - \theta_e(t)\}]\, dt \qquad (11\text{-}51)$$

Brown and Palermo show that the measure of resolution based on the radius of gyration of $|\chi_D|^2$ is approximated by

$$\sigma_\phi^2 = \frac{\int \phi^2 |\chi(\tau, \phi)|^2\, d\phi}{\int |\chi(\tau, \phi)|^2\, d\phi} + \sigma_{\theta_e'}^2 \qquad (11\text{-}52)$$

The first term of (11-52) is the second moment (or equivalent half-width squared) of the undistorted ambiguity function along the ϕ-axis direction. The second term represents the variance, or mean square frequency error, of $\theta_e'(t)$ the frequency modulation associated with $\theta_e(t)$, or

$$\sigma_{\theta_e'}^2 = E\{[\theta_e'(t)]^2\} - \{E[\theta_e'(t)]\}^2 = \text{mean square frequency error} \qquad (11\text{-}53)$$

where $E(y)$ is the expected value of y.

It is usually assumed that $E[\theta_e'(t)] = 0$, so that $\sigma_{\theta_e'}^2 = E\{[\theta_e'(t)]^2\}$. The expression given by (11-52) neglects the effect of the shift in the mean position of the distorted ambiguity function. This will ordinarily be zero for most random modulation error distributions for which $E[\theta_e'(t)] = 0$ for each sample function of the ensemble. For error distributions that have $E[\theta_e'(t)] = 0$ only in the aggregate the mean position error will tend to zero if the spectrum distribution of $\theta_e'(t)$ is relatively broad. Position errors at the output of the matched filter are normally associated with low frequency error components. It is entirely possible to have a random distribution of low frequency phase modulation errors for which $E[\theta_e'(t)] = 0$ over the long term, but not during each transmitted pulse. This will give the effect of a pulse-to-pulse shift in the location of the matched-filter output about a mean position, whereas the broad spectrum random phase error function will give the effect of a smeared matched-filter output for each received pulse.

For the case of a random phase error in the matched filter $\beta(\omega)$, and $E[\beta'(\omega)] = 0$, the expression for the ambiguity function spread is approximated by

$$\sigma_\tau^2 = \frac{\int \tau^2 |\chi(\tau, \phi)|^2 \, d\tau}{\int |\chi(\tau, \phi)|^2 \, d\tau} + \sigma_{\beta'}^2 \tag{11-54}$$

where $\sigma_{\beta'}^2 = E\{[\beta'(\omega)]^2\}$ is the mean square time delay error. Equations (11-52) and (11-54) have the form of variances, so that the spread of the ambiguity function is $2\sigma_\phi$ along the direction of the frequency axis, and $2\sigma_\tau$ along the direction of the time axis. σ_ϕ^2 can be obtained for any value of range shift τ by using (11-52), and σ_τ^2 can be obtained for any value of ϕ using (11-54).

An alternate measure of resolution, based on Woodward's time and frequency resolution constants discussed in Chapter 4, is the equivalent rectangle of range resolution, defined by

$$T_r = \int |g_d(t)|^2 \, dt/|g_d(0)|^2 \tag{11-55}$$

where $g_d(t)$ is the distorted matched-filter output signal.

This definition is based on equating the energy in an equivalent rectangular pulse to that of $g_d(t)$, as illustrated by Fig. 11.20. In practice, the assumption of random distortion errors may not be realistic for most types of matched-filter systems for radar application.[1] It may be more realistic to assume, once a particular matched-filter system has been built, that the associated errors are deterministic, with measurable characteristics. Using this assumption, Cheng has defined an upper bound on the equivalent rectangle of resolution for complicated phase error functions [12]. This approach is outlined below.

An arbitrary complex matched-filter phase error function is shown in Fig. 11.21. The maximum value of $\beta(2\pi f)$ with respect to the zero phase reference is m_p radians. The distorted matched-filter output signal $g_d(t)$ is described by

$$g_d(t) = \int |S(f)|^2 \exp[j\{2\pi ft + \beta(2\pi f)\}] \, df \tag{11-56}$$

where $|S(f)|$ is the spectrum of transmitted signal. Using Parseval's theorem,

---

[1] Random phase errors can also be caused by propagation through a medium with time or space varying transmission properties. In this case the medium can be treated as a filter for which the analysis of Brown and Palermo is applicable.

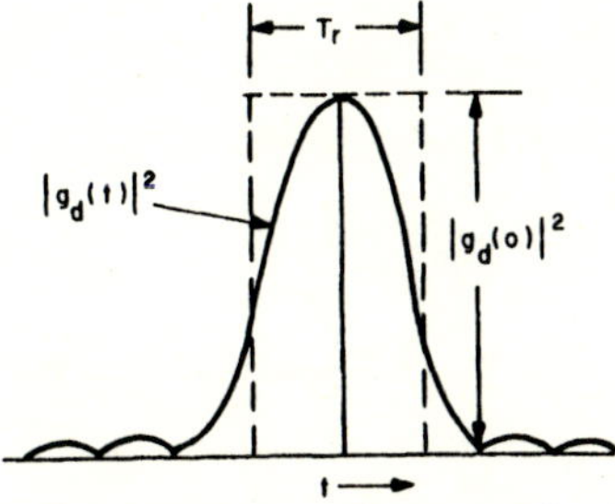

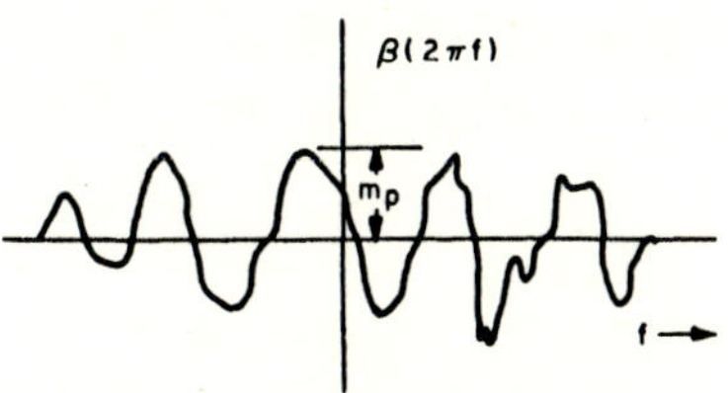

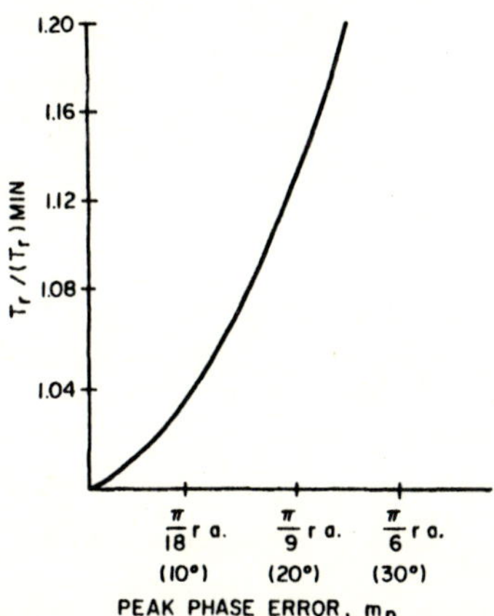

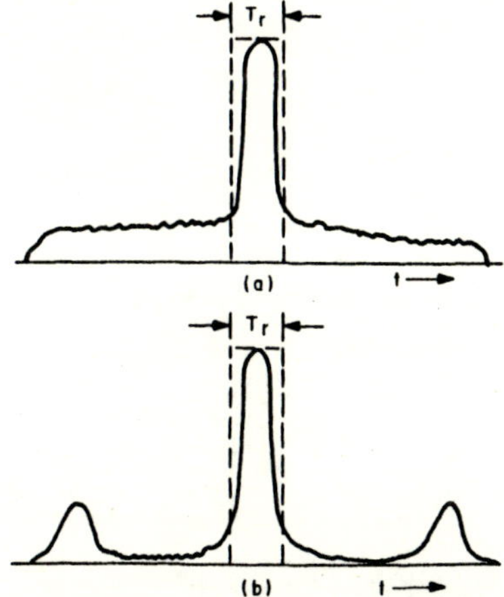

Fig. 11.20   Equivalent rectangle of resolution.

Fig. 11.21   Arbitrary matched-filter phase error function (after Cheng [12]).

Fig. 11.22   Upper bound on equivalent rectangle of resolution as a function of peak phase error, m_p.

Fig. 11.23   Distorted matched-filter waveforms for which the equivalent rectangle of resolution is a poor measure.

the energy of $g_d(t)$ is given by

$$\int |g_d(t)|^2 \, dt = \int \{|S(f)|^2 \exp[j\beta(2\pi f)]\}^2 \, df = \int |S(f)|^4 \, df \qquad (11\text{-}57)$$

This result is implicit in the fact that the energy of the signal is not affected by the phase distortion, except for its distribution in time. The peak value of $g_d(t)$ is obtained from (11-56) by setting $t = 0$, or

$$|g_d(0)|^2 = \left| \int |S(f)|^2 \exp[j\beta(2\pi f)] \, df \right|^2 \qquad (11\text{-}58)$$

If the phase errors are small the following approximation can be made:

$$\exp[j\beta(2\pi f)] = 1 + j\beta(2\pi f) - \tfrac{1}{2}\beta^2(2\pi f) + \cdots \qquad (11\text{-}59)$$

and (11-58) becomes

$$\left|\int |S(f)|^2 \exp[j\beta(2\pi f)]\,df\right|^2 = \left\{\int |S(f)|^2[1 - \tfrac{1}{2}\beta^2(2\pi f)]\,df\right\}^2$$
$$+ \left\{\int |S(f)|^2\beta(2\pi f)\,df\right\}^2 \qquad (11\text{-}60)$$

Noting that $|\beta(2\pi f)|_{\max} = m_p$, one obtains by invoking the mean value theorem for integrals

$$\left[\int |S(f)|^2[1 - \tfrac{1}{2}\beta^2(2\pi f)]\,df\right]^2 \geqslant \left[1 - \frac{m_p^2}{2}\right]^2 \left[\int |S(f)|^2\,df\right]^2 \qquad (11\text{-}61)$$

In addition,

$$\left[\int |S(f)|^2\beta(2\pi f)\,df\right]^2 \geqslant 0 \qquad (11\text{-}62)$$

and thus

$$|g_d(0)|^2 \geqslant [1 - m_p^2/2]^2 \left[\int |S(f)|^2\,df\right]^2 \qquad (11\text{-}63)$$

By substituting (11-57) and (11-63) into (11-55) the result given by Cheng is obtained:

$$T_r \leqslant \frac{\int |S(f)|^4\,df}{[1 - m_p^2/2]^2 \left[\int |S(f)|^2\,df\right]^2} \qquad (11\text{-}64)$$

Equation (11-64) represents an upper bound on the equivalent rectangle of resolution, being a minimum, with equality, when $m_p = 0$. Applying this result to the linear FM signal of unit amplitude, for which

$$|S(f)|^2 \simeq T/2\,\Delta f, \qquad -\Delta f/2 < f < \Delta f/2 \qquad (11\text{-}65)$$

results in

$$T_r = \frac{(T^2/4\,\Delta f^2)\,\Delta f}{[1 - m_p^2/2]^2(T^2/4\,\Delta f^2)\,\Delta f^2} = \frac{1/\Delta f}{[1 - m_p^2/2]^2} \qquad (11\text{-}66)$$

If a weighting function $W(f)$ is used to reduce the range sidelobes of the matched-filter signal then the factor $|S(f)|W(f)$ replaces $|S(f)|^2$ in the development above. If cosine weighting is used with the linear FM signal (-23 db sidelobes), $(T_r)_{\min} = 1.23/\Delta f$ compared to the 3 db pulse width of $1.34/\Delta f$; for $(\text{cosine})^2$ weighting $(T_r)_{\min} = 1.5/\Delta f$ compared to the 3 db pulse width of $1.6/\Delta f$. Since the factor

$$(T_r)_{\min} = \int |S(f)|^4\,df \Big/ \left[\int |S(f)|^2\,df\right]^2$$

applies to all matched-filter signals, or mismatched signals as defined above, the relationship

$$\frac{T_r}{(T_r)_{min}} = \frac{1}{[1 - m_p^2/2]^2} \tag{11-67}$$

represents a universal upper bound for the case of matched-filter phase errors. Equation (11-67) is plotted in Fig. 11.22. For the case of modulation phase errors a similar analysis yields an equivalent rectangle of Doppler frequency resolution where

$$(F_\phi)_{min} = \int |u(t)|^4 \, dt \Big/ \left[\int |u(t)|^2 \, dt\right]^2, \quad |u(t)| = \text{transmitted envelope} \tag{11-68}$$

and

$$F_\phi/(F_\phi)_{min} = 1/[1 - m_a^2/2]^2 \tag{11-69}$$

The upper bound expressions given by (11-64) and (11-68) are not applicable to all types of phase error functions. As an example, if the distortion function $\beta(2\pi f)$ considered above has many different high frequency components the distorted matched-filter output could have the form shown in Fig. 11.23(a). If one of the high frequency distortion components is dominant the distorted output signal might take on the form shown in Fig. 11.23(b). In these examples the effect of the phase distortion is to create interfering clutter signals or a major ambiguous response without having a particular smearing effect on the main matched-filter response. The major effect of the distortion in these cases is to degrade the capability to detect smaller signals, rather than to cause loss of resolution among relatively large signals. Thus, the use of the equivalent rectangle of resolution criterion must be applied with care, and would appear to be most useful when the distortion functions are composed of low frequency components. Here again, however, some care must be taken in interpreting results. The resolution factor given by (11-67) does not change if sidelobe reduction weighting is used, whereas it was seen for the case of low frequency quadratic phase errors considered in Chapter 6 that the use of the weighting function reduced the spread of the distorted linear FM matched-filter output signal. For these reasons it would appear to be more useful to attempt to define, where possible, the major components of the distortion functions in order to obtain a more accurate picture of the effect of a particular distortion function on the matched-filter signal under consideration, and to provide a basis for systematic distortion compensation at the receiver.

In a number of situations the distortions may arise from time variable factors associated with the system implementation. These would include power supply fluctuations and timing jitter in the transmitter, and also in the receiver if an active correlator is used in place of a matched filter. Here, the systems engineer may apply, with judicious care, the measures of loss of resolution discussed above. If the levels of distortion are more than can be tolerated, then the techniques of active distortion compensation discussed in the next section should be considered.

11.9 Distortion Compensation

Despite the best efforts of the system designer, obtaining the required level of allowable distortion in the transmitter by sheer brute force control over the individual components may become prohibitively expensive. When this is the case the best procedure to follow is one of designing distortion compensation circuits that introduce an equal and opposite amount of distortion to cancel the distortion that is generated in the high power transmitter chain. In practice, complete cancellation of the distortion components cannot be achieved, but reduction of these components by factors of up to 10 are realizable in operating systems.

Distortion correction can be based on open loop or closed loop error compensation techniques. General forms of both of these approaches have been discussed by DiFranco and Rubin[6], and their block diagram representations are shown in Figs. 11.24 and 11.25. The open loop technique shown in Fig. 11.24 is based on detecting the amplitude ripple on the

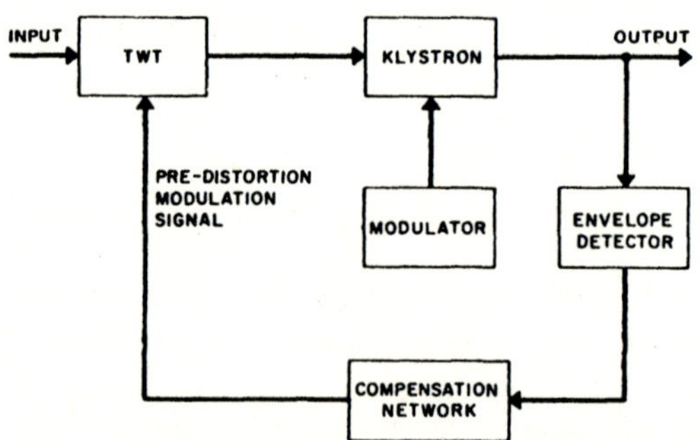

FIG. 11.24 An open-loop compensation technique (courtesy of J. V. DiFranco and W. L. Rubin [6]).

output of the final high power stage, and converting this amplitude error data into a phase predistortion signal that is applied to a stage prior to the final tube. This technique is applicable when the amplitude distortion associated with a phase modulation distortion is detectable but does not of itself produce distortion sidelobes above the required level. However, this approach, which is primarily based on compensating for the effects introduced by the final stage pulse modulator, will not be easy to implement when the phase modulation distortion components have little, or no, associated amplitude modulation. An alternate approach to an open loop correction system can be implemented if the modulation errors are known and repeatable from pulse-to-pulse, or if it is desired to compensate for some undesirable and deterministic characteristic of the normal transmitted matched-filter signal. In this situation the desired correction, or predistortion, function can be synthesized, synchronized with the time occurrence of the signal, and applied at a suitable point in the transmitter to accomplish the desired result. The linear FM predistortion function to reduce the Fresnel spectrum ripple and the associated paired-echo sidelobes described in Chapter 7 is an example of this latter type of open loop distortion compensation method [15].

The closed loop distortion correction shown in Fig. 11.25 is based on sampling the output of the final high power stage and comparing the phase

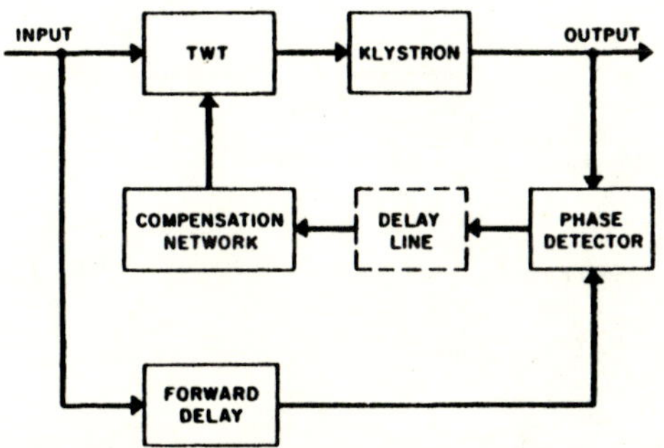

FIG. 11.25   A closed-loop compensation technique (courtesy of J. V. DiFranco and W. L. Rubin [6]).

error deviations of this signal to an undistorted reference signal obtained from the low level matched-filter signal generator (this may be an active or passive generator as discussed in Chapter 6). The two signals are brought into correct time and phase alignment by delay equalization of the path of the reference signal so that it matches the delay encountered in the high power chain. When this delay is small compared to the period of the highest distortion component to be corrected, then distortion correction during the transmitted pulse becomes feasible, and is the most desirable arrangement. If the forward delay does not meet this requirement, the distortion correction signal can be delayed for one repetition interval and used to correct the next pulse, provided that any pulse-to-pulse changes in the distortion frequency relationships do not change significantly in the interpulse period. Such changes could arise from time jitter in the pulse synchronizing signal, or from the effective time jitter in a passive pulse expansion generation technique caused by frequency drift. This method can be applied if the point of error correction in the transmitter occurs after the point where the distortion information is obtained. This method will not work if the error correction signal is applied in a feedback arrangement that reduces the distortion at the point where the distortion information is extracted. DeLorenzo and DeAngelis [13] consider the necessary requirements for techniques to measure the phase modulation error function and to generate an error correction signal for an intrapulse closed loop distortion compensation technique for high fidelity radar applications. Figure 11.26 illustrates one of the techniques described to generate the

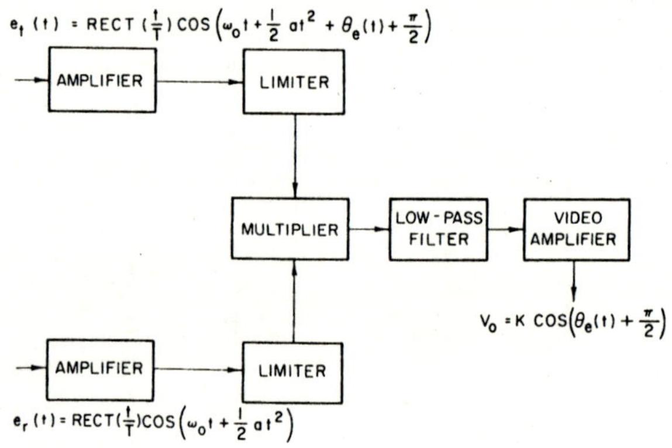

FIG. 11.26 Typical real-time phase comparator (courtesy of J. DeLorenzo and D. DeAngelis [13]).

error signal. The signals used for comparison are the undistorted matched-filter signal

$$s(t) = \cos[\omega_0 t + \theta(t)] \qquad (11\text{-}70)$$

and the distorted transmitter output signal

$$s_d(t) = \cos\left[\omega_0 t + \theta(t) + \theta_e(t) + \frac{\pi}{2}\right] \qquad (11\text{-}71)$$

where $\theta(t)$ is the coded modulation function and $\theta_e(t)$ the phase modulation distortion function.

After amplification and limiting, and the filtering out of high frequency terms, the output of the video amplifier in Fig. 11.26 is

$$v_0(t) = K \cos\left[\theta_e(t) + \frac{\pi}{2}\right] = K \sin[\theta_e(t)] \qquad (11\text{-}72)$$

For small values of $\theta_e(t)$ (i.e., less than $\pm 10°$), $v_0(t)$ can be approximated by

$$v_0(t) = K\theta_e(t) \qquad (11\text{-}73)$$

This represents the required error correction voltage for application to a linear phase modulation device. The measurement technique developed by DeLorenzo and DeAngelis may also be used to measure the frequency domain phase distortion of networks and amplifiers if the test signal has a linear FM modulation function having a relatively large time-bandwidth product, and if the expected phase errors are less than ± 10 degrees. This type of test setup is shown in Fig. 11.27. The linear FM output from the

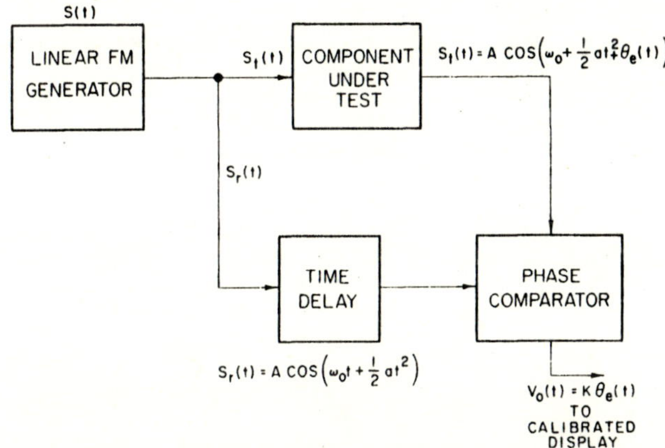

FIG. 11.27   Typical measurement system (courtesy of J. DeLorenzo and D. DeAngelis [13]).

device under test will contain a phase error modulation vs time characteristic that is functionally almost identical to the phase error vs frequency of the component. With a properly calibrated display, the modulation error voltage as derived from the circuit shown in Fig. 11.26 will serve as a fairly exact measure of the component frequency domain phase distortion for the conditions cited above. Tests using these techniques have been made on devices such as traveling wave tubes, microwave components and i.f. amplifiers with measurement accuracies of 0.1 degree, and a reduction of phase modulation distortion on an intrapulse basis of better than 12 db has been achieved.

Rademacher and Randise [14] have investigated the requirements and design parameters for the feedback loop necessary to apply the error correction signal to one of the possible voltage-to-phase transducers in the high power chain, as to a traveling wave tube for instance. Figure 11.28 illustrates the general configuration of the type of transmitter chain feedback loop considered for phase error modulation correction. The individual

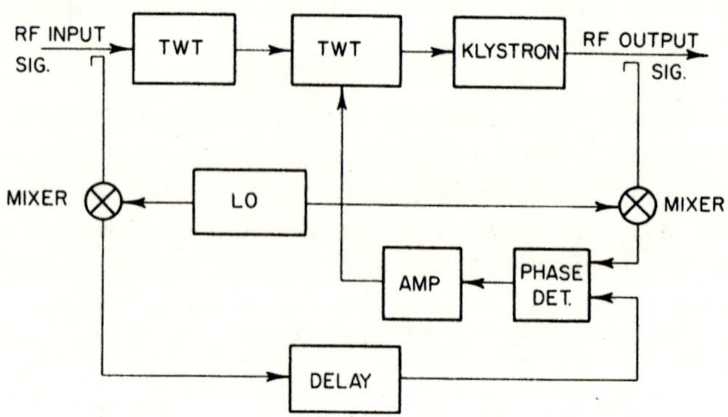

FIG. 11.28 Microwave system with feedback (courtesy of P. Rademacher and D. Randise [14]).

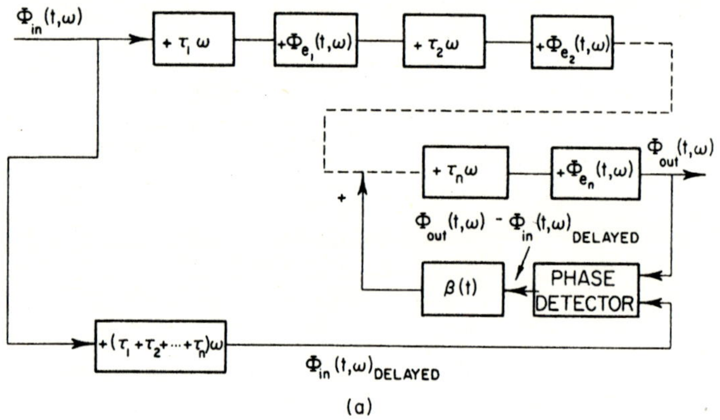

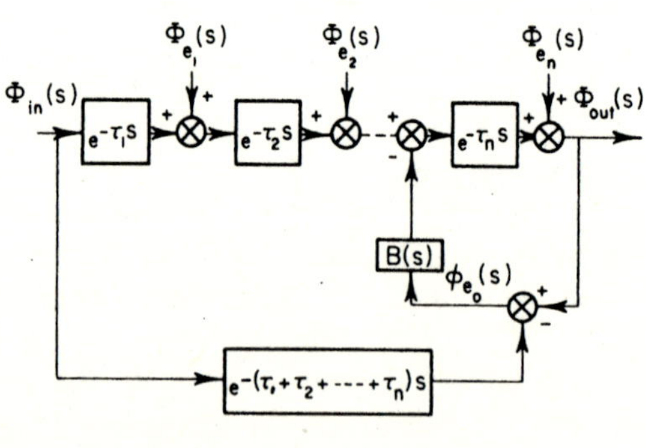

FIG. 11.29 Feedback loop characteristics. (a) Phase transfer characteristics of total feedback system. (b) Classical representation of feedback system. (Courtesy of P. Rademacher and D. Randise [14].)

sources of error and the forward delay requirements are identified in Fig. 11.29(a). The τ_i are the delays associated with the linear phase portion of each device phase function and the $\Phi_i(t, \omega)$ are the respective phase errors as a function of time and frequency. When frequency is a known function of time, as it is for the FM matched-filter signals, the $\Phi_i(t, \omega)$ can be represented as $\Phi_i(t)$, and by application of the Laplace transform the feedback system can be represented in the classical form shown in Fig. 11.29(b).

If the Laplace transform of the phase distortion in the absence of feedback is $\Phi_{e_t}(s)$ and the phase distortion with the application of feedback is $\Phi_{e_0}(s)$, then for the case shown in Fig. 11.29, the ratio of these two terms is, from considerations of basic feedback theory,

$$\frac{\Phi_{e_0}(s)}{\Phi_{e_t}(s)} = \frac{1}{1 + B(s)\exp[-\tau_n s]} \tag{11-74}$$

where τ_n is the delay from point of feedback to the output and $B(s)$ the transform of the feedback amplifier characterstic.

The requirements on the feedback amplifier characteristic are that the gain magnitude $|B(s)|$ must fall below 0 db before the phase term becomes greater than 180 degrees. The difference between 180 degrees and the actual amplifier phase when $|B(s)| = 0$ db is the phase margin of the feedback loop, and the level of gain below 0 db when the phase is 180 degrees is the gain margin of the feedback system. The need to have each of these margins sufficiently large to insure stability of the feedback operation restricts the gain-bandwidth product of the system and thus the effectiveness in reducing distortion. In this regard Rademacher and Randise point out the importance of the time delay τ_n, for this factor introduces an additional phase term to

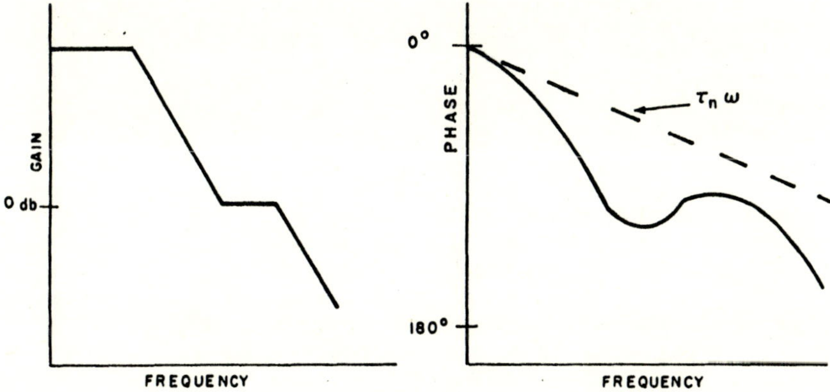

FIG. 11.30   Gain and phase characteristics of the feedback amplifier.

the feedback characteristic. This is illustrated for the feedback amplifier characteristics shown in Fig. 11.30. To keep τ_n small it is necessary to apply the feedback signal at a point fairly close to the final output stage. This means that the feedback must be introduced at a relatively high power point in the system, thus increasing the complexity and cost of the feedback amplifier design. Introduction of the feedback at a lower power point will generally increase τ_n and reduce the phase margin of the system. Results cited by Rademacher and Randise for a feedback amplifier designed to operate in a high power system show a phase margin of 95 degrees at the gain cross-over point when $\tau_n = 0$. However, when $\tau_n = 0.35$ microsecond the phase margin is reduced to only 13 degrees. To compensate for this the gain factor should be reduced if τ_n is large. Again, results cited for $\tau_n = 0.1$ microsecond showed a reduction in transmitter chain phase error of about $10:1$, but when $\tau_n = 0.3$ microsecond the error reduction factor was decreased to $5:1$ because of the necessary reduction of the gain characteristic to avoid system instability.

REFERENCES

1. B. Franklin, private correspondence to M. Leroy, 1789.
2. L. A. MacColl, unpublished manuscript referred to by C. R. Burrows, Discussion of paired-echo distortion analysis, *Proc. IRE (Correspondence)* **27**, 384 (1939).
3. H. A. Wheeler, The interpretation of amplitude and phase distortion in terms of paired echoes, *Proc. IRE* **27**, 359–385 (1939).
4. P. M. Liebman, The effect of distortion on pulsed radar residue level, unpublished report, Sperry Gyroscope Company (1957).
5. C. E. Cook, Effects of phase modulation distortion on pulse-compression signals, *IRE Intern. Conv. Record, Pt.* 4, 174–184 (1962).
6. J. V. DiFranco and W. L. Rubin, Signal processing distortion in radar systems, *IRE Trans.* **MIL-6**, 219–225 (1962).
7. J. Reed, Long line effect in pulse-compression radar, *Microwave J.* **4**, 99–100 (1961).
8. C. E. Cook, A general matched-filter analysis of linear FM pulse compression, *Proc. IRE (Correspondence)* **49**, 831 (1961).
9. C. E. Cook, Transmitter phase modulation errors and pulse-compression waveform distortion, *Microwave J.* **6**, 63–69 (1963).
10. W. M. Brown and C. J. Palermo, Theory of coherent systems, *IRE Trans.* **MIL-6**, 187–196 (1962).
11. W. M. Brown and C. J. Palermo, Effects of phase errors on the ambiguity function, *IEEE Intern. Conv. Record, Pt.* 4, 118–123 (1963).
12. D. K. Cheng, Degradation effects of arbitrary phase errors on high-resolution radar performance, *Proc. IEE (London)* **111**, 1375–1378 (1964).
13. J. DeLorenzo and D. DeAngelis, Real time measurement and correction of phase distortion in high fidelity radars, *IEEE Intern. Conv. Record, Pt.* 8, 96–103 (1963).
14. P. Rademacher and D. Randise, An automatic phase-correction system, *IEEE Intern. Conv. Record, Pt.* 3, 179–186 (1963).
15. J. F. Cerar and C. E. Cook, Electronic means for suppressing range side lobes of a compressed-pulse signal, U.S. Patent No. 3,281,842, October, 1966.

CHAPTER 12

The Design of Dispersive Delay Functions—I

12.1 Introduction

It was shown in Chapter 3 that the implementation of the general FM pulse-compression matched filter to process the signal

$$s(t) = a(t)[\cos \omega_0 t + \theta(t)] \qquad (12\text{-}1)$$

has, as an integral part, a dispersive group delay characteristic $-\Phi'(\omega)$ that has an inverse relationship to the FM function, $\theta'(t)$. The required dispersive delay function may be obtained in a variety of ways, ranging from the use of waveguide at microwave frequencies to ultrasonic delay lines at moderately low frequencies. From the information presented in Chapter 11 on distortion effects, it can be seen that the degree to which the actual delay matches $-\Phi'(\omega)$ can be quite critical in those applications that call for very low levels of range sidelobes associated with the compressed pulse. Three basic approaches are available to the pulse-compression system designer who must specify an accurately controlled dispersive delay function. These are:

(1) Lumped constant filter techniques based on the use of the bridged-T equivalent of the all-pass constant resistance lattice network.

(2) Ultrasonic delay lines designed to operate in a dispersive mode, while suppressing nondispersive modes. This approach provides an ultrasonic equivalent of the bridged-T approach.

(3) Optical signal processing techniques that operate on the space domain–frequency domain analog that can be achieved by optical Fourier transformations. This permits the dispersive delay function to be designed as a spatial, rather than a frequency, function.

Each of these methods has its advantages and disadvantages. The bridged-T approach developed in this chapter requires the use of many such networks to synthesize the necessary dispersive delay. Since these networks react critically to alignment errors, a filter composed of a great number of bridged-T sections requires great care in controlling the tolerance of the component L's and C's, as well as in the alignment procedure. Despite this, the first real success of pulse-compression system applications was achieved with such network designs, and this method is still applicable to systems that require bandwidths greater than a few megahertz. In addition, the basic design approach and lumped constant networks described in this chapter are useful for distortion compensation of other types of pulse-compression matched-filter design techniques. Thus, despite the fact that the bridged-T approach to dispersive filter design is being supplanted in many applications by the ultrasonic and optical methods described in the following chapters, familiarity with the basic properties of the bridged-T network can be of benefit to the system designer.

12.2 The All-Pass Time Delay Network

Several reference texts are available that describe the basic properties of the bridged-T networks that are used in pulse-compression matched filters. The development here essentially follows that given by O'Meara [1] and that derived on programs associated with systems at the Sperry Gyroscope Company [2].

The pulse-compression matched filter requires a dispersive time delay characteristic. That is, one that varies as a function of frequency and exhibits a minimum of attenuation properties in the useful band of frequencies. This rules out most standard types of filter sections, which have maximum dispersiveness in frequency regions near cutoff. A starting point for developing the desired type of network is shown in the basic lattice section shown in Fig. 12.1. The image impedance of this network is

$$Z_i = \sqrt{Z_a Z_b} \tag{12-2}$$

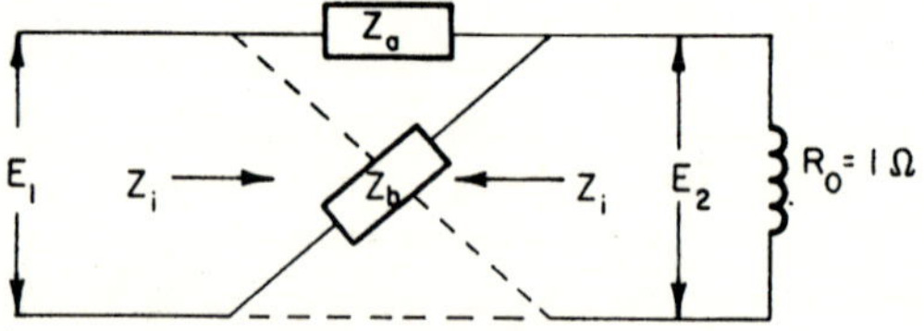

FIG. 12.1 Basic lattice section.

and the transfer function is

$$\frac{E_2}{E_1} = \frac{1 - \sqrt{Z_a/Z_b}}{1 + \sqrt{Z_a/Z_b}} \tag{12-3}$$

If the condition is imposed that

$$Z_a Z_b = 1 \qquad \text{or} \qquad Z_b = 1/Z_a$$

then $Z_i = 1$, a pure resistance (normalized), and

$$\frac{E_2}{E_1} = \frac{1 - \sqrt{Z_a^2}}{1 + \sqrt{Z_a^2}} = \frac{1 - Z_a}{1 + Z_a} = \exp[\alpha - j\beta] \tag{12-4}$$

where α is the attenuation function and β the transfer phase function. If Z_a is further restricted to be a pure reactance, so that

$$Z_a = jX_a, \qquad Z_b = jX_b$$

then

$$\frac{E_2}{E_1} = \frac{1 - jX_a}{1 + jX_a} = \frac{\sqrt{1 + X_a^2}\,\exp[-j\tan^{-1}X_a]}{\sqrt{1 + X_a^2}\,\exp[+j\tan^{-1}X_a]}$$

$$= \exp[-2j\tan^{-1}X_a] \tag{12-5}$$

It can be seen that the above transfer function has no attenuation (i.e., $\alpha = 0$) and a transfer phase characteristic

$$\beta = 2\tan^{-1}X_a \tag{12-6}$$

This type of network is designated as a constant resistance all-pass time delay network.

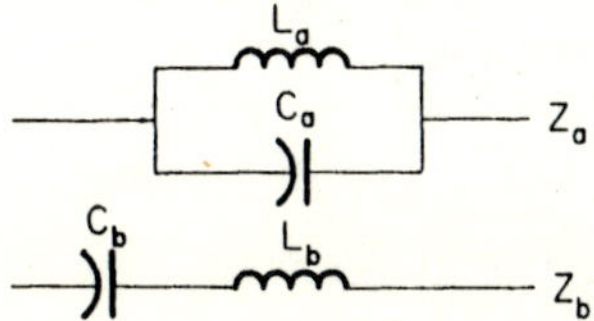

FIG. 12.2   Reactance arms of second-order all-pass lattice.

The most common type of all-pass section used as a basis for pulse-compression network design is the second order lattice where Z_a and Z_b are as shown in Fig. 12.2. If we let

$$1/C_a = 2\sigma_0 \tag{12-7}$$

and

$$1/L_a C_a = 1/L_b C_b = \bar{\omega}_0^2$$

then

$$L_a = \frac{2\sigma_0}{\bar{\omega}_0^2}, \qquad L_b = \frac{1}{2\sigma_0}, \qquad C_b = \frac{2\sigma_0}{\bar{\omega}_0^2} \tag{12-8}$$

The equations for L_b and C_b follow from the requirement

$$L_a/C_b = L_b/C_a = 1$$

Thus

$$Z_a = \frac{j\omega/C}{\bar{\omega}_0^2 - \omega^2} = \frac{2j\sigma_0\omega}{\bar{\omega}_0^2 - \omega^2} \tag{12-9}$$

and

$$X_a = \frac{2\sigma_0\omega}{\bar{\omega}_0^2 - \omega^2} \tag{12-10}$$

From (12-5) and (12-10) one obtains

$$\beta = 2\tan^{-1}\left[\frac{2\sigma_0\omega}{\bar{\omega}_0^2 - \omega^2}\right] \tag{12-11}$$

The matched-filter signals to be processed are i.f., or band-pass, signals, therefore the proper delay characteristic is group time delay. This is expressed as

$$T_D = \frac{d\beta}{d\omega} = \frac{d}{d\omega}\left[2\tan^{-1}\frac{2\sigma_0\omega}{\bar{\omega}_0^2 - \omega^2}\right] \tag{12-12}$$

$$= \frac{4\sigma_0(\bar{\omega}_0^2 + \omega^2)}{(\bar{\omega}_0^2 - \omega^2)^2 + 4\sigma_0^2\omega^2} \tag{12-13}$$

The following section derives an approximate value for T_D that is quite accurate under certain conditions, and is easier to handle.

12.3   All-Pass Network Delay Approximation

Using the expression for Z_a given by (12-9) in the all-pass transfer function (12-5), and letting

$$s = j\omega, \qquad s^2 = -\omega^2, \qquad 1/C_a = 2\sigma_0$$

one obtains

$$\frac{E_2}{E_1} = H(s) = \frac{1 - \dfrac{2s\sigma_0}{\bar{\omega}_0^2 + s^2}}{1 + \dfrac{2s\sigma_0}{\bar{\omega}_0^2 + s^2}} \tag{12-14}$$

Equation (12-14) reduces to

$$H(s) = \frac{s^2 - 2s\sigma_0 + \bar{\omega}_0^2}{s^2 + 2s\sigma_0 + \bar{\omega}_0^2} = \frac{(s - s_1)(s - s_3)}{(s - s_2)(s - s_4)} \tag{12-15}$$

where s_1, s_3 are the zeros of $H(s)$ and s_2, s_4 the poles of $H(s)$.

Use the quadratic root formula to solve for the poles and zeros of the transfer function; this yields

$$s_{1,3} = \sigma_0 \pm \sqrt{-\omega_0^2} = \sigma_0 \pm j\omega_0$$

$$s_{2,4} = -\sigma_0 \pm \sqrt{-\omega_0^2} = -\sigma_0 \pm j\omega_0$$

(12-16)

where $\bar{\omega}_0^2 - \sigma_0^2 = \omega_0^2$. The pole-zero plot in the complex frequency plane is shown in Fig. 12.3. Given a pole-zero plot, as above, a network theorem

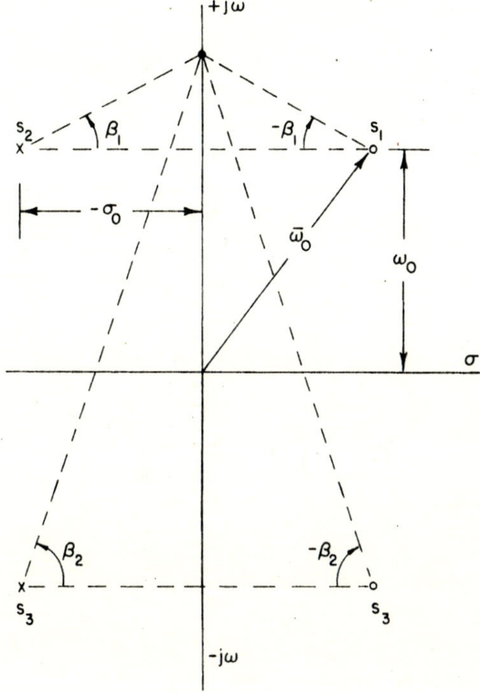

FIG. 12.3   Pole-zero plot of second-order all-pass lattice.

relates the total transfer phase of the network to the indicated angles by

$$\beta = \sum \beta_x - \sum \beta_0$$

(12-17)

where β_x is the vector direction from a pole to a general real frequency point, ω, and β_0 the vector direction from a zero to a general real frequency point, ω.

For the all-pass case $\beta = 2 \Sigma \beta_x$. Referring to the pole-zero plot, it can be seen that

$$\beta_{s_1} = \tan^{-1} \frac{\omega - \omega_0}{\sigma_0}, \qquad \beta_{s_3} = \tan^{-1} \frac{\omega + \omega_0}{\sigma_0} \tag{12-18}$$

and

$$\beta = 2 \left[\tan^{-1} \frac{\omega - \omega_0}{\sigma_0} + \tan^{-1} \frac{\omega + \omega_0}{\sigma_0} \right] \tag{12-19}$$

The following parameter assumptions are made by O'Meara:

$$\omega_0 / \sigma_0 > 3$$

$$|\omega_0 - \omega| < 0.3\omega_0 \quad (\pm 30\% \text{ bandwidth})$$

For the case of $\omega_0 / \sigma_0 = 3$ the second term takes on the following values:

$$\omega = 0.7\omega_0, \qquad \tan^{-1} \frac{1.7\omega_0}{\sigma_0} = \tan^{-1} 5.1 = 78° \, 55'$$

$$\omega = \omega_0, \qquad \tan^{-1} \frac{2\omega_0}{\sigma_0} = \tan^{-1} 6.0 = 80° \, 33' \tag{12-20}$$

$$\omega = 1.3\omega_0, \qquad \tan^{-1} \frac{2.3\omega_0}{\sigma_0} = \tan^{-1} 6.9 = 81° \, 47'$$

Compared to the variations of the first term in (12-19) around the frequency $\omega = \omega_0$, the second term can be approximated by a linear function, so that for the parameter assumptions given above β may be expressed as

$$\beta = 2 \tan^{-1} \frac{\omega - \omega_0}{\sigma_0} + k\omega + C \tag{12-21}$$

The group delay then becomes

$$T_D = \frac{d\beta}{d\omega} = \frac{2}{1 + \left(\dfrac{\omega - \omega_0}{\sigma_0} \right)^2} \frac{1}{\sigma_0} + k \tag{12-22}$$

The values tabulated above show that k is rather small so that

$$T_D \doteq \hat{T}_D = \frac{2/\sigma_0}{1 + \left(\dfrac{\omega - \omega_0}{\sigma_0} \right)^2} \tag{12-23}$$

For purposes of normalization $\omega - \omega_0 = x$ and

$$\hat{T}_D = \frac{2\sigma_0^{-1}}{1 + (x\sigma_0^{-1})^2} \tag{12-24}$$

where $2\sigma_0^{-1}$ represents the peak time delay obtained at $x = 0$. Equation (12-24) is the O'Meara approximation for the group delay of an all-pass, constant resistance lattice section.

For the sake of clarification the following are pointed out:

$\bar{\omega}_0$ = frequency at which impedances of the lattice arms resonate;

ω_0 = frequency at which maximum delay occurs (using the approximation formula for T_D), and $\bar{\omega}_0 > \omega_0$.

Values of $\hat{T}_D$ as given by O'Meara are plotted in Fig. 12.4 for normalized parameters of σ_0 and $\omega - \omega_0 = x$. Specific design data can be obtained from these curves. The procedure for this is described in Section 12.5.

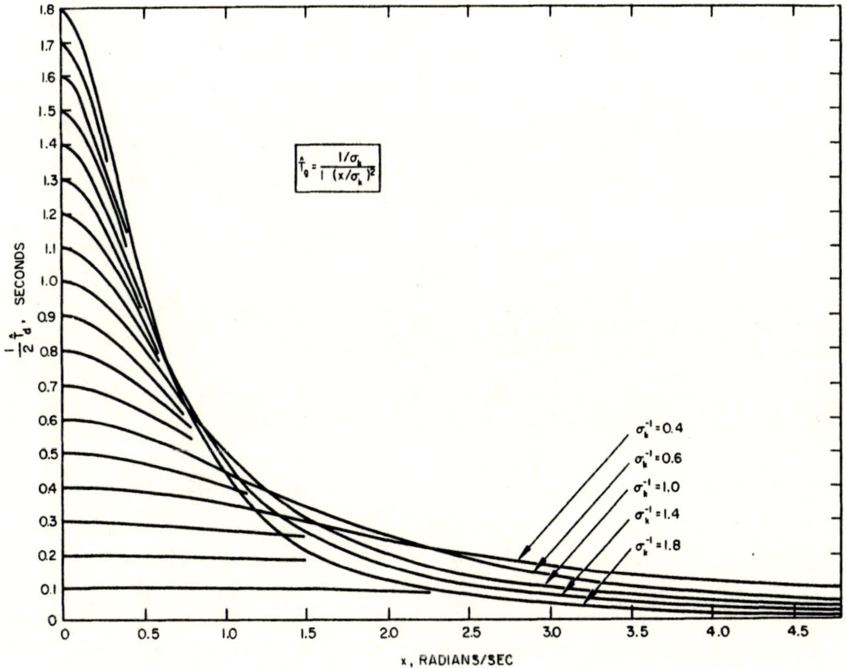

FIG. 12.4 Delay curves for O'Meara's approximation to second-order bridged-T network (courtesy of T. R. O'Meara [1]).

12.4   Lattice Conversion to Unbalanced Form

The lattice type of network shown in Fig. 12.1 is not generally desirable in itself, since it is an unbalanced network with respect to ground. The lattice has no common potential reference at any point, and it is quite critical to any mismatches in the impedances of the lattice arms. In order to obtain a more useful network the lattice must be converted to an unbalanced equivalent.

Figure 12.5 illustrates the general form of the conversion to a bridged-T network. The conditions of the conversion are

$$\frac{1}{2Z_a} = \frac{1}{Z_a'} + \frac{1}{2Z_c}, \qquad \frac{Z_b}{2} = Z_b' + \frac{Z_c}{2} \tag{12-25}$$

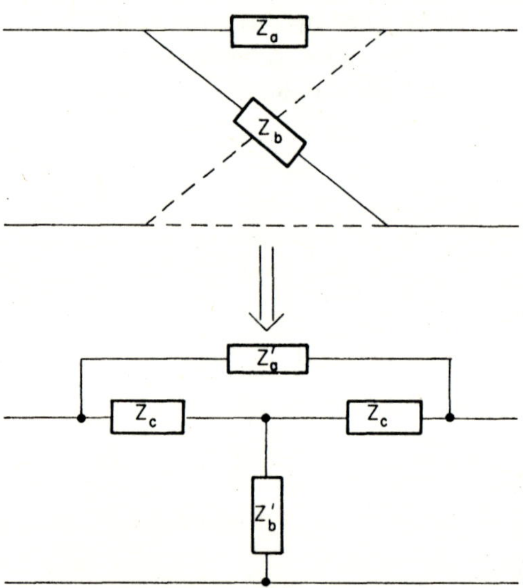

FIG. 12.5   Lattice to bridged-T conversion.

For the case of the second order lattice arms shown in Fig. 12.2 the minimum element bridged-T configuration is shown in Fig. 12.6. This particular form of the lattice/bridged-T conversion results in the element values shown. The capacity in the shunt arm to ground will determine whether this transform results in a realizable set of component values. This capacity may be written as

$$C_s = \frac{2C_a}{(C_a/C_b) - 1} \tag{12-26}$$

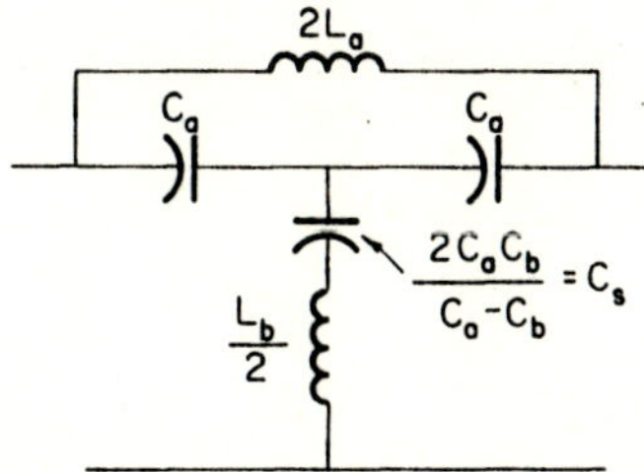

FIG. 12.6 A minimum element bridged-T.

The denormalized values of C_a and C_b are

$$C_a = \frac{1}{2\sigma_0 R_0} \tag{12-27a}$$

$$C_b = \frac{2\sigma_0}{R_0 \bar{\omega}_0^2} \tag{12-27b}$$

where R_0 is the desired impedance of the network. Thus

$$C_s = \frac{2C_a}{(\bar{\omega}_0^2/4\sigma_0^2) - 1} \tag{12-28}$$

and it can be seen that the condition for realizability is

$$\bar{\omega}_0/\sigma_0 > 2 \tag{12-29}$$

It is also possible to use the nonminimum element network shown in Fig. 12.7, having the element values as indicated. In this case the shunt arm capacity is

$$C_s = \frac{2kC_a}{(k\bar{\omega}_0^2/4\sigma_0^2) - 1} \tag{12-30}$$

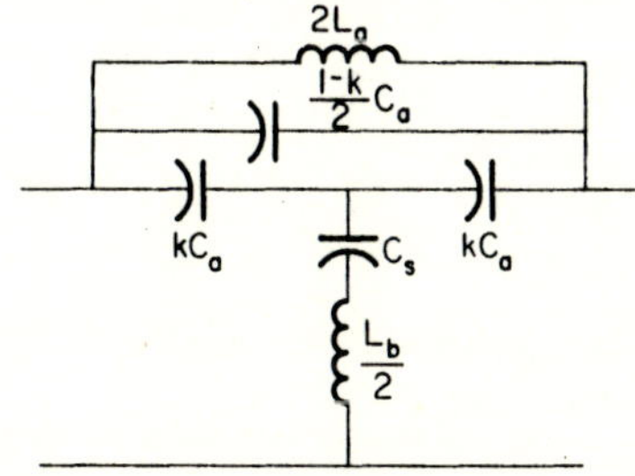

FIG. 12.7 A nonminimum element bridged-T.

and the realizability condition is

$$\bar{\omega}_0/\sigma_0 > 2/\sqrt{k} \tag{12-31}$$

Since $k < 1$, this is a more restricted requirement than that of (12-29).

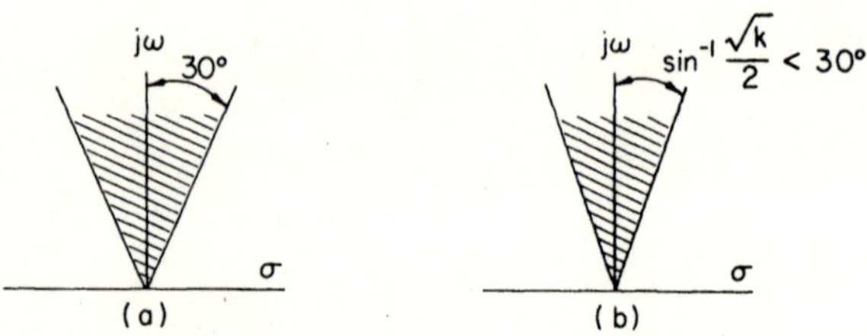

FIG. 12.8   Regions of realizability for bridged-T conversion. (a) Minimum element bridged-T. (b) Nonminimum element bridged-T.

Figure 12.8 shows, in the shaded regions of the complex frequency plane, the areas in which the zeros and poles (Eq. (12-16)) of the network transfer function (Eq. (12-15)) must fall for each of the network forms given in Figs. 12.6 and 12.7. The nonminimum element network can be useful when the minimum element network leads to component values that are difficult to obtain or fabricate. The equivalent reactance arms of the minimum element and nonminimum element bridged-T all-pass networks are shown in Figs. 12.9 and 12.10.

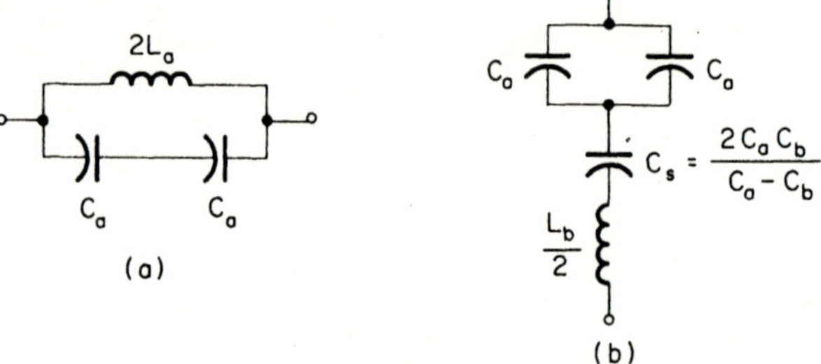

FIG. 12.9   Equivalent reactance arms of minimum element bridged-T. (a) Series arm. (b) Shunt arm.

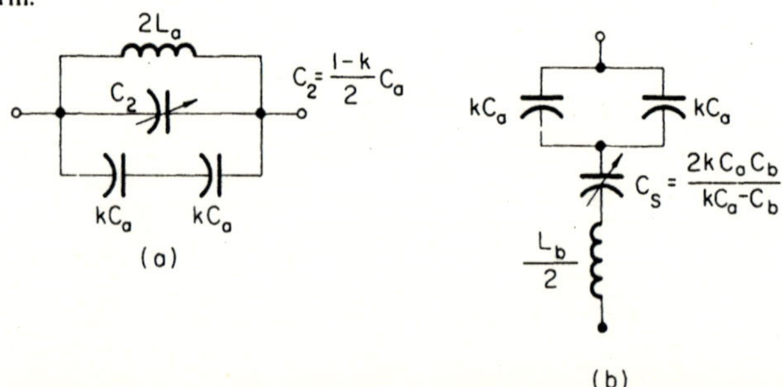

FIG. 12.10   Equivalent reactance arms of nonminimum element bridged-T. (a) Series arm. (b) Shunt arm.

12.5 Illustrative Design of a Linear Delay Network

The normalized bridged-T delay curves developed by O'Meara, and shown in Fig. 12.4, provide the basis for combining a number of such denormalized curves, staggered at different center frequencies, to obtain a composite dispersive delay characteristic. O'Meara has determined several such basic designs for linear time delay for combinations of 1, 2, 3, 4, and 6 bridged-T sections. The efficiency of a linear delay design has been defined by O'Meara as

$$\eta = \frac{T\,\Delta f}{2N} 100\% \tag{12-32}$$

where N is the total number of all-pass sections used. For linear delay curves designed to have approximately $\pm 1\%$ delay deviation from linearity the efficiency is roughly expressed as

$$\eta = 10n\%$$

when n is the number of all-pass networks in the basic design. The number of sections governs the spread of the values of L's and C's used, and the number of ripples in the time delay error curve (and hence influences the delay equalization procedure). Data supplied by O'Meara for a staggered triplet design will be used to illustrate a design example for a linear delay function.

In developing normalized delay curves the parameters σ_0^{-1} and x were used, where

$$2\sigma_0^{-1} = \text{peak delay at } x = 0$$
$$x = \omega - \omega_0, \quad \text{a normalized frequency}$$

For a design using three bridged-T sections the following relationships among these parameters are given by O'Meara as [1]:

$$x_1 = 0 \qquad \sigma_1^{-1} = 1.00$$

$$x_2 = 1.0 \qquad \sigma_2^{-1} = 0.65$$

$$x_3 = 2.6 \qquad \sigma_3^{-1} = 0.54$$

Figure 12.11 plots the normalized composite delay for O'Meara's triplet design. The linear delay portion of this curve extends from $x = 0.26$ to $x = 4.34$, being centered at $x = 2.3$. To denormalize this curve to fit a specific case the following must be known:

f_0,   the desired design center frequency

Δf,   the bandwidth over which the linear delay is to be maintained

Given these, a bandwidth denormalizing factor $2\pi\,\Delta f/\Delta x$ can be used to

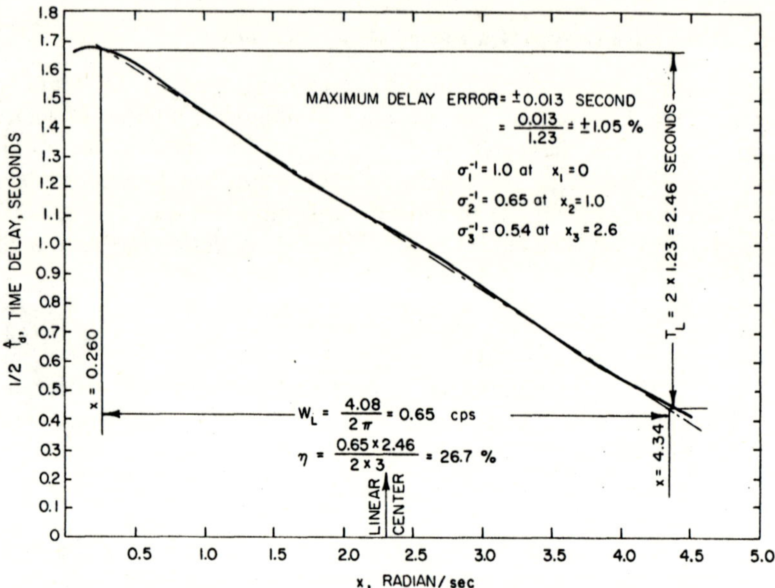

FIG. 12.11   Staggered-triplet approximation ($N = 3$) (courtesy of T. R. O'Meara [1]).

determine the actual delay of the triplet at any frequency about f_0 by the relationship

$$t_{D_{\text{actual}}} = \frac{\hat{T}_D}{2\pi\, \Delta f/\Delta x} \tag{12-33}$$

The position of the denormalized frequencies of peak delay relative to f_0 are obtained from

$$f_i = f_0 - \frac{(x_0 - x_i)}{2\pi}\frac{2\pi\Delta f}{\Delta x} = f_0 - (x_0 - x_i)\frac{\Delta f}{\Delta x} \tag{12-34}$$

As an example, the following values can be used:

$$f_0 = 5 \text{ MHz}, \qquad \Delta f = 1.6 \text{ MHz}$$

Since $\Delta x = 4.08$, and $x_0 = 2.3$ in the triplet design given by O'Meara, the frequencies of peak delay for each section are, from Eq. (12-34),

$$f_1 = 5 - (2.3 - 0)\frac{1.6}{4.08} = 4.098 \text{ MHz}$$

$$f_2 = 5 - (2.3 - 1.0)\frac{1.6}{4.08} = 4.49 \text{ MHz}$$

$$f_3 = 5 - (2.3 - 2.6)\frac{1.6}{4.08} = 5.112 \text{ MHz}$$

From the condition of Eq. (12-16)

$$\bar{f}_i^2 = f_i^2 + f_{\sigma_i}^2 \tag{12-35}$$

The denormalized values of the σ_i are given by

$$f_{\sigma_i} = \frac{\Delta f}{\Delta x \, \sigma_i^{-1}} \tag{12-36}$$

so that

$$f_{\sigma_1} = \frac{0.392}{1.0} = 0.392 \text{ MHz}$$

$$f_{\sigma_2} = \frac{0.392}{0.65} = 0.603 \text{ MHz}$$

$$f_{\sigma_3} = \frac{0.392}{0.54} = 0.726 \text{ MHz}$$

Substituting these into Eq. (12-35), the following alignment frequencies of the triplet are obtained:

$$\bar{f}_1 = 4.10 \text{ MHz}$$

$$\bar{f}_2 = 4.53 \text{ MHz}$$

$$\bar{f}_3 = 5.18 \text{ MHz}$$

With the information developed above, the element values of the basic triplet section can be worked out using Eqs. (12-7) and (12-8), and Eqs. (12-26)–(12-30) for the type of configurations given in Figs. 12.6 and 12.7. These values will be for unit impedance level, so that once an impedance level R_0 is specified, the actual component values can be determined from the design values by:

$$L_{\text{act}} = R_0 L_{\text{des}} \qquad C_{\text{act}} = \frac{C_{\text{des}}}{R_0} \tag{12-37}$$

The O'Meara triplet design will have a time delay error of approximately 1% of the total linear delay range. This is apparently a rather small error. However, as was shown in Chapter 11, the tolerable delay error is based on an absolute deviation from linear delay, and not a percentage deviation. Thus, as the total delay range is increased to obtain larger compression ratios, the percentage error must decrease in order to maintain the delay error and the paired-echo distortion sidelobes at a fixed value. At this point, many practical factors come into play such as the cumulative effect of impedance errors in a long delay line, the effect of stray L's and C's on the theoretical delay curve and the effects of using components that must be specified within a tolerance range. For these reasons, the next steps are not necessarily to refine the theoretical delay curve to reduce the delay error even further. An approach using the basic design discussed above, that has been used with success to obtain improved delay linearity, is described below.

Figure 12.12 gives measured time delay data for a combination of three triplets based on the design parameters worked out above. The delay error has been increased over that given by O'Meara because of the practical factors discussed above. In this design small changes in the values of σ_i and f_i were worked out in order to yield a more regular pattern on the delay curve. The reason for this was to permit the design of a second delay curve, based on frequency scaling techniques, having error ripples that are essentially 180 degrees out of phase with those of the first curve. This is illustrated by the dashed curve of Fig. 12.12, that gives the measured delay curve for three triplets of the basic design that has been scaled down in frequency by the factor 1.02. The combination of these two delay curves (measured data for six triplets) is also given in Fig. 12.12. The effect of this design technique in reducing the delay error is considerable. Despite this, the resultant delay error when a large number of combined delay designs are cascaded may still be too large, and even further steps may be necessary to obtain error reduction. This problem is discussed in the next section.

An estimate of the total number of sections required for a given application may be obtained from

$$N = \frac{\Delta T}{\Delta \hat{T}(\Delta x/2\pi\,\Delta f)} \tag{12-38}$$

where ΔT is the total delay difference required in the band Δf, $\Delta \hat{T}$ the normalized delay difference from O'Meara's data, and $2\pi\,\Delta f/\Delta x$ the denormalizing factor. Applying this to the triplet design in which a total delay of 50 microseconds is desired for the example discussed above, we obtain

$$N = \frac{50}{(3.34 - 0.88) \times \dfrac{4.08}{6.28 \times 1.6}} \doteq 50 \text{ triplet sections}$$

The considerations of a final design specification are discussed in the next section.

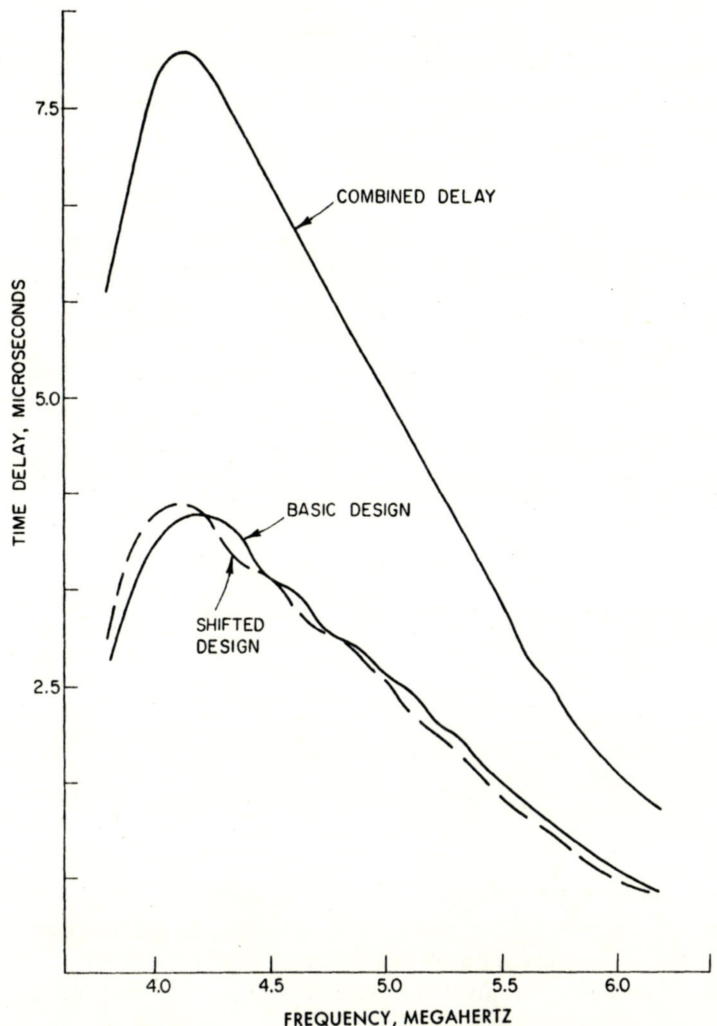

FIG. 12.12   Delay error reduction with frequency scaling of the basic design.

12.6 Matched-Filter Error Correction

The technique of group staggering outlined in the previous section can be quite useful in compensating for certain types of time delay errors. However, as the complete matched filter is built up the residual delay error may reach a level that exceeds the overall limit set for the system. In addition, other sources of delay error are present when a complete pulse-compression matched filter is assembled. Thus, final error correction networks may need to be incorporated into the design. One approach to designing the final error correctors is to establish an error curve for the assembled pulse-compression filter. This is done by obtaining an accurate delay measurement, and plotting the difference between the measured data curve and the desired theoretical curve. This is illustrated by the curve of Fig. 12.13(a) in which the error data of 30 triplet sections of the

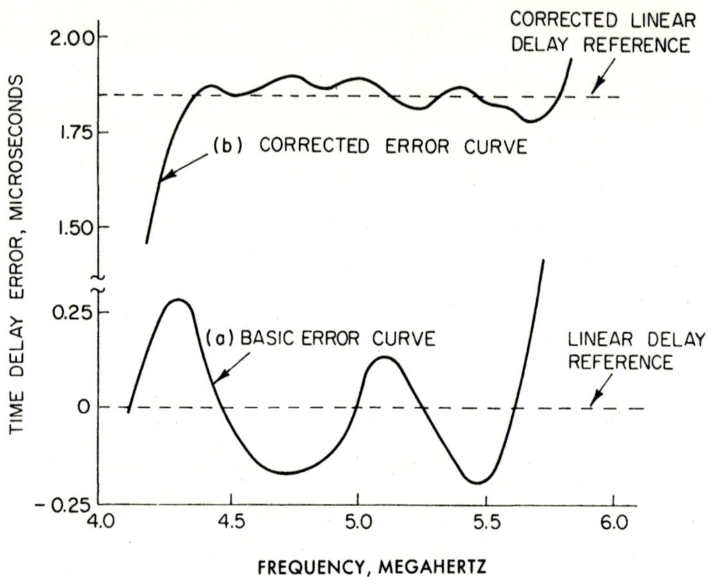

FIG. 12.13   Linear delay error data.

design example of the previous section is compared to the ideal linear delay curve. The ideal curve is represented by the dashed horizontal line through the zero coordinate of delay. The design of a delay correction network depends on employing a combination of a number of the delay curves as given by O'Meara. For the case being illustrated, a combination of four such bridged-T time delay networks was used. The individual and combined delay characteristics are shown in Fig. 12.14. The effect of adding this combination of delay correctors to the pulse-compression delay line is shown in Fig. 12.13(b). The new error reference line reflects the addition of fixed delay to the original reference line. However, as long as these two lines remain parallel, this has no effect on the matched-filter properties of the compression line, resulting only in an additional fixed delay of the signals at the filter output. The effect of these correctors in this example was to reduce the *delay* error by a factor of 4 : 1. However, an increase in the rate of the error ripple also resulted, so that there is an even

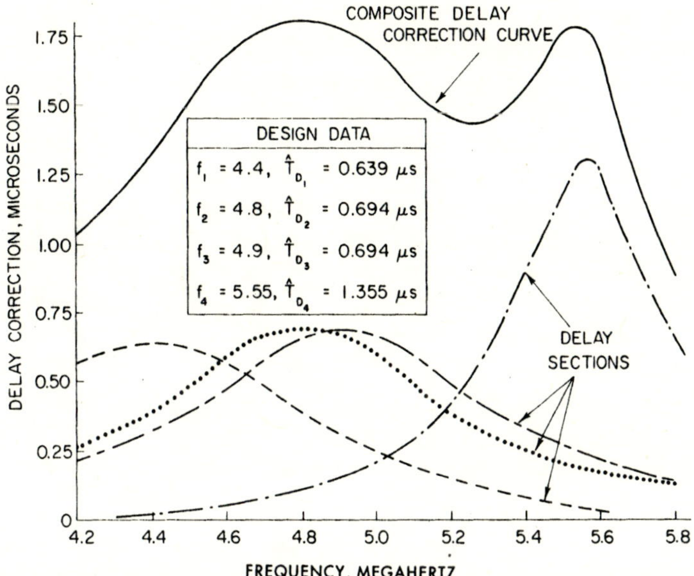

FIG. 12.14   Delay error correction design (courtesy of G. Barnocky, Sperry Piedmont Co., Charlottesville, Virginia).

greater reduction factor in the *phase* error, which is actually the most important criterion, as shown by the paired-echo analysis carried out in Chapter 11. If the phase error is represented by a sinusoid, such that

$$\beta_e = b_1 \sin C\omega \qquad (12\text{-}39)$$

then the associated time delay error is

$$t_e = \frac{d\beta_e}{d\omega} = b_1 C \cos C\omega \qquad (12\text{-}40)$$

From this it can be seen, that for a fixed peak delay error the peak phase error is inversely related to C, the number of cycles of the delay error function over the band Δf. In a specific practical development the group stagger approach and/or delay equalization may be required. However, the actual steps taken will depend largely on the engineer's evaluation of the error curve data.

In approaching the design of the delay error correction function it is more helpful to use *standardized*, rather than *normalized*, delay curves. From Eq. (12-27a) and the expression for the bridged-T all-pass peak delay (12-24) the relation between the bridged-T series arm capacity and the peak delay is

$$C_a = \hat{T}_{D_0}/4R_0 \qquad (12\text{-}41)$$

Since the peak delay determines all the other parameters of the delay curve, the center frequency of the design will not affect the shape of the curve once the peak delay has been chosen. Thus, one can determine from the delay error curve the approximate magnitude of peak delay contribution required from each bridged-T corrector, estimating the frequency at which the peak delay occurs. If, on examining the composite delay correction curve, further adjustments are required, the estimate of the center frequencies of peak delay can be altered. Because of the property just described, this will not change the shape of each contributing delay curve as long as the peak delays remain unchanged. To perform the design in this way requires the use of standard delay curves. Table 12-I and

TABLE 12-I

STANDARD DELAY DATA IN MICROSECONDS

					$|f_0 - f|$ MHz					
0.0	0.1	0.2	0.3	0.4	0.5	0.6	0.7	0.8	0.9	.1.0
0.25	0.247	0.244	0.238	0.226	0.216	0.204	0.191	0.179	0.166	0.155
0.50	0.488	0.455	0.408	0.360	0.310	0.265	0.224	0.190	0.168	0.145
0.75	0.712	0.610	0.500	0.394	0.315	0.251	0.202	0.163	0.135	0.114
1.00	0.910	0.720	0.530	0.380	0.290	0.220	0.170	0.140	0.110	0.092
1.25	1.025	0.775	0.525	0.362	0.256	0.187	0.134	0.115	0.093	0.076
1.50	1.220	0.788	0.502	0.326	0.228	0.166	0.126	0.098	0.078	0.064
1.75	1.340	0.788	0.472	0.298	0.205	0.147	0.110	0.086	0.068	0.056
2.00	1.440	0.760	0.440	0.280	0.194	0.130	0.098	0.077	0.060	0.049

					$|f_0 - f|$ MHz					
0.0	1.1	1.2	1.3	1.4	1.5	1.6	1.7	1.8	1.9	2.0
0.25	0.142	0.131	0.120	0.112	0.105	0.095	0.089	0.084	0.078	0.072
0.50	0.125	0.110	0.095	0.085	0.076	0.070	0.062	0.055	0.050	—
0.75	0.098	0.083	0.071	0.063	0.055	0.049	0.044	0.039	—	—
1.00	0.077	0.065	0.056	0.049	0.043	0.038	—	—	—	—
1.25	0.064	0.054	0.046	0.042	—	—	—	—	—	—

Fig. 12.15 present data for several such standard delay curves. Once the proper combination of peak delays and their associated frequencies have been set, as shown in the example of Fig. 12.14, then the components of each bridged-T section can be worked out in accordance with Eqs. (12-7) and (12-8). Since the design approach leads to values of σ_i and f_i, the frequency of peak delay, the alignment frequency of each section can be obtained from Eq. (12-35).

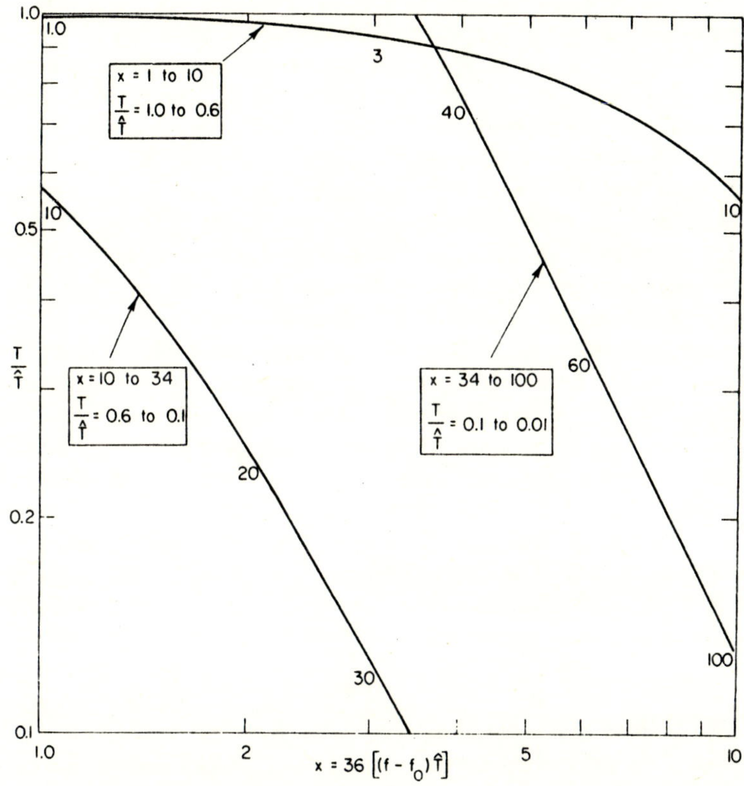

FIG. 12.15 Normalized time delay curve for a second-order constant resistance lattice (O'Meara approximation) (courtesy of R. Bruch and C. E. Cook [2]).

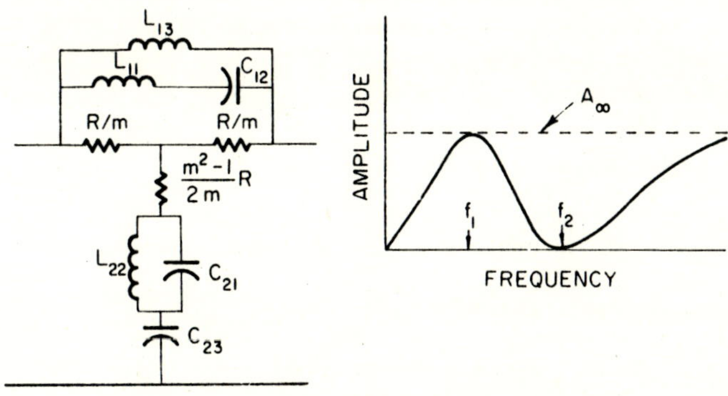

FIG. 12.16 Constant resistance amplitude equalizer.

The data given in Table 12-I were worked out for the values of peak delay necessary for use in the design example considered. If a set of delay characteristics are needed that extend over a wide frequency band, this standard delay data can be converted to yield delay correction functions with smaller (or larger) peak values and wider (or narrower) frequency extent. This is done by multiplying the peak delay, and thus C_a, by the desired factor and dividing the frequency increments by the same factor. As an example, if one half the peak delay is needed in a wider band application, each delay value in the desired row of the table is multiplied by one-half, and the frequency increments become 0.2 Mc rather than 0.1 Mc. This technique can be regarded as a variation of O'Meara's method, implemented for a specific application. The use of the standard delay curves has been found to be a useful way to set up the delay error correction problem.

The second major contribution to a nonideal realization of the matched filter is the nonflat amplitude response of the compression filter. This arises from the use of nonideal components (i.e., finite Q's) in which there are incidental dissipation losses. This type of loss can become quite large in a compression delay line that is composed of many sections. In general, the effect of the incidental dissipation takes the form developed by Bode [3]

$$\Delta A = \frac{\omega}{Q} \frac{d\beta}{d\omega} \tag{12-42}$$

where ΔA is the amplitude loss associated with incidental dissipation, Q the Q of all reactive elements (assumed equal), and $d\beta/d\omega$ the filter time delay function.

It is often assumed that the ω/Q factor remains relatively constant in a fairly narrow band, so that the added loss function is related to the values of the time delay curve. For the case of a linear delay function the loss should be approximately a linear function of frequency, being largest where the delay is a maximum. This is illustrated by the loss characteristic of the design example network shown in Fig. 12.17. To correct for this type of loss an amplitude response corrector, or equalizer, can be added to the design. A constant resistance amplitude correction network as shown in Fig. 12.16 is often used. The insertion loss of this type of circuit is given by

$$\exp[2A] = \frac{1 + (1 + m)^2 y^2}{1 + (1 - m)^2 y^2} \tag{12-43}$$

where $m = \coth \frac{1}{2} A_x$. A_x is the maximum attenuation, and

$$y = \frac{a_1 f - a_3 f^3}{1 - b_2 f^2}$$

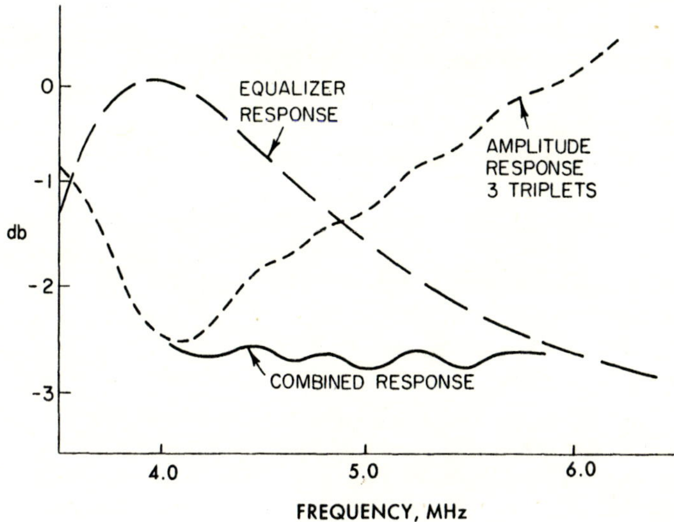

FIG. 12.17 Amplitude equalization data.

The form of the insertion loss function is shown in Fig. 12.16. Choice of m governs the maximum insertion loss, and the choice of a_1, a_3, and b_2 determines the critical frequencies f_1 and f_2, and thus the general shape of the curve. The element values of the bridged-T network are related to these parameters by

$$L_{13} = \frac{a_1 R}{\pi} \qquad C_{12} = \frac{\frac{a_1}{a_3}b_2 - 1}{4\pi^2\frac{a_1}{a_3}L_{13}} \qquad L_{11} = \frac{L_{13}}{\frac{a_1}{a_3}b_2 - 1} \qquad (12\text{-}44)$$

$$\frac{L_{13}}{C_{23}} = \frac{L_{11}}{C_{21}} = \frac{L_{22}}{C_{12}} = R_0^2 \qquad (R_0 = \text{characteristic resistance})$$

In practice, the parameters of the insertion loss function are adjusted so that a correcting amplitude response is obtained in either the region about f_1 or about f_2. The combination of the loss function of this type and the amplitude response of a portion of the compression network of the design example is shown in Fig. 12.17. The combined function yields the desired flat loss in the interval Δf. It is customary to divide the total amplitude correction response among several of these equalizer networks, which can be distributed along the compression delay line. If the total insertion loss becomes large amplifiers can also be inserted at intervals along the line, and one practice has been to combine the amplitude corrector and buffer amplifier into a single design unit. It should be noted that all of these additional circuits will influence the delay error curve and will have to be taken into account in that portion of the error correction procedure. A wider variety of correction networks are described by Zobel [4] and Mead [5].

Sources of error in the design of a matched filter include not only the inexactness of the design approximation, but also the effects of component tolerances, impedance mismatches and alignment errors. It is these latter practical aspects that can plague the system designer. Unnecessary additional efforts in the correction phase of the design can often be avoided by the application of care, and the use of quality components and accurate testing techniques. The cumulative effect of component tolerances in a bridged-T design, as reflected in the paired-echo sidelobe levels of the compressed-pulse, is shown in Fig. 12.18. The tolerance cited in these curves represents the rms of the combined component tolerances [6].

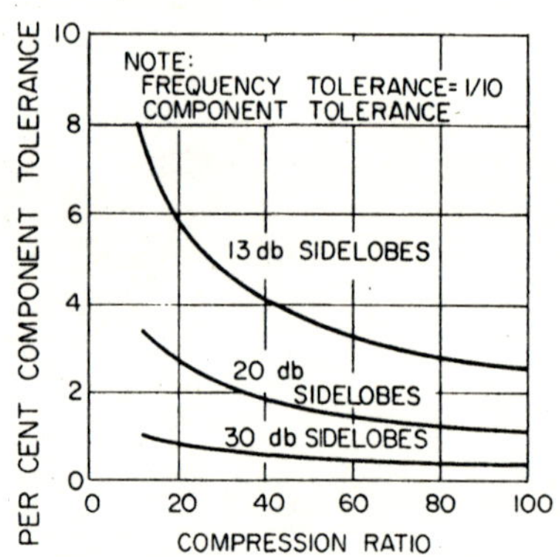

FIG. 12-18   Effect of component tolerance on distortion sidelobes (courtesy of S. E. Bogotch and R. Brackenbury [6]).

The techniques of dispersive delay design and error correction discussed in this and the previous sections have been based on the use of a linear delay design example. However, the general techniques plus the specific approach to error correction can be applied to any matched-filter design problem, although the basic delay function itself may have to be worked out on a computer if graphical approximation methods for nonlinear delay functions are not easily set up. The design problem of arbitrary group delay and phase functions has also been considered by O'Meara [7]. Other lumped constant dispersive delay line design techniques based on the use of all-pass networks are described by Brandon [8] and Steward [9].

12.7 Alignment of Bridged-*T* All-Pass Sections

In the lattice form of the all-pass time delay section the elements Z_a and Z_b are resonant at the frequency $\bar{\omega}_0$. The elements in the bridged-*T* section which are equivalent to Z_a and Z_b must also resonate at this frequency. These elements may be found by using the conversion equations (12-25):

$$\frac{1}{2Z_a} = \frac{1}{Z_a'} + \frac{1}{2Z_c} = \frac{1}{j\omega 2L_a} + j\omega \frac{C_a}{2} \tag{12-45}$$

At resonance Z_a approaches ∞ and the actual value of the resonant frequency is then

$$\bar{f}_0^2 = \frac{1}{(2\pi)^2(2L_a)(C_a/2)} \tag{12-46}$$

From this equation it may be seen that the elements equivalent to Z_a are the coil $2L_a$ in parallel with the capacitor $C_a/2$ (see Fig. 12.9).

The elements equivalent to Z_b are obtained in similar fashion from Eq. (12-25):

$$\frac{Z_b}{2} = Z_b' + \frac{Z_c}{2} = j\omega \frac{L_b}{2} + \frac{1}{j\omega C_s} + \frac{1}{2j\omega C_a} \tag{12-47}$$

At resonance $Z_b = 0$, and the actual value of the resonant frequency is

$$\bar{f}_0^2 = \frac{1}{(2\pi)^2 \dfrac{L_b}{2} \dfrac{C_s 2C_a}{C_s + 2C_a}} \tag{12-48}$$

The equivalent series and shunt elements in the minimum element bridged-*T* section are as shown in Fig. 12.9. The nonminimum element section is often used because provision can be made for alignment of each element to the resonant frequency. For this network section the equivalent series and shunt elements are as shown in Fig. 12.10.

Alignment of these elements to the resonant frequency can be made by observing the relative magnitude of the voltage at the output of this section when arranged in the test circuit shown in Fig. 12.19. A sweep generator is used to obtain an oscilloscope presentation of voltage vs frequency in which resonance is indicated by minimum output voltage, and the desired frequency, $\bar{f}_0$, by a crystal controlled frequency marker.

The series element can be aligned by disconnecting the inductance $L_b/2$ from the junction of the two capacitors (kC_a) and varying capacitor C_2 or inductance $2L_a$ to obtain a minimum output voltage at the specified frequency. However, the inductance, $L_b/2$ may be left connected if the

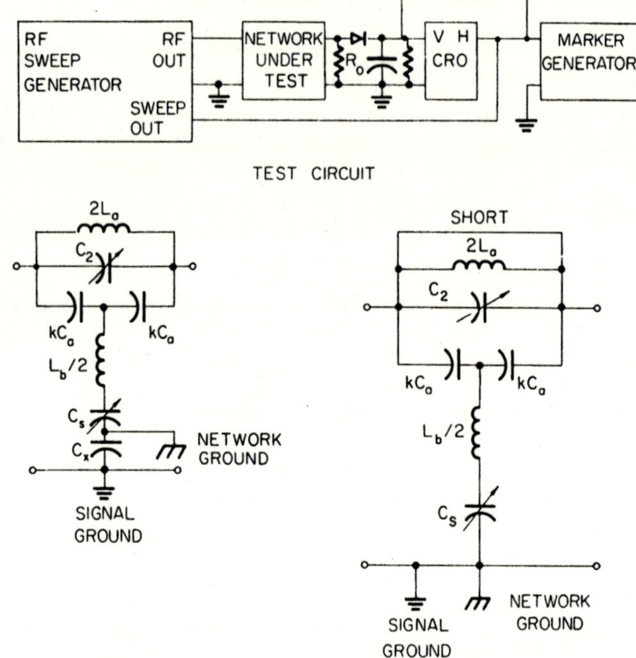

FIG. 12.19   Bridged-T alignment test circuit.

capacity between the section ground and the signal ground is low enough. Under this condition the equivalent network in the test circuit is given in Fig. 12.19. If the ratio C_a/C_x can be kept greater than $100:1$, then the net result on the error in the alignment frequency is less than 0.2%.

Alignment of the shunt element is accomplished by shorting the coil $2L_a$ and adjusting capacitor C_s or inductance $L_b/2$ for minimum output voltage at the specified frequency. The equivalent network for this test circuit is shown in Fig. 12.20.

The output voltage V_1 is given by

$$V_1 = iZ = \frac{\dfrac{iR_0}{2}\left(j\omega\dfrac{L_b}{2} + \dfrac{C_s + 2C_a}{2j\omega C_a C_s}\right)}{\dfrac{R_0}{2} + \left(j\omega\dfrac{L_b}{2} + \dfrac{C_s + 2C_a}{2j\omega C_a C_s}\right)} \tag{12-49}$$

$$= 0 \quad \text{when} \quad \omega^2 = \frac{C_s + 2C_a}{2C_s C_a (L_b/2)} = \bar{\omega}_0^2 \tag{12-50}$$

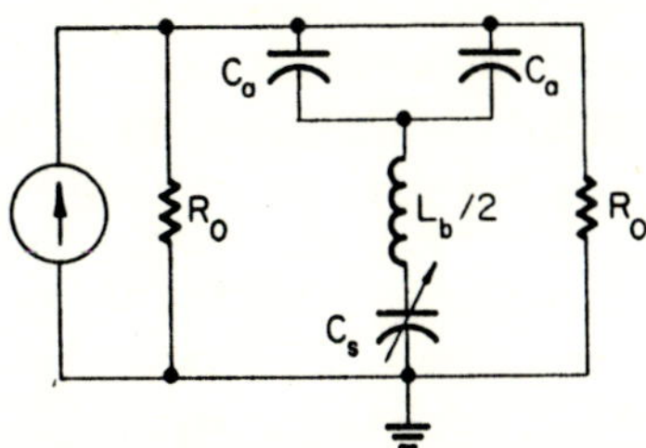

FIG. 12.20   Equivalent shunt element test circuit.

In this case the indicated frequency is the same as the actual resonant frequency of the shunt element. However, precautions must be taken to measure the voltage at the terminals of the shunt element using short connecting leads in order to eliminate errors caused by residual parameters of the measuring circuit. When a dispersive delay filter is comprised of many bridged-*T* sections, the test equipment of Fig. 12.19 can be instrumented by means of a test probe that can be applied in rapid succession to the individual sections before final connections are made. When these connections have been made, a further check on alignment accuracy is the

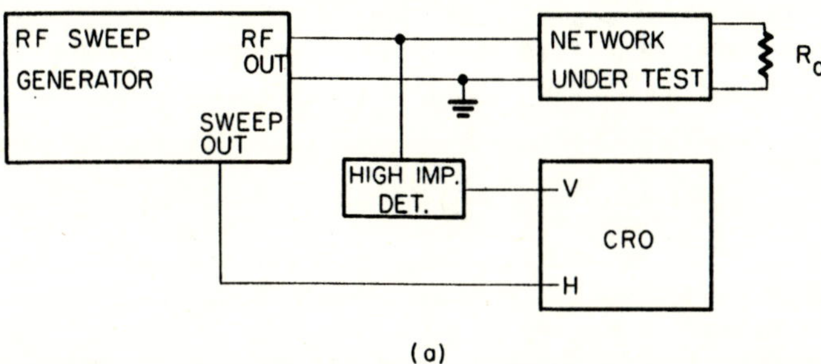

(a)

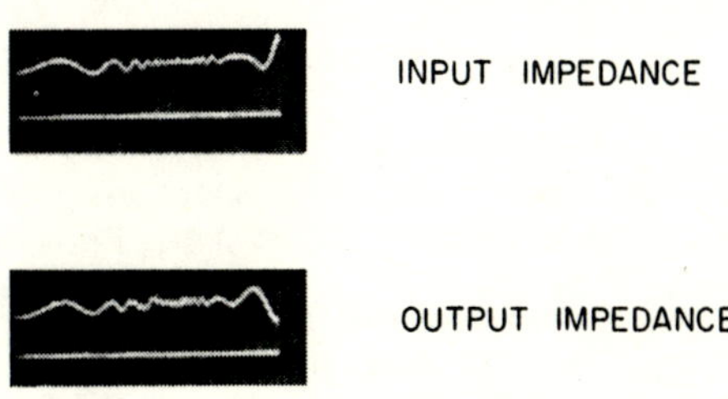

(b)

FIG. 12.21   Impedance test circuit for dispersive delay line. (a) Test circuit. (b) Oscilloscope waveforms.

behavior of the input and output impedances. The input impedance function can be obtained with the test setup shown in Fig. 12.21(a). The output impedance is obtained by reversing the input and output leads of the network in the test fixture. The impedance functions displayed on the oscilloscope will be as shown in Fig. 12.21(b). The VSWR can be obtained from this data, and should be essentially the same for both the input and output. The two way reflective loss can then be calculated (26 db for VSWR = 1.1; 32 db for VSWR = 1.05). The relative amplitude of the spurious outputs caused by reflection is obtained by adding to this loss the two way insertion loss and the effect of the linear delay mismatch suffered by the reflected signal by the time it has reached the output (see Section 6.6). VSWR's between 1.05 and 1.1 should suffice to keep these type of spurious responses 30 to 40 db below the desired signal. If the compression filter contains buffer amplifiers the reflection path length is that portion of the filter contained between two adjacent amplifiers.

At lower frequencies it also becomes feasible to include Q equalization of the series and shunt elements of each bridged-T. If $Q_{2L_a} < Q_{L_b/2}$ this is accomplished by inserting a small variable resistor in series with $L_b/2$ and adjusting it to obtain a flat response across the band at the output of the network under test. If $Q_{2L_a} > Q_{L_b/2}$, then the same adjustment can be made by placing a large variable resistor across $2L_a$. This procedure can reduce the cumulative effect of phase errors and mismatches.

<h2 style="text-align:center">REFERENCES</h2>

1. T. R. O'Meara, The synthesis of "band-pass", all-pass time delay networks with graphical approximation techniques (3rd ed.), Hughes Aircraft Co., Res. Lab., Malibu, California, Rept. 114 (February 1962).
2. R. Bruch and C. E. Cook, Summary of pulse-compression studies and filter techniques, unpublished design data, Sperry Gyroscope Co., Great Neck, New York (1961).
3. H. W. Bode, "Network Analysis and Feedback Amplifier Design." Van Nostrand, Princeton, New Jersey, 1945.
4. O. J. Zobel, Distortion correction in electrical circuits, *Bell Syst. Tech. J.* **7**, No. 3, 438–534 (1928).
5. S. P. Mead, Phase distortion and phase distortion correction, *Bell Syst. Tech. J.* **7**, No. 2, 195–224 (1928).
6. S. E. Bogotch and R. Brackenbury, private communication (1962).
7. T. R. O'Meara, Maximally flat approximations to arbitrary group delay and phase functions with Padé approximants, Hughes Aircraft Co., Res. Lab., Malibu, California, Rept. 159 (November, 1960).
8. P. S. Brandon, The design methods for lump-constant dispersive networks suitable for pulse-compression radar, *Marconi Rev.* **28**, 225–253 (1965).
9. K. W. F. Steward, A practical dispersive network system, *Marconi Rev.* **28**, 254–272 (1965).

CHAPTER 13

The Design of Dispersive Delay Functions—II Ultrasonic Delay Lines

13.1 Introduction

As noted in the preceding chapter, the design and construction of a filter having a dispersive group time delay characteristic can be achieved in a number of ways. When the design is based on the use of an iterated basic network composed of bridged-T all-pass sections, it was seen that a considerable amount of effort could be involved in the measurement and alignment procedures. In addition, there is the large number of inductances and capacitors necessary to build up the required number of sections. The example cited in Section 12.5 showed that approximately 50 bridged-T triplet sections (or 150 individual bridged-T sections) are required for a linear FM signal having a time-bandwidth product of 80. In the example, this turned out to be a somewhat optimistic number, since the specification of the desired time delay error reduced the usable bandwidth of the filter design.

As a result of the basic complexity and size of the lumped constant approach, alternate techniques have been sought to simplify the dispersive delay line design problem. One of the most successful of these to date, in terms of applicability to pulse-compression systems, has been the development of dispersive ultrasonic delay lines. The operation of these devices is based on the guided wave propagation properties of one class of cylindrical or rectangular solid bars. Ordinarily, these types of ultrasonic delay lines are thought of as yielding a fixed delay. However, in a restricted frequency region thin wires and strips have been known to exhibit dispersive properties for several different modes and types of wave propagation [1]. This is often referred to as a geometrical dispersion property, since it is a function of the geometry of the propagating medium, relative to the ultrasonic wavelength dimension, rather than an intrinsic property of the medium itself. It is the realization of these dispersive properties of wires and strips in practical forms that has resulted in the application of ultrasonic delay lines to pulse-compression radar systems.

In most cases, the pulse-compression system designer will not become involved in the actual fabrication of the ultrasonic dispersive delay line element, as he might in the case of a bridged-T design. However, the system designer will have the responsibility of determining the operational specifications that must be met by the ultrasonic delay line manufacturer, as well as of evaluating the result obtained. On this basis, a working understanding of the principles involved and of the significant parameters is desirable. The following sections cover the factors that should be of interest to the pulse-compression or matched-filter engineer. Basic papers in this area have been written by Meitzler [2–4], Meeker [5, 6], May [7], and Fitch [8, 9]. A thorough presentation of the theory of guided wave propagation in isotropic, homogeneous media is given by Meeker and Meitzler [10]. The design, realization, and applications of guided wave ultrasonic lines (dispersive and nondispersive) are discussed by May [11]. Extensive references and bibliographies are available in these latter two references. A more recent application of ultrasonic techniques to pulse-compression systems is based on the realization of a frequency dependent delay function with grating, or array, structures.

13.2 Parameters of Ultrasonic Waves in Isotropic Elastic Media

The propagation of ultrasonic waves in elastic media can take a number of forms. The two basic types of wave motion are longitudinal (compression and extension) waves and shear (transverse) waves. In longitudinal wave motion the displacement of the particles within the elastic medium is in the direction of the wave propagation; in shear wave motion the particle displacement is transverse to the direction of wave propagation. These are illustrated in Fig. 13.1. The displacement of the particles in the medium results from an applied ultrasonic stress, causing the type of deformations (or strains) within the medium, as shown. Electromechanical transducers that vibrate in either a shear mode or a longitudinal mode in response to an applied electrical signal can be used to achieve these two basic types of mechanical wave motion in ultrasonic delay lines that are used in electrical systems.

Other types of wave motion are also possible in elastic media, such as flexural (bending) waves in a plate or strip as illustrated in Fig. 13.1, or torsional (twisting) waves in a bar or wire. In the design of ultrasonic delay lines for pulse-compression applications these usually represent undesired modes of propagation, and great care is taken to suppress them, since

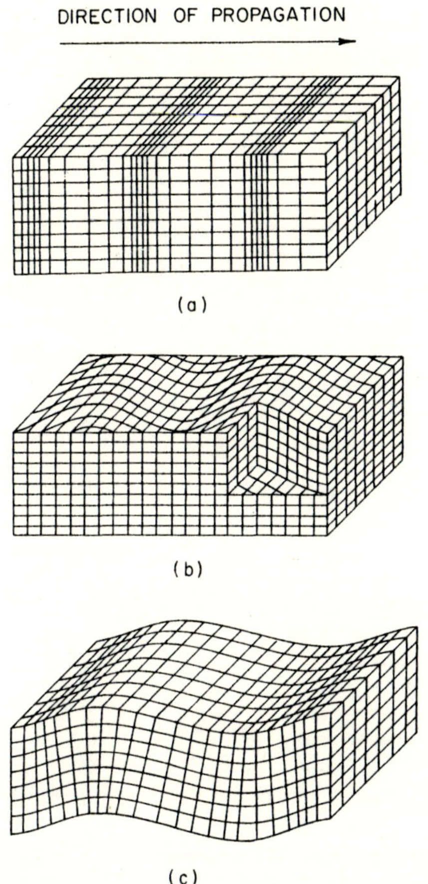

FIG. 13.1 Types of elastic wave propagation in a solid medium. (a) Longitudinal wave. (b) Shear wave. (c) Flexural wave. (After Krasil'nikov [30].)

signal energy travelling in these modes represents distortion terms at the output of the ultrasonic filter.

The propagation properties of elastic waves are related to a set of two independent elastic constants of the particular isotropic medium being considered. A well-known set of these is Young's modulus E and Poisson's ratio σ. A uniform cylindrical rod of length L and cross sectional area A is elongated by an amount ΔL under an applied extensional force F. Young's modulus, or the modulus of longitudinal elasticity, is the proportionality constant between the applied stress (F/A) and the resulting deformation or strain $(\Delta L/L)$, or

$$E = \frac{F/A}{\Delta L/L} \tag{13-1}$$

Poisson's ratio is related to the change in the cross sectional area when the rod is longitudinally deformed. If the rod diameter is d, and the change in diameter is Δd, Poisson's ratio is defined as

$$\sigma = -\frac{\Delta d/d}{\Delta L/L} \tag{13-2}$$

For most known materials $0.1 < \sigma < 0.4$. Both E and σ are constants in an isotropic homogeneous solid as long as the elastic limit of the medium is not exceeded. They are independent of the size and shape of the solid, and are sufficient to define the elastic properties of the isotropic, homogeneous solid. Poisson's ratio is a critical factor in determining the dispersive delay characteristics of ultrasonic guided wave delay lines. Typical values of σ are given in Table 13-I.

TABLE 13-I

POISSON'S RATIO FOR DIFFERENT MATERIALS

Material	Poisson's ratio, σ
Steel	0.28
Copper	0.35
Aluminum	0.32–0.35
Brass	0.37
Rubber	0.5
Glass (special types)	0.2
Fused quartz	0.17

Two other elastic constants of importance are the Lamé constants μ and λ. The constant μ is known as the shear modulus, and is the proportionality constant between an applied shear stress and the resulting shear deformation. In terms of Young's modulus and Poisson's ratio, μ is defined by

$$\mu = \frac{E}{2(1 + \sigma)} \tag{13-3}$$

The other Lamé constant is related to the above elastic constants by

$$\lambda = \frac{2\sigma\mu}{1 - 2\sigma} = \frac{E\sigma}{(1 + \sigma)(1 - 2\sigma)} \tag{13-4}$$

The elastic constants $(E, \sigma, \mu, \lambda)$ enter into the determination of the "free space" velocity of propagation for shear and longitudinal waves. In an infinite isotropic solid body the shear wave velocity is given by

$$V_s = (\mu/\rho)^{1/2} \qquad (\rho = \text{density of the medium}) \tag{13-5}$$

The velocity of longitudinal elastic waves in an infinite isotropic solid is

$$V_d = \left(\frac{E(1 - \sigma)}{\rho(1 + \sigma)(1 - 2\sigma)}\right)^{1/2} \tag{13-6}$$

In a rod the velocity of longitudinal waves at the low frequency limit is

$$V_b = (E/\rho)^{1/2} < V_d \tag{13-7}$$

V_s is the same in both an infinite medium or in a thin strip or rod. Making use of (13-3) and (13-5), we see that (13-6) becomes

$$V_d = V_s\left(\frac{2 - 2\sigma}{1 - 2\sigma}\right)^{1/2} \tag{13-8}$$

Equation (13-8) is the more usual form for relating the longitudinal and shear velocities in an infinite medium.

In the design of an ultrasonic dispersive delay line the parameters of greatest interest are the group velocity and the group time delay characteristics. It will be shown that these are related to the basic elastic constants defined above, as well as to V_s and V_d.

13.3 Dispersive Delay Characteristics of Shear Mode Propagation in Strips

The guided wave propagation of shear mode elastic waves in a thin plate of infinite extent is closely analogous to electromagnetic guided wave propagation between a pair of parallel conducting infinite plane surfaces. One important difference is that the strip can support a zero order mode, while the electromagnetic waveguide cannot. The theory of this type of ultrasonic wave propagation is described by Buchwald [12] and Meitzler [2]. Meitzler also describes the development of an experimental shear mode dispersive delay line. It was found that if the width of a thin strip of metal is 10 to 20 wavelengths wide, the measured propagation characteristics are in essential agreement with those predicted by theory for the thin plate. 5052-H32 aluminum alloy has found extensive use as a dispersive propagating medium, since it has a very low attenuation constant. In addition, aluminum alloy has the advantage of being formed easily into thin strips by means of a rolling process.

Figure 13.2 illustrates the geometry of the infinite thin plate. The transverse, or shear, wave function follows the classical wave equation

$$\frac{\partial^2 f_z}{\partial x^2} + \frac{\partial^2 f_z}{\partial y^2} = \frac{1}{V_s^2}\frac{\partial^2 f_z}{\partial t^2} \tag{13-9}$$

Given the constraint that the amplitude of the particle vibration in the solid is zero at the minor surfaces ($y = \pm\frac{1}{2}h$), the general solution of

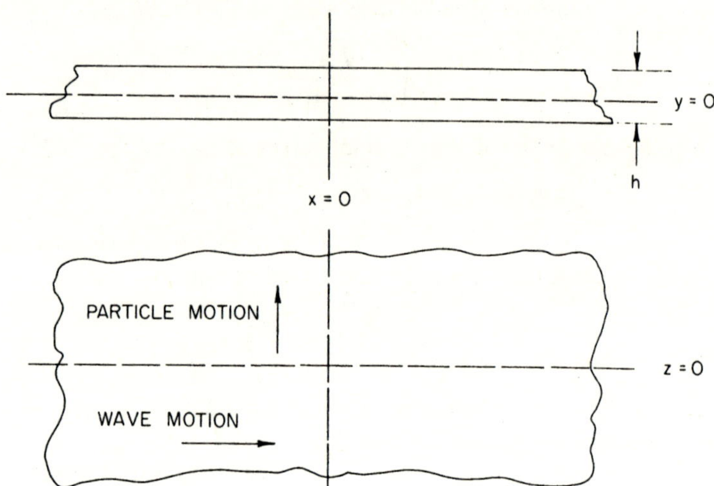

FIG. 13.2   Geometry of infinite thin plate for shear mode propagation.

(13-9) becomes

$$f_z(x, y, t) = \sum_n \left\{ \left(A_n \sin \frac{n\pi y}{h} + B_n \cos \frac{n\pi y}{h} \right) \exp[\, j(\gamma_n x - \omega t)] \right\} \quad (13\text{-}10)$$

where $A_n = 0$ for $n = 2, 4, 6, \ldots$, and $B_n = 0$ for $n = 1, 3, 5, \ldots$.

Equation (13-10) describes a wave traveling in the x-axis direction, in which the particle displacement along the z-axis is dependent on y. For $n \geqslant 1$, every mode of propagation is dispersive. A nondispersive mode exists for $n = 0$ [2]. The low frequency cutoff for each mode is given by

$$\left[\left(\frac{\omega_n}{V_s} \right)^2 - \left(\frac{n\pi}{h} \right)^2 \right] = 0 \quad (13\text{-}11)$$

resulting in $\omega_n = n\pi V_s/h$, or $f_n = nV_s/2h$.

In 5052-H32 aluminum alloy V_s is approximately 1.26×10^5 in./sec. Values of f_n in this material for different thicknesses are listed in Table 13-II. The phase velocity in the nth mode is given by

$$V_n = \frac{\omega}{\gamma_n} = V_s \left(1 - \frac{\omega_n^2}{\omega^2} \right)^{-1/2} \quad (13\text{-}12)$$

and the group velocity is $U_n = d\omega/d\gamma_n$. From (13-12) we have

$$\omega = (\omega_n^2 + \gamma_n^2 V_s^2)^{1/2} \quad (13\text{-}13)$$

TABLE 13-II

CUTOFF FREQUENCIES VS THICKNESS FOR A THIN ALUMINUM ALLOY STRIP

Thickness, h (in.)	First mode cutoff, f_1 (MHz)	Second mode cutoff, f_2 (MHz)
0.10	0.63	1.26
0.08	0.788	1.575
0.06	1.05	2.10
0.04	1.575	3.15
0.02	3.15	6.30

and

$$U_n = V_s^2 \left(\frac{\omega_n^2 + \gamma_n^2 V_s^2}{\gamma_n^2} \right)^{-1/2} = \frac{V_s^2}{V_n} \tag{13-14}$$

The phase and group velocities are identical for each mode when the frequency variable is normalized to the nth mode cutoff frequency. This is illustrated in Fig. 13.3. It can be seen that the dispersive delay characteristic, $1/U_n$, is nonlinear, with the greatest delay slope in the region near cutoff. The frequency of operation of a shear mode dispersive delay line should be between f_1 and f_2 in order to minimize energy in extraneous modes.

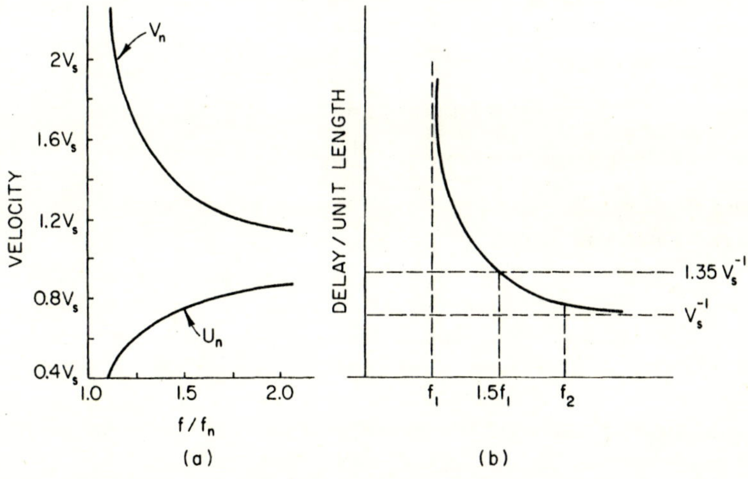

FIG. 13.3 Phase and group velocity of shear waves in thin plate. (a) Phase velocity V_n and group velocity U_n. (b) Group time delay. (After Meitzler [2].)

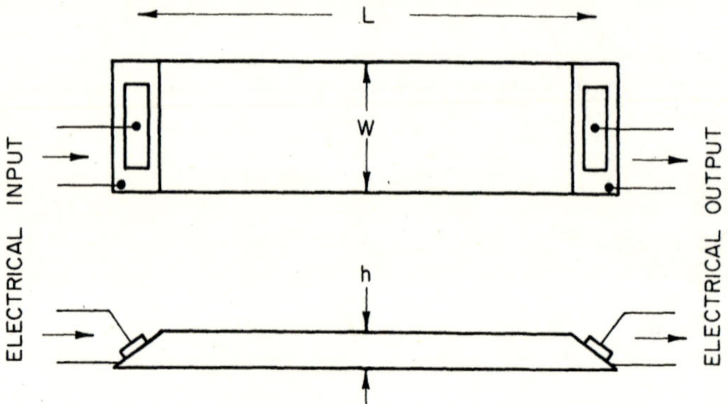

FIG. 13.4   Thin strip approximation to the infinite thin plate.

Figure 13.4 illustrates the type of thin strip shear mode delay line reported by Meitzler [2]. The transducers are poled to optimize the conversion of the electrical input signal into a shear mode ultrasonic wave. The ends of the strip are beveled to suppress the nondispersive zeroth mode, which can be kept in the order of 30 db below the desired signal (this does not include the effect of the compressed-pulse amplitude buildup). In order to suppress spurious modes caused by reflections at the minor surfaces it was found necessary to serrate these edges or to place tape around these edges. The latter approach has proved to be more effective, and has become a standard technique in the fabrication of strip delay lines. The application of a shear mode dispersive delay line would be confined to the type of nonlinear frequency modulation function shown in Fig. 13.5. A typical design could be based on the following specifications:

$$\text{normalized bandwidth,} \qquad \Delta f/f = 0.5$$

$$\text{normalized center frequency,} \qquad f_0/f_1 = 1.55$$

Using these specifications, and (13-12) and (13-14), one obtains a delay difference per unit length of 2.86 microseconds per inch and a midband delay of 10 microseconds per inch. These are both independent of the thickness of the strip. Assuming a bandwidth of 1 Mc, then $f_n = 2\,\Delta f$. Using (13-11) and $n = 1$, one obtains for the dimension of the strip $h = 0.0315$ inch. Also, $f_1 = 2.0$ Mc and $f_0 = 3.1$ Mc. For a required compression ratio of 200 (signal duration = 200 microseconds), the length of the line is given by $200/2.86 = 70$ inches, and the midband delay is 700 microseconds. In practice, the transducer characteristics, as well as losses in the material, would be important factors in determining the frequency band of interest.

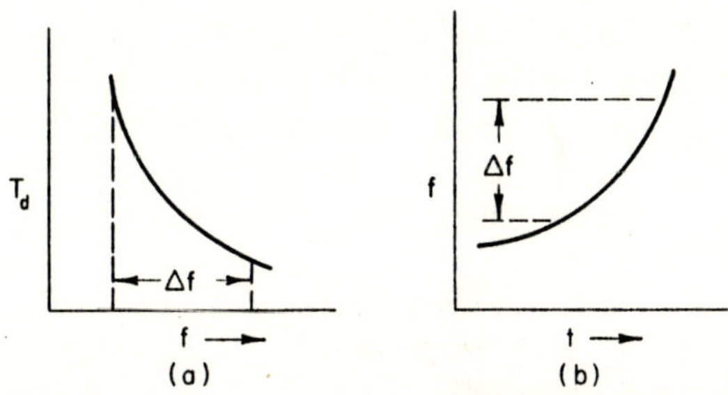

FIG. 13.5 Matching group delay and frequency modulation functions for optimum shear mode performance. (a) Delay function. (b) Frequency modulation.

13.4 Dispersive Delay Characteristics of Longitudinal Mode Propagation in Strips

The longitudinal modes of propagation of elastic waves in thin strips are related to the theory of longitudinal elastic wave propagation in an infinite thin plate. The relationship defining the longitudinal modes of propagation in an infinite plate is given by the Rayleigh–Lamb frequency equation [13, 14]. The standard form of this equation for longitudinal mode propagation is given by

$$\frac{\tan \frac{1}{2}\eta h}{\tan \frac{1}{2}\xi h} = -\frac{4\gamma^2 \xi \eta}{(\gamma^2 - \eta^2)^2} \qquad (h = \text{thickness of the plate}) \qquad (13\text{-}15)$$

Frequency appears implicitly in (13-15) through the constraints among the propagation constants

$$\gamma^2 + \xi^2 = \omega^2/V_d^2, \qquad \gamma^2 + \eta^2 = \omega^2/V_s^2 \qquad (13\text{-}16)$$

where V_s and V_d are given by (13-5) and (13-8), respectively. Since

$$\gamma = \omega/V \qquad (V = \text{phase velocity of the elastic wave}) \qquad (13\text{-}17)$$

(13-15) can be rewritten as

$$\frac{\tan\left\{\dfrac{\omega h}{2V_s}\left[1 - \dfrac{V_s^2}{V^2}\right]^{1/2}\right\}}{\tan\left\{\dfrac{\omega h}{2V_d}\left[1 - \dfrac{V_d^2}{V^2}\right]^{1/2}\right\}} = -\frac{4\dfrac{V^2}{V_d V_s}\left[1 - \dfrac{V_d^2}{V^2}\right]^{1/2}\left[1 - \dfrac{V_s^2}{V^2}\right]^{1/2}}{\left(2 - \dfrac{V^2}{V_s^2}\right)} \qquad (13\text{-}18)$$

The transcendental nature of (13-18) does not lead to a closed form solution relating ω and V, as was shown in the case of shear mode propagation in an infinite plate. The explicit appearance of V_s and σ (by reason of the definition of V_d) in (13-18) permits calculations of the behavior of V as a function of ω for various materials for which these elastic constants are known. Extensive analysis of elastic wave propagation in strips and cylinders is given by Holden [15].

The design of a dispersive delay line for pulse-compression applications requires a knowledge of the group velocity U_n of the desired mode, or the group delay per unit length $1/U_n$. The group delay characteristic is given by

$$\frac{d}{d\omega}(\gamma) = \frac{d}{d\omega}\left(\frac{\omega}{V}\right)$$

(13-19)

or

$$\frac{1}{V} - \frac{\omega}{V^2}\frac{dV}{d\omega} = \frac{1 - \dfrac{\omega}{V}\dfrac{dV}{d\omega}}{V}$$

Meeker [5, 6] has reported experimental data that show that a thin strip of thickness h has a delay characteristic quite close to that predicted by the Rayleigh–Lamb equation and (13-19), if the width of the strip is 10 to 20 wavelengths wide.[1] The general characteristic of the first longitudinal mode group delay is shown in Fig. 13.6. At low frequencies the delay per

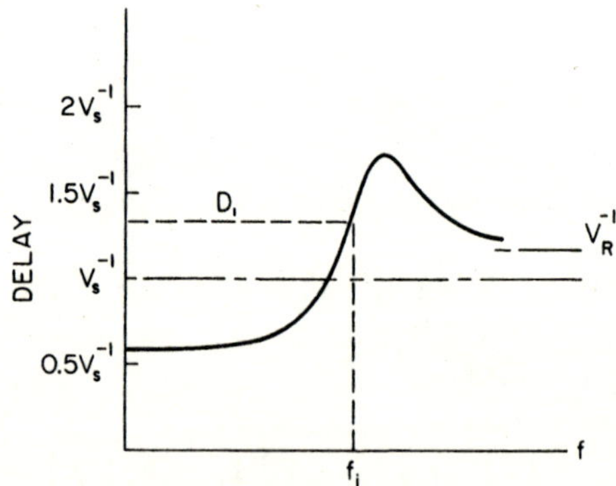

FIG. 13.6   Group delay characteristic for 1st longitudinal mode of infinite thin plate (after Meeker [5]).

---

[1] A larger width dimension permits use of a longer transducer, which has the advantage of being a more directive radiator. This reduces the chance of mode conversion resulting from reflection at the minor surfaces of the strip.

unit length approaches a constant $(V_p)^{-1}$, where V_p is the limiting low frequency velocity in a thin plate defined by

$$V_p = V_s[2/(1 - \sigma)]^{1/2} \tag{13-20}$$

For $\sigma = 0.35$, as an example, $V_p = 1.75V_s$. The high frequency limiting velocity in the lowest longitudinal mode is given by the Rayleigh surface wave velocity V_R, where V_R is approximately $0.9V_s$. Between these two regions the delay per unit length exhibits a frequency dependence as shown in Fig. 13.5, thus it is possible to choose a region on this curve that has a large amount of dispersive delay. The slope and peak delay of this dispersion region are a function of Poisson's ratio σ. Both of these quantities are larger for smaller values of σ. The dispersive region of the first longitudinal mode group delay curve contains a point of inflection at a frequency f_i. The inflection frequency dependence on Poisson's ratio and the thickness of the strip is shown in Fig. 13.7(a). The delay $D_i(f_i)$ at the inflection frequency is dependent only on the value of Poisson's ratio. This relationship is shown in Fig. 13.7(b). The measured dispersive delay region

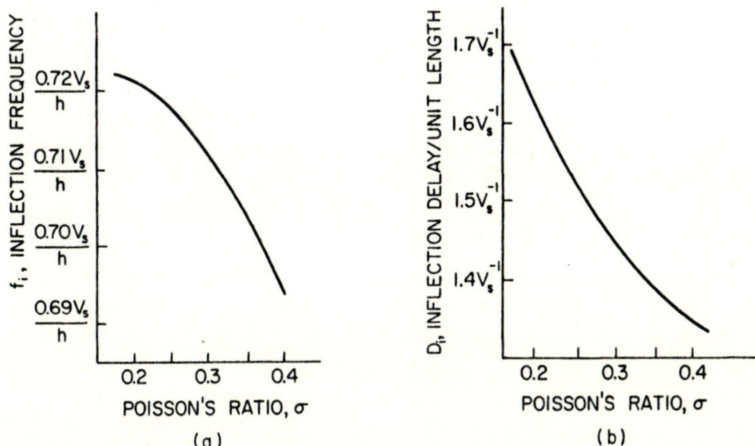

FIG. 13.7 Dependence of inflection frequency f_i and inflection delay D_i on Poisson's ratio (after Meeker [6]).

for a strip formed of 5052-H32 aluminum alloy ($\sigma = 0.35$) is shown in Fig. 13.8. For this value of σ, $f_i = 0.703V_s/h$. The choice of a desired operating center frequency (assuming operation about the inflection frequency) establishes the strip thickness, h. The signal bandwidth Δf yields the ratio $\Delta f/f_i$, and thus the dispersive delay difference Δt_d per unit length from Fig. 13.8, if aluminum alloy is used. The time duration T of the signal leads to the calculation of the overall length of the strip, $L = T/\Delta t_d$. From

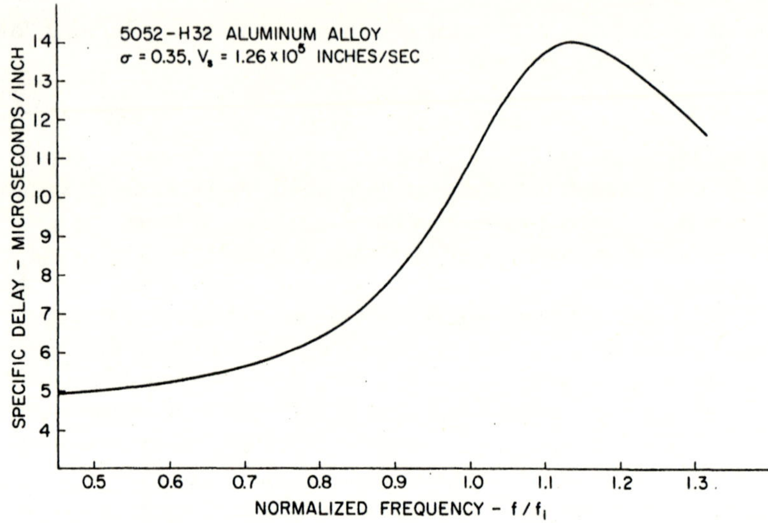

FIG. 13.8   Normalized delay vs frequency for 1st longitudinal mode in an aluminum strip (courtesy of H. C. Wing, Sanders Associates, Nashua, New Hampshire).

Fig. 13.8, the midband delay will be approximately $11 \times L$ microseconds for a line L inches long.

The existence of the inflection point in the dispersive delay characteristic makes operation in the longitudinal mode more desirable for approximating a linear delay vs frequency function than operation in the shear mode. However, as shown in Fig. 13.9, there exists deviations about the ideal linear delay curve (presuming use in a linear FM pulse-compression system). Data obtained by Meeker [6] show that the maximum deviation from linearity above the inflection frequency, δD_h, is greater than the maximum deviation from linearity below the inflection frequency, δD_l.

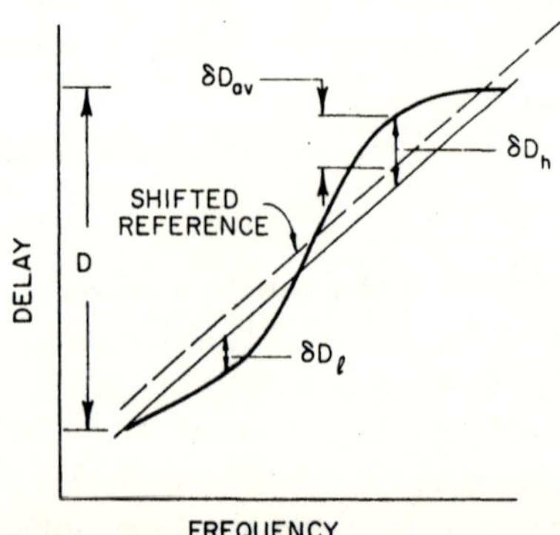

FIG. 13.9   Delay error of longitudinal mode group delay (after Meeker [6]).

These deviations become more symmetrical as the fractional bandwidth used, $\Delta f/f_i$, is decreased. For a 20% bandwidth ratio approximate values of δD_h and δD_l are 0.035 and 0.0175, respectively, in an aluminum alloy strip line. By defining a shifted reference delay, as in Fig. 13.9, it is possible to construct a nearly symmetrical delay error curve of approximately sinusoidal shape. Under this condition, the average delay from linearity vs percent bandwidth for $\sigma = 0.35$ is shown in Fig. 13.10. The average delay

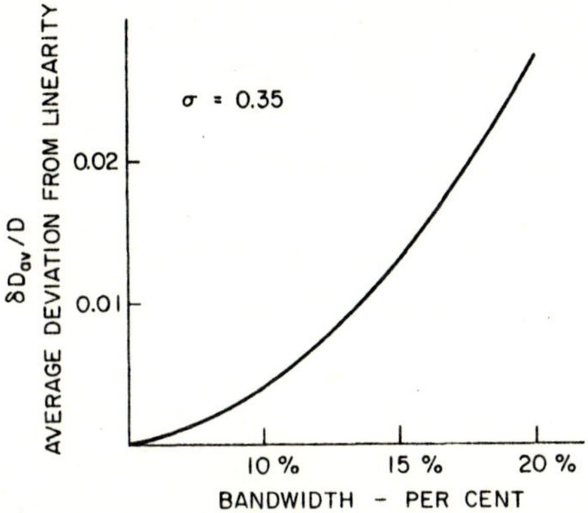

FIG. 13.10   Average delay error of longitudinal mode group delay.

error is larger for smaller values of σ [6]. As pointed out in Chapter 11, the waveform distortion effect created by this type of delay error depends only on the magnitude of the error, and not on the percent error relative to the dispersive delay. The actual criterion is the peak phase error in radians. For the longitudinal mode strip delay line, this phase error is approximated by

$$b_n = \frac{\delta D_{\mathrm{av}}\,\Delta f}{n} \tag{13-21}$$

where $n/\Delta f$ = error cycles/signal bandwidth. For the single thickness strip $n = 1$. Using the example of a linear FM signal with a 15% bandwidth factor and $\Delta f = 500$ kHz, $D = 200$ microseconds, then from Fig. 13.10 $\delta D_{\mathrm{av}}/D = 0.013$ and

$$b_n = \frac{0.013 \times 200 \times 10^{-6} \times 0.5 \times 10^6}{1} = 1.3 \text{ radians}$$

Phase errors of this magnitude are associated with large paired-echo distortion signals. As this example points out, there is an ultimate time-bandwidth product limitation in the use of a single thickness longitudinal mode delay line as a matched filter for a linear FM signal. This can be alleviated somewhat by use of the ultrasonic filter itself as a passive frequency generator, as described in Chapter 6. The frequency modulation obtained will be a nearly odd symmetrical (depending on percent bandwidth), slightly nonlinear FM function. Using frequency inversion techniques, the FM function can be time inverted, and after transmission and reception the signal can be processed through the same type of ultrasonic delay line to obtain the compressed-pulse signal. This approach will yield a better match between the transmitted signal and the receiver filter than obtained by the use of a strictly linear FM signal. The problem of large delay errors for the longitudinal mode dispersive line has been solved by changing the thickness of the strip along the length dimension, as described by Fitch [9]. The details of this approach are outlined below.

If several single thickness segments of different thicknesses h_1 through h_n and lengths L_1 through L_n are joined together, a composite delay function vs frequency is obtained. This is expressed by

$$D_T(f) = L_1 D_1(f) + L_2 D_2(f) + \cdots + L_n D_n(f) \qquad (13\text{-}22)$$

where $D_i(f)$ is the delay per unit length as a function of frequency for thickness h_i. Equation (13-22) provides a basis for synthesizing a linear delay function with less delay error than can be obtained with a single thickness line. Conceptually, this approach is analogous to that outlined by O'Meara [16] for the bridged-T network design discussed in the previous chapter. The design of the various thicknesses and lengths are obtained from a computer program described by Fitch [9]. In theory, any number of segments could be chosen, but in practice numbers of segments between 3 and 6 have proved adequate for the design of a linear delay function having a greatly improved delay error curve and a greater percent bandwidth. Once the design factors have been worked out the multiple thickness lines can be constructed by processing a single thickness strip through a rolling mill, in which the thickness can be adjusted according to the desired function, or by a chemical bath etching technique in which the thickness of any one segment is controlled by the amount of time the segment is immersed in the etching chemical. Figure 13.11 illustrates the thickness

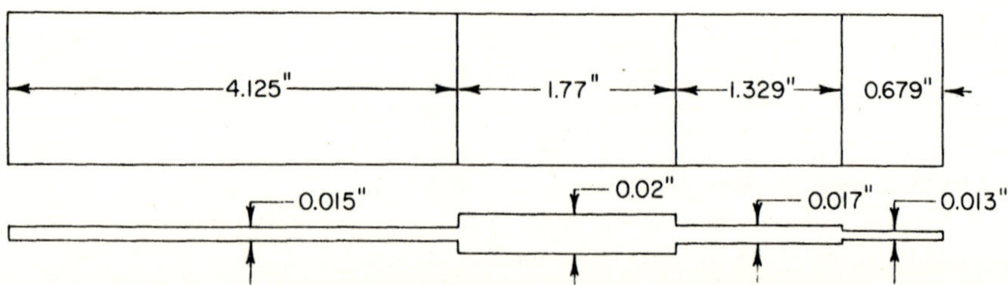

FIG. 13.11   Dimensions of four segment stepped thickness linear delay line (courtesy of R. D. Brew and Co., Concord, New Hampshire).

TABLE 13-III

INFLECTION FREQUENCIES OF STEPPED THICKNESS DELAY LINE

Thickness (in.)	Inflection frequency (MHz)
$h_1 = 0.02$	$f_1 = 4.43$
$h_2 = 0.017$	$f_2 = 5.21$
$h_3 = 0.015$	$f_3 = 5.90$
$h_4 = 0.013$	$f_4 = 6.81$

Figure 13.12 plots the delay curves of each individual segment and the composite delay curve. A linear delay curve has been synthesized that provides a dispersive delay of approximately 28 microseconds over a 1.25 MHz band centered at 5 MHz. The maximum compression ratio is 35. A multiplication of the length of each segment by a constant factor would increase the total dispersive delay and thus the compression ratio. The design can also be scaled up and down in frequency by multiplying each thickness by the same constant. This has the effect of increasing or decreasing the design bandwidth as the design center frequency is increased or decreased. Not only does the multiple segment design technique reduce the magnitude of the delay error, but it also results in an increased number of cycles in the delay error function. This also has the effect of reducing the associated phase error, as shown by Eq. (13-21).

Meitzler [3] has investigated the effect of tapering the width of the single thickness delay line as a function of the length variable. This was found to reduce the conversion of energy in the strip into undesired modes, thus improving the amplitude response characteristic. This technique was also found to result in a narrow band amplitude response characteristic that, with proper design, could be centered to coincide with the inflection frequency, thus providing a combination linear delay and weighted amplitude response. However, application of this technique has not been widespread, one of the reasons probably being the difficulty of controlling the exact nature of the narrow band response function. Single thickness lines have also been built of steel alloy to perform at higher frequencies (30 Mc and above). The steel alloy is used primarily because it is easier to

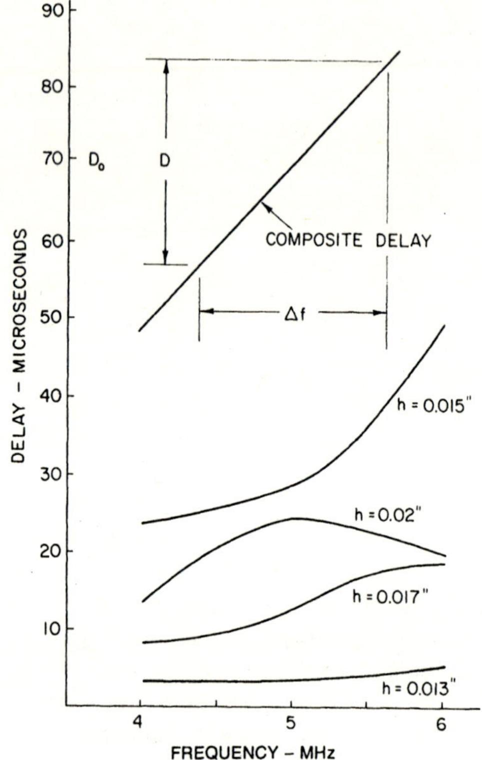

FIG. 13.12   Individual and composite delays of four segment stepped thickness line.

fabricate very thin ribbons with this material than with aluminum alloy. However, transducer fabrication is usually more difficult for these very thin, high frequency lines, and the characteristics of these ultrasonic dispersive delay lines exhibit a higher level of spurious signals caused by mode conversion, as well as by reflective mismatches. Mode conversion signals can occur as a result of improper orientation of the transducer and from interactions at the minor surfaces of the strip. Insertion losses for the high frequency lines can be in the order of 40 to 50 db, as compared to the 10 to 15 db for the aluminum lines operating at lower frequencies. Despite these problems the single thickness dispersive line has been used with success at higher frequencies, as shown by the waveforms in Fig. 13.13 obtained with a 30 MHz center frequency line having a compression ratio of 100. The operation of this line involved pulse expansion, gating, frequency inversion, and pulse compression as indicated in Fig. 6.15.

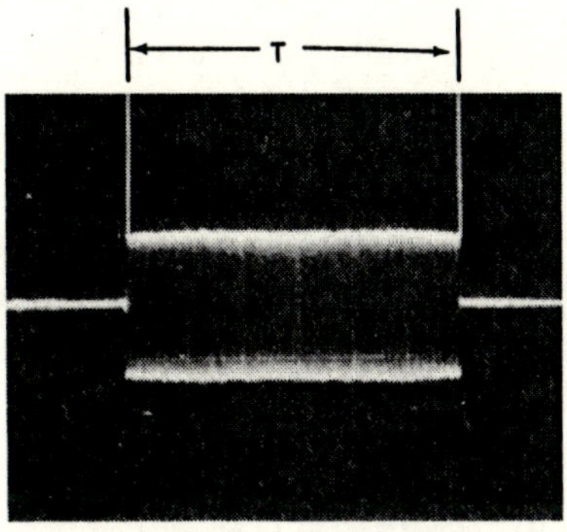

(a) EXPANDED PULSE (GATED)

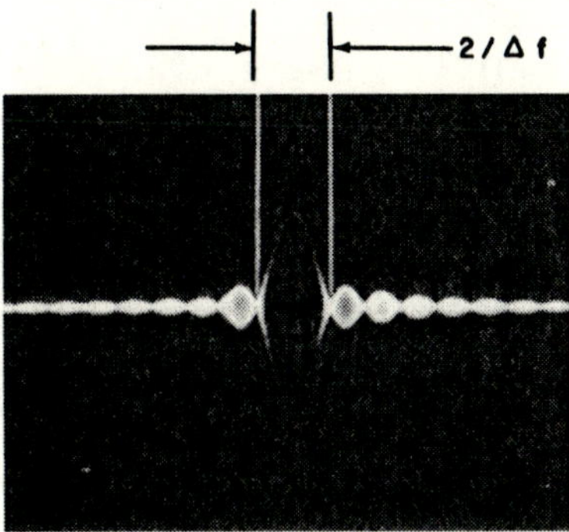

(b) COMPRESSED PULSE

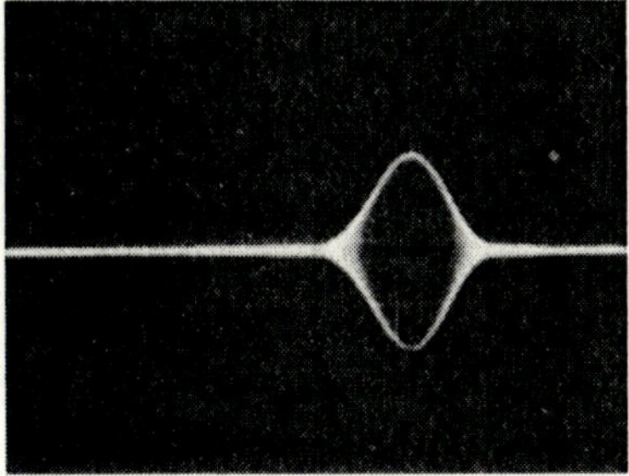

(c) WEIGHTED COMPRESSED PULSE-32 db SIDELOBE

FIG. 13.13   Waveforms for 30 MHz single thickness ultrasonic line, compression ratio = 100 (courtesy of S. E. Rittenburg, Andersen Lab., Bloomfield, Connecticut).

13.5 Wire Type Dispersive Delay Lines

The theory of propagation of ultrasonic waves in an infinite cylindrical bar was developed independently by Pochhammer [17] and Chree [18]. The Pochhammer–Chree frequency equation determines the allowable modes of propagation and frequency bands in an infinitely long cylindrical wire. These can be calculated if two elastic constants, for example Poisson's ratio and Young's modulus, are given. The Pochhammer–Chree equation for the cylinder serves the same function as the Rayleigh–Lamb frequency equation for the infinite plate. Bancroft [19] gives a simplified form of the Pochhammer–Chree equation as

$$(x - 1)^2 \phi(ha) - (\beta x - 1)[x - \phi(ka)] = 0 \tag{13-23}$$

where $\quad x = (V/V_b)^2(1 + \sigma)$

$$\phi(ha) = ha[J_0(ha)/J_1(ha)]$$
$$a = \text{radius of wire}$$
$$h = \gamma(\beta x - 1)^{1/2}, \qquad \beta = (1 - 2\sigma)/(1 - \sigma)$$
$$k = \gamma(2x - 1)^{1/2}$$

V, V_b, σ, and γ are as defined earlier.

Equation (13-23), as in the Rayleigh–Lamb case, is a transcendental function in which the phase velocity V can be calculated for specific materials. The first longitudinal mode $L(0, 1)$ of propagation exhibits a dispersive characteristic similar to that found in the strip. May [7] describes the development of a wire type dispersive delay line. Figure 13.14 shows the dispersive delay function calculated by May. The measured characteristics of a finite length wire delay line were found to agree substantially with the behavior predicted by the Pochhammer–Chree equation. Also shown in Fig. 13.14 are the first and second unwanted flexural modes, $F(1, 1)$ and $F(1, 2)$, of propagation as well as higher order flexural and longitudinal modes. There can exist an undesirable amount of mode conversion at frequencies where the phase velocity curves of different modes intersect. This results in the coupling of energy from the desired mode that causes spurious signals at the ultrasonic compression-filter output at certain frequencies. This imposes a limitation on the fractional bandwidth over which the linear delay characteristic can be achieved. Although this phenomenon exists in both wire and strip lines, it has been found to be less critical in strip lines operating in the first longitudinal mode. In addition, the transducers for a strip type of dispersive line are many times the area of a transducer for a wire type dispersive line having similar characteristics. This results in a lower impedance level for the strip transducer, making impedance matching over the desired band of frequencies less difficult. This has

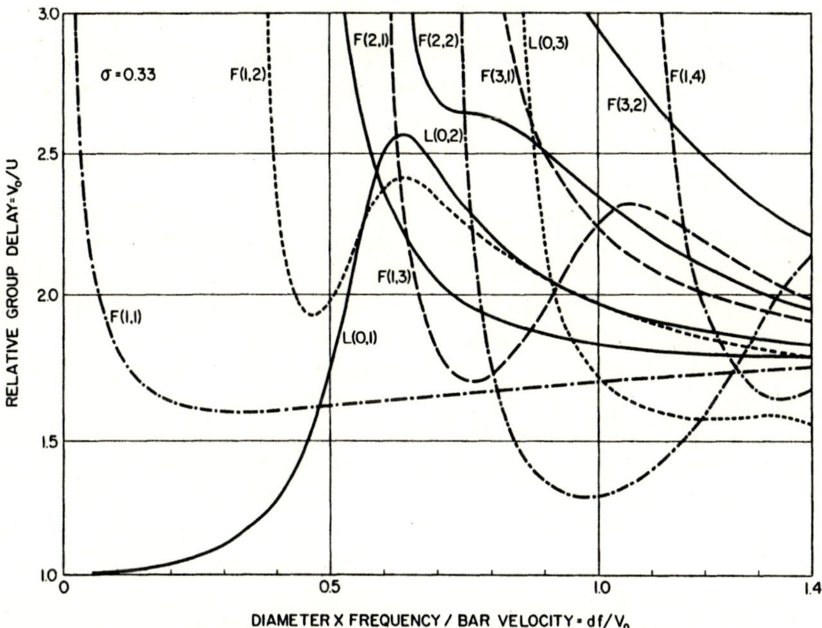

FIG. 13.14 Longitudinal and flexural modes of propagation in a wire (courtesy of J. E. May, Jr., Bell Telephone Lab., Murray Hill, New Jersey).

the effect of greater suppression of third time around reflections (triple travel) at the strip dispersive line output. For these reasons, for operation at megacycle frequencies in pulse-compression applications emphasis has been placed on the development of the thin strip as a dispersive delay line, rather than the thin wire.

13.6 Transducer Characteristics for Thin Strip Lines[1]

As previously mentioned, several of the modes of propagation in a thin strip could be used for a dispersive delay line. The first longitudinal mode, however, exhibits the most desirable shape for synthesizing a linear delay vs frequency curve, using the thickness tapering method described by Fitch [9].

The particular mode of propagation generated in the strip can be controlled by the transducer polarization and dimensions. With proper design, using these two parameters, it is possible to generate a pure first

---

[1] The material in this section was supplied by H. C. Wing, Sanders Associates, Nashua, New Hampshire.

longitudinal mode with other modes down 40 to 50 db. Asymmetries in the transducer configuration can lead to higher level spurious flexural modes. Meitzler [20] indicates that sources of these can be tilted transducer end faces, nonuniform bonding layers and nonuniform poling of the transducer material.

The amplitude characteristics of ultrasonic strip delay lines are determined by both the transducer and strip design. The absolute loss level is considered to be made up of three components: (1) mismatch loss, (2) material loss, and (3) beam spreading loss. The shape of the amplitude curve, however, is dependent on the above three mechanisms as well as on the mechanical Q of the transducers.

The bonded transducer, near resonance, has an equivalent circuit as shown in Fig. 13.15(a) [21, 22]. At the center frequency the series L and C are resonant, giving minimum loss. With the parallel capacity antiresonated at the center frequency by an external inductor, the general shape of the frequency response is that of a single series tuned circuit as the frequency is varied to either side of resonance. The value of the resistance is determined by the constants of the transducer material, the delay medium, and the type of bond between the two.

Because a component of the material loss in aluminum strip increases approximately as the fourth power of frequency, the maximum usable frequency in aluminum is limited to about 6 MHz. This rapid increase in

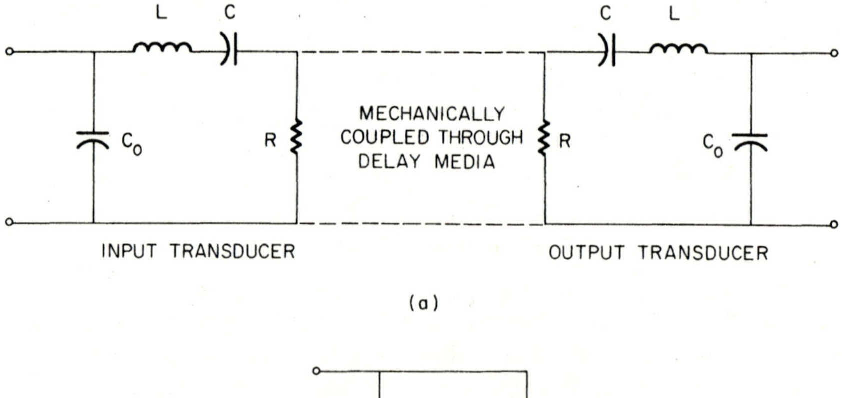

FIG. 13.15   Transducer circuit representations. (a) Equivalent circuit. (b) Effective parallel capacitance and resistance.

loss with frequency is to a great extent due to scattering of the acoustical signal at the grain boundaries of the material. This becomes very high as the wavelength in the material approaches the grain size.

On the short lines, usually used for pulse-compression applications, the beam spreading loss can be neglected if the transducer is made long compared to its width. Thus, the shape of the amplitude response is approximately the single tuned circuit response associated with the transducers.

With proper transducer and strip design, random spurious signals from a strip delay line can be maintained at 40 db below the desired output signal. The mismatch reflection from the end of the lines, however, is the major spurious problem in relatively short lines where material loss is small. This is commonly called triple travel since it occurs at three times the line delay from the input signal. It is effectively reduced by terminating the ends of the delay line with the complex conjugate of the transducer impedance. The equivalent circuit shown in Fig. 13.15(a) can be transformed into the parallel capacitance and resistance shown in Fig. 13.15(b). These are the terms that can be most easily measured with standard equipment. Note that the parallel elements are functions of frequency. This frequency function can be obtained from the admittances of the two circuits. For the circuit of Fig. 13.15(a)

$$Y = j\omega C_0 + \frac{1}{R + j(\omega L - 1/\omega C)} \tag{13-24}$$

And for Fig. 13.15(b)

$$Y = \frac{1}{R_E} + j\omega C_E \tag{13-25}$$

Equate the real and imaginary parts of the two admittance functions and solve for R_E and C_E; this yields

$$R_E = R + \frac{1}{R}\left(\omega L - \frac{1}{\omega C}\right)^2$$

$$C_E = C_0 - \frac{(\omega L - 1/\omega C)}{\omega(R^2 + [\omega L - 1/\omega C]^2)} \tag{13-26}$$

These are shown in Fig. 13.16. Excellent agreement with these theoretical curves has been obtained in practice, as shown by this figure.

For optimum triple travel rejection when working into a constant load impedance, a matching network is used. This network must match the resistance curve to a constant load value and tune the capacity curve across the frequency band of interest. The circuit shown in Fig. 13.17 has been

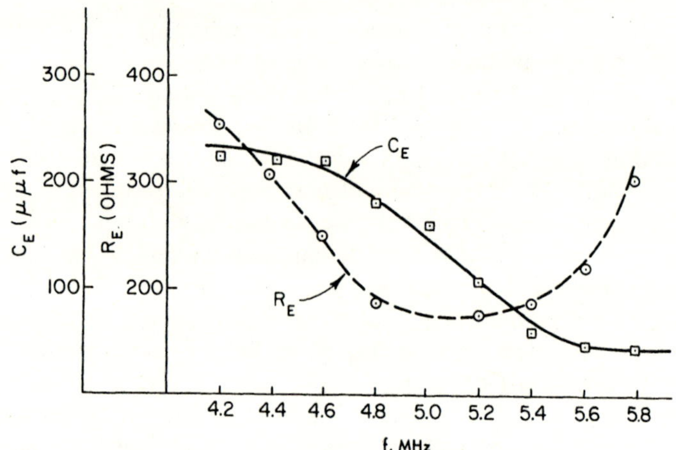

FIG. 13.16   Frequency dependent capacitance and resistance of transducer equivalent circuit. (After Meitzler [22].)

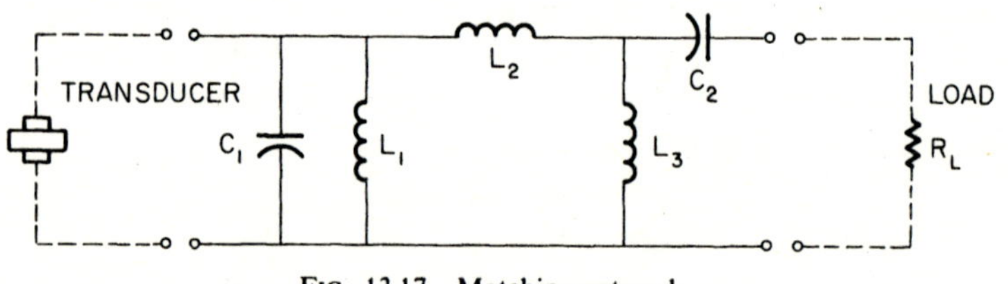

FIG. 13.17   Matching network.

used with good results for this purpose. It can be designed graphically, fitting the matching network curves to the measured transducer capacitance and resistance functions.

Since this network matches the transducer across the band, it has a negligible effect on the overall amplitude response. This method can be used for bandwidths up to 40% of the center frequency and yields compressed-pulse triple travel levels of 35 to 40 db on short strip delay lines.

13.7   System Specifications for Dispersive Ultrasonic Delay Lines

The pulse-compression systems engineer will, in most cases, purchase the ultrasonic delay line from one of several possible sources. This section summarizes some of the important characteristics that should be agreed upon by the supplier and user. Figure 13.18 illustrates the specification of delay and bandwidth. The ideal delay curve is derived from the bandwidth and time-bandwidth product. The specification should allow for some deviation of time delay slope from the slope of the ideal characteristic. This can be based on the considerations of linear delay mismatch described

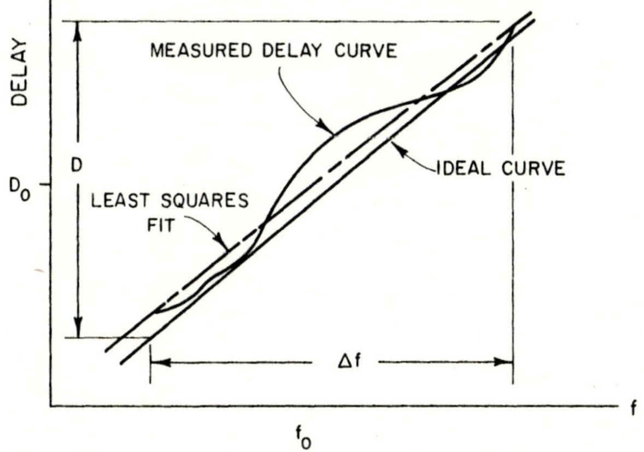

FIG. 13.18   Delay measurement specifications for ultrasonic dispersive delay line.

in Chapter 6. The delay error curve is the deviation from the best least squares fit to the measured time delay points. Specification should be made as to the number of points in the linear delay region at which the time delay will be measured. A technique such as described by May [23] can be used to measure the time delay, using a Gaussian envelope rf pulse with a bandwidth narrow compared to that of the delay line. Figure 13.19 shows a typical specification for the delay error curve and the linear delay slope mismatch. Since each of these can result in waveform deterioration at the compression-filter output, the systems engineer must determine which of these can be the most serious in his particular application, and what combinations of these two that he can tolerate. It has been found that attempting to obtain a very small amount of delay error in an ultrasonic

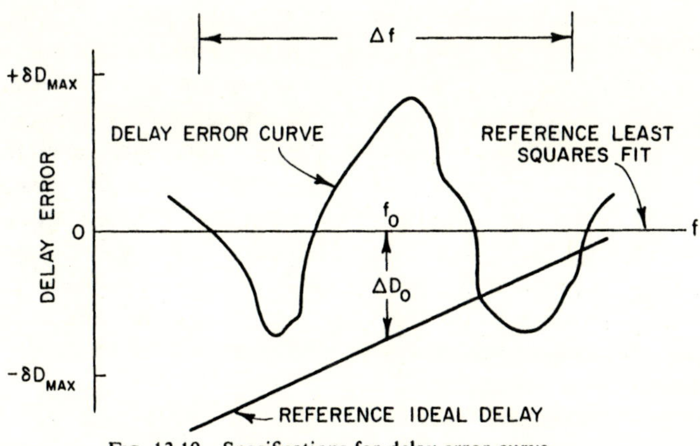

FIG. 13.19   Specifications for delay error curve.

line can lead to a costly item. It may be preferable to specify a delay error consistent with -25 db paired-echo sidelobes and then perform additional delay error correction with bridged-T equalizers, as illustrated in the previous chapter. Figure 13.20 compares the delay error curves of a bridged-T dispersive line before error correction and of a four segment stepped thickness ultrasonic dispersive line. This shows that they have similar error characteristics, and thus are amenable to the same error

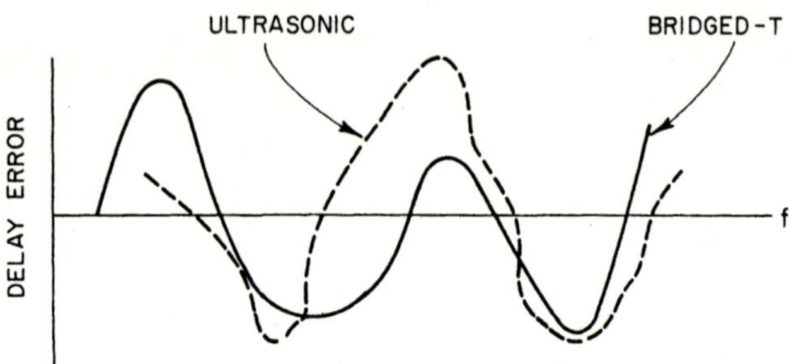

FIG. 13.20 Comparison of delay error curves for bridged-T and ultrasonic dispersive delay lines.

correction techniques. The value of nominal midband delay, D_0, is a function of the center frequency and time-bandwidth product of the design, and is not normally critical in itself. The variation in this delay may be critical when several ultrasonic lines are required to have matched characteristics. A specification for these parameters can allow a fairly broad variation in D_0, but a tighter variation in ΔD_0 once D_0 has been established.

The level of the undesired third time around signal should be at least 25 to 30 db below the desired signal output, when measured with a relatively long Gaussian envelope constant frequency pulse. If the specification is made in this way the third time around signal in a pulse-compression system will have added on to the specified difference level the pulse-compression ratio build-up factor ($10 \log T \Delta f$ in decibels) plus a further 6 db (approximately) resulting from the dispersion of the third time around signal as it travels back to the input and is then reflected back again to the output. Other spurious responses should be called out as being the desired number of db below the peak of the compressed pulse, and should be of the same order as the third time around signals when measured using a pulse-compression signal.

The amplitude response specification should allow for a nominal insertion loss plus variations about this level, as shown in the measured characteristic

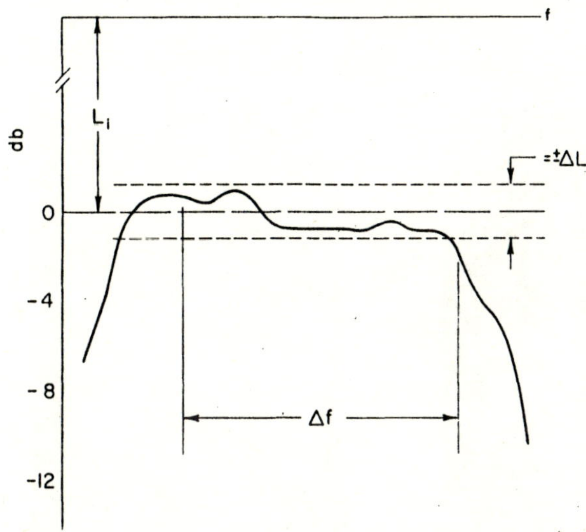

FIG. 13.21 Insertion loss and amplitude response specifications for ultrasonic pulse-compression filter.

in Fig. 13.21. Since amplitude ripple can also result in paired-echo side-lobes, allowable variations in the amplitude response should not be large, and certainly less than 1 db.

Figure 13.22 shows typical waveforms taken with an ultrasonic pulse-compression filter that illustrate some of the important characteristics described above. A dynamic test of the dispersive filter should be included as part of the specification. The dispersive delay line used for these waveforms was a Brew model 10400. This is a four segment stepped thickness line of the type described in Section 13.4. Figure 13.22(a) shows the linear FM uncompressed-pulse input and the compressed-pulse output delayed by $T/2 + D_0$ from the start of the uncompressed pulse. The peak of the compressed pulse relative to the peak of the uncompressed pulse should be 10 log $T\Delta f - L_i$, where L_i is the nominal insertion loss of the dispersive line. Figure 13.22(b) shows the compressed pulse in greater detail. If the input signal is set to the correct time duration and linear FM slope, the crispness of the nulls of the waveform is one indication of a good match in the slope of the time delay characteristic. If sidelobe reduction is not instrumented, the compressed pulse will have a $(\sin x)/x$ shape, with the first sidelobes 13.2 db below the waveform peak. Symmetry of the side-lobe structure and a clear definition of the sidelobes for several intervals on each side of the compressed pulse are an indication of an acceptable delay error curve. However, the effect of smaller paired-echo distortions will not show up on the $(\sin x)/x$ waveform. The width of the compressed

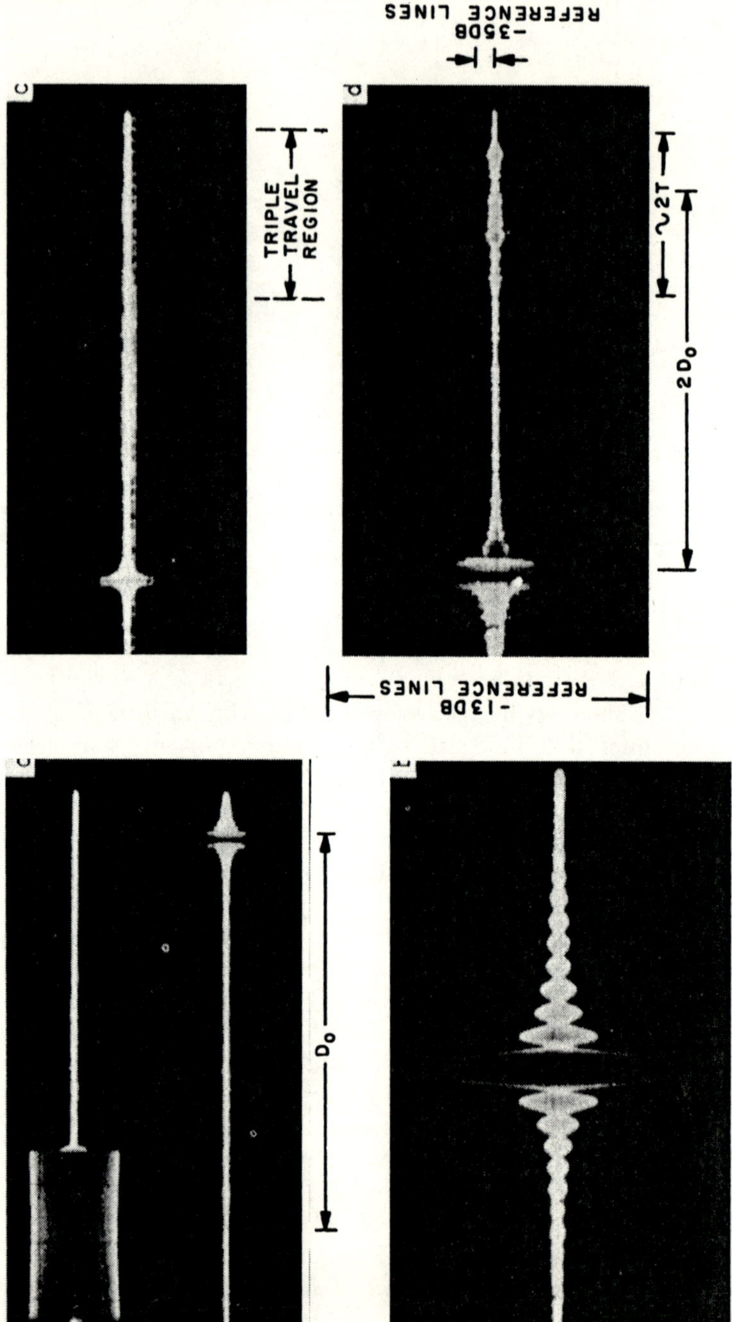

Fig. 13.22　Waveforms for ultrasonic strip dispersive delay line. (a) Input and compressed-pulse output. (b) Compressed pulse, expanded time scale. (c) Normal amplitude. (d) Expanded amplitude. (c) and (d) location and amplitude of triple travel signal relative to compressed pulse.

pulse should measure $1/\Delta f$ at the -4 db point, and $2/\Delta f$ between the first two nulls at the baseline. Figure 13.22(c) illustrates the region in which the third time around signal should be examined. This and other spurious responses, if based on an adequate specification that is met by the design, will not be discernible when the compressed pulse is viewed at normal amplitude levels. Figure 13.22(d) conveys the same information as Fig. 13.22(c), except at a much higher amplitude level. Here the triple travel signal shows up in clear detail, and is seen to be better than 35 db below the peak of the compressed pulse. The sidelobes around the compressed-pulse location should fall off as shown, and will extend over an interval $\pm T$ about the compressed pulse. The other signals between the regular sidelobes and the third time around region represent spurious signals and leakage from the signal generator used to obtain these waveforms.

13.8 Dispersive Diffraction Grating Devices

The ultrasonic dispersive delay lines discussed in previous sections operate on a guided wave principle. Recent investigations into the dispersive properties of ultrasonic diffraction grating devices have been described by Duncan and Parker [24] and Coquin and Tsu [25]. These devices operate on the basis of providing an effective delay path from an input array of transducers to an output array of transducers that is a function of frequency, although the wave propagation within the medium itself is a nondispersive shear mode. An example of this type of grating delay line is the perpendicular diffraction delay line (P.D.D.L.) described by Coquin and Tsu. The geometry of this device is illustrated by Fig. 13.23(a) showing an array of N radiating sources and M receiving elements arranged on two perpendicular sides of a propagating medium (aluminum, for instance). The N source elements are excited in phase, and the M receiver elements are connected in parallel, and $M > N$. The receiving elements are in the near field of the radiation pattern of the source array.

Coquin and Tsu describe the basic properties of the P.D.D.L. using interference analysis.[1] Referring to Fig. 13.23(b), the sources S_n and S_{n-1} on the x-axis are centered about the position $\bar{x}_n$ and separated by the distance d_n. The transmission path length difference between these two sources and the point $\bar{y}_m$ is approximated by (for $d_n \ll r_{m,n}$):

$$r_{m,n} - r_{m,n-1} \doteq d_n \cos \theta_{m,n} = \bar{x}_n d_n / \bar{r}_{m,n} \qquad (13\text{-}27)$$

Similarly, for the receivers R_m and R_{m-1}, the path length difference to the

---

[1] Silver [26] states that the mean value of the field intensity in the near zone resulting from a radiating array can be predicted quite accurately from considerations of geometrical ray theory propagation.

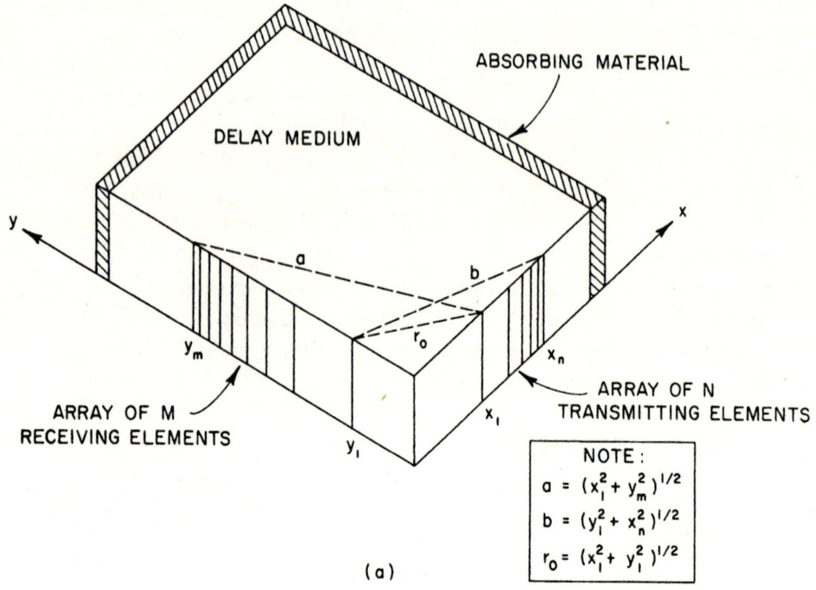

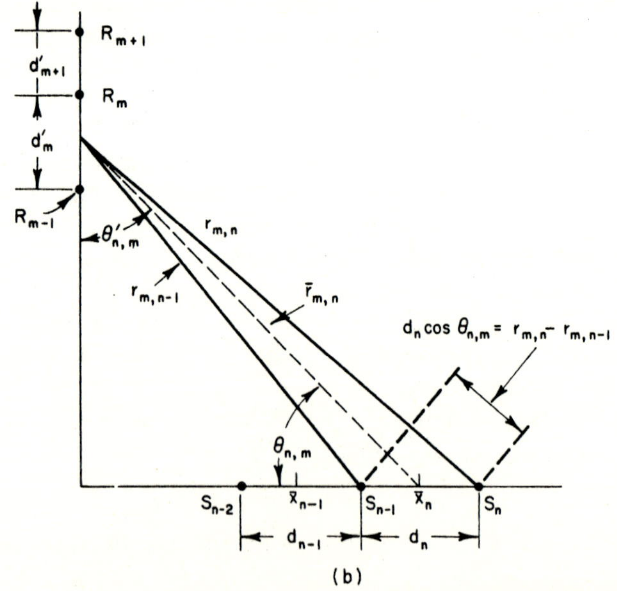

FIG. 13.23   The perpendicular diffraction delay line (P.D.D.L.). (a) Geometry of perpendicular diffraction delay line. (b) Interference relationships of P.D.D.L. (After Coquin and Tsu [25].)

point $\bar{x}_n$ is

$$r'_{m,n} - r'_{m-1,n} \doteq d'_m \cos \theta'_{m,n} = \bar{y}_m \, d'_m / \bar{r}_{m,n} \qquad (13\text{-}28)$$

For a frequency f_i transmitted from sources S_n and S_{n-1}, constructive interference at the receivers R_m and R_{m-1} occurs when the following condition is met:

$$\frac{\bar{x}_n d_n}{\bar{r}_{m,n}} = \frac{\bar{y}_m d'_m}{\bar{r}_{m,n}} = \frac{p V_s}{f_i} \qquad (13\text{-}29)$$

where p is the order of the interference and V_s the shear wave propagation velocity in the medium. If the set of sources and receivers are arranged so that $\bar{x}_n d_n = \bar{y}_m d_m = C$, a constant, so that as x and y increase the spacing between the transducers decreases, then all receiver source pairs separated by the distance r_i, where

$$r_i = C f_i / p V_s \qquad (13\text{-}30)$$

will transmit with maximum efficiency the incremental band of frequencies centered about f_i. For other source-receiver pairs, (S_{n-1}, S_{n-2}) and (R_m, R_{m-1}) for example, the geometry is such that for the shorter path length of this pair the path length *difference* $(d_i \cos \theta_{ij})$ is larger. Thus, transmission at a lower frequency is supported in accord with (13-30), yielding a linear delay vs frequency described by

$$t_d(f) = \frac{r}{V_s} = \frac{Cf}{p V_s^2} \qquad (13\text{-}31)$$

The factor $C/p V_s^2$ gives the delay slope for each order of transmission. The upper and lower frequency limits for each order are, respectively,

$$f_{2p} = \frac{p V_s}{C} \sqrt{x_1^2 + y_m^2} = \frac{p V_s a}{C} \qquad (13\text{-}32a)$$

$$f_{1p} = \frac{p V_s}{C} \sqrt{y_1^2 + x_n^2} = \frac{p V_s b}{C} \qquad (13\text{-}32b)$$

It is important that none of the orders of interest overlap in frequency. For the geometry shown in Fig. 13.23(b), this condition is met for the first p orders when

$$r_0 > \frac{C f_{2p}}{(p+1)V_s} = \frac{C}{(p+1)V_s} \frac{p V_s a}{C} \qquad \text{or} \qquad r_0 > \frac{pa}{p+1} \qquad (13\text{-}33)$$

For the case discussed of the sources and receivers being equally displaced about the general points $\bar{x}_n$ and $\bar{y}_m$, the following relationship holds for each m and n:

$$\bar{x}_n d_n = \tfrac{1}{2}(x_n^2 - x_{n-1}^2) = C$$

$$\bar{y}_m d_m = \tfrac{1}{2}(y_m^2 - x_{m-1}^2) = C \qquad (13\text{-}34)$$

From (13-34) is obtained the general relation for locating the source and receiver elements:

$$x_n^2 = x_1^2 + 2C(n - 1), \qquad 1 < n \leqslant N$$
$$y_m^2 = y_1^2 + 2C(m - 1), \qquad 1 < m \leqslant M \tag{13-35}$$

The properties of the P.D.D.L. device, based on the interference analysis, using (13-31)–(13-32b) and (13-35) are summarized as follows:

Center frequency

$$f_{0p} = \tfrac{1}{2}(f_{2p} + f_{1p}) = \frac{pV_s(a + b)}{2C} = \frac{M - N}{\Delta t_{d_p}} \tag{13-36}$$

Bandwidth

$$\Delta f_p = f_{2p} - f_{1p} = pV_s \frac{a - b}{C} \tag{13-37}$$

Dispersive delay

$$\Delta t_{d_p} = C\frac{\Delta f_p}{pV_s^2} = \frac{a - b}{V_s} = \frac{p(M - N)}{\tfrac{1}{2}[(f_{2p} + f_{1p})]} \tag{13-38}$$

Time-bandwidth product

$$\Delta t_{d_p} \Delta f_p = \frac{p(a - b)^2}{C} = p(M - N)\frac{\Delta f_p}{f_{0p}} \tag{13-39}$$

The factor C can be rewritten as $C = V_s^2 \Delta t_d/(f_2 - f_1)$, setting $p = 1$, and the lower frequency limit can be expressed as

$$f_1^2 = (V_s/C)^2[x_1^2 + y_1^2 + 2C(N - 1)] \tag{13-40}$$

Thus, given M, N, and C, the relationships given above define the significant parameters of the P.D.D.L. for use as a linear FM pulse-compression filter.

The condition for no orders overlapping, (13-33) and (13-40) yield

$$\tfrac{3}{4}(x_1^2 + y_1^2) > \tfrac{1}{2}C(M - 1)$$

From (13-36) $M = \tfrac{1}{2}(f_{2p} - f_{1p}) \Delta t_d + N$, and the bounds on N for none of the first p orders overlapping are

$$1 \leqslant N < 1 + \frac{\Delta t_d}{2(f_2 - f_1)}\left(f_1^2 - \frac{f_2^2}{4}\right) \tag{13-41}$$

In practice it has been found that an optimum value of N lower than the upper limit given above exists. This results from the fact that the modes do not cut off sharply enough for them to be completely separated if the upper limit for N (i.e., contiguous modes) is approached.

Coquin and Tsu give a more rigorous analysis based on diffraction theory that confirms the results derived from interference considerations. On the basis of their analysis typical response characteristics for the first order mode are as shown in Fig. 13.24, with the deviation from delay

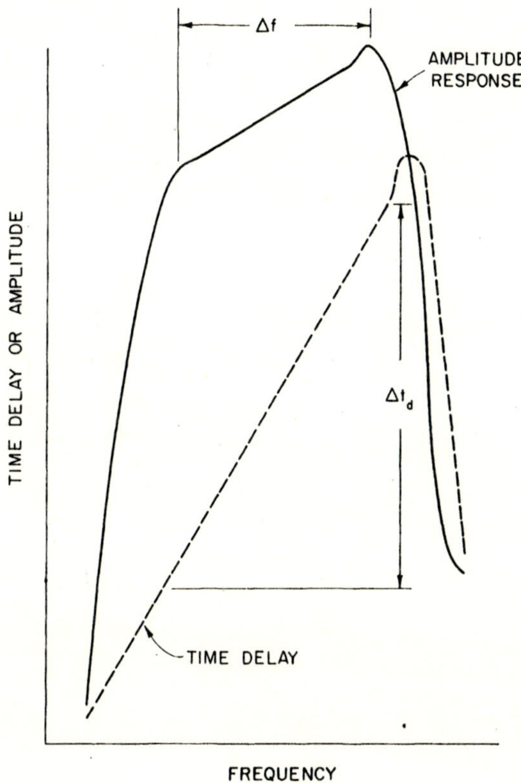

Fig. 13.24 Amplitude response and time delay of P.D.D.L. (courtesy of G. A. Coquin and R. Tsu [25]).

linearity shown in Fig. 13.25 for $N = 25$. For this type of design (large N) the percent bandwidth in each mode is the same, as is the overall dispersive delay (see Eqs. (13-36)–(13-38)). The delay error decreases as N increases up to an optimum value, above which the delay error increases again for the reason cited in the paragraph above. For the delay error curve given in

Fig. 13.25, the approximate phase error from Eq. (13-21) is 0.4 degrees, assuming $\delta D_{av} = 0.3$ microseconds and $\Delta f/n = 0.5 \times 10^{-6}/20$. This represents a significant improvement over the delay error obtainable with other ultrasonic or lumped constant design techniques, provided that spurious signals are also quite low. These can result from reflections off the other surfaces of the device. Test results on low frequency designs indicate that this level of delay error can be approached in practice. Insertion loss for the P.D.D.L. type of delay line can be about 30 db or greater.

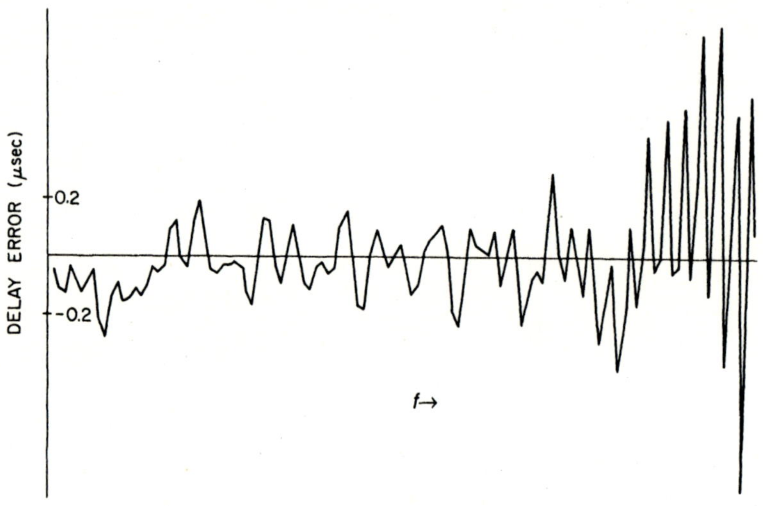

FIG. 13.25   Time delay error of a perpendicular diffraction delay line (courtesy of G. A. Coquin and R. Tsu [25]).

However, the general class of devices based on diffraction principles hold promise of operation at much higher frequencies and wider bandwidths than other ultrasonic devices, in addition to the lower level of delay distortion. This is illustrated by the waveform shown in Fig. 13.26, for which the signal bandwidth is 10 MHz and the time-bandwidth product is 100.

Further data on P.D.D.L. design is given by Eveleth [27], who cites a 60 db insertion loss for a 70 : 1 pulse-compression device ($T = 12.8\,\mu$sec, $\Delta f = 5.5$ MHz) operating at a 30 MHz center frequency. A spurious signal level of -45 db relative to the peak compressed-pulse output was obtained in

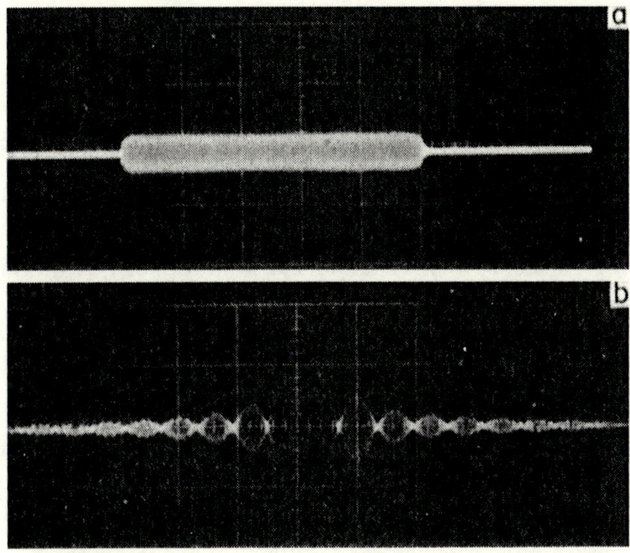

FIG. 13.26 Waveforms for P.D.D.L. expansion/compression filter; $T\Delta f = 100$, $f = 10$ MHz, center frequency = 45 MHz. (a) Gated expanded pulse, 2 μsec/div. (b) Unweighted compressed pulse, 0.2 μsec/div. (Courtesy of S. E. Rittenburg, Andersen Lab., Bloomfield, Connecticut.)

this design. A 400:1 design ($T = 20$ μsec, $\Delta f = 20$ MHz) centered at 45 MHz is also mentioned by Eveleth. Another type of grating device that has been successfully demonstrated takes the form of a triangular wedge [28]. An array of grating transducers is placed on the hypotenuse of the wedge, and a single long input transducer is placed on the base as shown in Fig. 13.27. The output array is again in the near field of the radiating transducer. This type of device does not make use of diffraction principles as does the

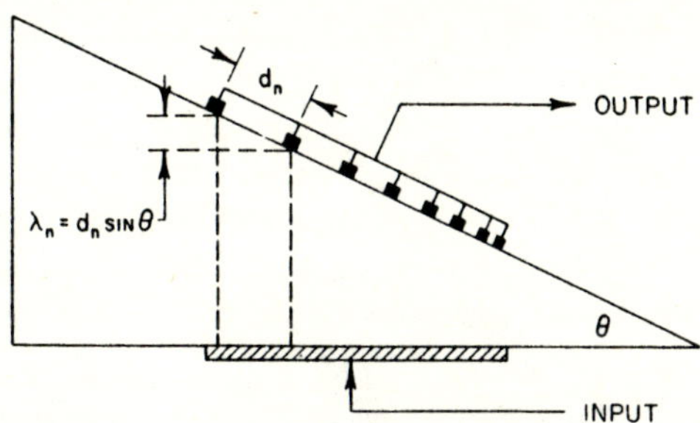

FIG. 13.27 The wedge grating array dispersive delay device.

P.D.D.L. The dispersive characteristic of the wedge depends on the fact that any two adjacent transducers act as a resonant pickup at a frequency for which their delay paths perpendicular to the source differ by one wavelength. Thus, a variable spacing of the component transducers in the array results in an overall delay vs frequency that is dispersive; that is, different pairs of transducers will respond to different segments of the spectrum of a wideband input signal. The wedge will provide either an expanded or compressed pulse, depending on whether a wideband impulse or matching frequency modulated signal is introduced at the input. This type of device can be designed for either a positive or negative slope delay function that is either linear or nonlinear. The delay error associated with the wedge is greater than that of the P.D.D.L. Eveleth discusses the results obtained with a matched pair of wedge dispersive lines for a 70 : 1 compression ratio having the parameters given above.

In applications where low level range sidelobes are of paramount importance the P.D.D.L. diffraction devices can solve the distortion problem for at least one critical component of the matched-filter radar system. In other applications in which range sidelobes in the order of -25 db are acceptable, the lower cost of the longitudinal mode thin strip delay lines (single or multiple thickness) would tend to make them the preferred component as long as the bandwidth requirement can also be met.[1] In many applications a single or tapered thickness line plus the use of bridged-T distortion equalizer networks, described in Chapter 12, will be an economical approach. Using this design method, the system engineer can lump in other receiver delay and amplitude distortions in the distortion correction design.

An additional point for consideration is the effect of the compression-filter insertion loss on the dynamic range and detection capabilities of the receiving system. In a lumped constant design the overall insertion loss can be offset by dividing the filter into smaller subunits and providing each subunit with its own loss compensation amplifier. No such incremental loss compensation can be as easily achieved with the ultrasonic dispersive delay lines, particularly since a large fraction of the loss derives from the input and output transducers. When the insertion loss is of the order of 40 to 60 db, as it can be with some of the ultrasonic devices, it may be necessary to compensate for a large part of this loss with a preamplifier before the compression filter. If the preamplifier gain is large then saturation may result with the stronger received signals, limiting the dynamic range of the receiver.

---

[1] The ultrasonic dispersive delay lines can be used in a parallel channel arrangement for wideband applications. Other grating devices under development, such as the surface wave grating device [29], may also be applicable to wideband pulse-compression implementations.

REFERENCES

1. W. P. Mason, "Physical Acoustics and the Property of Solids." Van Nostrand, Princeton, New Jersey, 1958.
2. A. H. Meitzler, Ultrasonic delay lines using shear modes in strips, *IRE Trans.* **UE-7**, 35–43 (1960).
3. A. H. Meitzler, Transmission characteristics of longitudinal-mode strip delay lines having asymmetrically tapered widths, *IRE Trans.* **UE-9**, 30–36 (1962).
4. A. H. Meitzler, Mode coupling occurring in propagation of elastic pulses in wires, *J. Acoust. Soc. Am.* **33**, 435–445 (1961).
5. T. R. Meeker, Dispersive ultrasonic delay lines using fhe first longitudinal mode in a strip, *IRE Trans.* **UE-7**, 53–58 (1960).
6. T. R. Meeker, The application of the theory of elastic waves in plates to the design of ultrasonic dispersive delay lines, *IRE Intern. Conv. Record, Pt.* 6, 327–333 (1961).
7. J. E. May, Jr., Wire-type dispersive ultrasonic delay lines, *IRE Trans.* **UE-7**, 44–53 (1960).
8. A. H. Fitch, A comparison of several dispersive ultrasonic delay lines using longitudinal and shear waves in strips and cylinders, *IRE Intern. Conv. Record, Pt.* 6, 284–292 (1960).
9. A. H. Fitch, Synthesis of dispersive delay characteristics by thickness tapering in ultrasonic strip delay lines, *J. Acoust. Soc. Am.* **35**, 709–714 (1963).
10. T. R. Meeker and A. H. Meitzler, Guided Wave Propagation, *in* "Physical Acoustics, Principles and Methods" (W. P. Mason, ed.), Vol. 1, Part A, pp. 111–167. Academic Press, New York, 1964.
11. J. E. May, Jr., Guided wave ultrasonic delay lines, *ibid.*, pp. 417–483.
12. V. T. Buchwald, Transverse waves in elastic plates, *Quart. J. Mech. Appl. Math.* **2**, 498–512 (1958).
13. Lord Rayleigh, On the free vibrations of an infinite plate of homogeneous, isotropic, elastic matter, *Proc. London Math. Soc.* **20**, 225–234 (1889).
14. H. Lamb, On waves in an elastic plate, *Proc. Roy. Soc. (London) A* **93**, 114–128 (1917).
15. A. N. Holden, Longitudinal modes of elastic waves in isotropic cylinders and slabs, *Bell Syst. Tech. J.* **30**, 956–969 (1951).
16. T. R. O'Meara, The synthesis of "band-pass" all-pass time delay networks with graphical approximation techniques, Hughes Aircraft Co., Res. Lab., Malibu, California, Res. Rept. 114 (June, 1959).
17. L. Pochhammer, Über die Fortpflanzungsgeschwindigkeiten kleiner Schwingungen in einem unbegrenzten isotropen Kreiscylinder, *J. reine angew. Math. (Crelle)* **81**, 324–336 (1876).
18. C. Chree, The equations of an isotropic elastic solid in polar and cylindrical coordinates, their solution and application, *Trans. Cam. Phil. Soc.* **14**, 250–269 (1889).
19. D. Bancroft, The velocity of longitudinal waves in cylindrical bars, *Phys. Rev.* **59**, 588–593 (1941).
20. A. H. Meitzler, Selective attenuation of elastic wave motions in strips of polycrystalline metals, *J. Acoust. Soc. Am.* **34**, 444–453 (1962).
21. R. A. Heising, "Quartz Crystals for Electrical Circuits." Van Nostrand, Princeton, New Jersey, 1948.
22. A. H. Meitzler, A procedure for determining the equivalent circuit elements representing ceramic transducers used in delay lines, *Proc. Electron. Components Symp.* pp. 210–219 (1957).
23. J. E. May, Jr., Precise measurement of time delay, *IRE Intern. Conv. Record, Pt.* 2, 134–142 (1958).

24. R. S. Duncan and M. R. Parker, Jr., The perpendicular diffraction delay line: a new kind of ultrasonic dispersive device, *Proc. IEEE (Correspondence)* **53**, 413–414 (1965).

25. G. A. Coquin and R. Tsu, Theory and performance of perpendicular diffraction delay lines, *Proc. IEEE* **53**, 581–591 (1965).

26. S. Silver, "Microwave Antenna Theory and Design." p. 170. McGraw-Hill, New York, 1948.

27. J. Eveleth, A survey of ultrasonic delay lines operating below 100 Mc/sec, *Proc. IEEE* **53**, 1406–1428 (1965).

28. M. R. Parker, U.S. patent application, filed June 11, 1964.

29. J. H. Rowen, U.S. patent application, filed Dec. 24, 1963.

30. V. A. Krasil'nikov, "Sound and Ultrasound Waves," third ed., (Transl.: N. Kaner and M. Segal). Israel Programs for Scientific Translations, Jerusalem, Israel, 1963.

CHAPTER 14

Microwave and Optical Matched-Filter Techniques

14.1 Introduction

The lumped constant and ultrasonic dispersive delay line designs considered in the two preceding chapters are applicable to the great majority of pulse-compression waveform implementations that require such devices. However, a certain number of radar applications involve either waveforms with extremely wide bandwidths or complex modulation functions for which the matched-filter design is not straightforward. For these cases processing with the usual narrower band dispersive devices is often not possible, although for some signal parameters the parallel channel design approach described in Chapter 6 permits the use of narrow band devices in wideband applications. Microwave pulse-compression filter design techniques, described in the literature, have wider bandwidths than either the lumped constant or ultrasonic designs, and can be used where there is the requirement for a real-time display of high resolution information. Most of the microwave techniques result in only a moderate amount of differential time delay, and thus may not be the best design technique when both long signal durations and wide bandwidths are required.

Several microwave dispersive delay design techniques are described in this chapter. The system designer can determine from the waveform and radar specification whether he should use a microwave technique with which he can obtain directly a wide bandwidth but small amounts of delay, or whether he should, at the expense of somewhat more complexity, rely on one of the combination techniques with which he can obtain both wide bandwidths and larger amounts of dispersive delay. The details of various microwave components that are used for a complete implementation are available in the literature [1, 2], and need not be considered here.

One of the more interesting areas for advances in matched-filter implementations is based on optical signal processing methods. The operation of the optical matched-filter device depends on the space-frequency analog between optical and electrical filter systems. Optical signal processing techniques generally can be considered to have the same level of bandwidth and time-bandwidth product capability that are available with the more advanced lumped constant and ultrasonic matched-filter designs. The unique advantages obtained through optical processing are that the equivalent of a matched-filter bank can be achieved in a compact arrangement, or that the optical processor can be designed to match rapid changes in the coded waveforms sent out by the transmitter. The availability of optical matched-filter signal processors expands the choices that accrue to the radar engineer, who will make his ultimate decision based on a combination of economic and technical factors, the relative weights of which will depend on individual judgement.

14.2   The Folded-Tape Meander Line

Several microwave dispersive delay line design techniques have been based on achieving a high frequency equivalent of the bridged-T all-pass network described in Chapter 12. The folded-tape meander line [3] is one of the microwave analogs of the lumped constant approach that has been used successfully to obtain a dispersive delay characteristic. A meander line design for pulse-compression applications has been developed by Dunn [4]. This design has the basic configuration shown in Fig. 14.1(a), where the folded conducting tape is positioned between two conducting planes. A dielectric material is used to hold the tape in position. The critical physical dimensions are shown in Fig. 14.1(b). The length L is large compared to the separation s, and the basic mode of propagation is a superposition of TEM waves that travel in the $\pm y$-direction. Requiring that the superposition of these waves match the boundary conditions at $y = \pm L/2$ yields the following relationship for the phase shift β between adjacent folds of the tape as a function of the general frequency $\omega = 2\pi f$ that is traveling in the x-direction:

$$\cot^2 \frac{\omega L}{c} = \frac{\gamma - \cos \beta - \gamma' \cos 2\beta + \cdots}{\gamma + \cos \beta - \gamma' \cos 2\beta - \cdots} \cot^2 \frac{\beta}{2} \qquad (14\text{-}1)$$

The parameters γ, γ', γ'', etc., are related to the static capacitances per unit length between different pairs of conductors of infinite length that have the same cross section dimensions as the folded tape. Dunn shows that for the geometry illustrated in Fig. 14.1 the parameters γ', γ'', etc., are negligible if the ratio $d/s \leqslant 1.5$. Inserting $\gamma' = \gamma'' = 0$ in Eq. (14-1) yields the following

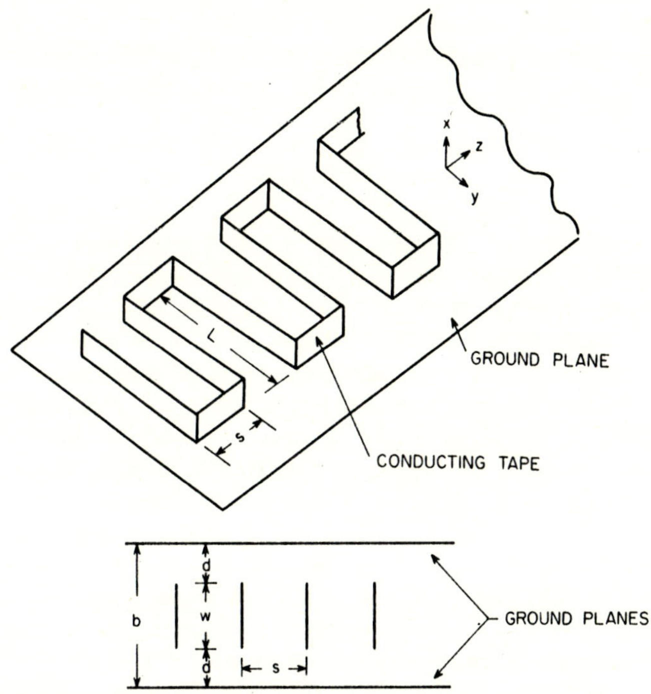

FIG. 14.1   Dimensions of folded-tape meander line (after Dunn [4]).

solution for β:

$$\beta = \cos^{-1}\left[\frac{1 - \gamma + \left[(\gamma - 1)^2 + 4\gamma \cos^2\dfrac{\pi f}{2f_0}\right]^{1/2}}{2\cos\dfrac{\pi f}{2f_0}}\right] \qquad (14\text{-}2)$$

where $f_0 = c/4L$, or $L =$ quarter wavelength at f_0. Differentiation of (14-2) with respect to ω yields the group time delay function per turn

$$T_D = \frac{d\beta}{d\omega} = \frac{\dfrac{1}{4f_0}\left|\tan\dfrac{\pi f}{2f_0}\right|\left[1 - \dfrac{1}{g(\gamma)}\right]}{\left\{2[g(\gamma) - 1] - \dfrac{4\cos^2\dfrac{\pi f}{2f_0}}{\gamma - 1}\right\}^{1/2}} \qquad (14\text{-}3)$$

where

$$g(\gamma) = \left[1 + \frac{4\gamma}{(\gamma - 1)^2}\cos^2\frac{\pi f}{2f_0}\right]^{1/2} \cong 1 + \frac{2\gamma}{(\gamma - 1)^2}\cos^2\frac{\pi f}{2f_0}, \qquad f \cong f_0$$

The general behavior of Eq. (14-3) is plotted in Fig. 14.2 for various values of γ. Equation (14-3) is indeterminate at $f = f_0$, and the value of peak delay can be obtained by L'Hospital's rule. The delay maximum at $f = f_0$ is

$$\hat{T}_D = \frac{1}{4f_0} \frac{\gamma}{\gamma - 1} \tag{14-4}$$

The curves shown in Fig. 14.2 are very similar to those obtained for the second order bridged-T all-pass section in Chapter 12. A major difference is the periodicity of the folded-tape delay function that makes this delay function differ from that of the all-pass network in the skirt regions.

The important factor in determining the dispersive delay characteristic of the folded-tape meander line is the parameter γ. A small value of γ indicates that the adjacent tape elements are tightly coupled, while a large value of γ indicates very loose coupling (for no coupling $\gamma = \infty$). From the curves of Fig. 14.2 it can be seen that small values of γ are desirable, and Dunn specifies a range of $1.05 \leqslant \gamma \leqslant 3.0$ as being easily obtainable and for which the parameters γ', γ'', etc., can be neglected. For the flat-tape design the dependence of γ on the dimensional ratios w/s and d/s are presented in Fig. 14.3. Normalized time delay curves for different values of γ are shown in Fig. 14.4. These are comparable to the bridged-T delay curves obtained by O'Meara and shown in Fig. 12.4.

The design relationships described above have been used by Dunn to obtain several equal ripple approximations to linear delay functions of both positive and negative delay slopes. Figure 14.5 is a typical example of such wideband designs. The negative slope delay curve is normalized to the lowest frequency f_L of the desired band. When the delay slope is negative the periodicity of the basic folded-tape delay characteristic limits the design to a 2 to 1 ratio of highest to lowest frequency. The values of the normalized design frequencies and γ are shown in the figure. Different designs must be worked out for every desired bandwidth ratio, although any particular design can be used over a narrower band than indicated. This is one way to avoid the larger delay errors that are usually present at the band edges. Dunn describes the design of a 50 : 1 pulse-compression ratio filter based on the data given in Fig. 14.5. The normalized design frequencies can be denormalized to obtain the desired bandwidth. This will determine the amount of actual dispersive delay per section, from which can be calculated the total number of sections needed. Additional design details are given in Dunn [4].

The meander line is not the exact analog of the lumped constant bridged-T all-pass network in that it is not a constant resistance network. The image impedance of the meander line is a function of the dimensional

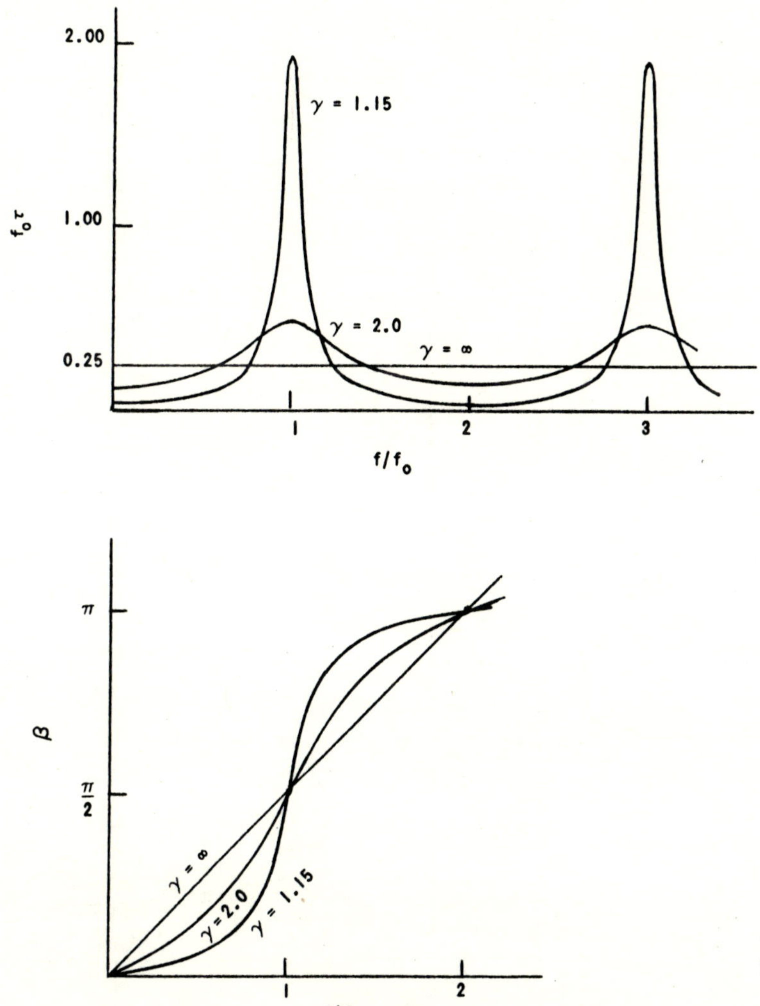

FIG. 14.2   Typical time delay and phase shift per turn of a meander line. Note the periodicity of the phase and time delay curves. (Courtesy of V. E. Dunn [4] and Stanford Electronics Laboratory, Stanford University, Stanford, California.)

ratios d/s and w/s, as well as of the dielectric material used. Thus, there can be mismatches between sections and between the first and last sections and the input and output load impedances. The quantitative effects of these on the meander line performance were not evaluated. However, satisfactory experimental results were obtained. The intended application for this particular design was in a rapid scanning receiver, or spectrum analyzer.

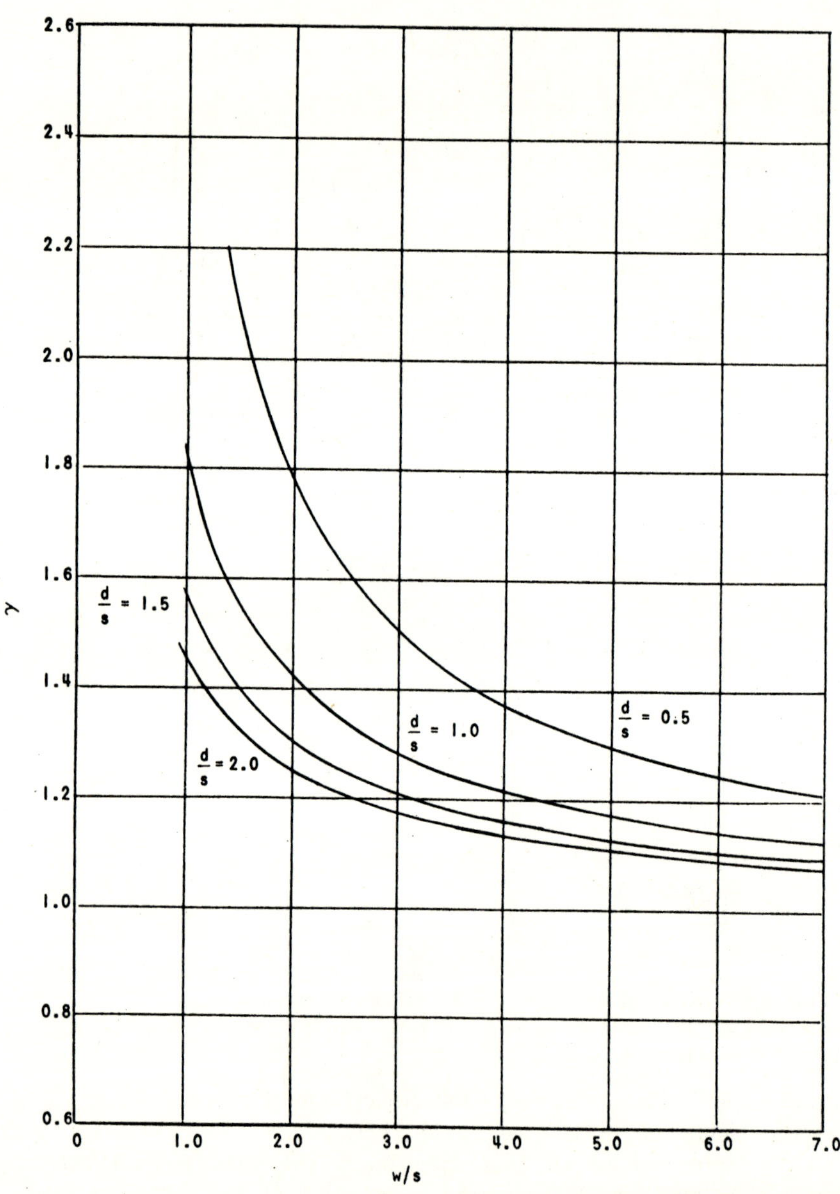

FIG. 14.3 γ as a function of w/s (courtesy of V. E. Dunn [4] and Stanford Electronics Laboratory, Stanford University, Stanford, California).

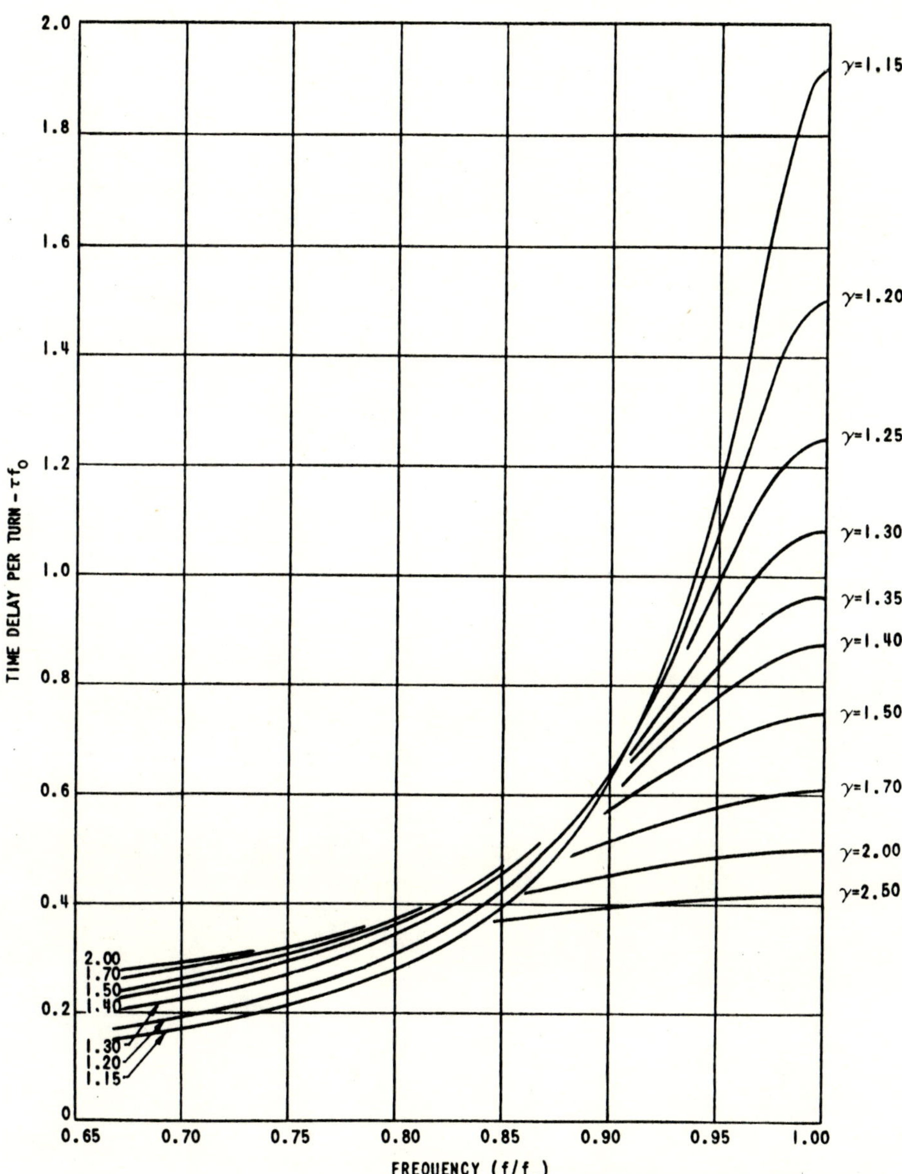

FIG. 14.4 Time delay per turn of a meander line as a function of frequency (courtesy of V. E. Dunn [4] and Stanford Electronics Laboratory, Stanford University, Stanford, California).

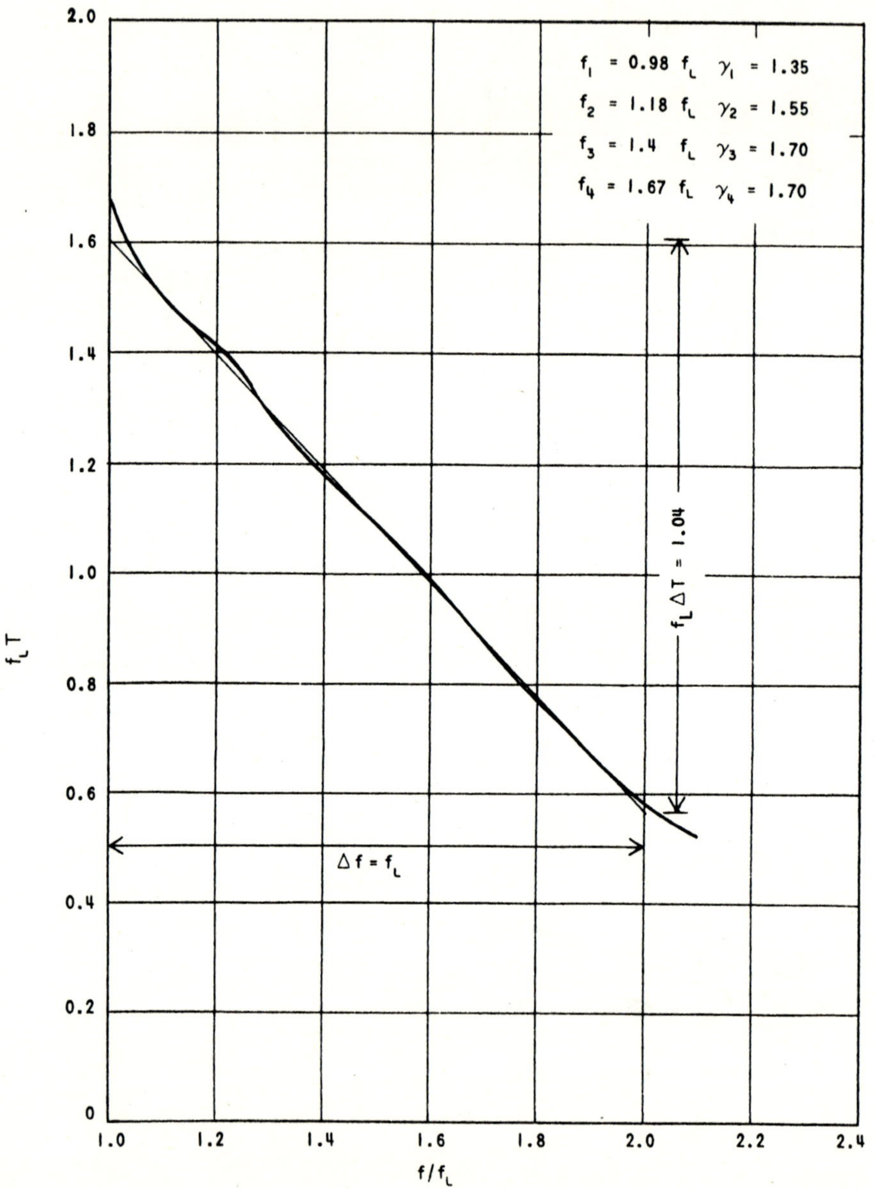

FIG. 14.5   Four-section approximation to a negative-slope characteristic over a 2.0-to-1.0 band (courtesy of V. E. Dunn [4] and Stanford Electronics Laboratory, Stanford University, Stanford, California).

The use of pulse compression in this type of equipment is described by Kincheloe [5].

14.3 The Hybrid Ring "All-Pass" Design

The basic folded-tape meander line is a slow wave structure whose normalized group delay function approximates that of the low frequency lumped constant all-pass section discussed in Chapter 12. A more direct microwave analog of the conventional all-pass section has been based on the use of the three-halves wavelength hybrid ring as shown in Fig. 14.6,

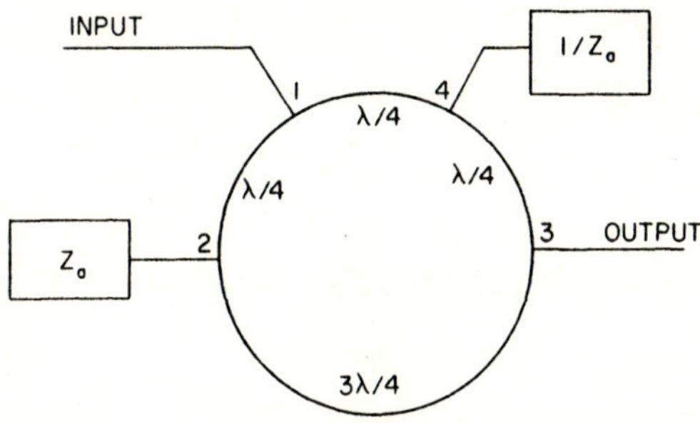

FIG. 14.6   The $3\lambda/2$ ring hybrid as a bandpass all-pass network.

although this design technique is not restricted to this particular configuration. Over the frequency range in which this structure acts as a hybrid the use of the dual impedances as shown results in a transfer function given by

$$\frac{E_2}{E_1} = \frac{1 - Z_a}{1 + Z_a} \tag{14-5}$$

When Z_a and Z_b are pure reactances (14-5) becomes identical to the lumped constant resistance all-pass transfer function given by (12-5), and

$$\frac{E_2}{E_1} = \exp[-j2\tan^{-1} X_a] = \exp[-j\beta(\omega)] \tag{14-6}$$

where $\beta(\omega)$ is the transfer phase characteristic.

At microwave frequencies the reactances jX_a and jX_b can be obtained by using sections of open or short circuit quarter-wave or half-wave resonant lines. If X_a is obtained with a shorted half-wave section (i.e., series resonant circuit) the associated reactance function is given by

$$jX_a = jZ_0 \tan \frac{\pi\omega}{\omega_0} \tag{14-7a}$$

$$= -jZ_0 \tan \pi\left(\frac{\omega_0 - \omega}{\omega_0}\right) \tag{14-7b}$$

In a relatively narrow frequency interval about ω_0 (14-7b) can be approximated by

$$jX_a = j\pi Z_0 \frac{\omega - \omega_0}{\omega_0} \tag{14-8}$$

and the transfer phase characteristic in the same frequency interval is approximated by

$$\beta(\omega) = 2 \tan^{-1} \frac{\omega - \omega_0}{\omega_0/\pi Z_0} \tag{14-9}$$

Recalling from (12-21) that O'Meara's approximation for the second order all-pass phase is

$$\beta(\omega) = 2 \tan^{-1} \frac{\omega - \omega_0}{\sigma_0} + \text{neglected terms}$$

it can be seen that (14-9) is identical to the expression above if the factor $\omega_0/\pi Z_0$ is defined as the equivalent microwave frequency σ_0. Thus, the hybrid ring structure can be used to synthesize microwave frequency dispersive delay functions based on low frequency lumped constant all-pass design data as long as the approximations given above are valid.

A practical realization of a microwave hybrid ring all-pass network is described by Ferguson and Barrett [6]. Their design takes the form of the stripline structure shown in Fig. 14.7. Each half-wave section shown is an

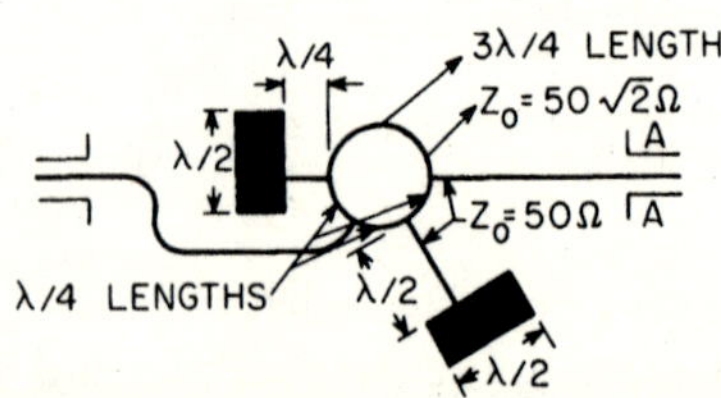

FIG. 14.7 Microwave all-pass network (courtesy of C. W. Barrett, Lockheed Electronics Co., Plainfield, New Jersey).

open circuit resonator. However, the quarter-wave connecting length at the jX_a location makes that half-wave section appear as its dual reactance, or half-wave short circuit section, at the hybrid ring junction no. 2. The width of the half-wave open circuit sections as well as the location of the tap points of the $\lambda/4$ and $\lambda/2$ connecting lines control the effective impedance level Z_0 of the resonant sections. This makes it possible to achieve the desired values of the equivalent σ_0's with which to scale one of O'Meara's all-pass designs to the microwave frequency region. Ferguson and Barrett describe the design of a staggered quadruplet of stripline all-pass sections at S-band. A symmetrical stripline section was formed by a photographic etching process. The geometry of this structure is shown in Fig. 14.8, with each of the indicated parameters having an effect on the characteristic impedance.

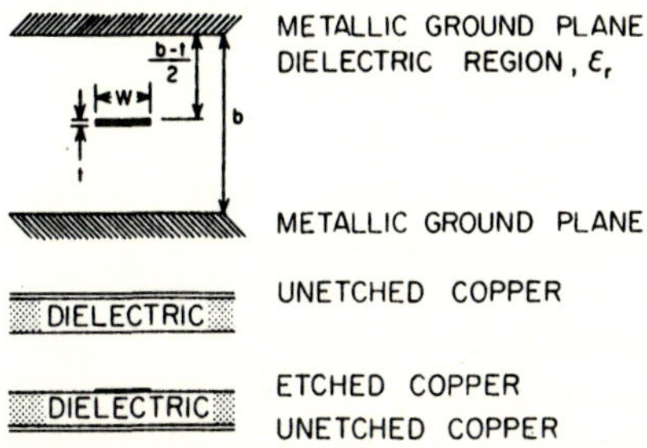

FIG. 14.8 Construction of symmetrical stripline (courtesy of C. W. Barrett, Lockheed Electronics Company, Plainfield, New Jersey).

Measurements made on the stripline all-pass sections indicate that the dispersive delay and amplitude response characteristics are similar to those obtained at low frequencies with the lumped constant designs. However, it was found that time delay errors and impedance mismatches were critically affected by deviations from the correct dimensions of the hybrid ring arms and the open circuit resonator sections. In this respect, the stripline sections suffer by comparison with the low frequency designs which are amenable to final tuning adjustments. However, this is offset by the fact that once an acceptable design is worked out the photographic etching technique is repeatable with high accuracy. As with the low frequency all-pass sections, the compression ratio can be built up by cascading the basic sections. A single hybrid ring section of the design cited above had a fixed loss and an additional dip due to incidental dissipation (see Eq. (12-42)) at the ring center frequency. Thus the quadruplet amplitude response would be similar to that shown in Fig. 12.6, requiring an overall equalization loss to correct the amplitude response.

In using the stripline design one is limited to a percentage bandwidth over which the hybrid ring relationships hold. In addition, the exact analogy to O'Meara's all-pass delay function exists only over the frequency regions where the resonant section impedance approximation given by (14-8) is valid. The latter restricts the bandwidth that can be obtained with this structure to about 15% when the microwave design is based essentially on a direct frequency scaling method. By comparison, the meander line designs derived by Dunn are based on exact time delay relationships and yielded much wider percentage bandwidths. Similar exact design data could also be obtained for the hybrid ring, as well as other microwave all-pass analogs.

14.4   Microwave Pulse-Compression Filters Using Tapped Delay Line Techniques

The use of microwave components that are approximate analogs of the low frequency lumped constant all-pass networks were described in Sections 14.2 and 14.3. These are directly applicable in the same sense as the low frequency components to the design of matched filters for the analog FM class of coded waveforms. Wideband implementations also exist for the class of discrete coded signals described in Chapter 8. In these applications the microwave versions of lower frequency tapped delay lines, bandpass filters and summation networks are used in a straightforward fashion, being complicated only by the problems of working at higher frequencies. Thus, microwave components could be used for obtaining a step-FM matched filter, or one of the tapped delay line configurations discussed in Chapter 8.

A somewhat different approach to pulse-compression filter design is the use of fixed delay components to construct a processor for an analog FM waveform. An example of this would be the use of a step-FM matched filter to compress a linear FM signal having the same sense and magnitude of frequency deviation as that of the matched step-FM signal. Obviously, this does not result in an exact matched-filter implementation, but can be a useful filter-design technique at frequencies at which nondispersive delay components may be easier to work with than the dispersive type of components. An example of this type of compression-filter design is described by Mueller and Goodwin [7]. Their design technique is presented below as illustrative of a general method not included in the matched-filter techniques mentioned above, as well as being an example of the use of nondispersive high frequency components applied to coded waveform processors. Although the example presented is a microwave design, it is clear that the same method can be used to obtain compression filters in the frequency ranges previously discussed for the lumped constant and ultrasonic dispersive lines.

The compression filter developed by Mueller and Goodwin is based on subdividing the ideal linear FM phase modulation characteristic into equal intervals as shown in Fig. 14.9. Each phase interval is associated with a different portion of the frequency band. The compression filter is a tapped delay line in which the respective delay spacing of the taps is matched to the midpoint of the delay associated with each phase subinterval, and is thus related to the phase modulation function. The minimum number of phase subintervals required for an adequate pulse-compression characteristic is the square root of the time-bandwidth product, or compression ratio.

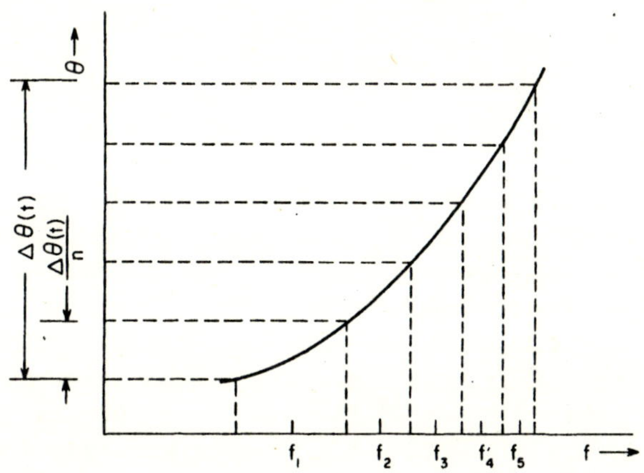

FIG. 14.9 Division of matched-filter phase characteristic into n equal segments. Note different frequency interval associated with each segment.

A larger number of delay line taps gives a better approximation to the ideal matched-filter time delay function. In essence, this technique is based on approximating a continuous delay vs frequency curve with discrete amounts of fixed delay. The phase modulation of the linear FM signal is given by

$$\theta(t) = \omega_0 t + \tfrac{1}{2}(\mu t^2), \qquad 0 < t < T \tag{14-10}$$

The representation of (14-10) eliminates consideration of negative delay in the compression filter. The total change in phase over the time interval T is

$$\Delta\theta(T) = \omega_0 T + \tfrac{1}{2}(\mu T^2) \tag{14-11}$$

For n equal phase subintervals the phase spacing of the delay line taps becomes

$$\theta_{\text{sp}} = \Delta\theta(T)/n \tag{14-12}$$

Locating the delay line tap at the midpoint of each phase interval yields the following expression for the cumulative phase at the tap points:

$$\theta_m = \frac{\Delta\theta(T)}{n}\frac{2m-1}{2}, \qquad 1 < m < n \tag{14-13}$$

The group delay associated with each tap point can be obtained from the phase modulation function (14-10), yielding

$$t_m = -\frac{\omega_0}{\mu} + \sqrt{\left(\frac{\omega_0}{\mu}\right)^2 + \frac{2\theta_m}{\mu}} \tag{14-14}$$

The midpoint frequency associated with each tap point is:

$$f_m = f_0 + \frac{\mu}{2\pi}t_m \tag{14-15}$$

The filter described by Mueller and Goodwin was designed to operate in the band from 300 to 400 MHz, and compress a linearly frequency modulated pulse of 1 microsecond duration, for a compression ratio of 100. The tapped delay line used had 20 taps ($n_{min} = 10$), and the overall delay was equally divided between the input and output branches, resulting in the ladder type of structure shown in Fig. 14.10. Directional couplers were used to tap off the signal at each input junction and to feed it back onto the output branch. The compression network parameters calculated from Eqs. (14-14) and (14-15), as given by Mueller and Goodwin, are tabulated in Table 14-I. The filter constructed by Mueller and Goodwin was designed to compress a one microsecond pulse swept from 300 to 400 MHz. The 302.9 MHz bandpass filter was placed at the largest delay increment relative to the input and output terminals. The incremental delays and cable lengths given in Table 14-I and 14-II are with reference to this filter, so that the synthesized delay decreases in a linear stepwise fashion in the interval from 300 to 400 MHz.

Type RG-213/U coaxial cable was used for the fixed delay elements. Included in the calculation of the delay length of each cable was the effect of the delays of the rf connectors and directional filters. The lengths of each cable are given in Table 14-II, remembering that each set of incremental delays is equally divided between the input and output branches of the tapped delay line. The total amount of RG-213/U cable used was approximately 640 feet. The bandpass filters working into and out of the directional couplers were of the two-pole Butterworth type. The bandpasses varied from 6.1 to 8 MHz, as given in Table 14-I. The response cross-overs were at the -1 db points relative to the response of each filter at its center frequency. Sharp skirts on the bandpass filter were not desired in order

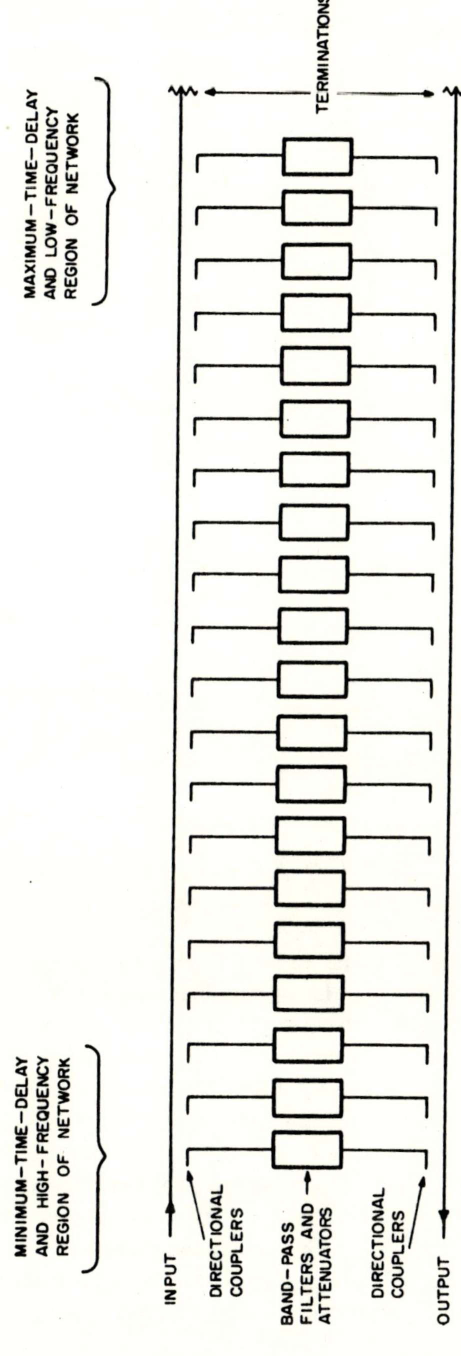

Fig. 14.10 Simplified diagram of compression network (courtesy of F. J. Mueller and R. L. Goodwin [7], and Airborne Instruments Laboratory. Melville, New York).

TABLE 14-I

COMPRESSION NETWORK PARAMETERS[a]

Filter center frequency[b] (MHz)	Filter bandwidth (MHz)	Differential time delay[c] (μsec)	Cumulative time delay[d] (μsec)
302.9	8.0	0.02903	0.02903
308.6	7.9	0.05723	0.08626
314.2	7.8	0.05619	0.14245
319.8	7.7	0.05521	0.19766
325.2	7.6	0.05426	0.25192
330.5	7.4	0.05338	0.30530
335.8	7.3	0.05253	0.35783
341.0	7.1	0.05172	0.40955
346.0	7.0	0.05094	0.46049
351.1	7.0	0.05021	0.51070
356.0	6.9	0.04950	0.56020
360.9	6.8	0.04882	0.60902
365.7	6.7	0.04817	0.65719
370.5	6.6	0.04754	0.70473
375.2	6.6	0.04694	0.75167
379.8	6.5	0.04636	0.79803
384.4	6.4	0.04580	0.84383
388.9	6.3	0.04526	0.88909
393.4	6.2	0.04474	0.93383
397.8	6.1	0.04423	0.97806

[a] Data from Mueller and Goodwin [7].

[b] $f_m = 300 + 100 t_m$, f_m in MHz, t_m in sec.

[c] $t_m = -3 + \sqrt{9 + 0.175(2n - 1)}\,\mu\text{sec}$ $(1 \leqslant m \leqslant 20)$.

[d] Cumulative time delay is in reference to the 300 MHz end of the network.

to prevent ringing in the filters. Finally, attenuators were inserted after and before each directional coupler to equalize the overall loss characteristic, as well as to help minimize the effect of reflections caused by mismatches. The attenuators can also be used for frequency response shaping to reduce sidelobes. The complete filter design is shown in Fig. 14.11. The amplitude response and insertion loss of the filter is shown in Fig. 14.12. The resonant frequencies indicated are those actually realized as compared to the values given in Table 14-I. The filter was packaged in three units, each of approximate dimensions 19 in. × 19 in. × 20 in. A weighting filter was used to shape the amplitude response, obtaining a range sidelobe level of about −20 db, and an increase in the compressed-pulse width from the optimum 10 nanoseconds to 15 nanoseconds, when the best operating conditions

TABLE 14-II
COMPRESSION NETWORK DELAY CABLE LENGTHS[a]

Differential delay between tap points (nsec)	Delay cable lengths exclusive of r.f. connectors[b,c] (inches)	Differential delay between tap points (nsec)	Delay cable lengths exclusive of r.f. connectors[b,c] (inches)
57.2	214.1	48.8	181.2
56.2	210.2	48.2	178.8
55.2	206.3	47.5	176.1
54.3	202.7	46.9	173.7
53.4	199.2	46.4	171.8
52.5	195.7	45.8	169.4
51.7	192.5	45.3	167.4
50.9	189.4	44.8	165.5
50.2	186.7	44.2	163.1
49.5	183.9		

[a] Data from Mueller and Goodwin [7].

[b] There are two delay cables at 214.1, 210.2, 206.3 inches, etc.

[c] Cable length $= \dfrac{\left[\dfrac{11.788 \text{ in.}}{\text{nsec}} \dfrac{t_d}{2}\right] - L_{dc} - L_c}{\sqrt{e}}$

where t_d is the differential delay (nsec);
L_{dc} is the electrical length of directional coupler, 12.17 inches;
L_c is the electrical length of two right angle r.f. connectors, 3.15 inches;
$\sqrt{e} = 1.503$, the square root of dielectric constant of RG-213/U cable;
11.788 is the cable length in air for 1 nsec of delay.

were realized. The same general type of design technique outlined above could also be applied to other types of FM functions, or to a step-FM matched filter [5]. The fact that, in general, the bandwidths of the band-pass filters can be all different means that care must be taken in combining these bandpasses responses to achieve an overall flat response.

14.5 Continuous Microwave Dispersive Structures

The microwave pulse-compression filters discussed in Sections 14.2 to 14.4 are based on essentially achieving a piecewise approximation to a prescribed matched-filter delay function. Specifically, linear delay charac-teristics were sought for, although these techniques have an inherent flexibility that would permit the design, within limits, of fairly arbitrary delay vs frequency functions. An alternate line of investigation has been concerned with the use of microwave structures that have monotonic, nonlinear dispersive delay characteristics, with the thought of matching

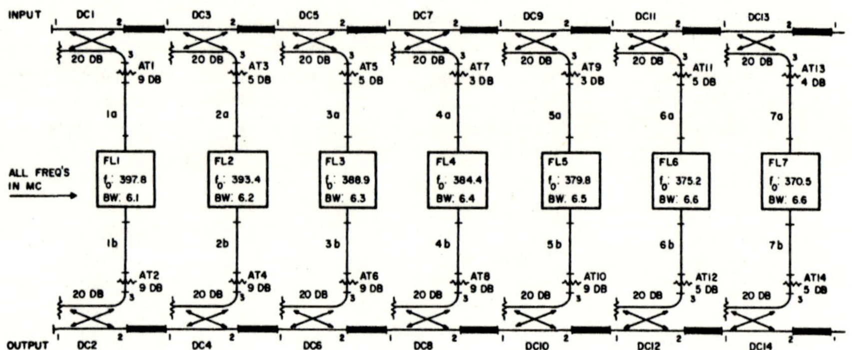

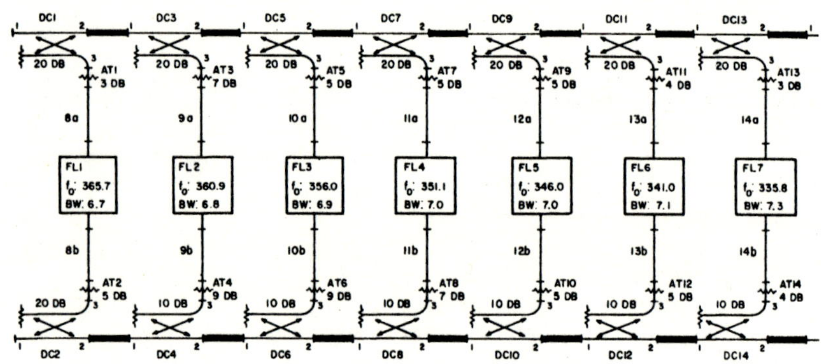

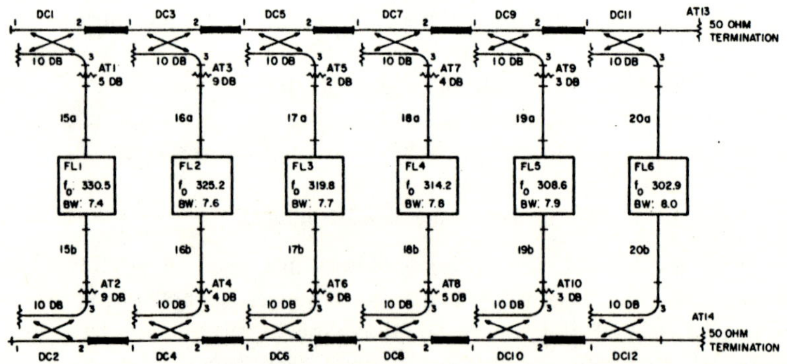

FIG. 14.11   Detailed schematic diagram of compression network (courtesy of F. J. Mueller and R. L. Goodwin [7]).

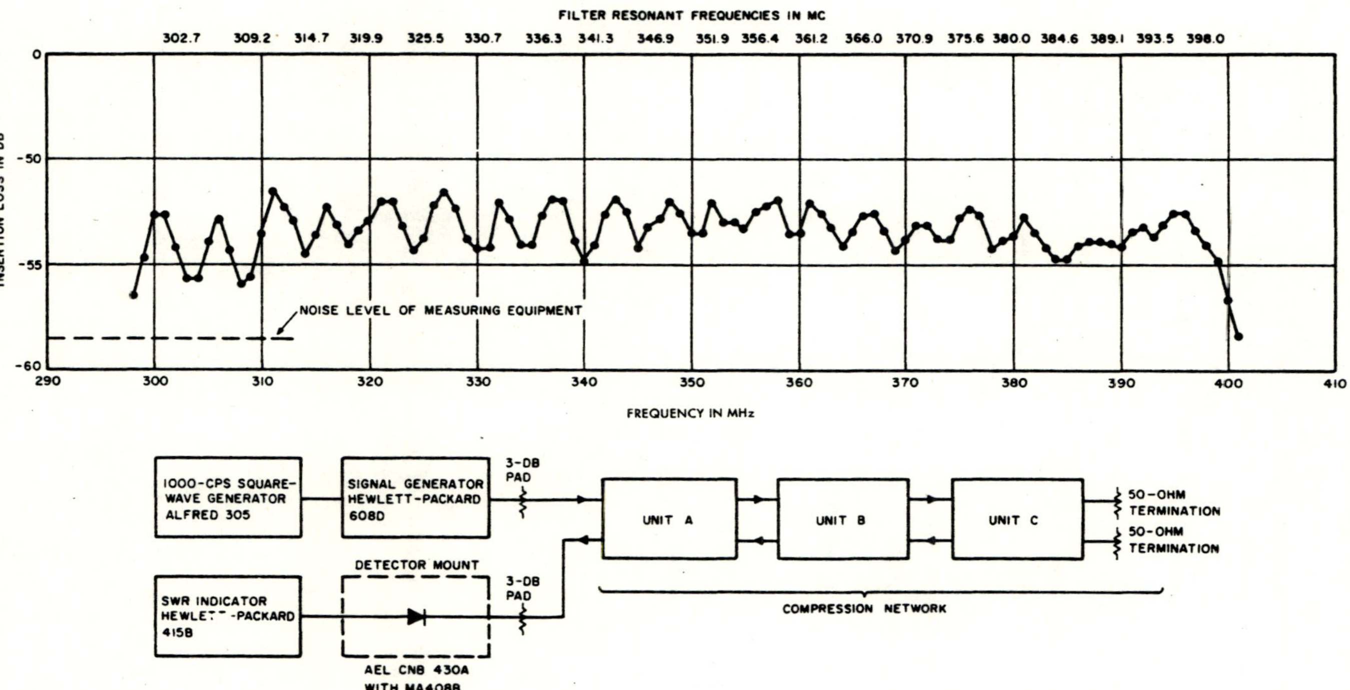

FIG. 14.12   Insertion loss vs frequency of complete compression network (courtesy of F. J. Mueller and R. L. Goodwin |7|).

the transmitted FM to the delay function of the structure. In some cases these delay functions are approximately linear over a narrow bandwidth, and thus could be used in linear FM systems. The nonlinear delay as a function of frequency of these types of microwave structures are such that they normally cannot be used in an expansion/compression matched-filter system, as depicted in Fig. 6.15, and are restricted to applications where the signal frequency modulation is generated by active techniques. This latter consideration makes the microwave structures discussed in this section less promising for general matched-filter applications, since each type of structure must be used with a specific form of nonlinear frequency modulation in order to achieve optimum pulse-compression results.

(a) Uniform Rectangular Waveguide

The most common type of microwave structure having nonlinear dispersive properties is the uniform rectangular waveguide operating in the TE_{10} mode. The phase function for this mode of propagation is

$$\beta(\omega) = \frac{L\omega}{c} \sqrt{1 - \left(\frac{\omega_c}{\omega}\right)^2} \qquad (14\text{-}16)$$

where c is the velocity of propagation, ω_c the cut-off frequency ($\lambda_c = 2a$, a being the width of waveguide), and L the length of waveguide. The group delay obtained from (14-16) is

$$T_D(\omega) = \frac{d\beta}{d\omega} = \frac{L}{c} \bigg/ \sqrt{1 - \left(\frac{\omega_c}{\omega}\right)^2} \qquad (14\text{-}17)$$

Equation (14-17) is plotted in Fig. 14.13. This delay function is similar to that of the shear mode of propagation in the thin strip ultrasonic delay line (see Section 13.3). The greatest amount of dispersive delay is realized in the region just above ω_c, and a lower limit ω_l of the band used is usually specified where $\omega_l/\omega_c > 1$. Ito [8] makes use of the principle of stationary phase to describe the relationship between the waveguide group delay given by (14–17) and the frequency modulation of a microwave signal at the waveguide output when the waveguide acts as a dispersive, or pulse expansion, filter for very narrow pulse inputs in the order of a nanosecond. Ito notes the application of this method to pulse-compression systems. In general, the frequency modulation of a pulse dispersed in a uniform waveguide filter will not be suitable for compression in the same waveguide because of the nonlinear delay characteristic of the waveguide, as shown by Figure 14.13. However, the frequency modulation function required to match the receiver delay given by (14-17) also can be obtained from the stationary

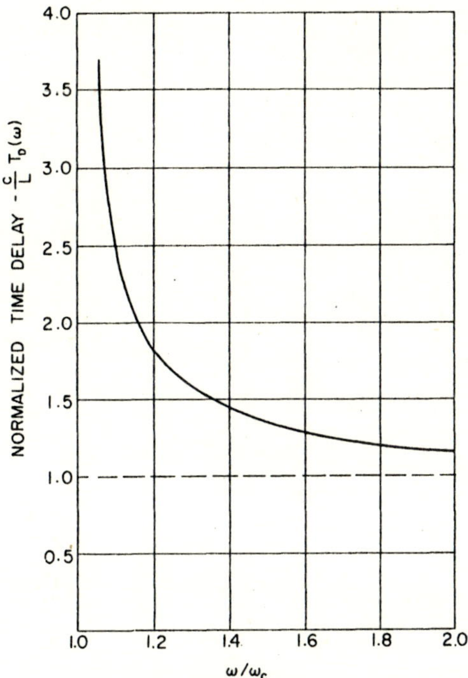

FIG. 14.13 Group delay of uniform rectangular waveguide.

phase method discussed in Chapter 3. It was shown there that if the matched-filter delay is $T_d(\omega)$, then the matching frequency modulation function $\omega(t)$ is approximately the inverse function $T_d^{-1}(t)$. For a specified lower frequency ω_l this gives as the optimum modulation function

$$\omega(t) = \frac{\omega_c}{\left\{ 1 - \left[\dfrac{1}{(c/L)T_D(\omega_l) - (c/L)t} \right]^2 \right\}^{1/2}} \tag{14-18}$$

where

$$t_{\max} = T_d(\omega_l) - T_d(\infty)$$

The modulation functions given by (14-18) are shown in Fig. 14.14 out to $t(2\omega_c)$. Ordinarily, the time duration of the transmitted signal would fall well within this limit unless extremely wide percentage bandwidths were desired. The modulation functions shown in Fig. 14.14 can be approximated by a linear frequency modulation in certain narrow band intervals. Such an application is related to the results described by Ohman [9]. However, most efficient use of this waveguide structure is obtained with the nonlinear FM

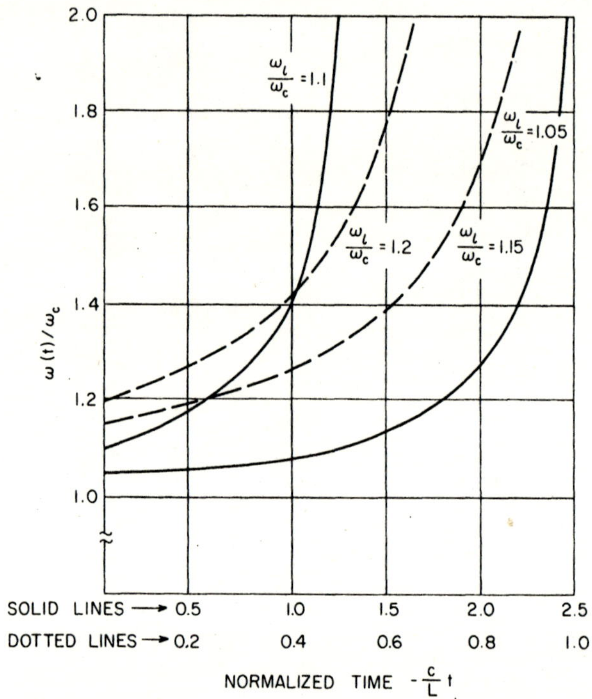

FIG. 14.14   Matching nonlinear FM function for uniform rectangular waveguide.

function described by (14-18). Generally, the best frequency region for use
with a linear FM waveform is obtained with high values of ω_l. This results
in a much longer length of waveguide for the same general pulse-compression
parameters than if the optimum modulation function is used in conjunction
with a value of ω_l that is much closer to ω_c. The ambiguity function properties
of a matched-filter waveform having a rectangular envelope and the frequency modulation (14-18) have not been studied extensively. However, it
can be assumed that they would be similar to the unidirectional parabolic
FM signal discussed in Chapter 9. Thus, the compressed-pulse waveform
would be approximated by a $(\sin x)/x$ pulse, being slightly wider and having
somewhat higher sidelobe levels (see Fig. 4.14). This type of compressed
pulse was obtained by Walther [10] in a water filled acoustic waveguide for
which the dispersive group delay was functionally equivalent to that of the
uniform microwave waveguide. The peak response of the matched-filter
output as a function of normalized Doppler shift would be similar to that of
the unidirectional parabolic FM signal shown in Fig. 9.11. However, since
the microwave signal could be extremely wideband, for all practical purposes
the behavior as a function of Doppler shift would not be too different from

that of a linear FM signal with very small Doppler shifts relative to the signal bandwidth.

Waveguide loss can be as high as 8 db per 100 feet of length. Once the significant waveguide parameters (ω_l/ω_c, Δf and ΔT_d) have been chosen the expected insertion loss can be readily calculated from the known loss characteristics of the various types of waveguide. The overall loss can be kept to a minimum by operating as close to ω_c as possible without encountering the larger losses occasioned by actually running into ω_c, and by using the proper nonlinear FM function.

(b)   Tapered Waveguide

The differential delay per unit length in a waveguide structure can be increased by using tapered waveguide. An example is the 2θ taper which has the same linear rate of taper with length in both the width and height aspects. The phase characteristic of this type of structure is

$$\beta(z) = \frac{\omega L}{c} \sqrt{1 - \left(\frac{\omega_c(z)}{\omega}\right)^2} \tag{14-19}$$

where $\omega_c(z)$ is the cutoff frequency at each z position.

The group delay function obtained from (14-19) is

$$T_D(\omega) = \frac{z_0}{c}\left\{\left[1 - \left(\frac{\omega_c(0)}{\omega}\right)^2\right]^{1/2} - \left(1 - \frac{L}{z_0}\right)\left[1 - \left(\frac{\omega_c(0)}{\omega}\right)^2 \frac{z_0^2}{(z_0 - L)^2}\right]^{1/2}\right\} \tag{14-20}$$

Equation (14-20) is plotted in Fig. 14.15 as a function of L/z_0 for different values of $\omega_l/\omega_c(0)$. Since the cutoff frequency of this high pass structure increases with the length of the guide, L, the choice of ω_l limits the maximum value of L/z_0. This is indicated by the cutoff contour in Fig. 14.15. The nonlinear frequency modulation that matches the group delay given by (14-20) is[1]

$$\omega(t) = \frac{\omega_c(0)}{\left[1 - \left\{\dfrac{1}{(c/L')T_D[\omega_c(0)] - (c/L')t}\right\}^2\right]^{1/2}}, \quad L' = L(1 - L/2z_0) \tag{14-21}$$

This expression is seen to be identical to (14-18), substituting L' for L. Thus, the same functional form of nonlinear frequency modulation is

---

[1] This expression, as well as (14-18), is based on unpublished work of L. Susman of the Sperry Gyroscope Company, Great Neck, New York.

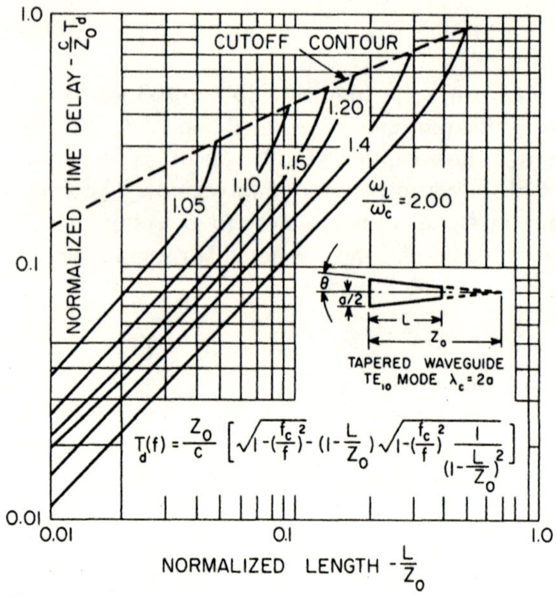

FIG. 14.15   Time delay vs length for tapered waveguide (courtesy of L. Susman, Sperry Gyroscope Co., Great Neck, New York).

required to match the dispersive delay characteristics of both the uniform and tapered rectangular waveguide.

When tapered waveguide is used a significant reduction in length can be achieved. Using (14-17) and (14-20), it can be shown that for a signal swept over the interval $1.05\omega_c$ to $1.20\omega_c$ (15% bandwidth normalized to cutoff) the uniform-taper waveguide length required is one fourth of the length of the uniform cross section rectangular waveguide. An application of tapered waveguide to pulse-compression or dispersion correction systems is described by Albersheim [11]. Tang [12] discusses the design of a nonuniformly tapered waveguide that yields a large linear dispersive delay. Clarricoats *et al.* [13] discuss the application of uniform and tapered dielectric loaded circular waveguides, operating in a backward wave mode at X-band, to pulse-compression radars. Techniques of phase and delay equalizer design using tapered waveguide are given by Torgow [14].

(c) *The Helix Structure*

The helix structure is most commonly known for its use in traveling wave tubes. In the frequency ranges of interest for this application the propagation properties of this structure result in a phase vs frequency characteristic that is essentially linear, and therefore nondispersive at these frequencies. However, at lower frequencies there is a region in which the phase characteristic is approximately parabolic, and the group delay function approximately linear. The phase response of the helix along the axial direction is [15]

$$\beta(\omega) = (\omega/v)\sqrt{(\omega/c)^2 + \gamma^2} \tag{14-22}$$

where v is the phase velocity and γ the radial propagation constant.

Evaluation of γ permits the calculation of the group delay function in the usual manner. Unfortunately, the expression for γ is not in closed form, being derived by Pierce [15] as

$$\gamma \frac{I_0(\gamma a)K_0(\gamma a)}{I_1(\gamma a)K_1(\gamma a)} = \frac{\omega}{c}\cot\psi \tag{14-23}$$

where a is the radius of helix, ψ the helix pitch angle, I_0, I_1 the modified Bessel functions of the first kind, and K_0, K_1 the modified Bessel functions of the second kind.

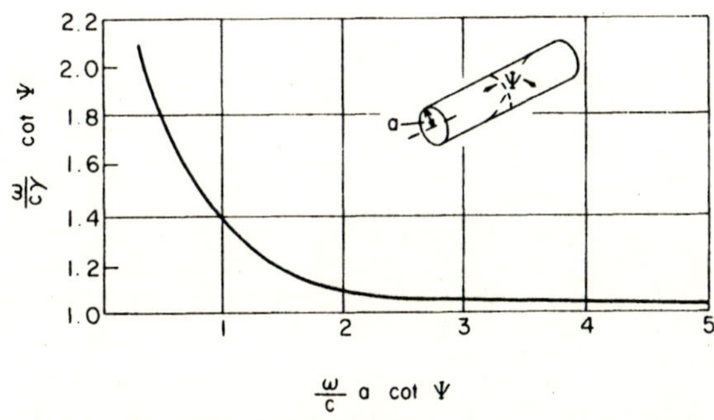

FIG. 14.16 Helix radial propagation constant γ (from J. R. Pierce [15], copyright 1950, D. Van Nostrand Company, Inc., Princeton, New Jersey).

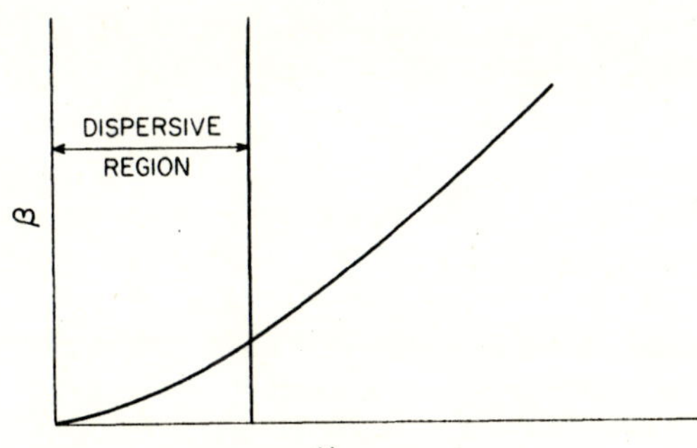

FIG. 14.17 Helix phase characteristic.

Figure 14.16 plots the relationship among ω, a, γ, and ψ. The phase characteristic of the helix is shown in Figure 14.17. In the dispersive region the delay increases with frequency, and thus the helix can be used as a compression filter for a signal that has a negative sweep rate. The design of a helix compression filter is described by Dunn [16]. This filter has an approximately linear delay, as shown in Fig. 14.18. The use of this type of pulse-compression filter in a scanning receiver is described by Kincheloe [5].

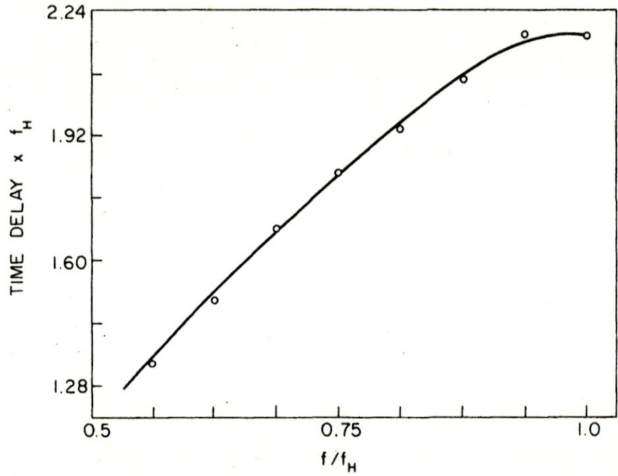

FIG. 14.18   Time delay vs frequency of the helix dispersive filter (courtesy of W. R. Kincheloe [5] and Stanford Electronics Laboratory, Stanford University, Stanford, California).

(d)   Other Microwave Structures and Devices

Many other types of microwave structures offer possibilities for use as wideband compression filters. An example is the Karp structure, a form of open faced ridged waveguide. A five foot length of this type of waveguide has been used to achieve a compression ratio of 16 at L-band [17].

As the examples in this section point out, these structures in general do not provide the flexibility in dispersive delay design that can be achieved with the microwave analogs of the bridged-T design technique or with the tapped delay line methods. In addition, the waveguide structures tend to be relatively large when differential delays of any sizable magnitude are required. The solution to this problem may very well rest with newer techniques that are coming to the fore. The dispersive properties of yttrium-iron-garnet (YIG) material when used as a propagation medium for magnetoelastic waves have been cited as being applicable to the design of dispersive delay lines in the microwave region [18, 19]. Experiments reported by Collins and Neilson [19] achieved compression of a one microsecond pulse to 14 nanoseconds, using a YIG dispersive delay line operating at a center frequency of 2280 MHz, and having a delay time that increased as a function of frequency. The frequency deviation was 125 MHz, yielding an input signal time-bandwidth product of 125. The 14 nanosecond compressed-pulse width was attributable to the weighted frequency response of the YIG apparatus. Range sidelobes were below the level of the $(\sin x)/x$ matched-filter compressed pulse. The YIG device described by Collins and Neilson was a rod 10 mm long and 3 mm wide, axially magnetized in a field of 1050 oersteds.

14.6 Optical Signal Processing—Spatial Filtering

Radar applications requiring simultaneous extraction of range and velocity information normally are implemented with a bank of matched filters, as pictured in Fig. 9.1. In some cases Doppler compensation following a single matched filter can yield equivalent results over a range of Doppler frequencies (an example for a binary phase coded waveform is discussed in Section 8.7). Nevertheless, the electrical filters described earlier have time as the only independent variable, and many filters that differ in center frequency are needed to process the received waveform information in time and frequency. Optical filters, on the other hand, can be designed to have two independent variables, and with proper implementation these can be realized as time and frequency. Thus, the optical filter has the potentialities of achieving a matched filter bank in one compact structure. The relation of optical filtering to communication and radar systems theory has been discussed by several authors [20–26]. O'Neill [22] gives several examples of filtering of optical signals in noise.

Lenses exhibit, to a first order approximation, a Fourier transform property that is applicable to matched-filter synthesis. A two-dimensional transform of an object placed in the front focal plane of such a lens can be obtained at the back focal plane [27]. This is illustrated in Fig. 14.19. The lenses shown are assumed to be of good quality and used in the region near the optical axis where best performance is obtained. The lens L_c serves to collimate monochromatic light from a point source (mercury arc reimaged through a pinhole, etc.) that illuminates an object (e.g., a grating pattern) located at the front focal plane of lens L_1. An aperture stop can be used to restrict the collimated light beam to the central region of the lens. The light crossing this plane is spatially coherent; that is the relative spatial phases of the light waves in the optical system are not functions of time. In the back focal plane of L_1 the amplitude and phase distribution of the light in the x-direction, assuming that the diffraction angle of the lens

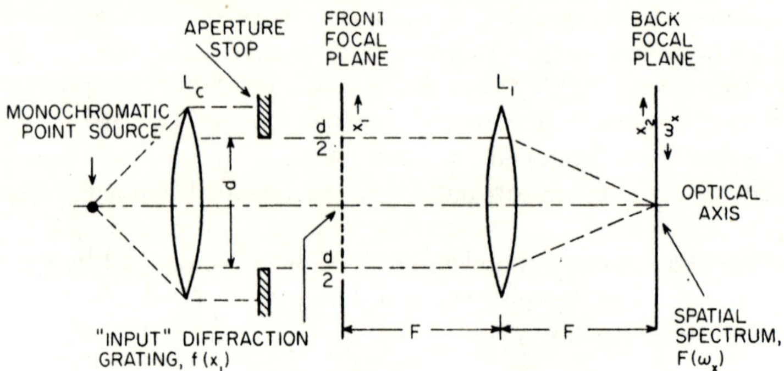

FIG. 14.19   Lens system for optical Fourier transform.

is small, is given by

$$F(\omega_x) = \int_{-d/2}^{d/2} f(x_1) \exp[-j\omega_x x_1]\, dx_1 \qquad (14\text{-}24)$$

where $|f(x_1)|^2$ is the transmission function in the x-direction of the object in the front focal plane, $\omega_x = 2\pi x_2/F\lambda_L$, the spatial "frequency" in the back focal plane, F the focal length of lens, and λ_L the wavelength of the monochromatic light.

Equation (14-24) can be recognized as the Fourier transform of the "input" function of duration d in the front focal plane. As an example, if $f(x_1)$ is a constant function with a uniform phase over the length d, then $F(\omega_x)$ is given by

$$d\left[\frac{\sin \tfrac{1}{2}\omega_x d}{\tfrac{1}{2}\omega_x d}\right] \qquad (14\text{-}25)$$

When the input function is described in both dimensions of the front focal plane then the Fourier transform is given by

$$F(\omega_x, \omega_y) = \int\!\!\int_{-d/2}^{d/2} f(x_1, y_1) \exp[-j(\omega_x x_1 + \omega_y y_1)]\, dx_1\, dy_1 \qquad (14\text{-}26)$$

For $|f(x_1, y_1)|^2$ a uniform transmission across a circular cross section of diameter d the Fourier transform obtained is[1]

$$F(\omega_x, \omega_y) = \frac{\pi d^2}{2\lambda_L F} \frac{J_1[(d/2)\sqrt{\omega_x^2 + \omega_y^2}]}{(d/2)\sqrt{\omega_x^2 + \omega_y^2}} \qquad (14\text{-}27)$$

---

[1] Equations (14-25) and (14-27) must be finite functions as a result of the physical limitations of the lens system and the approximations involved in (14-24). In the context of matched filtering this is equivalent to assuming both time limited and bandlimited signals. This is not of serious consequence if the extent of the frequency plane is sufficient to include the major portion of the spectrum.

Light detectors actually respond to the light intensity, so that the functions observed by physical devices would be $|F(\omega_x)|^2$ and $|F(\omega_x, \omega_y)|^2$, respectively, for (14-24) and (14-26). These two equations can be considered as descriptive of the "spatial frequency" spectrum of either one-dimensional or two-dimensional input functions.

As with the spectra of time functions, these spatial spectra can be modified by inserting a filter $H(\omega_x, \omega_y)$ in the back focal plane of L_1, or by a "modulation" function placed in the front focal plane. For example, in the one-dimensional case an x-axis weighting function, of the types discussed in Chapter 7, inserted into the front focal plane will cause the spectral distribution in the back focal plane to broaden and have lower sidelobe levels. After the spectrum is modified the amplitude and phase distribution in the "frequency" plane is given by $S(\omega_x, \omega_y) = F(\omega_x, \omega_y) \times H(\omega_x, \omega_y)$. A second lens L_2 added to the configuration, as shown in Fig. 14.20, yields a Fourier transform in the back focal plane of L_2 (output plane):

$$s(x_3, y_3) = \frac{1}{4\pi^2} \iint S(\omega_x, \omega_y) \exp[-j(\omega_x x_3 + \omega_y y_3)]\, d\omega_x\, d\omega_y \quad (14\text{-}28)$$

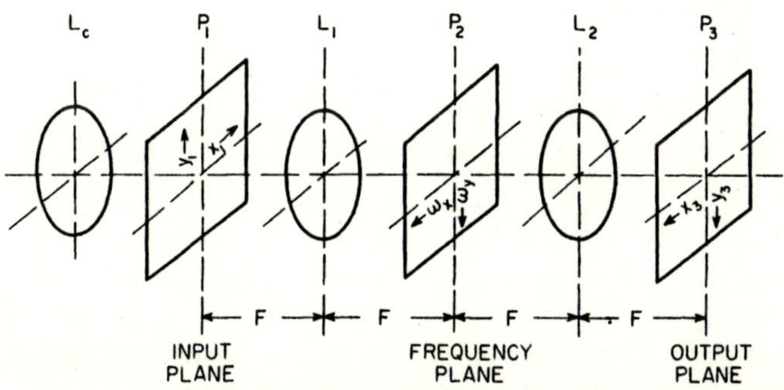

Fig. 14.20   Lens system for obtaining successive Fourier transforms.

Equation (14-28) is not the usual inverse Fourier transform realized in electrical filters, but is an additional direct Fourier transform. However, by recognizing the coordinate inversion that takes place between the input and output planes one can redefine (14-28) as an inverse transform. This is merely the normal image inversion obtained with a lens imaging system. If more than one input signal is present then the output is the superposition of the outputs caused by each individual input, provided that saturation does not occur.

Matched filtering of the input function can be achieved with the type of optical system described above by inserting a filter function in the spectral plane that meets the criterion [23, 28]:

$$H(\omega_x, \omega_y) = \frac{kF^*(\omega_x, \omega_y)}{N(\omega_x, \omega_y)} \tag{14-29a}$$

where $N(\omega_x, \omega_y)$ is the spatial noise density distribution and k the matched-filter normalizing constant. When $N(\omega_x, \omega_y)$ is white and Gaussian then

$$H(\omega_x, \omega_y) = k'F^*(\omega_x, \omega_y) = k'|F^*(\omega_x, \omega_y)| \exp[j\beta(\omega_x, \omega_y)] \tag{14-29b}$$

One of the advantages of an optical implementation is that the amplitude and phase factors of $H(\omega_x, \omega_y)$ can be realized separately; this is not the usual case for electrical filters. $|H(\omega_x, \omega_y)|$ is obtained by controlling the transmissivity of a transparency placed in the frequency plane, and $\beta(\omega_x, \omega_y)$ is obtained by varying the thickness of the transparency.

A second method of realizing the optical matched filter is to place a reference function $h(x_1, y_1)$ in the input plane such that [23]

$$h(x_1, y_1) = f(-x_1, -y_1) \tag{14-30}$$

The matched filtering is achieved by physically sliding $h(x_1, y_1)$ or $f(x_1, y_1)$ past the other function in the front focal plane of L_1. Taking the output function on the optical axis (i.e., $\omega_x = 0$, $\omega_y = 0$) in the back focal plane results in, from (14-26)

$$g(x_1', y_1') = \int\limits_{-d/2}^{d/2}\!\!\!\int f(x_1, y_1)f(x_1' - x_1, y_1' - y_1)\,dx_1\,dy_1 \tag{14-31}$$

where x_1', y_1' are the displacement variables. Equation (14-31) is the usual convolution integral formulation of matched filtering. As the input or reference function is slid by the stationary function a continuous readout of $|g(x_1', y_1')|^2$ can be obtained by a light detector positioned on the optical axis. This becomes equivalent to a matched filter followed by a square-law detector. This latter approach is the basis for an important implementation that makes use of ultrasonic light modulators.

The implementations described above appear to offer the exact analogs of electrical matched filtering if the space coordinates in the input plane can be converted into equivalent time units. Since time is required in only one dimension this is accomplished by causing an electrical input signal to "write" a continuous grating function on a recording film that is being transported in the x-direction, as shown in Fig. 14.21. When this is done both techniques described above yield outputs that are also functions of time.

A bank of matched filters is realized by using the y-direction to represent Doppler shift, and performing Fourier transforms only in the x-direction

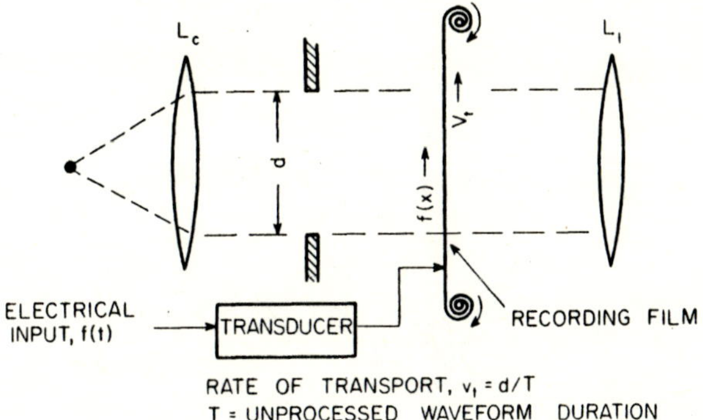

FIG. 14.21   Conversion of temporal waveform to spatial waveform (after Lambert [25]).

[23]. This is done with the combination of spherical and cylindrical lenses shown in Fig. 14.22. Since the filtering operations involve only one-dimensional Fourier transforms a line source, rather than a point source, of light can be used. One can think of the input function being sampled, as shown in Fig. 14.23, with each sample controlling the transmissivity of a grating line produced on the recording film being transported through the input plane. The bank of matched-filter functions in the frequency plane are arranged as shown. The cylindrical lens results in a one-dimensional Fourier transform. Thus, the combination of a spherical and cylindrical lens as shown in Fig. 14.22 causes a double Fourier transform in the y-direction (or no transformation, except for an inversion of the y-axis), but only a single Fourier transformation in the x-direction. The transform of

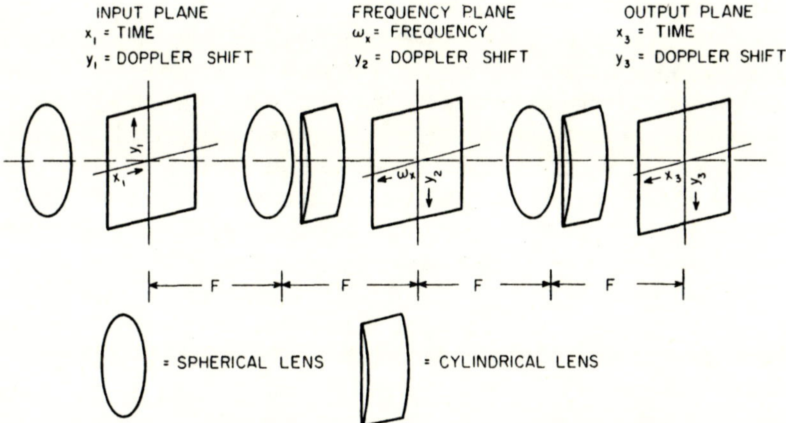

FIG. 14.22   Lens system for implementing a bank of matched filters.

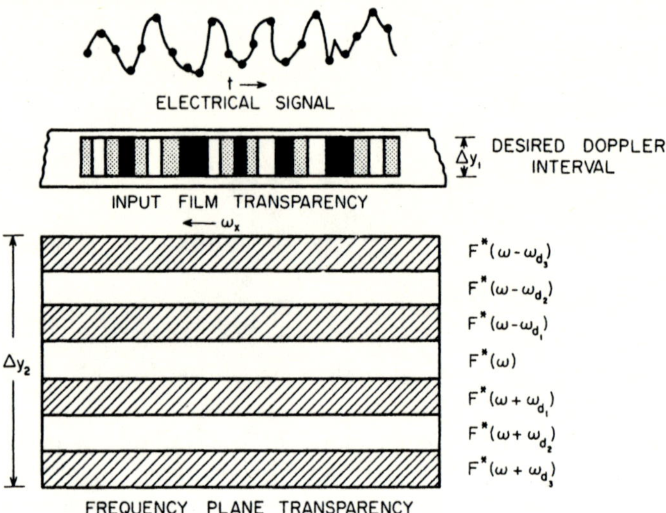

FIG. 14.23 Sampling of input signal to produce input plane film recorded function, and frequency plane matched-filter bank.

a particular Doppler shifted input signal is incident on all of the filter functions in the frequency plane, but its phase and amplitude distribution are matched to only one of the filters in the bank. Light detectors stacked in the y-direction at the output plane will give the time outputs from each filter in the bank as the recording film is transported through the input plane, achieving maximum output when the signal function fills the input plane. One of the detectors will yield the correct autocorrelation output, whereas the others will yield cross correlation, or off-Doppler, outputs. Thus, this operation is similar to that of the electrical matched filter bank of Fig. 9.1. An alternate configuration is to stack the desired set of Doppler shifted reference function gratings in the input plane, and to read out the multiple matched-filter signals with light detectors appropriately positioned behind a vertical slit placed along the y-axis in the back focal plane of L_1, this being an extension of the one-dimensional case of convolution filtering.

For most radar applications it is important that the recording of the input signals be accomplished in essentially real time. That is, no more delay in this function can be tolerated than the maximum time during which the recorded input signals are still useful as information outputs of the radar system. In some cases this may be less than a few seconds. Photographic film has been used as an input signal recording medium in applications where the developing time of the film is not restrictive. Cutrona *et al.* [23] cite the filming of the input functions directly off of the face of a cathode ray tube display. Thermoplastic films offer the possibility of essentially real time recording and use of the input data [29]. However, the transport velocities possible with most tape recording techniques are such that these methods are best suited to relatively long duration waveforms and relatively narrow bandwidths if large time-bandwidths products are desired. Lambert states that film transport velocities of 200 inches per second are possible at present. The use of film and tape has the advantage of permitting the use of many reference functions or matched-filter channels since the resolution capability of these signal recording techniques is typically between 5000 and 25,000 lines per inch [25]. However, the instabilities or time jitter of the transport mechanism may limit the practical resolution capability, or lines per inch, for the larger transport velocities. Ultrasonic light modulators can be used to obtain a real time matched-filter implementation, and are discussed further in Section 14.8.

Inherent in the discussion presented above has been the assumption that optical signals and filters can be treated as exact analogs of the electrical signals and filters discussed earlier. This is not precisely true, since optical transmission functions must always be positive. How this effects the implementation of the optical matched filter is briefly described in the following section.

14.7 Optical Matched Filters for Radar Application

It is assumed that the input signal to the optical matched filter is a pulsed signal with a constant carrier frequency described by

$$f(t) = a_1 \cos \omega_0 t, \qquad -T/2 < t < T/2 \qquad (14\text{-}32)$$

Equation (14-32) represents the waveform to be recorded on the light modulator mechanism that is being transported across the input plane. Since optical transparencies can only record positive functions this signal cannot be directly transposed as a grating mask function. One method of circumventing this restriction is to add a bias to the input signal so that all portions of the transmission function are positive. Thus, (14-32) can be rewritten as

$$f(t) = a_0 + a_1 \cos \omega_0 t, \qquad a_0 > a_1 \qquad (14\text{-}33)$$

The electrical to optical transducer, such as the film transport discussed above, converts this input signal to a spatial function $f(x_1)$ (Fig. 14.21), and the spatial spectrum appears in the back focal plane of L_1. This spectral function has a dc term a_0 plus two sidebands of amplitude $a_1/2$, as shown in Fig. 14.24. The intensity distribution would be, of course, a_0^2 for the dc term and $a_1^2/4$ for the sideband terms. The sideband spectral functions are

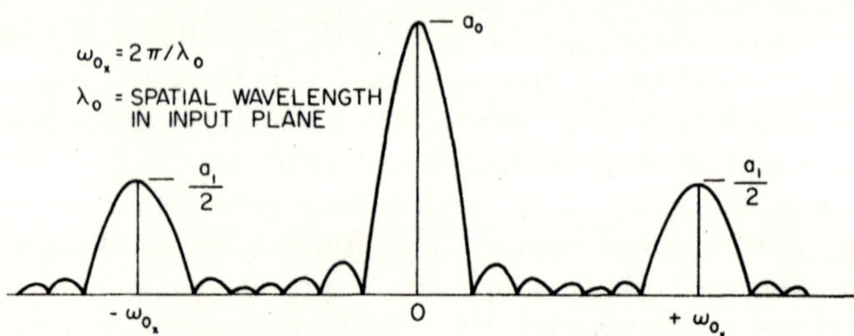

FIG. 14.24   Amplitude distribution of light in the frequency plane resulting from biased input function.

designated as diffraction orders in optics, and for the case discussed above are the -1st order, zero order and $+1$st order. Only one of the sidebands is required to be representative of the original signal (14-32), and light stops, or "bandpass" filters, can be inserted in the frequency plane to block the zero order and -1st order light. When this is done the light incident on the following transform plane is proportional to $f(x_1)$ in its complex form (magnitude and phase). A reference function can be placed in this next transform plane (a time plane) in order to perform the convolution operation, the final filter output being observed at $\omega_x = 0$ in a succeeding transform plane as illustrated in Fig. 14.25. The reference function will

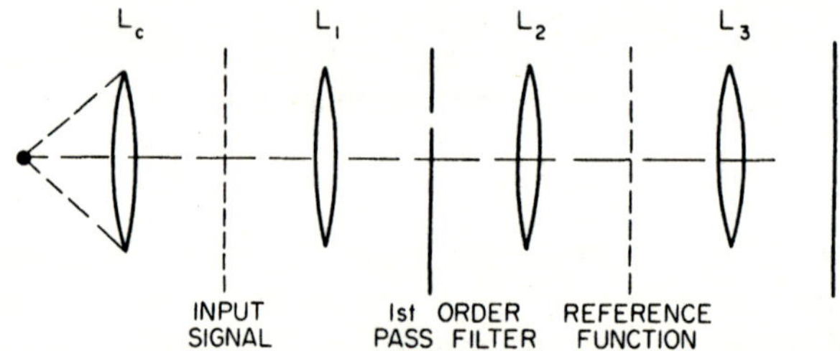

FIG. 14.25   Lens system and frequency plane stop for correlating input signal with reference function.

also have to be written with a bias attached to it. The light emerging from the reference function plane is thus, in general spatial coordinates,

$$f(x_1 - x_1')[b_0 + b_1 g(x)] \tag{14-34}$$

where $b_0 + b_1 g(x)$ is the biased reference function. Insertion of (14-34) into the convolution integral (14-31) yields the output amplitude as $f(x_1)$ is transported across the input plane. Since this output is observed at the zero frequency point in the output plane the convolution term arising from $b_0 f(x_1 - x_1')$ will only be objectionable if $f(x)$ has a dc component that is superimposed on the desired output. If this should happen to be the case then this additional light can be removed from the $\omega_x = 0$ position by rewriting the input and reference functions with a spatial carrier frequency such that they become

$$a_0 + a_1 f(x_1 - x'_1)\cos\omega_c x \qquad \text{and} \qquad b_0 + b_1 g(x)\cos\omega_c x$$

Multiplication of the above in the convolution integral, after blocking the a_0 term, yields

$$b_0 f(x)\cos\omega_c x + a_1 b_1 f(x_1 - x'_1)g(x)[\tfrac{1}{2} + \tfrac{1}{2}\cos 2\omega_c x] \qquad (14\text{-}35)$$

Now, only the term related to $\tfrac{1}{2}a_1 b_1 f(x_1 - x')g(x)$ will appear at the $\omega_x = 0$ position in the output plane, the other terms being displaced from this point in proportion to ω_c and $2\omega_c$. Examples of this type of process are discussed by Cutrona *et al.* [23]. A somewhat simpler implementation places both the input and reference functions in the input plane, with the order stops as before in the frequency plane. If either $f(x)$ or $g(x)$ are given a spatial carrier frequency ω_c, then observing the light at $\omega_x = \omega_c$ in the frequency plane of Fig. 14.19 will, from (14-24) (or (14-26) for the two dimensional case) yield a convolution integral output similar to (14-31). This output will appear at the shifted location in the frequency plane as the input function is transported past the reference function.

In most cases it will be easier to record the reference function on a film transparency as a set of black and white transmission levels, or grating lines, spaced in accordance with the modulation function, rather than as a continuous shading between these two extremes. Kozma and Kelly [30] show that this type of "clipped" reference function causes a loss of about 1.1 db relative to the matched-filter gain expected from the use of an exact replica function.

As an alternate to performing the convolution integration to obtain the matched-filter output, one can insert the appropriate matched filter in the first frequency plane. However, care must again be taken to position the filter transparency so that it operates on the appropriate sideband distribution while the other spectral components are filtered out. Particularly, the shot noise component of large zero order light terms that come through at the output will reduce the contrast, or dynamic range, of the signals that can be observed. Although the examples presented above were in terms of a simple pulse function it is obvious that the general approaches are equally suited to more complex types of waveform functions. Thus, in the case of a linear FM signal a linear FM reference function would be required, and the spectrum amplitude would be the Fresnel distribution discussed in earlier chapters. The spatial phase distribution in the frequency plane would be essentially square-law, requiring a parabolic curvature lens in this plane to provide the necessary phase retardation of the spatial light front if the matched filtering approach is adopted.

One of the chief difficulties in performing matched filtering in the spatial frequency plane is associated with the formation of the necessary phase retardation lenses if the signal modulation function is nonlinear or pseudorandom in phase or frequency. A technique for achieving a complex matched-filter function has been described by Vander Lugt [28]. This technique is based on the use of a Mach–Zehnder interferometer to inscribe the matched-filter function on photographic film, which of course can record only positive real functions. The configuration used by Vander Lugt is shown in Fig. 14.26. The light incident on the output plane from the monochromatic reference beam is

$$R(\omega_x, \omega_y) = |R(\omega_x, \omega_y)| \exp[j(b\omega_x + c\omega_y)] \qquad (14\text{-}36)$$

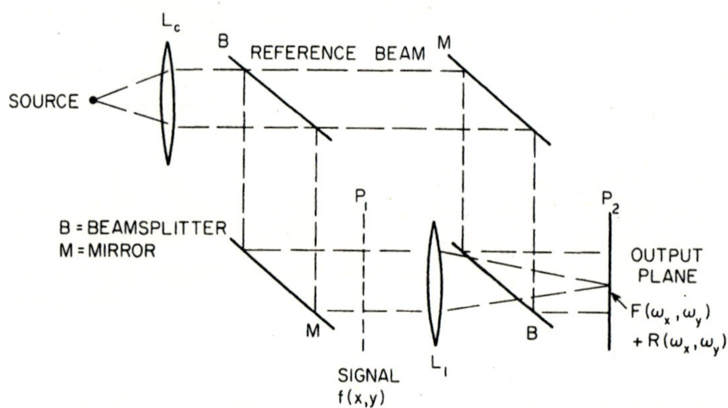

FIG. 14.26   Modified Mach–Zehnder interferometer for recording complex spatial filter functions (after Vander Lugt [28]).

The phase factor of (14-36), which is introduced by a phase plate in the reference beam, is linear in both the x and y dimensions of the output plane. The signal function $f(x, y)$, subject to the constraints discussed earlier, is inserted in the front focal plane of lens L_1, so that the light amplitude distribution in plane P_2 from the signal is

$$F(\omega_x, \omega_y) = |F(\omega_x, \omega_y)| \exp[j\beta(\omega_x, \omega_y)] \qquad (14\text{-}37)$$

The total observed output is thus:

$$|G(\omega_x, \omega_y)|^2 = |R(\omega_x, \omega_y) + F(\omega_x, \omega_y)|^2$$

$$= |R(\omega_x, \omega_y)|^2 + |F(\omega_x, \omega_y)|^2$$

$$+ R^*(\omega_x, \omega_y)F(\omega_x, \omega_y) + R(\omega_x, \omega_y)F^*(\omega_x, \omega_y) \qquad (14\text{-}38)$$

For the point source of monochromatic light $R(\omega_x, \omega_y) = \text{constant}, k$. Recalling that the matched-filter condition for white Gaussian noise is $H(\omega_x, \omega_y) = kF^*(\omega_x, \omega_y)$, (14-38) yields:

$$|G(\omega_x, \omega_y)|^2 = A(\omega_x, \omega_y) + H^*(\omega_x, \omega_y)\exp[-j(b\omega_x + c\omega_y)]$$
$$+ H(\omega_x, \omega_y)\exp[j(b\omega_x + c\omega_y)] \qquad (14\text{-}39)$$

where $A(\omega_x, \omega_y) = |R(\omega_x, \omega_y)|^2 + |F(\omega_x, \omega_y)|^2$. The last term of (14-39) contains the required matched-filter function. The entire expression given by (14-39) is recorded on film and then inserted in the frequency plane of an optical system of the type shown in Fig. 14.20, with the signal function placed in the input plane. The output function is, from (14-38) and (14-39),

$$g(x, y) = \frac{1}{4\pi^2}\int\!\!\int_{-\infty}^{\infty} F(\omega_x, \omega_y)A(\omega_x, \omega_y)\exp[j\{\omega_x x + \omega_y y\}]\, d\omega_x\, d\omega_y$$

$$+ \frac{1}{4\pi^2}\int\!\!\int_{-\infty}^{\infty} F(\omega_x, \omega_y)H^*(\omega_x, \omega_y)$$

$$\times \exp[j\{(x - b)\omega_x + (y - c)\omega_y\}]\, d\omega_x\, d\omega_y$$

$$+ \frac{1}{4\pi^2}\int\!\!\int_{-\infty}^{\infty} F(\omega_x, \omega_y)H(\omega_x, \omega_y)$$

$$\times \exp[j\{(x + b)\omega_x + (y + c)\omega_y\}]\, d\omega_x\, d\omega_y \qquad (14\text{-}40)$$

The last term of (14-40) represents the desired matched-filter output, for the recorded signal $f(x, y)$, centered at $x = -b$, $y = -c$. Vander Lugt states that for convenience $c = 0$, so that the desired term can be separated from the others by a light detector located at $x = -b$ if $b > x_{\text{in}}$, x_{in} being the total spatial duration of the input signal.

An alternate interferometer configuration for writing complex filter functions on film is described by Vander Lugt $et\ al.$ [31]. This technique makes use of the Rayleigh interferometer shown in Fig. 14.27. The lens L_r focuses the collimated light at a point $x = b$ in the front focal plane of lens L_1. This light amplitude is represented by the delta function $R\,\delta(x - b, y)$. The signal function $f(x, y)$ is also present in the front focal plane, so that

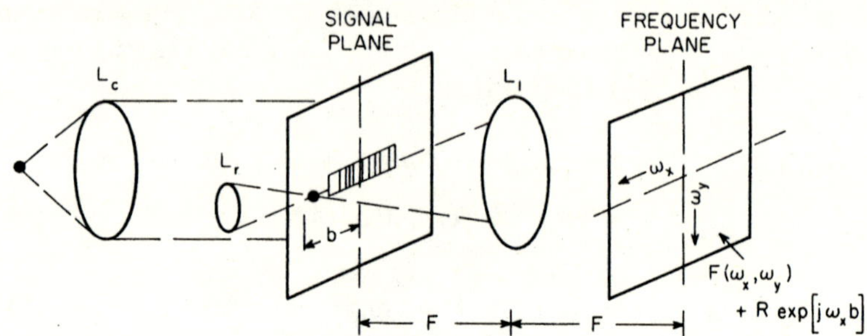

FIG. 14.27   Modified Rayleigh interferometer for recording complex spatial filter functions (after Vander Lugt *et al.* [31]).

the spatial spectrum in the back focal plane of L_1 is[1]

$$G(\omega_x, \omega_y) = \int\int_{-\infty}^{\infty} [f(x, y) + R\,\delta(x - b, y)]\exp[j(\omega_x x + \omega_y y)]\,dx\,dy$$

$$= F(\omega_x, \omega_y) + R\exp[j\omega_x b] \tag{14-41}$$

Photographic film located in the back focal plane of L_1 records

$$|G(\omega_x, \omega_y)|^2 = |F(\omega_x, \omega_y)|^2 + R^2 + R\exp[-j\omega_x b]F(\omega_x, \omega_y)$$

$$+ R\exp[j\omega_x b]F^*(\omega_x, \omega_y) \tag{14-42}$$

The last term of (14-42) contains the matched-filter function multiplied by a constant and a linear phase term. As in the case of the Mach–Zehnder interferometer, the recorded film function is placed in the frequency plane of Fig. 14.20 and, following the development given above, the desired output is centered at $x = b$ in the output plane, with the same condition being imposed on the magnitude of b as before.

14.8   The Ultrasonic Light Modulator

The use of photographic film to record and transport signals through the input plane of the optical matched filter usually does not yield outputs that can be classified as real time signals. A class of optical radar matched filters that operates in real time has been based on the use of ultrasonic light modulators (ULM) to achieve the equivalent of a moving signal grating function in the input plane. The ultrasonic light modulator is a

---

[1] The limits of integration in both (14-40) and (14-41) are finite, being controlled in practice by the finite dimensions of the lenses, the collimated light beam and the deviation of the input signal.

light transparent ultrasonic delay line (typically water, quartz, or glass) placed transverse to the direction of the collimated light beam, as shown in Fig. 14.28. The electrical signal that represents the input waveform excites the electromechanical transducer at the ULM input. The signal is usually converted to a longitudinal mode ultrasonic wave that propagates down the ULM. As the wave progresses down the delay line the resulting changes in the optical density of the line cause a variation in the index of refraction of the delay medium that is proportional to the exciting signal [32]. This in turn results in the spatial phase modulation of the collimated light passing through the ULM. This phase modulation is in direct proportion to the changes in the index of refraction, and thus to the

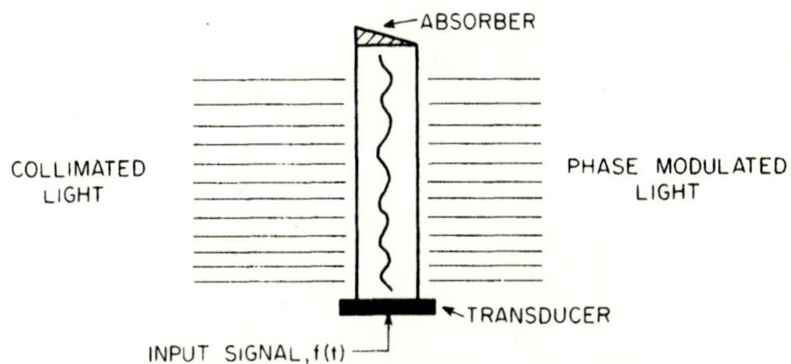

FIG. 14.28   Ultrasonic light modulator (ULM).

variations in the signal function as it passes a fixed point in the delay line. For the general input function the spatially phase modulated light can be described by:

$$f(x, t) = \exp[-ja_0 \cos\{\omega_c(x - v_s t) + \theta(x - v_s t)\}] \qquad (14\text{-}43)$$

where a_0 is the peak phase modulation, $\theta(x - v_s t)$ the modulation function of the signal propagating down the ULM, v_s the velocity of ultrasonic propagation in the ULM medium, and $-D/2 < x - v_s t < D/2$ where D is the spatial duration of the ultrasonic signal.

Equation (14-43) can be expressed as diffraction orders whose amplitudes are given by the usual Bessel function expansion (see Chapter 11). The frequency of the light contained in the respective diffraction orders is shifted by $\pm\omega_c$, $\pm 2\omega_c$, etc. When the spatial phase modulation is small then only the undiffracted light (J_0 term) and the first "sideband" orders (J_{-1}, J_1 terms) are significant. The two orders of diffracted light travel as plane waves at an angle $\pm\lambda/\Lambda$ to the zero order light (λ is the light wavelength, Λ the sound wavelength). This, plus the light frequency shift in each order, allows a region beyond the ULM to be found where the phase modulation is converted to spatial amplitude modulation that can then be correlated against an amplitude grating function, as described earlier for the example of the time signal written on a moving film.

Several techniques for utilizing the ULM in radar matched filters have been discussed in the literature. Reich and Slobodin [33, 34] have applied the ULM in the general type of system illustrated by Fig. 14.28. Lambert *et al.* [35] and Arm *et al.* [36] and Slobodin [37] describe uses of the ULM in a convolution type processor that contains the matched reference function in the input plane, as shown in Fig. 14.29. The reference grating function

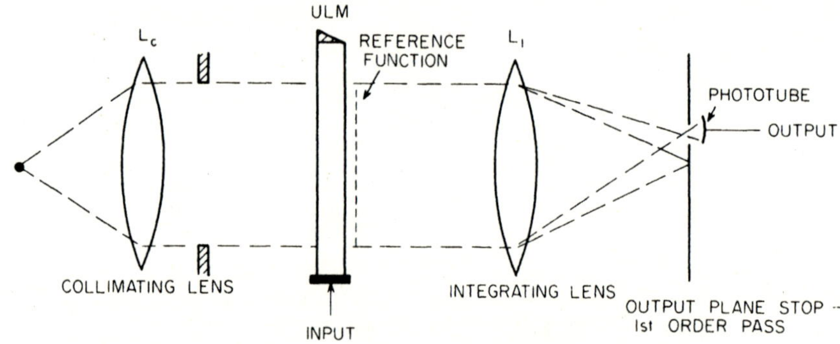

FIG. 14.29   Implementation of the basic ULM signal processor.

can be placed within or adjacent to the delay medium, which was water in the implementations discussed. In the optical filter described by Slobodin linear FM and pseudorandom FM signals with time-bandwidth products of 120 were obtained with a signal duration of 60 microseconds and a bandwidth of 2 MHz. The compressed-pulse width was 0.5 microsecond. The reference grating function was obtained by a photographic technique, and the optical system was used to form the coded waveform by impulsing the ULM input. Figure 14.30(a) shows the expanded linear FM pulse obtained by Slobodin, and Fig. 14.30(b) the associated compressed pulse. The tapering of the expanded pulse was caused by the overall bandpass characteristics of the processor. Figures 14.31(a) and 14.31(b) illustrate the expanded and compressed pulses for the pseudorandom FM signal. The compressed pulses at the photo tube output contained the ultrasonic carrier, so that further detection is required for video presentation of the signal. Arm *et al.* [36] discuss the use of a frequency offset in the reference function in order to achieve a greater separation of the desired first order

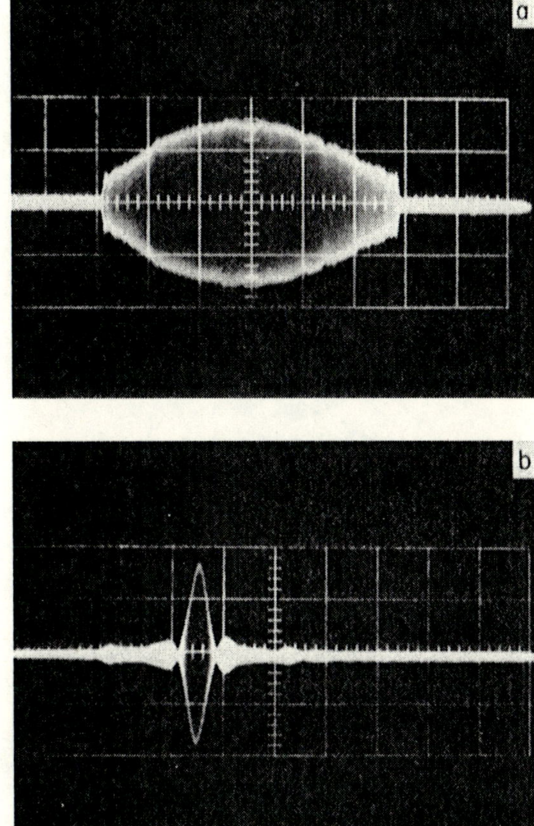

FIG. 14.30 Expanded- and compressed-pulse waveforms for linear FM signal obtained with a ULM optical correlator. (a) Expanded pulse (scale $= 10\,\mu\text{sec/div}$). (b) Compressed pulse (scale $= 2\,\mu\text{sec/div}$). (Courtesy of L. Slobodin [37], and Lockheed Electronics Company, Plainfield, New Jersey.)

light from the other light components if the interference among them is excessive. This results in an improved dynamic range capability at the processor output, with signal ranges of 40 to 60 db being cited as possible.

For the special case of the linear FM signal it is possible to use the ULM as a pulse-compression filter in a much simplified version that does not require the use of the reference grating function. Hoefer [24] notes that the first order light diffracted from the linear "FM" grating is convergent as shown in Fig. 14.32. This is also discussed by Gerig and Montague [38] who propose the implementation illustrated in Fig. 14.33 based on this phenomenon. The lens L_2 focuses the zero order light on the optical axis.

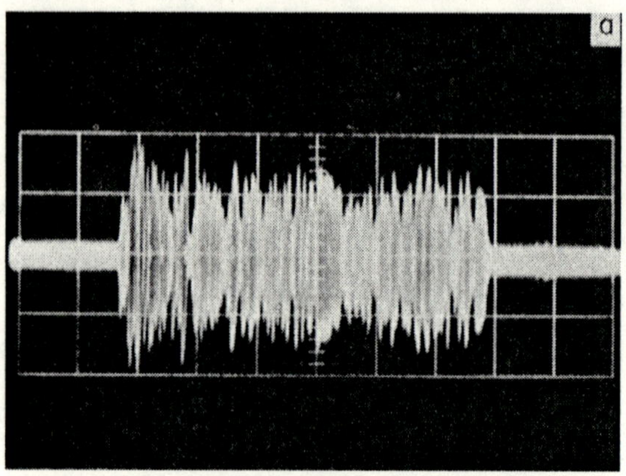

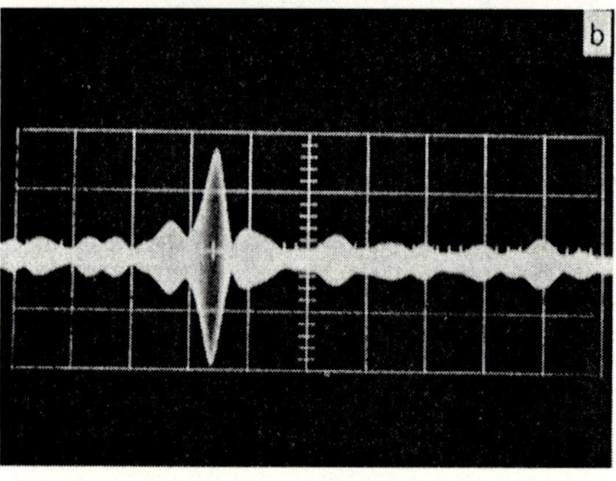

FIG. 14.31 Expanded- and compressed-pulse waveforms for pseudorandom FM signal obtained with a ULM optical correlator. (a) Expanded pulse (scale = 10 μsec/div). (b) Compressed pulse (scale = 2 μsec/div). (Courtesy of L. Slobodin |37|, and Lockheed Electronics Company, Plainfield, New Jersey.)

The convergent property of the first order light causes it to come to a focus at a point off the optical axis and at an equivalent focal length F_{eq}, described by

$$\frac{1}{F_{eq}} = \frac{1}{F_2} + \frac{1}{X_0} \tag{14-44}$$

where $X_0 = kc^2 T/\Delta\omega$ is the focal distance of the first order light in the absence of lens L_2, k the radian wave number of the light, c the acoustic velocity in the ULM, T the signal duration, and $\Delta\omega$ the radian bandwidth of the linear FM signal.

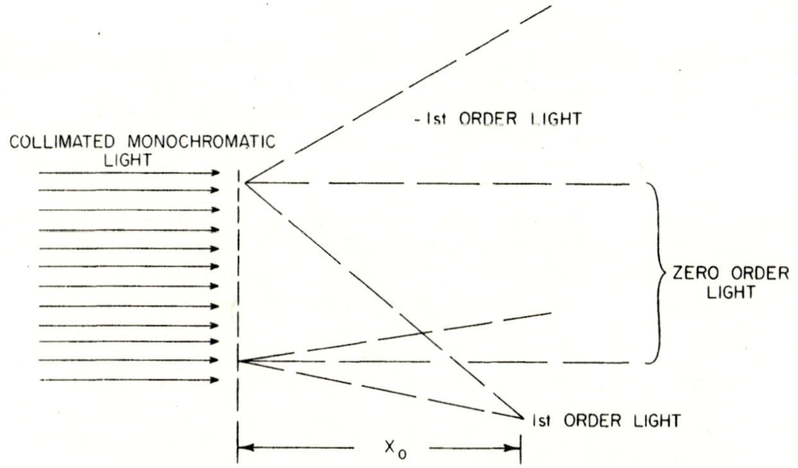

FIG. 14.32 Diffraction of light by a linear FM grating (after Hoefer [24]).

As the linear FM signal enters the ULM the convergence point of the first order light moves down the plane S_2, being coincident with the focal point shown when fully entered into the delay line. A slit placed in a stop in plane S_2 at this point will transmit the convergent light to the photodetector located behind the slit, converting the received light intensity to a voltage output. The first order convergence point sweeps by the slit in a time interval given by $\tau = 2\pi/\Delta\omega$. Thus, the duration of the peak response from the photodetector will have the correct dimensions of the compressed pulse.

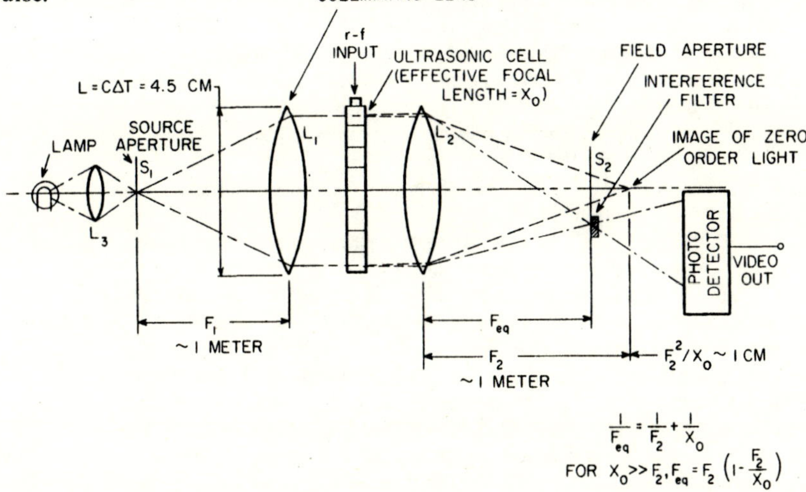

FIG. 14.33 Linear FM optical processor based on diffraction properties of a linear FM grating (courtesy of J. S. Gerig and H. Montague [38]).

REFERENCES

1. C. G. Montgomery, R. H. Dicke, and E. M. Purcell, "Principles of Microwave Circuits." McGraw-Hill, New York, 1948.

2. G. L. Matthaei, L. Young, and E. M. T. Jones, "Microwave Filters, Impedance Matching Networks, and Coupling Structures." McGraw-Hill, New York, 1964.

3. A. LeBlond and G. Mourier, Investigation of lines with periodic bar structure for UHF valves, Parts I and II, *Ann. Radioélec.* **180**, 311 (1954).

4. V. E. Dunn, Realization of microwave pulse-compression filters by means of folded-tape meander lines, Stanford Univ., Stanford Electron. Lab., Stanford, California, Rept. TR-557-3 (October, 1962).

5. W. R. Kincheloe, Jr., The measurement of frequency with scanning spectrum analyzers, *ibid.*, Rept. TR-557-2 (October, 1962).

6. J. G. Ferguson and C. W. Barrett, Stripline all-pass networks for pulse compression, *Proc. Symp. Pulse-Compression Tech. Appl.* TRR-62-580, Rome Air Develop. Center, New York, (1963).

7. F. J. Mueller and R. L. Goodwin, A wideband microwave compressive receiver, *IRE Intern. Conv. Record, Pt.* 3, 103–119 (1962).

8. M. Ito, Dispersion of very short microwave pulses in waveguide, *IEEE Trans.* **MTT-13**, 357–364 (1965).

9. G. P. Ohman, Getting high range resolution with pulse-compression radar, *Electronics* **28**, 53–59 (October 7, 1960).

10. K. Walther, Pulse-compression in acoustic waveguide, *J. Acoust. Soc. Am.* **28**, 681–686 (1961).

11. W. J. Albersheim, Electronic wave equalization system, U.S. Patent No. 2,863,127 (December 2, 1958).

12. C. C. H. Tang, Delay equalization by tapered cutoff waveguides, *IEEE Trans.* **MTT-12**, 607–615 (1964).

13. P. J. B. Clarricoats, C. D. Hannaford, and A. K. Bunn, Proposed microwave delay lines for pulse-compression radar, *Electronics Letters* **1**, 175–176 (1965).

14. E. N. Torgow, Equalization of waveguide delay distortion, *IEEE Trans.* **MTT-13**, 756–762 (1965).

15. J. R. Pierce, "Traveling Wave Tubes." Van Nostrand, Princeton, New Jersey, 1950.

16. V. E. Dunn, A time-compression filter employing a microwave helix, Stanford Univ., Stanford Electron. Lab., Stanford, California, Rept. TR-557-1 (October, 1960).

17. A. Kiis, Microwave pulse-compression techniques, M.S.E.E. Thesis, Polytechnic Inst. Brooklyn, New York (1962).

18. R. W. Damon and H. Van de Vaart, Spin waves and magnetoelastic waves in YIG, *Proc. IEEE* **53**, 348–354 (1965).

19. J. H. Collins and G. G. Neilson, Microwave pulse compression utilizing an yttrium-iron-garnet delay line, *Electronics Letters* **1**, 234 (1965).

20. P. Elias, Optics and communication theory, *J. Optical Soc. Am.* **43**, 229–232 (1953).

21. T. P. Cheatham and A. Kohlenberg, Optical filters—their equivalence to and differences from electrical networks, *IRE Intern. Conv. Record, Pt.* 4, 6–12 (1954).

22. E. L. O'Neill, Spatial filtering in optics, *IRE Trans.* **IT-2**, 56–65 (1956).

23. L. J. Cutrona, E. N. Leith, C. J. Palermo and L. J. Porcello, Optical data processing and filtering systems, *IRE Trans.* **IT-6**, 386–400 (1960).

24. W. G. Hoefer, Optical processing of simulated IF pulse Doppler signals, *IRE Trans.* **MIL 6**, 174–178 (1962).

25. L. B. Lambert, Optical Correlation, *in* "Modern Radar: Analysis, Evaluation and System Design" (R. S. Berkowitz, ed.), pp. 245–273. Wiley, New York, 1965.

26. J. W. Goodman, The role of coherent optics in electrical engineering, *Proc. WESCON* (1965).

27. J. Rhodes, Analysis and synthesis of optical images, *Am. J. Physics* **21**, 337–343 (1953).

28. A. Vander Lugt, Signal detection by complex spatial filtering, *IEEE Trans.* **IT-10**, 139–145 (1964).

29. W. E. Glenn, Thermoplastic recording, *J. Appl. Phys.* **30**, 1870–1873 (1959).

30. A. Kozma and D. L. Kelly, Spatial filtering for detection of signals submerged in noise, *Appl. Optics* **4**, 387–392 (1965).

31. A. Vander Lugt, F. B. Rotz, and A. Klooster, Character-reading by optical spatial filtering, *in* "Optical and Electro-optical Information Processing" (J. T. Tippet *et al.*, eds.), pp. 125–141. M.I.T. Press, Cambridge, Massachusetts, 1965.

32. A. B. Bhatia and W. J. Noble, Diffraction of light by ultrasonic waves, *Proc. Royal Soc.* **A220**, 356–385 (1953).

33. A. Reich and L. Slobodin, Optical pulse expansion/compression, presented at *Natl. Aerospace Electron. Conf.*, Dayton, Ohio (May, 1961).

34. L. Slobodin and A. Reich, Expansion and compression of electronic pulses by optical correlation, U.S. Patent No. 3,189,746 (June 15, 1965).

35. L. Lambert, M. Arm, and I. Weissman, A radar technique using an electro-optical two-dimensional filter, presented at *Natl. Conv. on Military Electron.*, Washington, D.C., 1960.

36. M. Arm, L. Lambert, and I. Weissman, Optical correlation techniques for radar pulse compression, *Proc. IEEE (Correspondence)* **52**, 842 (1964).

37. L. Slobodin, Optical correlation technique, *Proc. IEEE (Correspondence)* **51**, 1782 (1963).

38. J. S. Gerig and H. Montague, A simple optical filter for Chirp radar, *Proc. IEEE (Correspondence)* **52**, 1753 (1964).

Bibliography

The following list includes items of interest for supplemental reading, as well as important references not used directly in the text that will provide the basis for advanced study.

1. S. Applebaum and P. W. Howells, Waveform design for tomorrow's radar, *Space/Aeronaut.* **32**, 186–190 (1959).

2. E. D. Banta, A generalized ambiguity criterion, *Proc. IEEE (Correspondence)* **52**, 976 (1964).

3. M. Bernfeld and C. E. Cook, Matched filtering and pulse compression, *Sperry Eng. Rev.* **15**, 44–53 (1963).

4. M. Bernfeld and C. E. Cook, Measurement of radar parameters, *Electro-Technol., Sci. Eng. Ser.* **78** (March, 1967).

5. S. E. Bogotch and C. E. Cook, The effect of limiting on the detectability of partially time-coincident pulse-compression signals, *IEEE Trans.* **MIL-9**, 17–24 (1965).

6. E. Brookner, Optimum clutter rejection, *IEEE Trans.*, **IT-11**, 597–599 (1965).

7. E. Brookner, Effect of ionosphere on radar waveforms, *J. Franklin Inst.*, **280**, 1–22 (1965).

8. F. C. Bequaert, Some limiting factors in radar measurement accuracy, *Proc. Natl. Conv. on Milit. Electron.*, 79–82 (1963).

9. C. R. Cahn and J. E. Stalder, Bounds for correlation peaks of periodic digital sequences, *Proc. IEEE (Correspondence)*, **52**, 1262–1263 (1964).

10. W. J. Caputi, A technique for the time-transformation of signals and its application to directional systems, *Radio Electronic Eng.*, **29**, 135–142 (1965).

11. J. J. Connolly, Fundamental accuracy in the simultaneous measurement of transverse and radial trajectories of radar targets, *Proc. 1963 Natl. Conv. Milit. Electron.*, 64–68.

12. J. DeLorenzo and J. V. DiFranco, Response of a linear FM matched filter to gated noise, *Proc. IEEE (Letters)*, **54**, 905–906 (1966).

13. A. Detape, Codage et décodage en phase d'impulsions longues affectées d'effet Doppler, *Ann. Radioelectricité* **20**, (79) 3–13 (1965).

14. J. V. DiFranco, Closed form solution for the output of a finite-bandwidth pulse-compression filter, *Proc. IRE (Correspondence)* **49**, 1086 (1961).

15. C. W. Erickson, Clutter cancelling in autocorrelation functions by binary sequence pairing, U.S. Navy Electron. Lab. Res. Rept. 1047 (June, 1961).

16. E. B. Felstead, Progress report on an optical matched filter for radar signal processing, Canad. General Electric Co., Ltd., Toronto, Canada, Tech. Memo. RQ65EE37 (September, 1965).

17. S. M. Fomemko, Signal processing system, U.S. Patent No. 3,211,898, October, 1965.

18. R. A. Ford, A transversal filter for sidelobe reduction, *Proc. IEEE (Letters)* **54**, 703 (1966).

19. D. Fryberger, On the use of pulse compression for the enhancement of radar echoes from diffuse targets, *Proc. IRE (Correspondence)* **50**, 1993–1994 (1962).

20. D. A. George, Matched filters for interfering signals, *IEEE Trans. (Correspondence)* **IT-11**, 153–154 (1965).

21. W. Gersch, The limit of radar resolution for a distributed target, *Proc. IEEE (Correspondence)* **51**, 1787–1788 (1963).

22. N. R. Gillespie, J. B. Higley, and N. MacKinnon, The evolution and application of coherent radar systems, *IRE Trans.* **MIL-5**, 131–139 (1961).

23. M. Golay, Complementary series, *IRE Trans.* **IT-7**, 82–87 (1961).

24. R. O. Harger, An optimum design of ambiguity function, antenna pattern and signal for side-looking radars, *IEEE Trans.* **MIL-9**, 264–278 (1965).

25. H. S. Hewitt, A design procedure for tapped-delay-line compression filters, Stanford Electronics Lab., Stanford, California, Tech. Rept. 1965-1 (September, 1965).

26. N. F. Izzo, Optical correlation technique using a variable reference function, *Proc. IEEE (Correspondence)* **53**, 1740–1741 (1965).

27. R. Kronert, Impulsverdichtung, parts I and II, *Nachrichtentechnik* **7**, 148–152, 162 (1957); 305–308 (1957).

28. R. M. Lerner, Signals with uniform ambiguity functions, *IRE Intern. Conv. Record, Pt. 4*, 27–36 (1958).

29. R. M. Lerner, Means for counting "effective" numbers of objects or durations of signals, *Proc. IRE (Correspondence)* **47**, 1653 (1959).

30. R. M. Lerner, Signals having good correlation functions, *Proc. 1961 WESCON Conv.* paper 9/3.

31. E. S. Lurin, Digital pulse compression using polyphase codes, *Proc. IEEE (Correspondence)* **51**, 1262–1263 (1963).

33. T. Makimoto and T. Sueta, A three-dimensional radar of frequency-modulated pulse system, *Proc. IEEE (Correspondence)* **52**, 727–728 (1964).

34. D. E. Miller and M. R. Parker, Dispersion by plane wave excitation of piezoelectric transducer arrays, *Proc. IEEE (Letters)* **54**, 891–892 (1966).

35. H. Miyakawa and M. A. Husayn, Design of transistorized flat and linear delay circuits, Faculty of Eng., Univ. Tokyo, Tokyo, Japan (March, 1965).

36. F. E. Nathanson, Time sidelobes in a combined phase-code and pulse-Doppler system, *Proc. IEEE (Correspondence)* **53**, 1775–1776 (1965).

37. N. J. Nillson, On the optimum range resolution of radar signals in noise, *IRE Trans.* **IT-7**, 245–253 (1961).

38. T. R. O'Meara, Some applications of matched-filter concepts to seismic reflection profiling, Hughes Res. Lab., Malibu, California, Res. Rept. 304 (April, 1964).

39. T. R. O'Meara, The resonator-circulator all-pass network, part I, Hughes Res. Lab., Malibu, California. Res. Rept. 334 (May, 1965).

40. K. Milne. The combination of pulse compression with frequency scanning for three-dimensional radars, *Radio Electronic Engr.* **28**, 89–106 (1964).

41. C. A. Palmieri, Radar waveform design, *Sperry Eng. Rev.* **15**, 32–43 (1962).

42. W. A. Penn, Signal fidelity in radar processing, *IRE Trans.* **MIL-6**, 2-4–218 (1962).

43. A. W. Rihaczek, Doppler-tolerant signal waveforms, *Proc. IEEE* **54**, 849–857 (1966).

44. W. L. Root, Radar resolution of closely spaced targets, *IRE Trans.* **MIL-6**, 197–204 (1962).

45. W. L. Rubin and J. V. DiFranco, Analytic representation of wide-band radio frequency signals, *J. Franklin Inst.* **275**, 197–204 (1963).

46. F. C. Schweppe, On the accuracy and resolution of radar signals, *IEEE Trans.* **AES-1**, 235–245 (1965).

47. F. C. Schweppe and D. L. Gray, Radar signal design subject to simultaneous peak and average power constraints, *IEEE Trans.* **IT-12**, 13–26 (1966).

48. R. A. Simonelli, Generation of a phase-shift-keyed coded pulse signal, Eng. Res. Lab., Cornell Univ., Ithaca, New York, Res. Rept. EERL 52 (June, 1966).

49. G. K. Strother, Note on the possible use of ultrasonic pulse compression by bats, *J. Acoust. Soc. Am.* **33**, 696–697 (1961).

50. C. A. Stutt, Some results on real-part/imaginary-part and magnitude-phase relations in ambiguity functions, *IEEE Trans.* **IT-10**, 321–327 (1964).

51. S. M. Sussman, Least square synthesis of radar ambiguity functions, *IRE Trans.* **IT-8**, 246–254 (1962).

52. R. G. Sweet and W. R. Kincheloe, Jr., A real-time scanning spectrum analyzer using a tapped sonic delay-line filter, Stanford Electron, Lab., Stanford Univ., Stanford, California, Tech. Rept. 1967-1 (June, 1964).

53. P. Swerling, Parameter estimation accuracy formulas, *IEEE Trans.* **IT-10**, 302–314 (1964).

54. E. Titlebaum, A generalization of a two-dimensional Fourier transform property for ambiguity functions, *IEEE Trans. (Correspondence)* **IT-12**, 80-81 (1966).

55. D. W. Tufts, Matched filters and intersymbol interference, Cruft Lab., Harvard Univ., Cambridge, Massachusetts, Rept. 345 (July, 1961).

56. D. W. Tufts, Nyquist's problem—the joint optimization of transmitter and receiver pulse amplitude modulation, *Proc. IEEE* **53**, 248–259 (1965).

57. D. W. Tufts and D. A. Schnidman, Optimum waveforms subject to both energy and peak value constraints, *Proc. IEEE* **52**, 1002–1007 (1964).

58. H. van de Vaart, Pulse compression using X-band magnetoelastic waves in YIG rods, *Proc. IEEE (Letters)* **54**, 1007–1008 (1966).

59. R. A. Wainwright, On the potential advantage of a smearing-desmearing filter technique in overcoming impulse-noise problems in data systems, *IRE Trans.* **CS-9**, 362–366 (1961).

60. M. K. Ward, Low-pass linear-slope delay filters for compression, *Proc. IEEE (Correspondence)* **53**, 497–498 (1965).

61. M. K. Ward, Matched scan rate pulse-compression analysis, *Proc. IEEE (Letters)* **54**, 706–708 (1966).

62. G. R. Welti, Quaternary codes for pulsed radar, *IRE Trans.* **IT-6**, 400–408 (1960).

63. W. J. Williams and R. B. Rutledge, A generalized distortion analysis of a cross-correlation radar, *Proc. IEEE (Correspondence)* **53**, 200–201 (1965).

64. M. J. Withers, Method of measuring phase errors in linear frequency-modulated and pulse-compression radar systems, *Electronics Letter* **2**, 50–51 (1966).

65. M. J. Withers, Matched filter for frequency-modulated continuous-wave radar systems, *Proc. IEE (London)* **113**, 405–412 (1966).

66. J. M. Wozencraft and I. M. Jacobs, "Principles of Communication Engineering." Wiley, New York, 1965.

67. R. C. Yost and E. B. Smith, Application of transversal equalizer to radar receiver with emphasis on Doppler sensitivity, Mass. Inst. Technol., Lincoln Lab., Lexington, Massachusetts, Technical Report No. 420 (June, 1966).

68. *IRE Trans.* **IT-6**, entire issue (1960).

69. N. Zierler, Several binary sequence generators, *Proc. Am. Math. Soc.* **7**, 675–681 (1956).

Author Index

Numbers in parentheses are reference numbers and indicate that an author's work is referred to although his name is not cited in the text. Numbers in italic show the page on which the complete reference is listed.

Subject Index

The Artech House Radar Library

David K. Barton, Series Editor

Active Radar Electronic Countermeasures, Edward J. Chrzanowski

Adaptive Signal Processing for Radar, Ramon Nitzberg

Airborne Early Warning System Concepts, Maurice Long, et al.

Airborne Pulsed Doppler Radar, Guy V. Morris

AIRCOVER: Airborne Radar Vertical Coverage Calculation Software and User's Manual, William A. Skillman

Analog Automatic Control Loops in Radar and EW, Richard S. Hughes

Aspects of Modern Radar, Eli Brookner, et al.

Aspects of Radar Signal Processing, Bernard Lewis, Frank Kretschmer, and Wesley Shelton

Bistatic Radar, Nicholas J. Willis

CARPET (Computer-Aided Radar Performance Evaluation Tool): Radar Performance Analysis Software and User's Manual, developed at the TNO Physics and Electronic Laboratory by Albert G. Huizing and Arne Theil

Cryptology Yesterday, Today, and Tomorrow, Cipher Deavours et al.

Cryptology: Machines, History, and Methods, Cipher Deavours et al.

DETPROB: Probability of Detection Calculation Software and User's Manual, William A. Skillman

Designing Automatic Control Systems, W. L. Rubin

Detectability of Spread-Spectrum Signals, Robin A. Dillard and George M. Dillard

Electronic Countermeasure Systems Design, Richard Wiegand

Electronic Homing Systems, M. V. Maksimov and G. I. Gorgonov

Electronic Intelligence: The Analysis of Radar Signals, Richard G. Wiley

Electronic Intelligence: The Interception of Radar Signals, Richard Wiley

Electronic Warfare Receiving Systems, Dennis D. Vaccaro

Engineer's Refractive Effects Prediction Systems—EREPS, Herbert Hitney

EREPS: Engineer's Refractive Effects Prediction System Software and User's Manual, developed by NOSC

High Resolution Radar, Donald R. Wehner

High Resolution Radar Cross-Section Imaging, Dean Mensa

Interference Suppression Techniques for Microwave Antennas and Transmitters, Ernest R. Freeman

Introduction to Electronic Defense Systems, Fillippo Neri

Introduction to Electronic Warfare, D. Curtis Schleher

Introduction to Sensor Systems, S. A. Hovanessian

IONOPROP: Ionospheric Propagation Assessment Software and Documentation, Herbert Hitney

Kalman-Bucy Filters, Karl Brammer

Laser Radar Systems, Albert V. Jelalian

Lidar in Turbulent Atmosphere, V. A. Banakh and V. L. Mironov

Logarithmic Amplification, Richard Smith Hughes

Machine Cryptography and Modern Cryptanalysis, Cipher A. Deavours and Louis Kruh

Millimeter-Wave Radar Clutter, Nicholas Currie, Robert Hayes, and Robert Trebits

Modern Radar System Analysis, David K. Barton

Modern Radar System Analysis Software and User's Manual, David K. Barton and William F. Barton

Monopulse Radar, A. I. Leonov and K. I. Fomichev

MTI and Pulsed Doppler Radar, D. Curtis Schleher

Multifunction Array Radar Design, Dale R. Billetter

Multisensor Data Fusion, Edward L. Waltz and James Llinas

Multiple-Target Tracking with Radar Applications, Samuel S. Blackman

Multitarget-Multisensor Tracking: Advanced Applications, Volumes I and II, Yaakov Bar-Shalom, editor

Over-The-Horizon Radar, A. A. Kolosov, et al.

Principles and Applications of Millimeter-Wave Radar, Charles E. Brown and Nicholas C. Currie, editors

Pulse Train Analysis Using Personal Computers, Richard G. Wiley and Michael B. Szymanski

Radar and the Atmopshere, Alfred J. Bogush, Jr.

Radar Anti-Jamming Techniques, M. V. Maksimov, et al.

Radar Cross Section Analysis and Control, A. K. Bhattacharyya and D. L. Sengupta

Radar Cross Section, Second Edition, Eugene F. Knott, et al.

Radar Electronic Countermeasures System Design, Richard J. Wiegand

Radar Engineer's Sourcebook, William Morchin

Radar Evaluation Handbook, David K. Barton, et al.

Radar Meteorology, Henri Sauvageot

Radar Polarimetry for Geoscience Applicatons, Fawwaz Ulaby

Radar Propagation at Low Altitudes, M. L. Meeks

Radar Reflectivity Measurement: Techniques and Applications, Nicholas C. Currie, editor

Radar System Design and Analysis, S. A. Hovanessian

Radar Technology, Eli Brookner, editor

Radar Vulnerability to Jamming, Robert N. Lothes, Michael B. Szymanski, and Richard G. Wiley

RGCALC: Radar Range Detection Software and User's Manual, John E. Fieldng and Gary D. Reynolds

SACALC: Signal Analysis Software and User's Guide, William T. Hardy

Signal Detection and Estimation, Mourad Barkat

Small-Aperture Radio Direction Finding, Herndon Jenkins

Solid-State Radar Transmitters, Edward D. Ostroff, et al.

Space-Based Radar Handbook, Leopold J. Cantafio, editor

Spaceborne Weather Radar, Robert M. Meneghini and Toshiaki Kozu

Statistical Signal Characterization, Herbert Hirsch

Statistical Signal Characterization: Algorithms and Analysis Programs Software and User's Manual, Herbert Hirsch

Statistical Theory of Extended Radar Targets, R. V. Ostrovityanov and F. A. Basalov

Surface-Based Air Defense System Analysis, Robert Macfadzean

Surface-Based Air Defense System Analysis Software and User's Manual, Robert Macfadzean and James M. Johnson

Synthetic-Aperture Radar and Electronic Warfare, Walter W. Goj

VCCALC: Vertical Coverage Plotting Software and User's Manual, John E. Fielding and Gary D. Reynolds

For further information on these and other Artech House titles, contact:

Artech House   Artech House
685 Canton Street   6 Buckingham Gate
Norwood, MA 01602   London SW1E6JP England
(617) 769-9750   +44(0)71 630-0166
Fax:(617) 762-9230   +44(0)71 630-0166
Telex: 951-659   Telex-951-659